The New England States

(and Long Island)

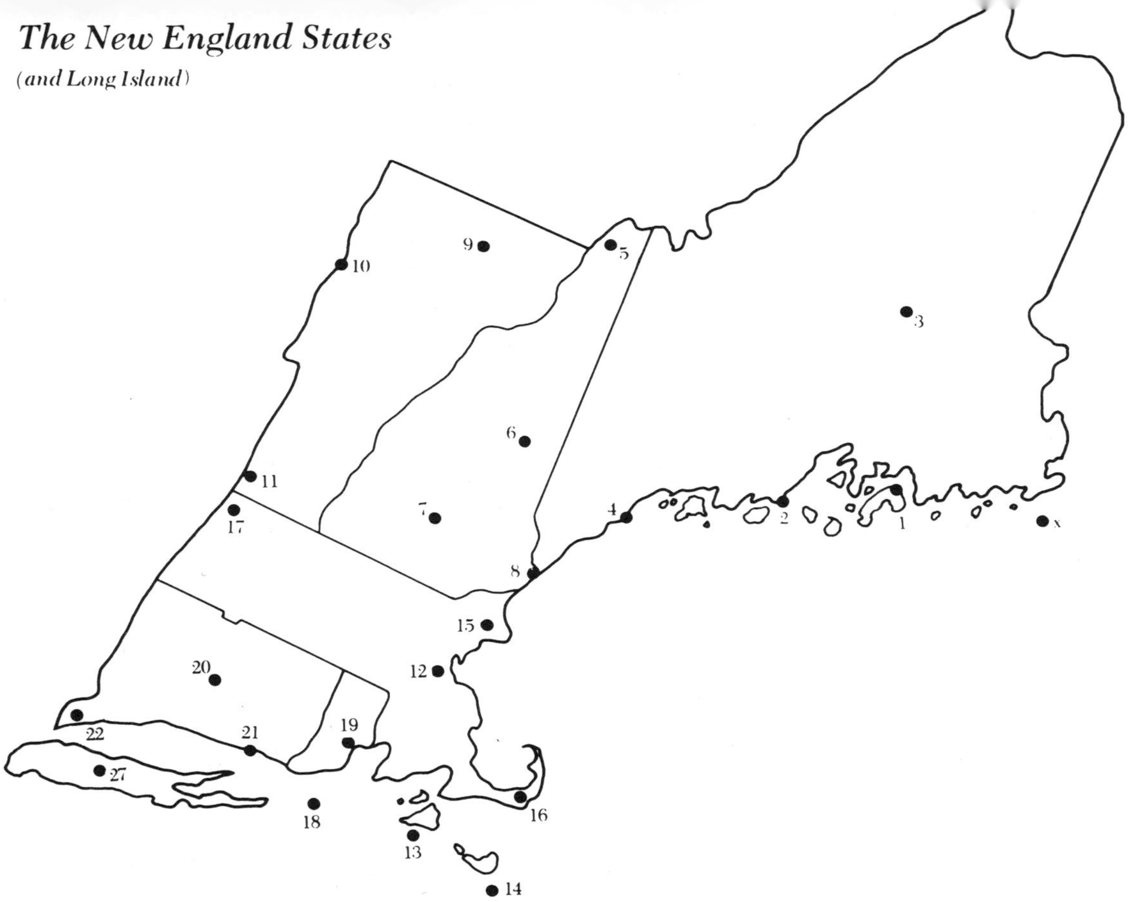

MAINE
1 Bar Harbor
2 Rockland
3 Millinocket
4 Portland
x Grand Manan (N.B., Canada)

NEW HAMPSHIRE
5 Pittsburg
6 North Conway
7 Concord
8 Portsmouth

VERMONT
9 Newport
10 Burlington
11 Arlington

MASSACHUSETTS
12 Boston
13 Martha's Viney
14 Nantucket
15 Newburyport
16 Wellfleet
17 North Adams

RHODE ISLAND
18 Block Island
19 Providence

CONNECTICUT
20 Hartford
21 New London
22 Greenwich

NEW YORK
27 Long Island

COULTER LIBRARY ONONDAGA COMM. COLL.
3 0418 00229047 4

Sidney B.
Coulter
Library
ONONDAGA COMMUNITY COLLEGE
SYRACUSE, NEW YORK 13215

A Guide to

Bird Finding EAST OF THE MISSISSIPPI

Second Edition

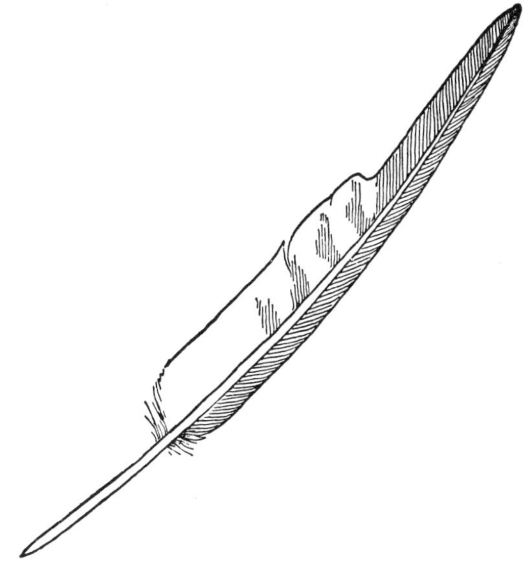

OLIN SEWALL PETTINGILL, JR.

with illustrations by GEORGE MIKSCH SUTTON

New York · OXFORD UNIVERSITY PRESS · *1977*

LIBRARY OF CONGRESS CATALOGUE CARD NUMBER: 76–9253
PRINTED IN THE UNITED STATES OF AMERICA
9 8 7 6 5

FOR ELEANOR

Preface to the First Edition

One clear mid-October day in 1945 my wife and I climbed to the observation promontory on Hawk Mountain in eastern Pennsylvania and for the first time in our lives watched the hawks, eagles, and falcons loafing southward on the updrafts along the Kittatinny Ridge. We were there at the right time of the year; we had no difficulty in finding our way. Indeed, the trip was enormously rewarding—thanks to our good friends, Mr. and Mrs. Raymond J. Middleton of Norristown, who were our guides.

How often in our trips to find birds have we not been so fortunate as to have the Middletons with us! Learning of some bird attraction in a part of the country strange to us, we have attempted to locate it ourselves, all the while losing precious time and, in the end, perhaps not finding it at all. How often, too, have we visited a part of the country only to learn later that we had been near an attraction but had missed it altogether because we did not know about it at the time.

Immediately after our trip to Hawk Mountain I decided to prepare a guide to the outstanding places for birds in the United States. In the weeks that followed I drew up plans for the project and proceeded, early in 1946, to accumulate information.

Although I have toured through all the states of the Union and have enjoyed bird watching in most of them, I realized at the outset of this project that I neither possessed in my own notes nor could acquire in a lifetime of field work the special type of information most greatly needed—namely, precise locations of ornithological attractions and specific directions for reaching them. Moreover, I realized that, despite the extensive literature on American ornithology, this information was lacking in published form. And so I resorted to a nationwide correspondence with persons qualified to provide the facts. My usual procedure on receiving the data was to work them up in my own style, then submit them to the donors for checking.

The organization of the book was troublesome. Though several plans were considered, only one proved feasible—a book divided into chapters according to states. A book with chapters on physiographic regions (e.g. coastal plains, mountain systems) or biological communities (coniferous forests, prairie grasslands), would have been more sound scientifically, but it would have caused confusion to the average bird finder, who, like most other Americans, pictures his country in terms of states. Recognizing nonetheless the scientific value of presenting United States birdlife and bird-finding places as related to regions and communities, I prepared each chapter with an introductory section in which this approach was emphasized.

I wrote most of the text during a period extending from January 1949 to June 1950. Occupied with my duties as an instructor at Carleton College and (in the summers) at the University of Michigan Biological Station, I accomplished the actual writing in the evenings, on vacations, and on class-free days, except from 1 September 1949 to 1 February 1950, when I was granted a leave of absence from Carleton College to devote my full time to the project.

After the text was well under way, it became apparent that it would be too long for publication in one volume; so it was decided to bring it out in two volumes. The states lying entirely east of the Mississippi River are covered here; the remaining states will be taken up in a second volume, the text of which is close to completion.

In preparing the chapters in this volume, I frequently referred to ornithological literature and felt free to draw upon information as needed. Papers published in *The Auk*, *The Wilson Bulletin*, *Bird-Banding*, *Audubon Magazine* (including its predecessor *Bird-Lore*), and numerous local journals often proved useful. *Audubon Field Notes* frequently contained pertinent items. Occasionally, good sources of information were the *Naturalist's Guide to the Americas*, edited and compiled by Victor E. Shelford, and the American Guide Series, compiled and written by the Federal Writers' Project, or the Workers of the Writers' Program, of the Works Progress Administration.

In a large sense this volume represents a co-operative undertaking. I am immensely indebted to the following individuals who prepared entire chapters, faithfully conforming to my specifications in regard to organization and style: Burt L. Monroe (Kentucky), Robert E. Stewart and Chandler S. Robbins (Maryland and the District of Columbia), Alexander Sprunt, Jr. (South Carolina), Maurice Brooks (West Virginia). I am similarly indebted to Mr. Robert S. Arbib, Jr., for preparing the account of New York City.

Over 300 persons contributed information. Time and again I was deeply impressed by their willingness to meet requests and, in many instances, to go out of their way to gather data not readily available to them. (Some of the persons who helped immeasurably I have never met.) Most of the individuals who contributed information are listed at the ends of the chapters in which their information is used. Of those listed, a few stand out as either having contributed a substantial amount of data, or having taken time to read critically one or more chapters. These persons are Francis H. Allen, Stanley C. Ball, Fred W. Behrend, Thomas D. Burleigh, Roland C. Clement, Ben B. Coffey, Jr., Edward F. Dana, Albert F. Ganier, Alfred O. Gross, Joseph C. Howell, Thomas A. Imhof, Douglas L. Kraus, Robert B. Lea, Peggy Muirhead MacQueen, Karl H. Maslowski, Joseph James Murray, A. W. Schorger, Wendell P. Smith, Herbert L. Stoddard, Phillips B. Street, George Miksch Sutton, Wendell Taber, Edward S. Thomas, W. E. Clyde Todd, Milton B. Trautman, Josselyn Van Tyne, Francis M. Weston, and Ralph E. Yeatter.

I owe a special word of gratitude to Dr. J. Speed Rogers, Director, and Dr. Josselyn Van Tyne, Curator of Birds, in the University of Michigan Museum of Zoology. Through their generosity the Museum's facilities, including an office in which to work, were placed at my disposal during my leave of absence from Carleton College. In this five months' interval I particularly appreciated Dr. Van Tyne's counsel.

Standing squarely behind this work is the encouragement and aid of my associate of long standing, Dr. George Miksch Sutton. Many times when vexing problems arose, I appealed to him for possible solutions and invariably obtained them. His illustrations in this volume speak for themselves, but I must pay special tribute to the fine spirit with which he undertook to combine certain ideas I wished to have shown with the subject matter as seen through his own eyes.

It is my sincere wish that this volume and the one to follow will bring to persons the country over a greater awareness of the endless opportunities for bird finding in every state of the Union. If it serves to enrich their appreciation of birdlife in their own sections of the country, or if it leads them to more enjoyable vacations and more satisfactory methods of relaxation between business engagements in metropolitan centers, so much the better.

A work of this sort necessarily has its limitations. Many an ornithologist will be disappointed that his choice spot for birds is insufficiently treated, if it is treated at all. More than one person will doubtless discover that a particular bird-finding area does not match the description. It must be remembered that any environment is in a constant state of change. As the late Charles A. Urner once remarked, it is 'risky to depend tomorrow on today's description of good birding places.' For a future edition of this initial effort, corrections and suggestions will be most welcome.

OLIN SEWALL PETTINGILL, JR.

Carleton College
Northfield, Minnesota
May 1951

Preface to the Second Edition

For the first edition of this work, published in 1951, I began assembling the material in 1946 and virtually completed the manuscript in 1950. In the quarter-century since its publication, the changes in the twenty-six states I have found staggering.

Take, for example, the changes in the distribution of birds. The Peregrine Falcon no longer breeds east of the Mississippi. The Cattle Egret, not even mentioned in the first edition, is widespread. The House Finch, mentioned on Long Island where it was introduced, has expanded its range in all landward directions. The Northern Mockingbird and Northern Cardinal, living up to their names, have moved into northern New England, Michigan, and Wisconsin, as have the Tufted Titmouse and Carolina Wren. And those wading birds the Snowy Egret and Glossy Ibis, so commonly associated in our memory with southern swamps and bayous, are nesting on the coast of Maine.

The changes in environment are similarly staggering. On the outskirts of expanding cities, the marshes that once yielded hosts of birds now feature human creations from parking lots to high-rise apartments. Clean, remote sand beaches and dunes on the Atlantic barrier islands are accessible by bridge to all manner of human activities and their litter. One now customarily looks for shorebirds foraging among discarded beverage containers, bottle carriers, and plastic bags, and too often searches in vain for nesting Wilson's Plovers and Little Terns.

There is, however, a brighter side, thanks to the expansion and proliferation of parks, refuges, sanctuaries, preserves, and nature centers, embracing some of the finest places for bird finding, frequently with the benefit of interpretive services. But again, in the case of the big parks, no matter how well managed, there is a disadvantage—too many people. Here the bird finder must now adjust, pursuing his interest in the early morning, before and after the vacation season, along the trails that are least used. In this edition I have included many more publicly owned and publicly managed areas, at the same time excluding areas that may be 'developed' to the detriment of occupant birds. Nothing is more heartbreaking than to drive many miles to a particular wild area in anticipation of some special bird, only to find a new shopping center.

In keeping up with these changes, I had to face an Interstate Highway System and the consequent relocating and renumbering of many local routes. Recognizing the convenience of the Interstates for bird finders who may have to travel a great distance to reach a particular area in country unfamiliar to them, I have routed many places from the nearest Interstate exits rather than from the centers of cities and towns.

Adjusting to these changes was difficult. No less so was selecting English names for birds to denote precisely different kinds, species or subspecies, that are readily identifiable by sight or sound. Unfortunately a number of English names authorized by the Committee on Classification and Nomenclature of the American Ornithologists' Union are inadequate for the thousands of people who travel countrywide—and beyond—to see different birds. For example, the names Cardinal for *Richmondena cardinalis* and Mockingbird for *Mimus polyglottos* lack modifiers to distinguish the species from others that bear 'cardinal' and 'mockingbird' in their English names. The names American Golden Plover and European Tree Sparrow misrepresent the extent of the native ranges of the species. In certain names the modifier 'common' is used when a more suitable one is available.

Realizing these shortcomings, the Checklist Committee of the American Birding Association, after far-reaching deliberation, published its own list of English names (A.B.A. *Checklist: Birds of Continental United States and Canada*, 1975). This I am following with a few exceptions (*see* The Plan of This Book). I am using Traill's Flycatcher for the two species, Willow and Alder, because in many instances I was unable to determine whether one species or the other occupied the locality in question. I am using Yellow-shafted Flicker, Myrtle Warbler, Baltimore Oriole, and Slate-colored Junco—the long-established names of these familiar birds in eastern North America—even though each of these birds and its western counterpart (or counterparts) is considered one species and was consequently given a new 'umbrella' name, viz., Common Flicker, Yellow-rumped Warbler, Northern Oriole, and Northern Junco (Dark-eyed Junco, new A.O.U. name). Such names fail to indicate obvious differences among geographical populations, and are thus unsuitable in a handbook of this sort.

In the first edition I cited colleges, universities, and other educational institutions offering courses dealing with birds; I also mentioned museums, libraries, art collections, and zoos of ornithological interest, and called attention to state and local ornithological societies. Now that there are so many deserving of recognition, I decided to omit them all in this edition, using the space they might have occupied for more bird-finding places.

In preparing this edition I frequently consulted articles and reports in *American Birds, Birding, The Chat* (Journal of the Carolina Bird Club), and *The Migrant* (Journal of the Tennessee Ornithological Society). I also found very useful the following publications: *Where To Find Birds and Enjoy Natural History in Florida* (Florida Audubon Society), *A Birder's Guide to Georgia* (Georgia Ornithological Society), *Bird Finding in Illinois* (Illinois Audubon Society), and *Wisconsin's Favorite Bird Haunts* (Wisconsin Society for Ornithology).

Much of the information on New York State exclusive of the New York City area is taken, with updating, from *Enjoying Birds in Upstate New York* (Cornell Laboratory of Ornithology), written by Sally Hoyt Spofford and myself. I have also used, after updating them, some of my *Audubon Magazine* columns appearing from 1957 through 1968, and some area descriptions from *Enjoying Maine Birds* (Maine Audubon Society).

This edition, like its predecessor, is a cooperative undertaking. So many people have generously supplied information that I cannot begin to acknowledge all of them. And when I say 'all of them' I refer to the persons who helped me with the first edition as much as with the second. Indeed, the second edition was built on the first and would not exist without it. All the people on whom I, or my authors, relied for information in either edition or both are listed in this book at the end of the general introduction to the chapter containing the information they contributed.

I am most grateful to Maurice Brooks for revising his West Virginia chapter; Burt L. Monroe, Jr., for revising the Kentucky chapter written by his late father; Chandler S. Robbins for revising the Maryland chapter that he wrote jointly with Robert E. Stewart; Dennis M. Forsythe and Sidney A. Gauthreaux, Jr., who collaborated on a new chapter on South Carolina to replace that prepared by the late Alexander Sprunt, Jr., for the first edition; and R. Dudley Ross, who, with the cooperation of Seal T. Brooks, Lloyd L. Falk, and Winston J. Wayne of the Delmarva Ornithological Society, wrote a separate chapter on Delaware. I am grateful also to Robert S. Arbib, Jr., who with the assistance of Thomas H. Davis, Jr. revised his account of the New York City area, drawn largely from the more recent *Birds Around New York City* (Houghton Mifflin), of which he was the lead author.

Among the many authorities, I single out for special mention with my appreciation the following who provided extensive information and/or read critically parts or all of specific chapters: Alabama, Thomas A. Imhof; Florida, Margaret H. Hundley and Louis A. Stimson; Illinois, Jean W. Graber and Richard R. Graber; Indiana, Henry C. West and Jon E. Rickert; Massachusetts, Ruth P. Emery; Mississippi, Jerome A. Jackson; New Hampshire, Tudor Richards; Ohio, Karl H. Maslowski and Milton B. Trautman; Pennsylvania, Phillips B. Street; Tennessee, Fred J. Alsop III, Ben B. Coffey, Jr., Kenneth H. Dubke, and Lee R. Herndon; Vermont, Elizabeth Downs; Virginia, Curtis S. Adkisson; Wisconsin, Samuel D. Robbins, Jr.

Altogether eighty pen-and-ink drawings by George Miksch Sutton enliven the pages of this edition. Four—of the Northern Cardinal, Swallow-tailed Kite, Connecticut Warbler, and Black-backed Three-toed Woodpecker—were created especially for this edition. Three—of the Great Cormorant, Wilson's Storm Petrel, and Whimbrel—were first reproduced in *Fundamentals of Ornithology* by the late Josselyn Van Tyne and Andrew J. Berger (New York: John Wiley & Sons, 1959, 1976). For permission to use these three drawings, I thank the artist, junior author, and publisher.

I owe a special word of appreciation to Philip A. DuMont of the United States Fish and Wildlife Service for sending me, as soon as they were issued, the many descriptive folders and bird checklists pertaining to the National Wildlife Refuges.

Finally, I could not have brought this book through all its stages without my wife. Always my 'pilot' with road map in hand on our long drives for birds, she is likewise the pilot of all persons using this book. For it is she who largely worked out the route directions with the required patience and exactitude. All bird finders who reach the fine places for birds described in this book can thank Eleanor Rice Pettingill.

OLIN SEWALL PETTINGILL, JR.

Wayne, Maine
June 1976

Contents

The Plan of This Book

Area Covered. This book covers the twenty-six states east of the Mississippi River. In three states—Maine, New York, and Michigan—adjacent Canadian areas are included. Places for bird finding in each state are chosen to show (1) the widest variety of regular species; (2) seasonal concentrations of birds and migratory movements; and (3) the best representation of birdlife in the vicinities of metropolitan areas and leading vacation centers. Included whenever they offer productive bird finding are National Parks and National Wildlife Refuges, many state and municipal parks, refuges, and preserves, and numerous public or privately owned sanctuaries and nature centers.

Birds Covered. Attention is given to all species residing or appearing regularly in the twenty-six states. Subspecies or races receive attention if their populations are readily identifiable in the field.

Terminology. The English vernacular names of birds follow the *A.B.A. Checklist: Birds of Continental United States and Canada* (American Birding Association, 1975), with the following exceptions: Traill's Flycatcher is used for both the Willow and the Alder Flycatchers; Yellow-shafted Flicker is used for Common Flicker, Myrtle Warbler for Yellow-rumped Warbler, Baltimore Oriole for Northern Oriole, and Slate-colored Junco for Northern Junco. In some instances, where a subspecies is recognizable by sight or sound, the English name of the subspecies precedes the name of the species in parentheses—for example, (Bicknell's) Gray-cheeked Thrush, (Ipswich) Savannah Sparrow.

Frequently the following terms are used to indicate groups of species:

Waterfowl: swans, geese, and ducks

Waterbirds: herons, egrets, bitterns, storks, ibises, spoonbills, cranes, limpkins, and all swimming birds other than waterfowl

Wading birds: herons, egrets, bitterns, storks, ibises, spoonbills, cranes, and limpkins
Shorebirds: oystercatchers, plovers, turnstones, woodcock, snipes, curlews, sandpipers, godwits, avocets, stilts, and phalaropes
Alcids: razorbills, dovekies, murres, guillemots, and puffins
Pelagic birds: chiefly albatrosses, shearwaters, and petrels; other waterbirds that habitually prefer the open sea to inshore waters
Landbirds: vultures, hawks, eagles, falcons, gallinaceous birds, doves, cuckoos, owls, swifts, hummingbirds, kingfishers, woodpeckers, and passerine birds
Parulids: all wood warblers
Fringillids: grosbeaks, finches, sparrows, and buntings

Names for physiographical features follow in most cases those used in *Physiography of Eastern United States* by Nevin M. Fenneman (New York: McGraw-Hill, 1938).

Organization of the Chapters. Each chapter consists of an introduction followed by a series of bird-finding places.

The introduction presents the birdlife of the state with relation to its physiographical regions and associated habitats, to migration, and to the winter season. After the lead paragraph, which usually points up one or more ornithological features of the state, comes a sequence of paragraphs describing briefly the regions and their habitats, with mention of characteristic breeding birds. In the case of 'farmlands'—a general category for fields (cultivated and fallow, including pastures), wet meadows, brushy lands (including forest edges and hedgerows), orchards, and dooryards (lawns, ornamental shrubbery, shade trees)—and the predominant forests, the characteristic breeding species are listed. The lists, however, are incomplete. Some species are omitted because they are commonplace, others because their habitat preferences are either too specialized or too broad, still others because their presence in the state during the breeding season is scarce or questionable. Below are species consistently left out of such lists, for one or more of the above reasons, in all the chapters.

Cooper's Hawk
Red-tailed Hawk
Bald Eagle
Northern Harrier

Osprey
American Kestrel
Ring-necked Pheasant
Wild Turkey
Killdeer
American Woodcock
Upland Sandpiper
Rock Dove
Barn Owl
Great Horned Owl
Long-eared Owl
Short-eared Owl
Common Nighthawk
Chimney Swift
Ruby-throated Hummingbird
Belted Kingfisher
Traill's Flycatcher
Bank Swallow
Rough-winged Swallow
Cliff Swallow
Purple Martin
American Crow
Sedge Wren
American Robin
Veery
Cedar Waxwing
European Starling
Black-and-white Warbler
Golden-winged Warbler
Blue-winged Warbler
House Sparrow
Red-winged Blackbird
Brown-headed Cowbird

Following the paragraphs concerned with regions and their habitats are at least two paragraphs. One deals with the peculiarities of migration, pointing out the principal migration lanes (if any) through the state, and concludes with a timetable giving the dates within which one may expect the peak flights of waterfowl, shorebirds, and landbirds. The second paragraph deals with winter birdlife, emphasizing regular visitants from out of the state.

The introduction concludes with the names of authorities from whom bird-finding information was obtained for both the first edition and this one. If there are any available publications of recent date that will be helpful to bird finders visiting the state, they are cited after the authorities.

The accounts of bird-finding places after the state introduction are usually presented under the nearest cities or towns that are indexed in the *Rand McNally Road Atlas.* If there is a publication pertinent to the birds in a place described, it is cited at the end of the account.

Index. There is one index incorporating both (1) bird-finding places and the states, cities, and towns under which they are described, and (2) the bird species mentioned in connection with the bird-finding places. Species cited in the introductions to chapters are indexed in a few instances.

Endpapers. For instant reference to the states covered in this Guide, the endpapers display condensed maps of (1) the New England states; (2) the northern states exclusive of New England; and (3) the southern states. Map 1 appears on the front endpapers, Maps 2 and 3 at the rear. On these maps many cities or towns near important sites for birds are mentioned. Then, in anticipation of the adoption of the metric system in the United States, a conversion table for measurements commonly used in this book has been prepared for the first endpaper.

Suggestions for Bird Finders

When Wanting To Know Locations for a Bird Species. Look up the name of the species in the Index, then consult the pages given. On each page the bird is mentioned in connection with a place. Reading this information will determine how to reach the bird from the city or town under which the place is listed, what to expect as to terrain, vegetation, and other birds.

When Visiting a State. Consult the chapter devoted to the state, reading first the introduction, which gives the principal features of the state's birdlife. Then read the accounts of bird-finding places. A road map of the state and its index will readily reveal the cities or towns from which the places may be reached. Usually helpful is the official highway map issued by the state, as it shows local routes and gives information about state parks, campgrounds, picnic areas, and other recreational facilities.

When Visiting a National Park. Consult the Index to find where the Park is described with directions for reaching it. Either at the entrance or at headquarters, obtain a a map and general information about the Park and its interpretive services. If there is a visitor center, stop for supplementary information.

When Visiting a National Wildlife Refuge. The National Wildlife Refuges, whose boundaries and directional signs bear the figure of a flying goose, are federally owned and supervised by the Fish and Wildlife Service of the United States Department of the Interior. Because they embrace some of the finest places for birds, almost all Refuges having resident personnel and permitting access to the public are described in this book. Consult the Index. Directions for reaching the headquarters of each Refuge are included, with route directions from the nearest city or town.

Never visit a National Wildlife Refuge without first going to headquarters and making known the purpose of the visit. This is important. While bird finders are always welcome, there will doubtless be regulations and restricted areas that they should know about. Also, there may be available maps, a checklist of birds of the Refuge, and advice on the best vantage points for viewing birds. Many such Refuges have self-guiding car tours and foot trails laid out especially for bird finding.

When Visiting Other Parks and Refuges, Sanctuaries, and Nature Centers. This book describes, and gives directions to, a great number of parks and refuges owned and operated by states, counties, and municipalities. In many cases they are patterned after those of the Federal Government, and many have excellent interpretive centers, museums, self-guiding nature trails, and naturalists in residence (usually during warmer months in the northern states). Also listed is a considerable number of sanctuaries and nature centers, both publicly owned and privately owned, with a resident staff and interpretive services. Make a point of consulting headquarters *first* when visiting any of these places, but especially in the case of refuges, sanctuaries, and nature centers, since there may be restrictions, as well as opportunities to be gained, in finding out precisely where to see certain birds.

When in Quest of Particular Birds. When looking for shorebirds on the mud flats and sand flats in tidal estuaries and bays, consult the tide schedules in local newspapers. When the tide is low, the shorebirds feed on the flats, widely scattered and often far from shore. As the tide comes in, the birds move closer to shore—and closer for viewing.

Along the Appalachian Mountains and the Great Lakes there are 'flight lanes' that fall-migrating hawks follow in great numbers, depending on the physiographical peculiarities of the area and—very importantly—the weather. Some of the best lookouts for observing flights and the conditions governing them are described briefly in this book. Bird finders desiring extensive information on this spectacular phenomenon, as well as on all known lookouts, can do no better than to consult *Autumn Hawk Flights. The Migrations in Eastern North America* (New Brunswick: Rutgers University Press, 1975) and *A Guide to Eastern Hawk Watching* (University Park: Pennsylvania State University Press, 1976) by Donald S. Heintzelman.

Searching for owls after sundown can be facilitated by the use of a portable cassette recorder. Thus, when visiting woods where the sought-after owls reside, come equipped with a cassette on which their particular vocalizations are pre-recorded, and play them. If the owls are there, chances are that they will respond and possibly draw closer to the source.

The cassette recorder is useful also during the breeding season, for bringing small landbirds into view. On hearing the recorded songs of their species, males mistake them for those of rival males and consequently approach, revealing themselves. Pre-recorded songs help to turn up males that are not vocalizing at the time of the visit. If the males are already singing, although out of sight, and a view of them is desired, one may record the songs and play them back, to achieve the same result. *But heed this urgent warning:* Repeated broadcasts of the songs may be disruptive, causing the males to desert the area or distracting attention from their nests. Males as well as females of some species play a significant role in the incubation of the eggs and care of the young.

Such birds as pelicans, cormorants, many wading species, and gulls and terns gather in colonies for nesting. Because their aggregations are impressive, the locations of some of the more accessible ones are given in this book. Bird finders must bear in mind, however, that a walk into any colony can drive the adults from their nests and seriously expose the eggs or young, or drive older young from their nests and subject them to starvation or to attacks by neighboring adults or by predators. Therefore, observe these precautions: *Circle the colony but never go into it.* Do not stay near the colony longer than a few minutes. If the air temperature is exceptionally cool or the sun's rays are uncomfortably hot, do not approach the colony at all. Instead, postpone the visit to a time when the weather is more suitable.

Miscellaneous Advice and Precautions. Although this book gives directions, where possible, to all areas described, highways are sometimes relocated and renumbered, and privately owned areas are 'developed'—to the exclusion of the desired birds. Failing to find the areas or their birds, contact local people knowledgeable about birds, in cities and larger towns through the chamber of commerce, in the smaller communities through the postmaster.

Before entering farmlands or any areas fenced off or posted, determine the ownership and ask permission for entry. Violation of property rights can mean the subsequent exclusion of all bird finders from choice areas that are highly productive.

Finally, always be alert for hazards and even personal danger. In all large metropolitan areas—in parks, preserves, anywhere—beware of muggers. Walk with another person, preferably with a group of people, but in any case, *never* alone! When exploring forests without well-marked trails, carry a good map and a compass and consult both before entering. When exploring any wild areas, be informed as to the prevalence of venomous snakes and poisonous plants and be able to recognize them. Wear clothing that will serve as a protection from noxious insects.

Aids to Identification of Birds

Birds of North America. A Guide to Field Identification. By Chandler S. Robbins, Bertel Bruun, and Herbert S. Zim. New York: Golden Press, 1966.

Audubon Bird Guide: Small Land Birds, and *Audubon Water Bird Guide: Water, Game, and Large Land Birds.* By Richard H. Pough. Garden City, N.Y.: Doubleday, 1946 and 1951. Both books cover eastern North America.

A Field Guide to the Birds. By Roger Tory Peterson. 2d ed. Boston: Houghton Mifflin, 1947. Covers eastern North America.

A Field Guide to Bird Songs. Either two cassettes or two 12-inch phonograph records to accompany, page by page, *A Field Guide to the Birds* (Peterson). Available from Cornell Laboratory of Ornithology, 159 Sapsucker Woods Road, Ithaca, N.Y. 14853.

A Guide to Bird Songs. Descriptions and Diagrams of the Songs and Singing Habits of Land Birds and Selected Species of Shore Birds. By Aretas A. Saunders. Rev. ed. Garden City, N.Y.: Doubleday, 1951.

Aid to Keeping Records of Birds

The Bird Finder's 3-Year Note Book. By Paul S. Eriksson. New York: Paul S. Eriksson Inc., 1976. Contains two parts: a life-list index of more than 700 species of North American birds, with space after each name for logging the date when first found, and a three-year diary for entering details of the observation under the same date.

Information about Birds and Their Study

Bird finders wishing to be informed about birds or be in touch with others who enjoy finding, watching, and studying birds should subscribe to *American Birds* and join the American Birding Association.

American Birds is a bimonthly journal published by the National Audubon Society (950 Third Avenue, New York, N.Y. 10022). One issue each year reports the Christmas bird counts taken in all fifty states and the ten Canadian provinces. The other five issues contain, besides seasonal reports from both countries, descriptions of sites for bird watching, articles on techniques of bird finding, photography, and sound-recording, and reports on various studies of birds.

The American Birding Association (Box 4335, Austin, Texas 78765) meets every other year in a different area of North America where the birdlife is particularly enticing, and publishes, six times a year, the magazine *Birding*, replete with various subjects pertinent to bird finding in North America and throughout the world.

Bird finders eager to increase their knowledge of birdlife generally can begin at home by taking the *Seminars in Ornithology* offered by the Cornell Laboratory of Ornithology. This is a college-level home-study course, readable in style and profusely illustrated, dealing with the way birds are structured and fly, how they live and behave from the time they are hatched, their migrations and how they determine direction, where they nest and how they rear their young. For further details on the course and how to enroll, write the Laboratory at 159 Sapsucker Woods Road, Ithaca, N.Y. 14853.

A GUIDE TO BIRD FINDING

EAST OF THE MISSISSIPPI

Second Edition

Alabama

BACHMAN'S SPARROW

Alabama, in the heart of Dixie, is commonly pictured as a land of heavy human populations, of cottonfields, of antebellum plantations, and sharecroppers' shacks. In reality, no part of the state conforms to this popular conception. Even in the so-called Black Belt, a relatively small area in central Alabama where the soil is black and fertile, the raising of cattle has replaced the production of cotton, once the state's most valuable crop. Though there is agriculture in sections of the state other than the Black Belt, Alabama has vast, sparsely settled areas that produce no crops of any sort. Some of these areas are particularly inviting to bird finders—namely, the scenic northeastern ridges and deep valleys with their cascading streams; the big forests of the Warrior Tableland near Birmingham; the trackless swamps covering the delta of the Mobile and Tensaw Rivers; the beaches, marshes, and bays on the Gulf Coast.

Sloping gradually from the southern extremities of the Appalachians down to the Tennessee Valley in the northwest and to the Gulf Coast in the south, Alabama has four major natural divisions: the Mountain Region, the Piedmont Plateau, the Black Belt, and the Coastal Plain. All except the Black Belt, considered by geologists to be part of the Coastal Plain, were once heavily forested.

The Coastal Plain constitutes the southernmost division, or roughly that part of the state south of the Fall Line from Phenix City westward through Wetumpka, then northwestward to Tuscaloosa and north into Tennessee. It is rolling country throughout, especially northward, where elevations reach 500 feet. The southeastern and central sections have fertile soils, farmed extensively. In the few forests now existing on the uplands, longleaf pine is the predominant tree growth, mixed with loblolly pine and oak; on the lowlands along the streams, hardwoods such as oaks and gums are the principal trees. The southwestern section, generally less suitable for farming, is distinguished by several areas:

1. *The delta,* formed by the Mobile and Tensaw Rivers as they approach Mobile Bay, is relatively narrow from east to west, but extends northward from the Bay for 50 miles and is primarily swampland, heavily forested with giant cypresses, black gums, red maples, water oaks, and other trees, all festooned with moss. The American Anhinga, Snowy Egret, Louisiana Heron, Purple Gallinule, and Fish Crow nest here commonly. US 43 going north from Mobile parallels the delta swampland for many miles.

2. *The coastal pinelands, or flatwoods,* characterized by open forests of longleaf and slash pines, frequently have a dense, shrubby undergrowth. The ground is boggy in spots, but here and there in undisturbed places elevated tracts, or hammocks, support magnolias, laurel oaks, and live oaks. One may expect Wild Turkeys here. In the flatwoods along State 59 from Summerdale south through Foley (*see under* **Mobile**) there is always a good chance of finding Red-cockaded Woodpeckers and Bachman's Sparrows, and, south of Foley, a pair or two of Sandhill Cranes.

3. *The coastal marshes and beaches.* The best marshes are at the head of Mobile Bay around the outlets of the Mobile and Tensaw Rivers, crossed by the Cochran Bridge and Causeway (*see under* **Mobile**). Clapper Rails, Marsh Wrens, Red-winged Blackbirds, and

Boat-tailed Grackles are common breeding birds in all coastal marshes; Seaside Sparrows are more restricted, appearing only in isolated colonies. A typical coastal beach is in Gulf State Park (*see under* **Mobile**), where Wilson's Plovers, Little Terns, and a few pairs of Snowy Plovers regularly nest. Dauphin Island, connected with the mainland by bridge and causeway, also has sandy beaches backed by tracts of pine and salt marshes.

The Mountain Region comprises the state's northernmost division, extending from the Tennessee boundary south to points near Heflin, Ashland, Columbiana, Centreville, Tuscaloosa, Fayette, and Hamilton. In its eastern part rise the southernmost spur ridges of the Appalachians, most of them quite rocky, with thin soil supporting a somewhat stunted growth of pine, red cedar, oak, and hickory. Nearly all these ridges run in a northeast-southwest direction, with deep valleys between them. Cheaha Mountain (2,407 feet), highest point in the state, rises from a ridge south of Anniston.

In the northern part of the Mountain Region the Tennessee River flows across the state from the northeastern to the northwestern corner. Dams built across the River by the Tennessee Valley Authority have created three large reservoirs: Guntersville, Wheeler, and Wilson Lakes. The broad, rolling Tennessee Valley is largely agricultural land, though badly eroded by careless use. South of the Tennessee Valley and composing the remaining part of the Mountain Region lies the Warrior Tableland (sometimes called the Coal Region), an area of low hills constituting the southern end of the Cumberland Plateau. In most of the valleys between, streams pass through deep gorges with precipitous walls. Warrior Tableland has some of Alabama's finest forests, principal trees being loblolly, shortleaf, and Jersey pines, chestnut oaks, hickories, and black gums.

The Piedmont Plateau is a triangular area extending south from the spur ridges of the Appalachians and meeting the Coastal Plain. Its eastern boundary is the Georgia line; its southwestern boundary extends roughly from Phenix City through Wetumpka to Centreville; its northwestern boundary passes from Centreville through points near Columbiana and Ashland to Heflin. Within the region are Alexander City, Opelika, and Wedowee. Much of the Piedmont, which is moderately hilly, is under cultivation, but there are rougher sections not suitable for crop production, containing forests of pines, red gum, blackjack oak, and red oak among the predominant trees.

Beginning a few miles southwest of Phenix City, the Black Belt, 20 to 25 miles wide, crosses the state south of Montgomery to the Mississippi line. Originally grassy prairie with timber growing only on the river bottomlands and higher elevations, it was once the state's best cotton-growing section. In recent years most of the cotton plantations have become stock farms.

In the Mountain Region—or in some instances the Piedmont Plateau or Black Belt—the following species reach the southern limits of their ranges:

Broad-winged Hawk
Whip-poor-will (*higher elevations*)
Eastern Phoebe
White-breasted Nuthatch
Bewick's Wren
Gray Catbird
Black-and-white Warbler
Blue-winged Warbler
Yellow Warbler
Black-throated Green Warbler
Ovenbird (*higher elevations*)
Scarlet Tanager
Dickcissel
American Goldfinch
Grasshopper Sparrow
Chipping Sparrow

Throughout Alabama, in the pine and hardwood forests and on the farmlands (fields, brushy lands, orchards, and dooryards), the following species breed regularly, except as indicated:

PINE AND HARDWOOD FORESTS

Turkey Vulture
Black Vulture
Red-shouldered Hawk
Yellow-billed Cuckoo
Common Screech Owl
Barred Owl
Chuck-will's-widow
Yellow-shafted Flicker
Pileated Woodpecker
Red-bellied Woodpecker
Hairy Woodpecker
Downy Woodpecker
Red-cockaded Woodpecker (*pine forests*)
Great Crested Flycatcher
Acadian Flycatcher (*except near coast*)
Eastern Pewee

Blue Jay
Fish Crow (*mainly near coast*)
Carolina Chickadee
Tufted Titmouse
Brown-headed Nuthatch (*pine forests, mainly south of Mountain Region*)
Wood Thrush
Blue-gray Gnatcatcher
Yellow-throated Vireo
Red-eyed Vireo (*except near coast*)
Prothonotary Warbler
Northern Parula Warbler (*uncommon in Mountain Region*)
Yellow-throated Warbler
Pine Warbler (*mainly pine forests*)
Louisiana Waterthrush (*except near coast*)
Kentucky Warbler
Hooded Warbler
American Redstart
Summer Tanager

FARMLANDS

Common Bobwhite
Mourning Dove
Ground Dove (*Coastal Plain*)
Red-headed Woodpecker
Eastern Kingbird
Barn Swallow
Carolina Wren
Northern Mockingbird
Brown Thrasher
Eastern Bluebird
Loggerhead Shrike
White-eyed Vireo
Prairie Warbler
Common Yellowthroat
Yellow-breasted Chat
Eastern Meadowlark
Orchard Oriole
Common Grackle
Northern Cardinal
Blue Grosbeak
Indigo Bunting
Rufous-sided Towhee
Bachman's Sparrow
Field Sparrow (*except near coast*)

Small numbers of shorebirds always migrate through the state; if the water level of the TVA reservoirs is low in late summer, they stop to rest and feed on the mud flats. Spring and fall movements of small landbirds on the Alabama coast show the same peculiarities as those on the Mississippi coast (*see* **Mississippi** chapter). The following timetable applies to the main flights through Alabama:

Shorebirds: 15 March–15 May; 15 August–20 October
Landbirds: 15 March–1 May; 15 September–10 November

Bird finding in Alabama during winter can be productive. Christmas counts made at Birmingham usually total 80 or more species, and on the coast usually exceed 120. Wintering populations of blackbirds, fringillids, and other passerine species such as Golden-crowned and Ruby-crowned Kinglets, Water Pipit, and Myrtle Warbler are often impressively large, as are the populations of many waterbirds and waterfowl. Large assemblages of loons, grebes, pelicans, cormorants, ducks, gulls, terns, and skimmers remain in Mobile Bay and Bon Secour Bay and on the outlying waters of the Gulf of Mexico. The reservoirs in the interior (e.g., Wheeler Lake) are havens for geese as well as for loons, grebes, cormorants, ducks, and gulls. American Coots are often abundant on the Coastal Plain, where there are stretches of open water.

Authorities

William Calvert, Jr., Lawrence S. Givens, Thomas A. Imhof, Francis X. Lueth, Henry M. Stevenson, Francis M. Weston.

Reference

Alabama Birds. By Thomas A. Imhof. 2d ed. University: University of Alabama Press, 1976.

BIRMINGHAM
Lake Purdy | Bayview Lake

Alabama's largest city and the South's leading iron- and steel-producing center, Birmingham lies in a broad valley between the southernmost spurs of the Appalachians. Beyond the suburbs much of the terrain is rugged and forest-covered, generally unsuited to agriculture, and surprisingly wild. Bird finding possibilities the year round are varied and seemingly limitless.

1. An outstanding place for birds between mid-October and mid-April is **Lake Purdy** and its environs, reached by going southeast from Birmingham on US 280. After crossing the Cahaba River about 5 miles from the city, continue for 1.6 miles to a crossroad and turn left on County 27. For 2 miles this road passes through farming country—fields, brushy pastures, roadside and dooryard hedges—where may be found American Kestrels, Common Bobwhites,

Mourning Doves, Eastern Phoebes, Winter and Bewick's Wrens (occasionally), Northern Mockingbirds, Brown Thrashers, Eastern Bluebirds, Loggerhead Shrikes, Eastern Meadowlarks, Northern Cardinals, and Savannah and Vesper Sparrows.

The road then crosses Cox's Creek, a locality for Pied-billed Grebes, Great Blue Herons, American Coots, and several species of ducks. Killdeers, Common Snipes, and small flocks of Water Pipits (winter only) feed on the exposed mud flats. In August and September these mud flats and the adjoining shallows attract a few transient shorebirds as well as egrets and other wading birds. After crossing Cox's Creek, turn left. Ahead is the main body of Lake Purdy, a huge storage reservoir for the Birmingham Water Works, formed by damming the Cahaba River. For the next 2 miles—to the eastern end of the reservoir—there are good vantage points for observing wintering birdlife: Pied-billed Grebes, American Coots, and several species of ducks, especially Ring-necked Ducks, Lesser Scaups, and Ruddy Ducks.

Another spot for winter bird finding in the Lake Purdy area may be reached by recrossing Cox's Creek and taking the first turn to the right, about 0.25 mile west of the stream. This road passes over a low ridge, then skirts the western end of Lake Purdy, and, after 1.5 miles, leads to the dam. Here one may see diving ducks to good advantage and explore the adjoining stretches of open pine woods, where Red-cockaded Woodpeckers, Brown-headed Nuthatches, and Pine Warblers are permanent residents, and Yellow-throated Warblers and Bachman's Sparrows are present from mid-March to mid-October.

2. West of Birmingham as far as the Warrior River (12 to 15 miles distant) extends rough coal-mining country with scrub pine-covered bluffs and deep, narrow, beech-covered ravines—areas particularly good between the first of April and July for breeding birds, especially warblers. To reach this area, drive southwest from the center of Birmingham on US 11 for 4.5 miles to the Alabama State Fair Grounds (or exit from I 59 on Ensley Avenue), turn right on Ensley Avenue for about a mile, and then left on 20th Street. After about 0.5 mile, 20th Street crosses a viaduct over the Ensley Steel Plant and becomes County 61, which, after 3 miles, crosses the southern end of **Bayview Lake** by bridge.

Just before the bridge, County 61 crosses a north-south road running the length of Bayview Lake and worth exploring for birds. Pied-billed Grebes maybe seen on the Lake every month except June. At its southern end is a 40-acre cattail marsh, where Virginia Rails, Soras, and Marsh Wrens are common transients and winter residents. Large numbers of Red-winged Blackbirds roost here from October to April. Barred Owls and Pileated Woodpeckers are residents in the surrounding oak-pine timber.

From the bridge at Bayview Lake, continue west on County 61, crossing two railroads on one bridge and stopping at the next (second) bridge, about 3 miles from the Lake. Descend to railroad level and walk north. Along the track for 8 miles stretches a splendid bottomland forest of beech, tulip tree, and gum; here one may expect the following breeding warblers: Black-and-white, Worm-eating, Black-throated Green, Cerulean, Yellow-throated, Pine, Prairie, Louisiana Waterthrush, Kentucky, Yellow-breasted Chat, Hooded, and American Redstart. Yellow-billed Cuckoos, Acadian Flycatchers, and Summer Tanagers are relatively abundant as well.

DECATUR
Wheeler National Wildlife Refuge

The 34,063-acre **Wheeler National Wildlife Refuge** in northern Alabama embraces Wheeler Lake (a reservoir created by damming the Tennessee River), and adjacent lands extending eastward from Decatur. Roughly half the Refuge is open water, much of it consisting of shallow backwaters or embayments near the mouths of creeks. Often the embayments are more than a mile wide. The land area supports bottomland hardwoods—oak, hickory, ash, maple, gum, hackberry—and plantations of loblolly and shortleaf pines. A portion of the land is used for crops of various kinds.

Even though Wheeler Refuge is far east of the great Mississippi waterfowl flyway and in the heart of a densely populated, highly agricultural section of the country, it has become an important wintering ground for waterfowl since its creation in 1938. At the winter peak the duck population represents all species using the Mississippi flyway. Mallards make up over half the population, and American

Black Ducks, Common Pintails, and American Wigeons are rivals for second place. The wintering population of Hooded Mergansers often exceeds 500. There are also thousands of Canada Geese and several hundred Snow Geese. Waterbirds, common on the Refuge in winter, include Common Loons, Pied-billed Grebes, Double-crested Cormorants, Great Blue Herons, American Coots, Herring Gulls, and Ring-billed Gulls.

Among breeding waterbirds and waterfowl on Wheeler Refuge or in the immediate vicinity are Great Blue Herons, Great Egrets, a few American Anhingas, a few Mallards and American Black Ducks, and many Wood Ducks. Dickcissels are common summer residents on both cultivated and abandoned farmlands. In late summer and early fall, Green Herons, Little Blue Herons, Snowy Egrets, Black-crowned Night Herons, and Yellow-crowned Night Herons frequently come in.

When Lake Wheeler's water is low between mid-August and September first, shorebirds appear on the mud flats. Common species are the Semipalmated Plover, Common Snipe, Spotted Sandpiper, Solitary Sandpiper, Greater and Lesser Yellowlegs, Pectoral Sandpiper, Least Sandpiper, and Semipalmated Sandpiper. The Black-bellied Plover, Whimbrel, Willet, Red Knot, White-rumped Sandpiper, and Stilt Sandpiper occasionally show up. American Woodcock occupy the shrubby thickets late in fall and through winter. In spring, because of high water, transient shorebirds are relatively scarce.

From mid-August to mid-October, transient waterbirds commonly include Common Terns and Black Terns. At this same time of year the first ducks from the north show up. Between mid-October and mid-November many Snow Geese stop on the Refuge. By the last two weeks of December the winter population is stabilized: all the waterfowl and waterbirds have arrived and the transients have moved southward.

To reach Refuge headquarters, drive south from Decatur on US 31 for 3 miles, then turn left on State 67 and proceed for 1.5 miles to the entrance, on the left, marked by a sign and a stone gate. In dry weather one can reach all parts of the Refuge by car; after heavy rains several of the clay roads are difficult to travel.

EUFAULA
Eufaula National Wildlife Refuge

Seven miles north of this southeastern Alabama city, the **Eufaula National Wildlife Refuge** (11,160 acres) encompasses both sides of the Walter F. George Reservoir on the Chattahoochee River, the state boundary, with 7,929 of its acres in Alabama and 3,231 in Georgia. Refuge headquarters are in Eufaula. Both US 431 and State 97, north from Eufaula, cross sections of the Refuge; State 165, which branches east from US 431 and runs north parallel to it, forms part of the Refuge boundary.

The Refuge, established in 1964, not only fulfills its primary purpose in providing still one more welcome for migrating and wintering waterfowl, but also attracts many other kinds of birds, including Great Blue Herons, Cattle Egrets, and Great Egrets in any season. Landbirds in variety stop over at the Refuge during migration. Common Snipes, Common Bobwhites, and Mourning Doves are plentiful during winter.

FLORENCE
Tennessee River Lowlands

Florence lies in the Muscle Shoals district of northwestern Alabama. Just south of the city, north of the Tennessee River, and west of US 43 are the **Tennessee River Lowlands,** with a large pasture where a few shallow ponds have been excavated. In spring and fall many species of surface-feeding ducks visit the ponds, and, if the water is low, a few shorebirds also come in. During the same seasons the marshy edges of the ponds occasionally attract visitors such as American Bitterns, Virginia Rails, Soras, Marsh Wrens, and Lincoln's Sparrows. Species nesting commonly in the pasture during late spring and early summer include the Eastern Meadowlark, Dickcissel, and Grasshopper Sparrow, and possibly a pair or two of Lark Sparrows.

Across US 43, east of highway and pasture, is a canal bordered by woods in which Warbling Vireos nest. West of the pasture is a bottomland woods around the mouth of Cypress Creek. Along this stream one may encounter breeding Blue-winged Warblers, Cerulean Warblers, and American Redstarts, and perhaps be so fortunate as to find a pair of summer-resident Swainson's Warblers in a dense thicket.

FORT PAYNE
De Soto State Park

On Lookout Mountain (elevation 1,500–1,900 feet) in northeastern Alabama is **De Soto State Park** (4,869 acres), one of the most scenic areas in the state. A highway, State 89, extends for 28 miles from the northernmost part, near Mentone, southward along the rim of the Little River Gorge to May's Gulf, a steep, 700-foot chasm into which the Little River cascades. A dense pine-hardwood forest borders the highway for a considerable distance. Rhododendrons, mountain laurels, and azaleas grow in profusion over Lookout Mountain and the upper reaches of the Gorge, and when they attain full bloom in April and early May, the color display is magnificent.

Both Chuck-will's-widows and Whip-poor-wills breed in De Soto Park, the latter species in forests at higher elevations. Among other species nesting here are the Wild Turkey, Pileated Woodpecker, Eastern Phoebe, Carolina Chickadee, Yellow-throated Vireo, Black-and-white Warbler, Worm-eating Warbler, Black-throated Green Warbler, Yellow-throated Warbler, Pine Warbler, Ovenbird, Scarlet Tanager, and Summer Tanager.

To reach De Soto Park from Fort Payne, drive east on State 35 to the top of Lookout Mountain and turn left on State 89 for 5 miles to headquarters; or drive right for 7 miles to May's Gulf, following directional signs; or drive north from Fort Payne on I 59 or US 11 to the Hammondville exit, go east on State 117 to Valley Head, and then south on State 89.

US 11 north from Fort Payne runs for 15 miles through a valley that is a migration route for swallows and excellent bird-finding country. Two miles north of Hammondville, a 300-acre pond on the edge of the highway attracts wading birds, surface-feeding ducks, and shorebirds.

JACKSON
Salt Springs Game Sanctuary | Choctaw National Wildlife Refuge

Thirteen miles south of this community in southwestern Alabama is the 5,500-acre **Salt Springs Game Sanctuary,** owned by the state and closed to the public. Lying along the Tombigbee River, the terrain is somewhat steeply rolling, with alternating hollows and ridges. Forests cover most of the area: stands of loblolly and longleaf pines in some places, mixed stands of pines and hardwoods (oaks and gums) in others. Where the forests are open, shrubs and vines form a dense ground cover. Although many species of birds characteristic of Alabama forests inhabit the Sanctuary, the large population of Wild Turkeys is a particular attraction.

Bird finders wishing to enter the area must first obtain permission and directions from the Alabama Department of Conservation (Montgomery, Alabama 36101).

For another good spot near Jackson, drive west on US 43 across the Tombigbee River, park the car where a ramp runs down to a

BACHMAN'S WARBLER

large pasture, and explore the bordering woods for a sample of bottomland birds. The Bachman's Warbler has been seen here in April.

North of Jackson and also on the Tombigbee River—on a section flooded by Jackson Dam at Coffeeville—lies the **Choctaw National Wildlife Refuge,** embracing 4,218 acres of land and water. Two small creeks divide the level river-bottomland into three sections, all broken by numerous lakes and sloughs, making it excellent habitat for migrating and wintering waterfowl. Mallards and Wood Ducks predominate among the wintering population, along with lesser numbers of American Black Ducks, Common Pintails, American Wigeons, Blue-winged Teal, Green-winged Teal, Ring-necked Ducks, and scaups.

Nesting on the Refuge are American Anhinga, Great Blue Heron, Little Blue Heron, Cattle Egret, Great Egret, Snowy Egret, and White Ibis. Wild Turkeys are common in winter. Also nesting commonly are Turkey Vultures, Black Vultures, Mourning Doves, Eastern Kingbirds, Purple Martins, Fish Crows, Northern Mockingbirds, Yellow-breasted Chats, Red-winged Blackbirds, Orchard Orioles, Northern Cardinals, and Indigo Buntings.

To reach the Refuge, drive north from Jackson on State 69 to Coffeeville; then left on US 84 for about 8 miles to a country store; turn right, following country roads to Womack Hill, and here drive eastward for a short distance. Refuge headquarters is in Jackson (Box 325, Jackson, Alabama 36545).

JACKSONVILLE
Boozer's Pond | Reed's Mill | Edwards Farm

The town of Jacksonville lies in a region of rolling hills and sheltered valleys at the end of the Appalachians in the northeastern part of the state. Even though it is off the main migration routes, in its environs there are a number of surprisingly good places for bird finding, all of them readily accessible.

Boozer's Pond, one of the best places for transient waterbirds and waterfowl, is reached by going west from the center of town for about 2 miles and turning right on a minor road that borders the Pond about one-third mile beyond. Having a surface of nearly 5 acres and being relatively shallow, with a fringe of cattails and other aquatic plants, the Pond is particularly attractive in early March and late October to such birds as Common Loons, Pied-billed Grebes, Mallards, American Black Ducks, Gadwalls, American Wigeons, Common Pintails, Blue-winged Teal, Northern Shovelers, Redheads, Ring-necked Ducks, Canvasbacks, Lesser Scaups, American Coots, and Common Gallinules.

In August and September, Great Blue Herons, Green Herons, and sometimes Little Blue Herons and Great Egrets appear. During winter the nearby meadows and fields, with their brushy edges, hold a variety of fringillids, including Savannah Sparrows, Vesper Sparrows, Chipping Sparrows, Field Sparrows, White-throated Sparrows, Swamp Sparrows, Song Sparrows, and occasionally a few Henslow's and Fox Sparrows.

Farther west, at **Reed's Mill,** about 9 miles from Jacksonville on State 204, the road cuts through a narrow pass between high hills and crosses a wide creek. Park the car near the bridge and follow a path starting on the right side of the road and running along the creek for about a mile. Except for a few fields at the water's edge, the creek is bordered by a dense wood consisting of oak, black birch, sweet gum, tulip tree, and beech, with a thick undergrowth of mountain laurel and azaleas. A walk along this path in late April and early May will undoubtedly yield the longest list of transient

WATER PIPIT

landbirds in the Jacksonville area. Among warblers nesting here in late spring and early summer are the Blue-winged, Yellow-throated, Kentucky, and Hooded.

Two miles north from the center of town on State 21 is the **Edwards Farm,** with the most extensive open country in the region. Grasshopper Sparrows nest in the wide, grassy fields; Blue Grosbeaks, Indigo Buntings, and Dickcissels, in low shrubs bordering both fields and roadsides. Horned Larks and Water Pipits frequently appear in large numbers during winter, and migrating hawks of several species—e.g., Red-tailed, Red-shouldered, Broad-winged, and Northern Harriers—pass here often during April and October.

MARION
Federal Fish Hatchery

Not far from Marion in west-central Alabama is a **Federal Fish Hatchery,** a reserve of 600 acres with about fifty ponds. Pied-billed Grebes, American Coots, and many ducks stay here from October to April. On the ponds that have been drained, shorebirds such as Solitary and Pectoral Sandpipers frequently gather during May, August, September, and sometimes the winter months. Plainly visible in a grove of trees east of the Hatchery is a Turkey Vulture and Black Vulture roost. Many individual birds use it, most often in winter.

To reach the Fish Hatchery, drive north from Marion on State 5 for about 6 miles.

MOBILE
Dauphin Island | Gulf State Park

Mobile, at the head of Mobile Bay in the southwestern corner of the state, has in its vicinity two remarkable bird-finding areas.

One of the favorite places for Mobile residents is **Dauphin Island,** a barrier strip about 14 miles long and a little over a mile at its widest point, between the Gulf of Mexico and Mississippi Sound, about 35 miles south of the city. Connected with the mainland by bridge and causeway, the Island has beaches on the Gulf side and marshes and lagoons on the Bay side. Despite the fact that it is thickly settled, with unbroken rows of cottages on its western two-thirds—only a long, narrow, barren sand spit—and many landscaped homes lining the tree-shaded streets on the eastern third, there are several areas where bird finding can be most productive.

To reach Dauphin Island from city center, drive west on I 10 a short distance and turn south on State 163. When coming from the west, leave I 10 at Theodore, go south on County 59, following signs to Bellingham Gardens, and turn right on State 163. Bellingham Gardens is worth a stop on the way to look for landbirds.

Just south of Alabama Port, State 163 cuts across a large marsh with fresh water on the east and salt water on the west. Stop here and look for Least Bitterns, King Rails, Common Gallinules, American Coots, and Marsh Wrens on the east, and for Clapper Rails and Seaside Sparrows on the west.

On a small island about 0.25 mile beyond the toll gate, Gull-billed Terns, Little Terns, and Black Skimmers nest in June. On the oyster reefs just south of the lift bridge, American Oystercatchers are permanent residents. The mud flats and shallow lagoons of Dauphin Island itself are a haven for shorebirds in winter and during migration.

After reaching Dauphin Island, continue south to Bienville Boulevard, turn left, and drive to Fort Gaines at the end of the road. In the spring—particularly late April and early May—explore the shrubs and lawns around the embankment. During migration this area may be alive with vireos, warblers, and other small passerines. Return on Bienville Boulevard for a short distance, turn left into the campground, and park on the far right side, near a gate to the eastern entrance to the Audubon Bird Sanctuary (164 acres). Enter and follow the trails beneath enormous live oaks and between tangles of bushes and vines and across the open grassy patches.

In this section of the Sanctuary, Mobile bird-banders set up their nets in May, and in one season often capture and band hundreds of warblers—Northern Parula, Cerulean, Kentucky, and Hooded—as well as orioles, tanagers, towhees, and other woodland birds. Continue west on Bienville Boulevard to Audubon Street and the western entrance to the Sanctuary, where there are stretches of open lawns and a pond.

During spring migration the bird finder need not limit his activities to the Fort and the Sanctuary on Dauphin Island. A walk up and down the shady streets anywhere in the east end may yield such birds as American Robins, Baltimore Orioles, Indigo Buntings, and Rose-breasted Grosbeaks—not by ones or twos, but by the dozen—on the lawns and in the shrubs around the homes. Although there is no public access to the shore on the west end, a drive west (still on Bienville Boulevard) to see the assemblage of swallows in fall—mostly Barn, Rough-winged, and Tree—should be worthwhile.

Also during fall, when many birds from western United States move eastward, there is always the possibility of sighting a Western Kingbird, Scissor-tailed Flycatcher, Vermilion Flycatcher, Bullock's Oriole, or Western Tanager. During winter the shorebirds gather to feed on the mud flats and reefs on the land-side of the Island, and diving ducks appear in surrounding waters.

Another excellent area not far from Mobile is **Gulf State Park** (6,000 acres) on a sandy peninsula separating Bon Secour Bay and part of Mobile Bay from the Gulf of Mexico. To reach the Park, drive east from the city on I 10, crossing Cochran Bridge and Causeway, and turn south on State 59 through Summerdale and Foley to Gulf Shores. Within the Park, turn right on State 180 for Fort Morgan, and left on State 182 for Alabama Point.

The marshes at the head of Mobile Bay, crossed by I 10 on the Bridge and Causeway, were once one of the most productive bird-finding areas on the coast—in the days when one could park and scan the lagoons, tidewater creeks, and open water for such residents as Marsh Wrens, Red-winged Blackbirds, and Boat-tailed Grackles after mid-May; for herons in late summer; for many shorebirds in May, August, and September; and for several species of diving ducks and possibly Bald Eagles from November first to April. At the present time, although the birds are still there, the traffic on this busy highway precludes any stopping or even slowing down. The only available parking area is in a vacant lot beside the Holiday Inn on the north side of the Causeway.

Along State 59 south of Summerdale, through farmland and cutover longleaf pine forest, Eastern Bluebirds on roadside fences are a common sight in winter. In any season, look for Red-cockaded Woodpeckers, Brown-headed Nuthatches, and Pine Warblers in stands of pine, and Bachman's Sparrow in open pine woods. About 2 miles south of Foley, Sandhill Cranes appear regularly on boggy land in open pine forests and are sometimes visible from the highway. A few may be permanent residents.

Gulf State Park, which extends for 27 miles east of Fort Morgan, has a wide beach and extensive dunes, three fresh-water lakes, and vegetation varying from forests of scrub oak and pine to areas of holly, myrtle, scrub palmetto, and grass. Both beach and dunes are remarkably white, thus producing a severe glare even in winter. Anyone spending any length of time in this area should wear sun glasses to guard against 'snow' blindness.

During spring, late summer, and fall, shorebirds of many species appear in fair numbers at Gulf Shores (at the end of State 59) on the beach and around the marsh pools and lagoons. Transient Black Terns, common in May, become very abundant in August and September. Brown Pelicans, Laughing Gulls, Royal Terns, Caspian Terns, and Black Skimmers are present offshore in any season, though more frequently in winter (December to April), when the following are also in evidence: Northern Gannet, Double-crested Cormorant, Herring Gull, Ring-billed Gull, Bonaparte's Gull, Forster's Tern, and Common Tern.

From Gulf Shores east to Alabama Point, State 180 follows along the beaches and dunes, providing opportunities to see waterbirds and an occasional Bald Eagle.

State 182 west to Fort Morgan (which in the days of wooden ships guarded the entrance to Mobile Bay) passes through well-forested country, worth investigating for birds found commonly in woods of southern uplands. To the left of the road is the site of a Great Blue Heron colony active from late January to mid-June. The direction taken by the big birds as they cross the road will guide the bird finder to the site.

The last few miles of the road afford good views of the outlying waters of the Gulf and of both Bon Secour Bay and Mobile Bay, where Common Loons, Horned Grebes, Lesser Scaups, Common Goldeneyes, and Red-breasted Mergansers stay in winter. Here also are small, marshy ponds around which transient shorebirds are sometimes common and near which Willets nest from late April to mid-June. On the extreme western end, another reward in any season may be the sight of a Snowy Plover, a pale little bird easily overlooked as it stands motionless on the light sand.

At Fort Morgan, Magnificent Frigatebirds sail overhead from July through September, sometimes no more than one or two, sometimes in circling flocks of as many as a hundred individuals. During late spring and summer a walk through the sally port of the old fort to one of the more recent fortifications leads to a large group of nesting Barn Swallows—a rarity in the South. At the same times of year a few pairs of Wilson's Plovers and small colonies of Little Terns may nest on the sand flats between the old fort and the beach. From the wharf at the end of the road, or from abandoned wharves along the bayside, the bird finder may always observe the continuous activity of many kinds of waterbirds.

TUSCALOOSA
Moody Swamp

Wooded **Moody Swamp,** southwest of Tuscaloosa in the west-central part of the state, has a rich variety of breeding birds, possibly including the Bachman's Warbler, a secretive bird that usually begins nesting during March or April and can best be found at that time, when it is in full song. By mid-May this scarce parulid becomes silent and is thus difficult to locate. Blackberry thickets, a favorite nesting habitat of the species, are abundant on the edges of Moody Swamp. Species nesting regularly, if not commonly, in Moody Swamp include the Pileated and Red-bellied Woodpeckers, Acadian Flycatcher, Cerulean Warbler, Kentucky Warbler, Hooded Warbler, American Redstart, and probably the Swainson's Warbler.

To reach Moody Swamp, drive southwest from the center of Tuscaloosa on US 11 to 32nd Avenue, turn left, and drive south. After about 2 miles the road begins to cross the Swamp. The best section for bird finding is on the right.

Connecticut and Rhode Island

GREATER SCAUP

While the seaworn headlands and open beaches of Maine, New Hampshire, and Massachusetts are being lashed by winter winds and whitecapped surges, rafts of Greater Scaups ride easily on the sheltered waters of Rhode Island's Narragansett Bay and Connecticut's New Haven Harbor. Even such hardy coastal birds as Oldsquaws, Common Eiders, Surf Scoters, and Dovekies seek the lee shores of Block Island, Sakonnet Point, and Point Judith, or the quieter surfaces of Long Island Sound. As one Rhode Island ornithologist remarked, when extolling the possibilities for winter bird finding on the coast of southern New England, 'It can get downright raw in winter here, but there are enough birds to offset the cold, even in January.'

The Connecticut coastline, the entire length of which faces Long

Island Sound, is typical of New England, showing a disorderly succession of rocky headlands, sandy beaches, harbors, bays, and estuaries. Salt marshes fringe many of the coastal indentations and lie behind the beaches. Rhode Island, on the other hand, boldly fronts on the Atlantic, with a series of salt ponds and occasional salt marshes separated from the open sea by a barrier of low bluffs and sand beaches. The principal indentation—big, gaping Narragansett Bay—cuts deeply into the state for over 28 miles, its many islands and points of land forming a labyrinth of coves and channels. Islands off the coast of the two states are relatively few, only Block Island southeast of Rhode Island being of any prominence or special ornithological interest.

Both the number and the variety of birds breeding on the Connecticut and Rhode Island coasts are limited. Common Terns, together with a few Roseate Terns, maintain small colonies on little islands close to shore. Herring Gulls and, in recent years, Greater Black-backed Gulls breed on the islands directly off Sakonnet Point in Rhode Island. Piping Plovers and Little Terns sometimes find suitable habitat on undisturbed beaches. Far more abundant are the birds of the salt marshes: the Clapper Rails near the northern limits of their range, the Sharp-tailed and Seaside Sparrows, whose ranges overlap, and the Marsh Wrens and Red-winged Blackbirds, which are characteristic also of fresh-water marshes. The best salt marshes are in Connecticut—e.g., in the Barn Island and Great Island Wildlife Management Areas near Stonington and Old Lyme, respectively.

In both states the terrain near the coast is low and slightly rolling. Northward the slopes become more pronounced. In Connecticut two distinct series of rugged hills comprise an eastern highland and a western highland, between which lies a central lowland with gentle contours—except near New Haven, where traprock ridges extend from the coast. The western highland has the greater elevation of the two, culminating in Bear Mountain (2,355 feet) in the northwestern corner. In Rhode Island the highest point (805 feet) is Durfee Hill in the northwest.

In former times Connecticut and Rhode Island were heavily covered with both deciduous and coniferous trees, their distribution varying according to the condition of soils and drainage. Chestnuts, oaks, hickories, tulip trees, sugar maples, white pines, and hemlocks were widespread; red spruces grew on the higher elevations of northwestern Connecticut; pitch pines on the sandy plains near the coast; elms, red maples, and white cedars in the swamps; cottonwoods and silver maples along the streams. The original forests have been decimated; nearly half the area of the two states is today covered with second-growth woods, since much of the land is unsuited to agriculture, and a large part of the population is confined to towns and cities. The composition of present-day forestation differs from the original mainly in the greater abundance of deciduous trees such as oaks and hickories.

The farms of Connecticut and Rhode Island are small for the most part, and produce a variety of commodities for nearby markets. Each farm has its hayfields, brushy pastures, orchards, fence rows, woodlots, and neat, tree-shaded buildings. In the central lowlands of Connecticut, however, where the loam is rich, the farms are devoted principally to tobacco production, and thus wide stretches of land are under cultivation.

Characteristic breeding birds of the Connecticut and Rhode Island deciduous woods and farmlands are the following:

DECIDUOUS WOODS

Red-shouldered Hawk
Broad-winged Hawk
Ruffed Grouse
Yellow-billed Cuckoo
Black-billed Cuckoo
Common Screech Owl
Barred Owl
Whip-poor-will
Yellow-shafted Flicker
Hairy Woodpecker
Downy Woodpecker
Great Crested Flycatcher
Least Flycatcher
Eastern Pewee
Blue Jay
Black-capped Chickadee
Tufted Titmouse
White-breasted Nuthatch
Wood Thrush
Yellow-throated Vireo
Red-eyed Vireo
Warbling Vireo
Black-throated Green Warbler
Ovenbird
American Redstart
Baltimore Oriole
Scarlet Tanager
Rose-breasted Grosbeak

FARMLANDS

- Common Bobwhite
- Mourning Dove
- Eastern Kingbird
- Eastern Phoebe
- Tree Swallow
- Barn Swallow
- House Wren
- Northern Mockingbird
- Gray Catbird
- Brown Thrasher
- Eastern Bluebird
- Yellow Warbler
- Chestnut-sided Warbler
- Common Yellowthroat
- Bobolink
- Eastern Meadowlark
- Common Grackle
- Northern Cardinal
- Indigo Bunting
- House Finch
- American Goldfinch
- Rufous-sided Towhee
- Savannah Sparrow
- Grasshopper Sparrow
- Vesper Sparrow
- Chipping Sparrow
- Field Sparrow
- Song Sparrow

Along the coast and in some lowlands extending northward, a few species of more southern affinities regularly breed in suitable habitats. These include the Acadian Flycatcher (mainly southwestern Connecticut), Carolina Wren, White-eyed Vireo, Black-and-white Warbler, Worm-eating Warbler (probably Connecticut only), Blue-winged Warbler, Prairie Warbler, Louisiana Waterthrush, Yellow-breasted Chat, Hooded Warbler, Orchard Oriole, and Henslow's Sparrow. In the Connecticut highlands, notably in the northwest, northern birdlife is represented by such summer residents as the Hermit Thrush, Solitary Vireo, Nashville Warbler, Black-throated Blue Warbler, and Canada Warbler.

For their small size, the two states have a goodly number and variety of watery environments. Hundreds of lakes and ponds dot their landscape, and there are many rivers, all passing southward, all broadening and deepening as they approach the sea, and nearly all forming wide mouths. Three Connecticut rivers predominate: the Connecticut, cutting through the central lowland from northern New England; the Thames with its tributaries, coming from the eastern highland; and the Housatonic, flowing down through the western highland. Many rivers in both states are bordered by fresh-water marshes and wooded swamps, some of which are impressive in extent and provide the bird finder with a rewarding area for exploration. One such marsh lies along the Connecticut River north of Hartford; another is near the mouth of the Quinnipiac River in the vicinity of New Haven. A fascinating area, quite unlike any other in southern New England, is the Great Swamp in southeastern Rhode Island near Kingston. Although ornithologically it has been only partially explored, enough is known of its birdlife to indicate that it has an exceptional combination of woodland- and marsh-inhabiting species.

On the whole, the water environments of neither state hold spectacular breeding aggregations. But this lack is readily compensated for in spring and fall, when vast numbers of warblers and other passerine species move in waves along the river valleys, shorebirds gather on beaches, and waterbirds and waterfowl abound in the bays and harbors. The larger valleys, especially the Connecticut River Valley, are migratory routes for many landbirds. In adjoining swampy thickets—e.g., that on the Connecticut River north of Hartford— bird finders have an excellent opportunity in September and early October to observe the Connecticut Warbler, a species discovered by Alexander Wilson when he was visiting Connecticut more than 175 years ago. A few waterfowl, waterbirds, and shorebirds follow the river valleys, but the majority pass along the coastline. In both Connecticut and Rhode Island, peak flights may be expected within the following dates:

Waterfowl: 15 March–15 April; 10 October–25 November
Shorebirds: 1 May–1 June; 1 August–1 October
Landbirds: 15 April–20 May; 25 August–20 October

In winter the coastal region is remarkably productive. If a tour is so arranged as to include woods, brushy fields, open fields, rivers, ponds, bays, and views of the open sea, it should easily yield a hundred or more species, very likely including the following: Common Loon, Red-throated Loon, Horned Grebe, Great Cormorant, American Wigeon, Greater Scaup, Common Goldeneye, Bufflehead, White-winged Scoter, Red-breasted Merganser, Red-tailed Hawk, American Kestrel, Purple Sandpiper, Greater Black-backed

Gull, (Northern) Horned Lark, Brown Creeper, Eastern Bluebird, Golden-crowned Kinglet, Cedar Waxwing, Myrtle Warbler, Evening Grosbeak, Purple Finch, Pine Siskin, Slate-colored Junco, American Tree Sparrow, White-throated Sparrow, Fox Sparrow. If the Short-eared Owl and (Ipswich) Savannah Sparrow are on the list, consider the trip a success!

Authorities

Stanley C. Ball, E. Alexander Bergstrom, Thomas H. Bissonnette, Martha M. Capizzano, Ralph Case, Edward C. Childs, Roland C. Clement, Paul Desjardins, Elizabeth Dickens, John G. Erickson, William John Frank, Harold N. Gibbs, Alfred L. Hawkes, John P. Holman, Douglas L. Kraus, Arroll L. Lamson, Edward B. Lang, Harold L. Madison, Thomas P. McElroy, Jr., Robert Moeller, Charles E. Mohr, S. Dillon Ripley, Eleanor H. Stickney, Ralph B. Wainright, Kenneth E. Wright, Thomas J. Wright.

Reference

Places to Look for Birds. An Area Guide to Birding. Hartford: Connecticut Board of Fisheries and Game, State Office Building, 1970.

BLOCK ISLAND, RHODE ISLAND

Block Island, 11 miles square, 12 miles southwest of Point Judith, and 12 miles east of Montauk Point on Long Island, lies along the coastal flyway and is thus of primary interest as a stopping place for migrating and wintering birds. Although the Island is mainly a summer resort, a number of people reside here permanently; hence it has regular plane and ferry service from the mainland the year round.

About 6 by 3.6 miles in greatest diameters, Block Island is surrounded by rocky coast, except for sand dunes at the north end and beaches on the northeast and west. Great Salt Pond, which all large boats must cross to land at New Harbor (the chief port), almost bisects the Island. There are low hills, a few marshes, and many fresh-water ponds fed by springs. On the south and east the clay bluffs of Mohegan and Clay Heads rise to 150 feet. Very few trees of the original forest remain; the principal cover now consists of grasses, thickets of low-growing shrubs (such as bayberry), which have overrun abandoned farmlands, and old apple orchards.

At different seasons the birds to be viewed on the Island or offshore include the following:

Late November through early March: Red-throated Loons, Rednecked Grebes, Northern Gannets, Great Cormorants, Common Eiders, White-winged Scoters, Surf Scoters, Black Scoters, Redbreasted Mergansers, Greater Black-backed Gulls, Herring Gulls, Ring-billed Gulls, Bonaparte's Gulls, and Purple Sandpipers, and, occasionally, an Iceland Gull, Black-headed Gull, Black-legged Kittiwake, or Dovekie

March, late October through early November: Canada Geese and other transient waterfowl

March through April, late September through early November: many transient fringillids

May, and mid-September: waves of warblers

May, and August through September: transient shorebirds

Good roads lead to good places for bird finding: the fishing docks at Old Harbor and New Harbor, for gulls; the sandy beaches, particularly those on the north side of Old Harbor, for shorebirds; the leeward side of the Island and the many ponds, for waterbirds and waterfowl; the thickets and fields around the ponds and along the edges of beaches on the north end, and the interior of the Island, at Center Crossroads, for warblers, fringillids, and other small landbirds. Center Crossroads is reached by taking Ocean Street south from New Harbor and turning right on Center Road. The area along Center Road between Beacon Hill Road and Conneymus Road is well worth investigating for landbirds.

Block Island is a remarkable 'trap' for small landbirds migrating south following a cold front in fall. At night they come down on this bit of land after being blown to sea by strong northwest winds, and here they stay, held for the succeeding day or longer—until the winds moderate and they have gained sufficient energy to return to the mainland and continue their flights.

If there is a 'best' time to visit Block Island, it is during the first ten days of October, during which period the Audubon Society of

Rhode Island traditionally—and appropriately—schedules its Annual Block Island Birding Weekend. Right after a cold front the Island literally swarms with small birds. Because the Island is almost treeless, the birds are in plain view, and many are consequently out of place ecologically. Brown Creepers, Red-breasted Nuthatches, and Golden-crowned Kinglets resort to foraging in shrubs, on driftwood, and even on the ground. They show up on porches and on the roofs of buildings. The species one sees in greatest abundance on a given day depends on the species that happened to be moving at the time the cold front set in. Sometimes, for example, the warblers may consist almost entirely of Myrtles and Palms, with no others in sight. What is so impressive about the aggregation of trapped birds is the quantity, not the variety, of the few species.

If at all possible, take the trip to Block Island by ferry rather than by plane, as there is always a good chance of seeing a shearwater, storm petrel, or some other oceanic species from the boat. For current information on ferry and flight service, as well as for a map of the Island and a list of accommodations, write the Block Island Chamber of Commerce (Block Island, Rhode Island 02807).

BRIDGEPORT, CONNECTICUT
Seaside Park

For convenience in observing waterbirds and waterfowl in winter, **Seaside Park** (210 acres) surpasses most places on the Connecticut shore. Although the species appearing here between November and April are not unusual, and in some cases are not equal in variety and number to those found at the mouth of the Connecticut River at Old Lyme, the fact that one may watch the birds from a warm car appeals to some bird finders.

From Exit 26 on the Connecticut Turnpike, drive east on Railroad Avenue and turn right on South Avenue, Park Avenue, or Main Street to Sound View Drive (also called Marine Boulevard), which runs along the sea wall for 2.5 miles. From this drive the bird finder will have good views of Long Island Sound and the sheltered waters of Black Rock Harbor.

CHARLESTOWN, RHODE ISLAND
Kimball Bird Sanctuary

Kimball Bird Sanctuary (29 acres) is on the southeastern shore of Watchaug Pond in the Burlingame Reservation, a state park and game refuge. Once a private estate but now owned by the Audubon Society of Rhode Island, the Sanctuary has a pond, oak woods with low undergrowth through which a trail winds, a small field, and a house and boathouse surrounded by plantings of conifers and shrubs. Gray Catbirds, Chestnut-sided Warblers, Prairie Warblers, and Rufous-sided Towhees nest commonly.

The pond, containing aquatic plants attractive to waterfowl, is one of the stops for a great number of Ring-necked Ducks, as well as American Black Ducks and Greater and Lesser Scaups. The birds arrive in October and stay until the ice forms; in the spring they reappear as soon as the ice is out of the pond. If ice fails to form during winter, they remain throughout the season. Canada Geese not infrequently share the pond with the ducks.

To reach the Sanctuary from the junction of US 1 and State 2 in Charlestown, proceed westward about 2 miles along US 1, turn right on an unnumbered road opposite Charlestown Naval Air Station, bear left at the first intersection, and continue to the end of the road, about 1.5 miles from US 1.

CLINTON, CONNECTICUT
Hammonassett Beach State Park

A number of roads traverse **Hammonassett Beach State Park** (1,015 acres), through groves of deciduous trees and shrubs in the interior, to a 5-mile sand beach on Long Island Sound, to extensive marshes toward Clinton Harbor, to Meigs Point and the breakwater.

The visitor is almost certain to find some birds of interest here except during the summer months, when the Park is crowded with people. The area is outstanding ornithologically by reason of the migrating warblers, sparrows, and other small landbirds that stop in May, the shorebirds in May, August, and September, and the enormous numbers of waterbirds and waterfowl that may be seen offshore in late fall and winter. Sometimes there are many Greater Scaups, together with Common Loons, Red-throated Loons, Horned Grebes, Oldsquaws, White-winged Scoters, Surf Scoters, Black Scoters, and Red-breasted Mergansers. The greatest concentrations of waterfowl appear off Meigs Point and breakwater: the shorebirds, on the east side of the Point; the landbirds, in the vegetation along the shore and in the interior.

From Exit 62 on I 95, drive straight south for about a mile and cross US 1 to the entrance, turn left at the first circle, and leave the car in the parking lot.

EAST GREENWICH, RHODE ISLAND
Goddard Memorial State Park

On Popowomut Peninsula, directly across Narragansett Bay from East Greenwich, is **Goddard Memorial State Park** (472 acres), once a private estate. The Park grounds contain a sandy beach on the Bay, backed by a narrow belt of oaks with some white pine and bordered by young Douglas fir, Norway spruce, and deciduous shrubs; two small ponds, one surrounded by white, Scotch, and Austrian pines, the other, by thickets of deciduous shrubs; large, sandy areas with bayberry and blackberry predominating; and one of the largest stands of white oaks in Rhode Island.

Birds nesting in the Park include the American Kestrel, White-breasted Nuthatch, Gray Catbird, Wood Thrush, Red-eyed Vireo, Yellow Warbler, Chestnut-sided Warbler, Prairie Warbler, American Redstart, Scarlet Tanager, and Grasshopper Sparrow. In summer, Black-crowned Night Herons, American Bitterns, and Belted Kingfishers frequent the small pond at the northern end. During fall and winter American Black Ducks and rafts of Greater Scaups that gather in the Bay may be viewed from the beach.

To reach Goddard Park, drive south from East Greenwich on US 1 for about a mile, bear left for a short distance, and left again into the Park.

GREENWICH, CONNECTICUT
Audubon Center in Greenwich | Audubon Fairchild Garden | Tod's Point

In the southwestern extremity of Connecticut, 35 miles from New York City, are the **Audubon Center in Greenwich** (352 acres) and the **Audubon Fairchild Garden** (125 acres), both owned and maintained by the National Audubon Society. At the Center the bird finder may explore grassy meadows, brushy pastures, deciduous woodlands, ponds, and streams along well-maintained trails. Among the eighty-five bird species known to nest in the area are the Ruffed Grouse, American Woodcock, Yellow-billed and Black-billed Cuckoos, Pileated Woodpecker, Great Crested and Traill's Flycatchers, Scarlet Tanager, Rose-breasted Grosbeak, and at least fourteen species of warblers.

Although the Center is worth visiting at any season, winter is the least productive. Early May, at the height of the warbler migration, and late May to mid-July, during the nesting season, are the most rewarding. The primary aim of the Center with its interpretive building, workship, and permanent staff, is to instill in all visitors, young and old, an appreciation of the natural environment and a sense of responsibility for maintaining it.

Fairchild Garden, a short distance from the Center, contains more than six hundred species of plants from trees to wildflowers. Pleasant trails lead through birch, beech, and maple woods, stands of hemlock, and damp meadows.

To reach the Audubon Center: From Merritt Parkway at Exit 28, drive north on Round Hill Road for 1.4 miles, turn left at the Round Hill Church on John Street, and continue for 1.5 miles to the entrance at the corner of John Street and Riversville Road. To reach the Fairchild Garden: From John Street at the Center entrance, turn left on Riversville Road, proceed for about 0.3 mile, then go left on Prochuck Road for about 0.5 mile to the entrance. Write Audubon Center, Greenwich, Connecticut 06803, for current information on scheduled bird walks and courses at the workshop.

Tod's Point (sometimes called Greenwich Point) extends into Long Island Sound south of Old Greenwich. Although a town park, and restricted to area residents between 30 May and 31 October, the

Point fortunately is open to the general public at the time when bird finding is at its best. About 1.7 miles long, and only 0.3 mile at its widest, the Point has gravelly shores with stretches of sand and grassy patches. There are two small ponds, as well as a small marsh along the Hammonassett River. Vegetation ranges from a small grove of pines and a few mature deciduous trees to open grasslands and stretches with tangled vines and brush.

In May, migrating flycatchers, warblers, and sparrows (especially Savannah Sparrows) are abundant. This is also a good time for wading birds—Great Blue Herons, Black-crowned Night Herons, Great Egrets, and, possibly, a few Yellow-crowned Night Herons. During fall, winter, and spring a variety of waterbirds and waterfowl assemble on the Sound or in the sheltered harbors: Great Cormorants, Canada Geese, Snow Geese, American Black Ducks, scaups, Common Goldeneyes, scoters, and mergansers. Others often sighted are Common Loons, Red-throated Loons, and Horned Grebes.

To reach Tod's Point, leave the Connecticut Turnpike at Exit 5, drive east on US 1 for 0.2 mile, go right on Sound Beach Avenue for 1.8 miles, and right again on Shore Road for 0.9 mile to the park entrance.

HARTFORD, CONNECTICUT

Connecticut River at South Windsor | Reservoir No. 1
Batterson Park Pond

Hartford, almost in the geographical center of Connecticut and on the west bank of the Connecticut River, lies on a major north-south migration route. The River, wet meadows, tobacco fields, wooded areas, and reservoirs in and around the city offer good stopping places for transient birds in both spring and fall.

An outstanding spot for bird finding is along the **Connecticut River at South Windsor,** 6 miles north of Hartford. To reach this site, drive east on I 84 across the River, go north on US 5 for about 1.5 miles, and, at McAuliffe Park, bear left on Main Street to South Windsor. In South Windsor, turn left on an unnumbered road (known locally as Station 43), leave the car, and walk down along the River.

In this area a diversity of habitats—e.g., a small marsh, shrubby thickets, open meadows, and the River itself—brings a corresponding diversity of birds. Least Bitterns, Virginia Rails, Soras, Common Gallinules, and Marsh Wrens nest in the marsh. Solitary Sandpipers are among the shorebirds that sometimes appear in late summer. Small landbirds are numerous in migration during May, August, and September; the Connecticut Warbler is present regularly in September and early October.

Another outstanding spot for bird finding is in the immediate vicinity of the Metropolitan District Commission's **Reservoir No. 1** in West Hartford. Although the Reservoir water attracts few birds, the surrounding mixed deciduous-coniferous woods provides habitat for many species. Ruffled Grouse and Pileated Woodpeckers reside the year round; Broad-winged Hawks, Brown Creepers, Golden-winged Warblers, and Louisiana Waterthrushes are among a wide array of summer residents.

Drive west from Hartford on I 84, turn off at Exit 41, and proceed north on Main Street to the Boulevard; turn left here, continue for a short distance, and turn left again on Farmington Avenue, which soon passes Reservoir No. 1 on the right.

More attractive to waterfowl for feeding and resting is **Batterson Park Pond** in the southwestern part of the metropolitan area, reached by continuing west from Hartford on I 84, turning off at Exit 37, crossing the freeway, and turning left on Batterson Park Road, which follows the north edge of the Pond. Ruddy Ducks and Hooded Mergansers are regular visitants in late October and November.

KINGSTON, RHODE ISLAND

Great Swamp | Worden Pond

For the hardy bird finder who likes to paddle and pole a canoe or skiff along brush- and vine-choked streams into territory that has unlimited possibilities for unusual birds and yet is almost impenetrable on foot, Rhode Island has the **Great Swamp,** covering 8 square miles and bordered on its southeastern edge by **Worden Pond.**

The composition of the Great Swamp is relatively complex. There are low open stretches with grass hummocks and clusters of shrubs, which in wet seasons are under 6 to 18 inches of water and in dry seasons are interspersed with pools of various sizes and often obscured by a thick cover of aquatic plants. There are quaking sphagnum bogs; other bogs covered with cedar and a heavy undergrowth of rhododendron; moist areas densely forested with red maple, tupelo, pin oak, and swamp white oak; and well-drained areas with black oak, white oak, and white pine. In sharp contrast are dry sandy 'islands' with open stands of pitch pine, where blueberries and other low shrubs grow.

It was in the Great Swamp that the late Harry S. Hathaway, well-known Rhode Island ornithologist, did much of his field work and discovered the first recorded Rhode Island nesting of the Louisiana Waterthrush. By late May and early June the following breeding species are fairly common in the Great Swamp: American Bittern, American Black Duck, Wood Duck, Yellow-shafted Flicker, Downy and Hairy Woodpeckers, Tree Swallow, Black-capped Chickadee, White-breasted Nuthatch, Gray Catbird, Wood Thrush, Veery, Black-throated Green Warbler, Chestnut-sided Warbler, Prairie Warbler, Ovenbird, Northern Waterthrush, Common Yellowthroat, Canada Warbler, American Redstart, Red-winged Blackbird, Scarlet Tanager, Rufous-sided Towhee, Swamp Sparrow, and Song Sparrow. Perhaps less common are the Red-shouldered Hawk, Broad-winged Hawk, Carolina Wren, Marsh Wren, White-eyed Vireo, Solitary Vireo, Blue-winged Warbler, Northern Parula Warbler, and Louisiana Waterthrush.

For the only approach to the Swamp accessible by car—and one that gives the bird finder a slight idea of the Swamp's possibilities—drive west from Kingston on State 138, then left on State 2 for 2.5 miles, and left again on a road where a sign gives directions to the site of the Great Swamp Fight. Follow this road for about a mile, park the car, and walk along a path to the monument commemorating the Great Swamp Fight on 19 December 1675, when the colonists destroyed the winter camp of the Narragansett Indians. Bird finding along this path can be worthwhile.

There are two canoe routes that follow meandering streams through the heart of the Swamp. (1) Take State 138 west from Kingston for about 2 miles to a bridge across the Chipuxet River. This stream, after wandering about for 4 miles, empties into the north side of Worden Pond. (2) Drive west on State 138 and south on State 2 for 2 miles to a bridge crossing the Queen's River (sometimes called the Usquepaug River). Paddle downstream about 2 miles on this River to its intersection with a larger one—Charles River, then upstream on the Charles to the west side of Worden Pond.

Worden Pond (1,400 acres), largest fresh-water pond in Rhode Island, has an average depth of only 3.5 feet and many submerged aquatic plants. The ducks most commonly seen during spring and fall are, in order of decreasing abundance, Lesser Scaups, American Black Ducks, Greater Scaups, Ruddy Ducks, Common Mergansers, Common Goldeneyes, and Buffleheads.

To look for ducks on Worden Pond, take State 138 west from Kingston for about 2 miles, go left on State 110 for 3.9 miles, and then right to a parking area on the shore of the Pond.

LITCHFIELD, CONNECTICUT
Litchfield Nature Center

Litchfield Nature Center in northwestern Connecticut embraces a 4,400-acre sanctuary and museum owned by the White Memorial Foundation. When visiting the Center, stop first at the museum near the entrance for a map of the area. Drive west from Litchfield on US 202 for 2.2 miles, and turn into the well-marked entrance on the left.

The property provides a good variety of habitats: (1) four small ponds, one of which—Little Pond—has a fine marsh accessible by a board walk; (2) a portion of the shoreline of Bantam Lake, the largest natural fresh-water lake in the state; (3) a part of the meandering Bantam River; (4) deciduous woods containing conifers, mostly pines; and (5) various wooded swamps, alder thickets, wet meadows, and old fields reverting to shrubs and trees. Also on the property close to Little Pond is the Litchfield Wild Garden, with plant species native to Connecticut. More than 30 miles of roads and trails wind through nearly all sections of the sanctuary, which is open to the public at all times.

In early May and in September, when warbler migrations are at their height, twenty-three species may be recorded in the Wild Garden. At the end of August and through September, transient shorebirds—notably Greater Yellowlegs and Solitary Sandpipers—appear at Little Pond; later in October and in November, migrating ducks such as Blue-winged and Green-winged Teal, American Wigeons, Canvasbacks, Redheads, and Ruddy Ducks stop off here. Species nesting in the marsh at Little Pond and its immediate vicinity are Pied-billed Grebes, American Black Ducks, Wood Ducks, American Bitterns, Green Herons, Virginia Rails, Marsh Wrens, Sedge Wrens, Red-winged Blackbirds, and Swamp Sparrows.

Species known to nest in other environments of the Wild Garden include the Ruffled Grouse, American Woodcock, Barred Owl, Mourning Dove, Black-billed Cuckoo, Hairy Woodpecker, Least Flycatcher, Black-capped Chickadee, Wood Thrush, Veery, Red-eyed Vireo, Black-and-white Warbler, Blue-winged Warbler, Black-throated Blue Warbler, Black-throated Green Warbler, Canada Warbler, American Redstart, Bobolink, Eastern Meadowlark, Bal-

BOBOLINK

timore Oriole, Rose-breasted Grosbeak, Indigo Bunting, Purple Finch, Rufous-sided Towhee, Grasshopper Sparrow, and Field Sparrow. Transient ducks stop on Bantam Lake from the last of September to mid-December and from early March until May; a nonmigratory flock of Canada Geese is here the year round.

MYSTIC, CONNECTICUT
Denison Pequotsepos Wildlife Sanctuary

Denison Pequotsepos Wildlife Sanctuary (125 acres), owned by the Denison Society, has a variety of habitats for landbirds. About two-thirds of the Sanctuary is covered with second-growth deciduous woods and brush; the remainder consists of fields, lowland swamps, a fresh-water marsh, and several ponds and streams.

Bird finding in the Sanctuary is particularly rewarding at the height of the warbler migration during the first week of May, and in the nesting season in late May and June. Some of the species nesting regularly are the Ruffed Grouse, Hairy Woodpecker, Great Crested Flycatcher, Least Flycatcher, White-breasted Nuthatch, House Wren, Wood Thrush, White-eyed Vireo, Yellow-throated Vireo, Black-and-white Warbler, Blue-winged Warbler, Chestnut-sided Warbler, Prairie Warbler, Hooded Warbler, Eastern Meadowlark, Baltimore Oriole, and Rufous-sided Towhee.

To reach the Sanctuary, take the Mystic Seaport Exit south from I 95 on State 27 to Coogan's Boulevard; follow this road, and at its end go right for 0.75 mile, then right again on Pequotsepos Road for 0.5 mile to the entrance on the left. Obtain a map at the museum before starting to look for birds.

NEW HAVEN, CONNECTICUT
East Rock Park | Sleeping Giant State Park | Quinnipiac Marshes | Konold's Pond | West Haven Sand Spit | New Haven Flats

Commanding a fine harbor at the mouth of the Quinnipiac River, New Haven has a wide variety of bird-finding places such as deciduous woods, marshes, and sand spits.

East Rock Park (3 square miles), northeast of the city, embraces a traprock ridge, locally known as 'The Rock,' with 150-foot cliffs rising from a steep, 150-foot talus. On the west and south the Rock is skirted by the Mill River, with its nearby cattail marshes, alder thickets, fields, and groves of oak, maple, and beech. On the east the Rock slopes down through a residential section to the Quinnipiac River. With the exception of the cliffs and some 4 acres of lawn and shrubs, the Rock is forested with deciduous woods and patches of hemlock.

The best part of the Park for birds—that bordering on the Mill River—is reached by taking State 10A from New Haven and turning right on East Rock Road. Just before crossing the bridge over the Mill River, park the car and take a footpath through the woods and over low, wet ground, where cattails, briars, and shrubs grow thickly. During the nesting season in late May and June, this walk should yield such species as the American Black Duck, Wood Thrush, Veery, Worm-eating Warbler, Blue-winged Warbler, Black-throated Green Warbler, Prairie Warbler, Louisiana Waterthrush, Hooded Warbler, American Redstart, Scarlet Tanager, and Indigo Bunting. A walk along this same footpath during early May will be rewarded by views of many transient warblers and other small passerines. In winter, Pied-billed Grebes and Common Mergansers can be seen on the Mill River, unless it freezes over.

Sleeping Giant State Park (1,246 acres), with a forest of mixed deciduous trees and some brushy thickets, is a fine area for warblers in early May. About 28 miles of trails meander through the area. Drive north on State 10A, cross the Wilbur Cross Parkway, continue on State 10, and turn right on Mt. Carmel Road, proceeding to the Park entrance on the right.

The **Quinnipiac Marshes,** which once covered 5 square miles along the Quinnipiac River, are gradually disappearing near the city, but there is still a section where one may find birds. For this section, drive north from New Haven on US 5, cross the River, and take the second road on the right.

Birds nesting in the Marshes include American Bitterns, American Black Ducks, Marsh Wrens, Red-winged Blackbirds, and Swamp Sparrows. Black-crowned Night Herons and other herons may be seen along the River in the late summer. During October and November thousands of Red-winged Blackbirds, Rusty Blackbirds, and Common Grackles roost in the cattails. In spring and fall, Soras and Yellow Rails are occasional transients, as are Short-eared Owls.

State 69 north from the center of New Haven passes West Rock Park—almost a twin to East Rock Park and a fine place for migrating warblers—and then skirts **Konold's Pond** (0.5 mile in diameter) at the foot of the West Rock talus. This small body of water is surrounded by marshes, fields, brushy areas, and forests of mixed deciduous trees and hemlocks. Birds such as Pied-billed Grebes, American Bitterns, American Black Ducks, Common Gallinules, Red-winged Blackbirds, and Swamp Sparrows are summer residents in the marshy areas, and numerous passerines nest nearby. In spring and fall, transient Canada Geese, American Wigeons, Redheads, Greater Scaups, and American Coots gather on the open water.

West Haven Sand Spit, 0.5 mile long and 60 yards wide, stretches out into the mouth of New Haven Harbor. Its only cover consists of scattered clumps of beach grass, and its only nesting species is the Spotted Sandpiper; but the Sand Spit becomes important ornithologically in September, when shorebirds, such as Ruddy Turnstones, Least Sandpipers, Dunlins, Short-billed Dowitchers, Semipalmated Sandpipers, Sanderlings, and, rarely, Northern Phalaropes, begin to appear in migration. Transient species seen from the Sand Spit from November to March include the Common Loon, Red-throated Loon, Horned Grebe, Greater Scaup (often in the thousands), Common Goldeneye, Oldsquaw, White-winged Scoter, Surf Scoter, Black Scoter, Greater Black-backed Gull, Herring Gull, Ring-billed Gull, and occasionally the Redhead and Bonaparte's Gull.

From Exit 43 on I 95, drive south on Campbell Avenue and go left on Beach Street to the mouth of the Harbor and the Sand Spit.

Gulls are always present on the **New Haven Flats;** shorebirds stop by in migration; ducks show up in winter. Take Frontage Street (Exit 46) from I 95 and drive east along the Flats, which parallel the waterfront.

NEW LONDON, CONNECTICUT
Connecticut Arboretum | Harkness Memorial State Park | Waterford Beach

The **Connecticut Arboretum** (340 acres within this city) includes a variety of natural habitats—hillside fields and deciduous woods, a lowland wooded swamp, fresh-water marshes, a fresh-water pond, and numerous brushy thickets. Owned by the State of Connecticut and maintained by nearby Connecticut College, the Arboretum is open to the public, and most of its habitats are readily accessible by footpaths from the Williams Street entrances. This is reached from I 95 by exiting on State 32 and driving north about 0.5 mile, then turning left on Williams Street. The entrance is 0.5 mile farther at the top of the hill on the left.

Among the wide variety of birds breeding in the Arboretum are the Broad-winged Hawk, Black-billed Cuckoo, Great Crested Flycatcher, Wood Thrush, Ovenbird, Louisiana Waterthrush, Common Yellowthroat, Yellow-breasted Chat, Hooded Warbler, Scarlet Tanager, Field Sparrow, and Swamp Sparrow. Nearly all these species are nesting in late May. Earlier in May, many warblers linger temporarily during their migratory passage to nesting grounds in northern New England and Canada. During winter there are always permanent residents such as Ruffed Grouse, Northern Mockingbirds, Tufted Titmice, and Northern Cardinals, together with numerous visitants—e.g., Brown Creepers, Cedar Waxwings, Myrtle Warblers, Slate-colored Juncos, American Tree Sparrows, and White-throated Sparrows—and in some winters, Common Redpolls, Pine Siskins, and crossbills.

Harkness Memorial State Park (231 acres), on Long Island Sound south of New London, attracts a different variety of birds to its open fields, mud flats, sandy beach, rocky shore, and offshore waters. Migrating shorebirds stop in May, August, and September; waterfowl begin showing up in early October, and many remain through the winter.

To reach the Park, exit from I 95 on State 213 and drive directly south for about 5 miles to the entrance. Just east of the Park is **Waterford Beach**, behind which is an extensive salt marsh that often attracts Short-eared Owls and, occasionally, a Snowy Owl in winter.

NEWPORT, RHODE ISLAND
Easton Pond

Easton Pond, a fresh-water reservoir with a big cattail bed in the southeastern corner, very seldom freezes over in winter. Consequently, it is visited quite regularly from November to April by such waterfowl as American Wigeons, Blue-winged Teal, Northern Shovelers, and Ruddy Ducks. Gadwalls and Ring-necked Ducks sometimes put in an appearance, as does, on rare occasions, a Eurasian Wigeon.

Easton Pond lies east of Newport and just north of the Municipal Beach. From atop the high embankment surrounding the Pond the bird finder may readily observe the ducks.

OLD LYME, CONNECTICUT
Smith's Neck | Great Island Wildlife Management Area | Poverty Island | Rocky Neck State Park | Tern Sanctuary

Take Exit 70 from the Connecticut Turnpike and drive south on State 156. After going under the railroad viaduct, turn right at the first crossroad on Smith's Neck Road, which runs across a barren upland called **Smith's Neck.** Here one may look down on the extensive salt marshes in the mouth of the Connecticut River, an outstanding place for ducks. Continue across Smith's Neck to the shore and the state-operated boat landing, where one may launch his own craft or hire one, for exploration of **Great Island Wildlife Management Area** (504 acres), owned and administered by the Connecticut Board of Fisheries and Game.

Great Island, just to the west offshore, is a flat, brackish marsh with many tidal channels and ditches. Grasses, bulrushes, and cattails comprise the principal cover. During May and June one may see several occupied Osprey nests, some in trees on the few scattered hummocks, others on heaps of driftwood and debris. Among species probably nesting in the marsh are the Clapper Rail, Marsh Wren, Savannah Sparrow, Sharp-tailed Sparrow, and Seaside Sparrow. From November until April, ducks such as Mallards, American

OSPREY

Black Ducks, Greater Scaups, Common Goldeneyes, and Buffleheads gather in the channels and on the offshore waters.

Return to State 156 and continue south. After crossing the Black Hall River, take the next two right turns onto a road to **Poverty Island.** The last part of this road is private, but permission to enter will be granted bird finders who wish to reach the sand spit at the mouth of the Connecticut River. Not only is the sand spit an excellent vantage point from which to view ducks, but it is also one of the best spots on the Connecticut shore for (Ipswich) Savannah Sparrows during fall and winter. Look for these pale, sand-colored birds in the beach grasses.

Return to State 156 and drive eastward to **Rocky Neck State Park**

(561 acres). A road running south from highway to beach parallels a long, narrow marsh along Brides Brook. Here, in fall and spring, may be sighted Great Blue Herons, Mallards, American Black Ducks, Green-winged Teal, and occasionally Common. Pintails, Wood Ducks, and Hooded Mergansers. At the end of the road is a rocky cove, a favorite wintering place for Common Loons, Red-throated Loons, Horned Grebes, Common Goldeneyes, White-winged Scoters, Surf Scoters, Black Scoters, and Red-breasted Mergansers.

Continuing eastward from the Park on Route 156, take the first right turn on a road leading to Black Point. Follow the road down the west side of Black Point until a small island known as **Tern Sanctuary** appears a few hundred yards offshore. Beginning in late May, Common Terns nest on its 3 acres of ledges and sandy areas with their scattered clumps of low vegetation. Permission to visit the island during the nesting season must be obtained from the Connecticut Board of Fisheries and Game (State Office Building, Hartford, Connecticut 06115). Boats serving the island are available for hire on Black Point.

POINT JUDITH, RHODE ISLAND
Point Judith Coast Guard Station | Galilee

One of the best sites from which to observe coastal birds is the **Point Judith Coast Guard Station** at the very end of Point Judith, where there is a low, gravelly, nearly treeless bluff overlooking the ocean. Among birds expected regularly from October to April are Common Loons, Red-throated Loons, Horned Grebes, Northern Gannets, Great Cormorants, Double-crested Cormorants (rare in winter), Common Goldeneyes, Oldsquaws, White-winged Scoters, Surf Scoters, Black Scoters, and Red-breasted Mergansers. Occasional visitants are Brant in November and March; Common Eiders, King Eiders, Dovekies, and Razorbills during winter.

Another place to see coastal birds, particularly gulls and terns, is reached by driving north from the Coast Guard Station for 1.25 miles and then left to the little village of **Galilee.** Follow the road past the docks. On its west side are sand flats where, at low tide,

Greater Black-backed Gulls, Herring Gulls, and Ringed-billed Gulls are in view the year round; Laughing Gulls from May to mid-December; Bonaparte's Gulls from mid-November through December; and Common Terns, Roseate Terns, and Little Terns from mid-May to mid-September. These are occasionally joined by Black Terns in August and September, Glaucous and Iceland Gulls from December through March, Forster's Terns in August and September, and Black Skimmers in summer.

On the east side of this same road are salt marshes and mud flats, where many species of transient shorebirds feed. Lesser Golden Plovers, Willets, Whimbrels, and Western Sandpipers are among the more unusual species to appear.

PROVIDENCE, RHODE ISLAND
Narragansett Bay

The mild climate of **Narragansett Bay** induces many ducks to remain throughout the winter. In the northern part of the Bay the wintering population of Greater Scaups is one of the largest on the Atlantic coast. For a good view of the scaup rafts, which remain from November to April, go to the Edgewood Yacht Club. In addition to the scaup rafts, the bird finder may see Horned Grebes, Common Goldeneyes, Buffleheads, and Red-breasted Mergansers. For the Yacht Club, drive south on US 1A from Providence for about 2 miles, continue on Narragansett Boulevard, and turn left on Ocean Street.

South about 10 miles, either on State 117 or on unnumbered roads nearer the shore, is Conimicut Point, a long sand bar extending into the Bay. The species in view here are the same as those seen from the Yacht Club, with the addition of transient or wintering Bonaparte's Gulls and an occasional Black-headed Gull.

If the bird finder continues down the shore to Warwick Neck and drives straight out to the lighthouse at the southern tip, there is always a good chance of seeing Oldsquaws, as well as the birds mentioned above, out in the Bay.

SHARON, CONNECTICUT
Sharon Audubon Center

The **Sharon Audubon Center** (540 acres), owned and operated by the National Audubon Society, lies in northwestern Connecticut's rolling hill country, much of it formerly farmland reverting to deciduous woods. From the clock tower on Sharon's main street, drive east on State 4 for 2.5 miles and watch for the well-marked entrance to the Center on the right.

On the property are two ponds—one near headquarters, the other about a half-mile away—a stream, swamp and marsh areas, fields, and a young woodland dominated by hemlock, tamarack, red maple, aspen, red oak, and chestnut oak. Well-kept trails give access to all parts of the area. Mallards, American Black Ducks, and Wood Ducks nest, and Common Pintails, Green-winged and Blue-winged Teal, American Wigeons, Ringed-necked Ducks, and Hooded and Common Mergansers frequent the ponds and wetlands in fall and early spring. Nesting landbirds include the Broad-winged Hawk, Ruffed Grouse, Pileated Woodpecker, Veery, Eastern Bluebird, Golden-winged and Blue-winged Warblers, Chestnut-sided Warbler, Ovenbird, Louisiana Waterthrush, American Redstart, and Scarlet Tanager.

STONINGTON, CONNECTICUT
Barn Island Wildlife Management Area

Probably the best place in southeastern Connecticut for marsh birds and transient shorebirds, waterbirds, and waterfowl is the **Barn Island Wildlife Management Area** (707 acres), owned and administered by the Connecticut Board of Fisheries and Game. Extensive salt, brackish, and fresh-water ponds and marshes (their water levels controlled by dikes), as well as mud flats and a few upland woods and fields bordered by brushy thickets, are attractive habitats for birds the year round. Through most of the Management Area, meandering trails provide ready access to good vantage points for observation.

Any time of the year is recommended for bird finding except

when the Area is open to hunting—usually from mid-October through January. Clapper Rails, Marsh Wrens, Sharp-tailed Sparrows, Seaside Sparrows, and, in drier spots, Savannah Sparrows begin nesting in late May. During summer one may expect to see wading birds: Great Blue Herons, Green Herons, Great and Snowy Egrets, Black-crowned and Yellow-crowned Night Herons, Least and American Bitterns, and, infrequently, Little Blue and Louisiana Herons, and sometimes a Glossy Ibis. In May and again in late August and September, numerous shorebirds appear, including Black-bellied Plovers, Ruddy Turnstones, and Whimbrels. Throughout winter, look for Short-eared Owls and an occasional Snowy Owl.

In Stonington, drive east on State 1 from its junction with State 1A for about a mile, then turn right on Palmer Neck Road and proceed to a boat-launching area 1.7 miles beyond. Leave the car in a parking lot just before reaching the launching area, and walk along a gated road that begins opposite the lot and leads west into the Management Area.

TIVERTON, RHODE ISLAND
Sakonnet Point | Warren Point | Brigg's Marsh

Rhode Island bird watchers seem agreed that the most exciting year-round spot for waterbirds and waterfowl in the state is on the coast south of Tiverton and Little Compton, in the area that includes **Sakonnet Point, Warren Point,** and **Brigg's Marsh.** The most direct route from Tiverton leads straight south on State 77. At Fort Rodman, where State 77 veers right to Sakonnet Point, the road straight ahead leads to Warren Point, passing Brigg's Marsh on the east. The Sakonnet Point Road passes Long Pond, with its big cattail beds, and Round Pond on the left. A longer but far more rewarding route is to follow the estuary border south of Stone Bridge at Tiverton, circle Nannaquacket Pond, and return to State 77. This is a good area for ducks in fall and winter.

Proceed on State 77 to Seapowet Avenue, turn right, skirt the shore and the Seapowet Marshes, and take the first left back to State 77. This stretch of shore has several Osprey aeries and nesting habitats for Grasshopper and Sharp-tailed Sparrows. In late summer,

Great Egrets and immature Little Blue Herons gather in the marshes.

Sakonnet and Warren Points are quiet, resort/fishing communities in semi-open and rolling coastal country, with abandoned fields containing herbaceous perennials and shrubs such as bayberry, rose, blackberry, and poison ivy. The shore has a beach backed by an abrupt bluff from 10 to 40 feet high. Off Sakonnet Point are three rocky islands and smaller rocky islets with scattered, thick clumps of shrubs. Brigg's Marsh (206 acres), to the north and east of Warren Point, is a bay-mouth lagoon of brackish water fringed by panic grass, quillreed, three-square bulrush, and spartina grass, with a good growth of eel grass in the shallow waters. The beach adjacent to this marsh is privately owned, and one must have permission to enter.

In June, Piping Plovers nest on the beach, Virginia Rails, Marsh Wrens, and Red-winged Blackbirds in the cattail marsh at Long Pond and in Brigg's Marsh. At the same time the islands off Sakonnet Point have nesting Herring Gulls, a few pairs of Common and Roseate Terns, and several pairs of Greater Black-backed Gulls.

In late September a minor hawk flight passes here. The birds, largely Cooper's Hawks, Red-shouldered Hawks, and Northern Harriers, pass westward about a mile inland and turn northward over the Sakonnet estuary. The flight, which lasts only two or three days, can best be viewed either from the golf course at Sakonnet Point or from open areas commanding a good northeast view of Brigg's Marsh.

Brigg's Marsh is one of the best spots on the coast for a view of wintering ducks. A few Canada Geese and several hundred American Black Ducks, American Wigeons, Common Pintails, Green-winged Teal, and Blue-winged Teal remain from October to May. The bluff at Sakonnet Point and Warren Point makes an excellent vantage point in winter to look for occasional Common and Red-throated Loons, Horned Grebes, Northern Gannets, Common Goldeneyes, Oldsquaws, Common Eiders, White-winged Scoters, Surf Scoters, Black Scoters, Red-breasted Mergansers, Razorbills, and Dovekies (when there is an onshore wind). Great and (occasionally) Double-crested Cormorants rest on the islands off Sakonnet Point.

Delaware

R. DUDLEY ROSS

SHOREBIRDS

Delaware, although in area the second smallest of the United States, offers abundant opportunities for bird finding throughout the year, in habitats varying from the tidal marshes along the coast, where waterfowl and shorebirds congregate in spring and fall, to the uplands in the north, where woodland and field birds pause in migration and many remain to nest. In fact, when considering the proportion of good bird-finding places in relation to size, Delaware surpasses many other states. The annual list of species seen in Delaware generally numbers about three hundred.

Delaware occupies the northeastern part of Delmarva Peninsula, which lies between Chesapeake Bay on the west and the Delaware River and Bay and the Atlantic Ocean on the east. Of Delaware's approximately 2,400 square miles—including 80 of inland water and

350 in Delaware Bay—approximately 180, or almost 8 per cent, are tidal wetlands, and state policy is to protect as many of these valuable wetlands as possible. They include the brackish and salt marshes occupying much of the Delaware Bay coast, where there are two national wildlife refuges, Bombay Hook (*see under* **Smyrna**), and Prime Hook (*see under* **Milford**), as well as two state wildlife areas, Woodland Beach (*see under* **Smyrna**) and Little Creek (*see under* **Dover**). Then on Little Assawoman Bay, which faces the Atlantic in the southeastern corner of the state, lies Assawoman Bay State Wildlife Area with its tidal marshes (*see under* **Bethany Beach**). Any sizable fresh-water marshes remaining today are along Dragon Creek (*see under* **Delaware City**) in New Castle County and in Sussex County.

Some of the birds breeding regularly in Delaware's marshes are:

Pied-billed Grebe
American Bittern
Least Bittern
American Black Duck
Gadwall
Blue-winged Teal
King Rail
Clapper Rail (*salt marshes*)
Virginia Rail
Common Gallinule
American Coot
Willet (*salt marshes*)
Marsh Wren
Sedge Wren
Boat-tailed Grackle
Sharp-tailed Sparrow (*salt marshes*)
Seaside Sparrow (*salt marshes*)
Swamp Sparrow

The southern two of Delaware's three counties (Kent and Sussex) and most of the northern county (New Castle) are within the Atlantic Coastal Plain. The northern part of New Castle lies on the Piedmont Plateau. Within the Piedmont the state rises to its peak elevation of 442 feet above sea level at Centerville, a few miles north of Wilmington, the largest city. The Fall Line, separating Piedmont from Coastal Plain, runs from the northeastern corner through Wilmington to the western boundary near Newark. I 95 roughly parallels this line. The state's Atlantic coastline is almost 25 miles long, 12 of these miles being within Cape Henlopen State Park (*see under* **Lewes**) and Delaware Seashore State Park (*see under* **Rehoboth**) and thus safe from damaging development. The 6,000-acre

Cypress Swamp (*see under* **Selbyville**) belongs to Delaware Wildlands, Inc., an organization devoted to acquiring and preserving unique natural areas.

Much of northern Delaware, particularly north of the Chesapeake and Delaware Canal, is under either cultivation or commercial development. In the southern part, numerous areas are still forested with second-growth timber. Half the land area in the southernmost county (Sussex) is still forested. Hickory, beech, maple, sweet gum, tulip tree, ash, sycamore, and various oaks are the principal trees, but pines become more prominent in the south.

The birds listed below breed more or less regularly on Delaware's farmlands (fields, wet meadows, brushy lands, orchards, and dooryards) and in deciduous woods:

FARMLANDS

Common Bobwhite
Mourning Dove
Eastern Kingbird
Eastern Phoebe
Tree Swallow
Barn Swallow
House Wren
Carolina Wren
Northern Mockingbird
Gray Catbird
Brown Thrasher
Eastern Bluebird
White-eyed Vireo
Yellow Warbler
Prairie Warbler
Common Yellowthroat
Yellow-breasted Chat
Bobolink
Eastern Meadowlark
Orchard Oriole
Common Grackle
Northern Cardinal
Indigo Bunting
House Finch
American Goldfinch
Rufous-sided Towhee
Grasshopper Sparrow
Vesper Sparrow
Chipping Sparrow
Field Sparrow
Song Sparrow

DECIDUOUS WOODS

Turkey Vulture
Red-shouldered Hawk
Broad-winged Hawk
Yellow-billed Cuckoo
Black-billed Cuckoo
Common Screech Owl
Barred Owl
Whip-poor-will

DECIDUOUS WOODS (*Cont.*)

Yellow-shafted Flicker
Pileated Woodpecker
Red-bellied Woodpecker
Hairy Woodpecker
Downy Woodpecker
Great Crested Flycatcher
Acadian Flycatcher
Eastern Pewee
Blue Jay
Carolina Chickadee
Tufted Titmouse
White-breasted Nuthatch
Wood Thrush
Blue-gray Gnatcatcher
Yellow-throated Vireo
Red-eyed Vireo
Warbling Vireo
Ovenbird
Kentucky Warbler
Hooded Warbler
American Redstart
Baltimore Oriole
Scarlet Tanager
Summer Tanager (*local*)

During breeding season a number of southern species nest in Delaware, among them the Purple Gallinule, Black-necked Stilt, Chuck-will's-widow, Brown-headed Nuthatch, Yellow-throated Warbler, Kentucky Warbler, Hooded Warbler, Boat-tailed Grackle, Summer Tanager, and Blue Grosbeak.

Migration seasons are very productive in Delaware. The spring migrations begin in late February or early March and peak during the first two weeks in May. The Delmarva Ornithological Society, in its 'Spring Roundup' by parties covering areas throughout the state in early May, lists between 210 and 220 species. The following timetable indicates when the heaviest flights may be expected:

Waterfowl: 20 February–1 April; 15 October–10 December
Shorebirds: 1 May–1 June; 1 August–5 October
Landbirds: 15 April–15 May; 1 September–25 October

Winter regularly finds certain species, such as Lapland Longspurs and Snow Buntings, close to their southern limits. Northern finches can be abundant, particularly in the northern third of the state, during 'outbreak' years. Along the coast, winter regularly brings Great Cormorants, Common and King Eiders, and Purple Sandpipers. Probably no other state has a greater proportion of its area within Christmas bird-count circles. From five separate counts, totals range from nearly 100 in the north, the Wilmington area, to the 120s to 140s from Bombay Hook southward.

The bird finder will find invaluable the free official Delaware highway map, which shows all United States highways and the major State roads, and three inexpensive county maps, showing other roads as numbered for maintenance. All four maps are available from the Division of Highways (P.O. Box 8, Bear, Delaware 19701). Any other road numbered is referred to in the text as 'Road,' regardless of the agency maintaining it; and one will see it indicated on the highway by a small white sign attached to a stop sign or by a regular road sign.

Authorities

Maurice V. Barnhill III, William Baxter, Jr., Harry S. Bristow, Jr., Seal T. Brooks, Elizabeth Dyer, Lloyd L. Falk, Violet L. Findlay, Clayton M. Hoff, Curtis O. Johnson, Jay G. Lehman, Stanley B. Speck, George P. Spinner, Phillips B. Street, Winston J. Wayne.

BETHANY BEACH
Assawoman Bay Wildlife Area

Just a few miles southeast of this summer resort is state-owned **Assawoman Bay Wildlife Area,** encompassing 1,500 acres of mixed deciduous and coniferous woodlands, of ponds, and of tidal marshes.

There are about 6.5 miles of roads in the Area, an excellent site for undisturbed observation of the region's birds. A fine variety of waterfowl spends the winter on the ponds, marshes, and adjoining Bay, which opens into the Atlantic. August and September bring a fair number of shorebirds and a good assortment of wading birds. By far the most rewarding time is the month of May, when the principal spring migration takes place. A morning's bird finding will produce at least a hundred species, including waterbirds, shorebirds, and a nice array of warblers and other landbirds.

From Bethany Beach, go west on State 26, take the first road on the left (Kent Avenue), and follow the signs to 'Camp Barnes' for 6.3 miles to the well-marked Wildlife Area entrance.

DELAWARE CITY
Dragon Creek | Dragon Run Park | Thousand Acre Marsh | State 9 south of Delaware City

From late spring until fall the fresh-water marsh along **Dragon Creek** in northern Delaware has a rich avifauna that includes the Pied-billed Grebe, American and Least Bitterns, King Rail, Common Gallinule, and American Coot. Wood Ducks nest in the surrounding woods. The marsh is a feeding area for Great Blue, Green, Little Blue, and Black-crowned Night Herons, for Cattle, Great, and Snowy Egrets, and for Glossy Ibises, which nest on Pea Patch Island nearby in the Delaware River.

A good bird-finding area in the nesting season, the marsh is even better during spring and fall migrations, when there are large populations of herons and egrets, ducks, Bobolinks, and Red-winged Blackbirds. During these periods, warblers, sparrows, and other landbirds congregate in the woods and thickets. In winter, large numbers of surface-feeding ducks inhabit the marsh and surrounding fresh-water ponds.

From the intersection of US 13 and State 72 just west of Delaware City, go east on State 72 for 1.5 miles, turn right (south) onto Road 378 for 0.8 mile to a bridge crossing Dragon Creek. This western portion of the marsh is a good place for nesting King and Virginia Rails.

Return to State 72, turn right onto State 9, and continue for 2 miles to a bridge crossing the outlet of Dragon Creek, where one can view the marsh from the south side of the road. Continue southeast on State 9 for 0.2 mile and turn right on a paved road to a parking lot in **Dragon Run Park.** The timid bird finder can search the low woods, hedgerows, and thickets here for landbirds; the bolder can view the entire marsh by piercing the thickets between the parking area and the marsh and taking a trail south along the edge.

After leaving Dragon Run Park, continue southeast on State 9 into Delaware City, turn right at the traffic light on Clinton Street, and drive about 0.8 mile until a marsh choked with spatterdock appears on the right. From the north side of this road there is an unobstructed view of the entire marsh. Here, Purple Gallinules nested until recently. From the high levee on the south side of the road a large fresh-water pond is in view. In fall, winter, and early spring, coots along with wigeons and other surface-feeding ducks and occasional diving ducks congregate in big flocks. The Eurasian Wigeon appears here irregularly in spring.

Return on Clinton Street to the traffic light and drive right on State 9 for 0.3 mile to Reedy Point Bridge. Immediately after crossing the bridge, turn right at the sign 'To Dutch Neck Road,' follow the paved road to the canal, and turn west on Dutch Neck Road, which runs along the canal. Continue for 0.3 mile to an open area on the south side of the road, where a fresh-water pond, **Thousand Acre Marsh,** comes into view. Late in a dry summer season, when the water is low, this is the best place in northern Delaware for shorebirds. Large numbers of Greater and Lesser Yellowlegs, Short-billed Dowitchers, Pectoral, Least, Stilt, Semipalmated, and Western Sandpipers, and a few phalaropes rest and feed here, and Marbled Godwits and Black-necked Stilts occasionally show up. Many gulls and sometimes Forster's and Caspian Terns appear during late summer and fall, even when the water level is high.

Continue west on Dutch Neck Road, keeping on the paved road to the left for 1.2 miles to a marsh with much spatterdock, where Least Bitterns can be heard and seen at close range.

From Delaware City, **State 9** continues southward, running parallel to the Delaware River and Bay until it merges with US 113 south of Dover Air Force Base. The areas adjacent to State 9, particularly those to the east—toward the water—are rich in birdlife. Some of the more thoroughly studied are mentioned below, in geographical sequence from north to south.

About 7.3 miles south of Port Penn, State 9 intersects State 299, and about a mile south of this intersection State 9 crosses a small stream called Hangman's Run, on a concrete bridge. In addition to the Blue Grosbeak, which nests here regularly, look for other passerines and hawks. Continue south on State 9 for 1.6 miles, bear right on Road 456 to where it crosses two branches of Blackbird Creek (about a half-mile) in moist, wooded valleys—the first branch just before the first house on the right, the second at a wooden bridge. Besides many passerines, one may expect vultures, hawks, Common Bobwhites, Common Screech Owls, Great Horned Owls, and Belted Kingfishers. Both Blackbird Creek crossings are particularly good for winter finches.

Return to State 9 and continue south about 9 miles to the intersection with Road 317. The deciduous woods on both sides of the next half-mile of State 9 is good for American Woodcock, as well as for flycatchers, vireos, warblers, and other passerines during spring migration.

DOVER
Little Creek Wildlife Area

From this capital city, take State 8 east for 4.5 miles, turn right (south) on State 9 for 0.4 mile, and then left on Road 89 (Port Mahon Road), which enters the **Little Creek Wildlife Area** (3,897 acres) between the Delaware Bay shore and State 9. At 0.9 mile a gravel road (right) leads to the top of a dike around a mosquito-control impoundment, part of the Little Creek Area. The amount of water in this large pool varies from almost complete coverage to intermittent pools with bare mud flats and extensive patches of marsh vegetation—to make it a fine place for waterfowl, wading birds, shorebirds, gulls, and terns, particularly during migration. Besides the usual species, rarer birds appearing fairly regularly include the Eurasian Wigeon, Red Knot, White-rumped and Baird's Sandpipers, Black-necked Stilt (a breeding bird in the Area), and Northern Phalarope. Little Gulls and Black-headed Gulls sometimes appear with Bonaparte's Gulls in April and May. The path on the dike leads completely around the pool.

Drive 1.2 miles farther along Port Mahon Road to a footbridge over the tidal gut, giving access to a dike leading either east to Delaware Bay or back to the dike around the pool. This road next crosses a gut draining into the Bay, where the mud flats, exposed at low tide, are superb for shorebirds, as is the debris-covered mud on the beach along the shore north toward Port Mahon.

The large tidal marsh north of Port Mahon Road is good for Rough-legged Hawks, Northern Harriers, and Short-eared Owls during the colder months; Clapper Rails and Sharp-tailed and Seaside Sparrows the year round; and Whimbrels during migration. Diving ducks and scoters come into the Bay during any season except summer.

There are two other access roads to the Little Creek Area. For the first, return to State 9 and drive south for about 1.5 miles to where a 'wildlife area' sign indicates a gravel road on the left. This passes through a farmyard, turns left, and winds its way for 1.2 miles to a parking lot from which a boardwalk on the right leads to an observation tower. The road straight ahead, closed to vehicles, is worth covering on foot for viewing another large impoundment, south of the Port Mahon Road pool. The birdlife is similar to that at the Port Mahon impoundment, but there is a better opportunity here for seeing Least and American Bitterns, Redheads, Ring-necked Ducks, Hooded Mergansers, and Boat-tailed Grackles. The hedgerows along the way are excellent for passerine birds; the multiflora rose hedge at the entrance off State 9 is a favorite spot for White-crowned Sparrows in winter.

For the second access road, continue south on State 9 for 0.9 mile and take the next left road (Road 349) to Pickering Beach, 2.1 miles distant. Search the holly woods on the left, the likeliest place in Delaware for a Saw-whet Owl in winter. Along this road, but before dawn, it is also possible to hear Common Screech, Great Horned, and Barred Owls, and, in summer, the Whip-poor-will. In early summer, listen for singing Henslow's Sparrows in the fields south of Road 349. On the north side of the Road, about 100 yards west of the 'wildlife area' sign, is a wet field where Sedge Wrens usually reside in May and June. At the sign the gravel road to the left winds back to a parking area at the corner of two pools. One can walk the dikes to the north and east all the way to the Bay. Throughout the colder months during migration, the Bay off Pickering Beach and the neighboring village of Kitts Hummock produces scoters and other sea ducks. Farther south, at the intersection of State 9 and Road 68 to Kitts Hummock, the field to the west within Dover Air Force Base has had Upland Sandpipers regularly in migration.

LEWES
Cape Henlopen State Park | Roosevelt Inlet

Cape Henlopen State Park (1,641 acres) lies south and east of Lewes and includes Cape Henlopen, which juts almost due north, with the Atlantic on its eastern side and Delaware Bay on its west.

Its northern portion consists of beaches and of dunes covered with a variety of plants such as marram grass, dusty miller, beach heather, and seaside goldenrod. Nesting birds include the Piping Plover, Common and Little Terns, Black Skimmer, Common Nighthawk, (Prairie) Horned Lark, and Savannah Sparrow. (Northern) Horned Larks, (Ipswich) Savannah Sparrows, and often Lapland Longspurs and Snow Buntings appear in winter in the dunes.

During spring and fall migrations a surprising assortment of passerines and other birds stop on the Cape. Both ocean and bayside beaches provide excellent bird finding the year round, with a good variety of shorebirds, terns—including the more unusual Royal, Caspian, Roseate, and Black—and several species of gulls—including occasional Black-headed and Little among large flocks of Bonaparte's.

Delaware Bay, throughout almost the whole year, has loons, grebes, and a good assortment of waterfowl. In winter the ocean just offshore is preferred by Red-throated Loons, Red-necked and Horned Grebes, and Red-breasted Mergansers. With easterly winds the shearwaters, storm petrels, Northern Gannets, and jaegers sometimes come close enough to shore for satisfactory observation. Two long stone breakwaters provide resting areas for a variety of birds—e.g., Double-crested Cormorants and, in winter, Great Cormorants.

In the southern part of the State Park, pitch pine, red cedar, sumac, bayberry, bittersweet, and other woody plants afford year-round habitat for Common Bobwhites and nesting habitat for Pine and Prairie Warblers. Some depressions between the dunes retain water, thus encouraging bog plants such as cranberry, blueberry, sundews, rushes, and club mosses. Huge shifting dunes, which slowly engulf the vegetation in their path, are another feature of the Park.

To reach the Park from the intersection of State 14 and State 18 east of Lewes, either drive east on State 18 to the Bay and turn right on a road direct to the Park, or go south for about a mile on State 14 to the Lewes–Cape May ferry cutoff, turn left, and, just before reaching the ferry, turn right.

After exploring the Park, drive back toward Lewes and stop at the fishing pier for waterfowl at close range, and at the ferry slip for gulls, among them, in winter, the lingering Laughing Gulls, as well as scoters, possibly an eider or two, and Purple Sandpipers.

Continue west along the shore through Lewes to **Roosevelt Inlet** and the entrance to the Lewes–Rehoboth Beach Canal. The jetty and small beaches offer further opportunities to see waterfowl and shorebirds in season. Across the Canal lies a favorite area for Boat-tailed Grackles, and a nearby clam-packing plant attracts hundreds of gulls.

MILFORD
Prime Hook National Wildlife Refuge | Redden State Forest

Prime Hook National Wildlife Refuge, southeast of Milford, extends some 8 miles along Delaware Bay from Slaughter Beach to Broadkill Beach. The Refuge's 10,700 acres include 7,300 acres of marsh and water, 1,200 of timber and brush, and 2,100 of pasture and croplands. In addition to all-weather roads for vehicles many foot trails and 15 miles of canoe waterways are provided. Although the Refuge is still being developed, bird finding should be similar to that at Bombay Hook National Wildlife Refuge (*see under* **Smyrna**).

Several roads going eastward from State 14, south of Milford, enter different parts of the Refuge. From Milford, drive east on State 36 to Slaughter Beach. After exhausting the possibilities on this beach, continue on State 36, which forms a loop and returns to State 14 at Argos Corner, 6 miles south of Milford. Turn left (south) on State 14 and 1, and watch for roads east to Fowler Beach and Primehook Beach; both roads cross the Refuge. Return to State 14 and 1, turn south to State 16, turn left and proceed for 1.2 miles, and left again for 1.6 miles to Refuge headquarters. Return to State 16, turn left (east) for 2.8 miles to Broadkill Beach.

A short distance south of Milford is **Redden State Forest** (3,330 acres in two sections). Drive south from Milford on US 113 for about 12.5 miles and turn right on Road 40 (Gravelly Run Road) toward Bridgeville. The wooded areas, primarily of pine and extending both north and south along this road, comprise the western portion of the Forest. This section is interlaced with dirt roads along which are good habitat for Summer Tanagers. The best bird-finding spots are near the intersections on Gravelly Run Road, and the wet, wooded roadsides.

Despite the current practice of selective cutting of the pine stands, with consequent reduction in birdlife, the area still attracts late April warblers, including the Yellow-throated. One can usually hear and sometimes see Pileated Woodpeckers. A pair of Red-shouldered Hawks are residents each year.

NEWARK
White Clay Creek Valley

North of Newark is **White Clay Creek Valley,** with birds typical of both the Coastal Plain and the Piedmont. Observations during a recent ten-year period produced more than 180 species, of which approximately 90 are breeding birds, among them Traill's Flycatcher, Veery, Yellow-throated and Warbling Vireos, and Northern Parula, Cerulean, and Yellow-throated Warblers.

From Newark, go north on State 72 for about 2 miles to Milford Crossroad, then left for 1.6 miles, and again left on Chambers Rock Road to the Creek. After searching the area around the bridge over the Creek, continue on Chambers Rock Road for about 0.1 mile to the intersection with Creek Road. To the left (south), Creek Road is passable for cars through wooded bottomlands for about 3 miles. To the right, it leads to steeper terrain, unsuitable for cars but passable on foot for about 2 miles.

REHOBOTH BEACH
Henlopen Acres | Gordon Pond Wildlife Area | Silver Lake | Delaware Seashore State Park | Indian River Inlet

The area of Rehoboth Beach abounds in birdlife, even in winter. The Christmas bird count, centered near Indian River Inlet, totals about 135 species each year.

On the north edge of the town of Rehoboth Beach, **Henlopen Acres,** an extensive development with a variety of plantings, provides year-round bird finding. From the center of town, drive west on State 14A to the last street on the right before the Lewes–Rehoboth Canal (Columbia Avenue), take a right for 0.7 mile to Second Street, then left for one block to Henlopen Avenue, entrance to Henlopen Acres. Spring transients, winter finches, and such year-round residents as the Brown-headed Nuthatch and Pine Warbler all find shelter in the trees and shrubs.

Return to Henlopen Avenue, turn left (east) for 0.3 mile, then left on Ocean Drive for 1.0 mile to a parking area and sign indicating **Gordon Pond Wildlife Area** (300 acres). Park the car and follow a dirt road for a few hundred yards to the southern end of Gordon Pond, which lies between the southern end of Cape Henlopen State Park (*see under* **Lewes**) and the Canal. The Pond, roughly a mile long and several hundred yards wide, is merely a depression normally covered by about a foot of water. Its edges support a heavy growth of grasses, rushes, and sedges, with bayberry and other woody plants on higher ground.

The Area is excellent for Pied-billed Grebes, Glossy Ibises, Mute Swans, Canada Geese, surface-feeding ducks, rails, gulls, and terns, the number and species varying with the season. In early summer, by walking along the sand road circling the Pond, one can find Least and American Bitterns, Common Gallinules, Willets, Marsh Wrens, Yellow Warblers, and Common Yellowthroats.

Return to Rehoboth Beach and drive south on State 14A (Bayard Avenue) across **Silver Lake,** a small fresh-water pond on the outskirts of town about a hundred yards from the Atlantic. Throughout the colder months Silver Lake entertains Canada Geese, both surface-feeding and diving ducks, coots, and gulls. Several hundred Canvasbacks winter here.

South of Rehoboth Beach, State 14A joins State 14, and the route continues south for 6 miles on a narrow strip of coastal barrier through **Delaware Seashore State Park** (2,020 acres) to **Indian River Inlet,** the link of both Rehoboth Bay and Indian River Bay with the Atlantic. Frequent short roads left and right of State 14 and 1 permit ready views of both Bays and the Atlantic for waterbirds throughout the year. At Haven Road, just before the Inlet, Little Terns nest in early summer.

At Indian River Inlet the stone breakwater jetties extending into the Atlantic provide good bird finding during colder months. Purple

Sandpipers feed on the rocks, and Red-throated Loons, Horned Grebes, Oldsquaws, Common Eiders, White-winged, Surf, and Black Scoters, and Red-breasted Mergansers favor the rough water at the end of the jetties. The Bonaparte's Gull flocks frequently include Black-headed and Little Gulls. Easterly winds increase the chance of close views of shearwaters, Northern Gannets, and jaegers. The tidal marshes and waterways north of the Inlet harbor Sharp-tailed and Seaside Sparrows and Clapper Rails the year round, and herons, egrets, and shorebirds most of the year.

On the south side of the Inlet, immediately beyond the bridge, a road (right) through a trailer park leads to a marina on Indian River Bay. Here the small beaches, marshy islands, and the Bay are worth viewing for birds in any season. During colder months look for Brant in the Bay. In warmer months Gull-billed Terns and other terns appear, along with numerous wading birds and marsh sparrows. Ospreys, American Oystercatchers, Willets, Common Terns, Black Skimmers, and Boat-tailed Grackles all nest in the area.

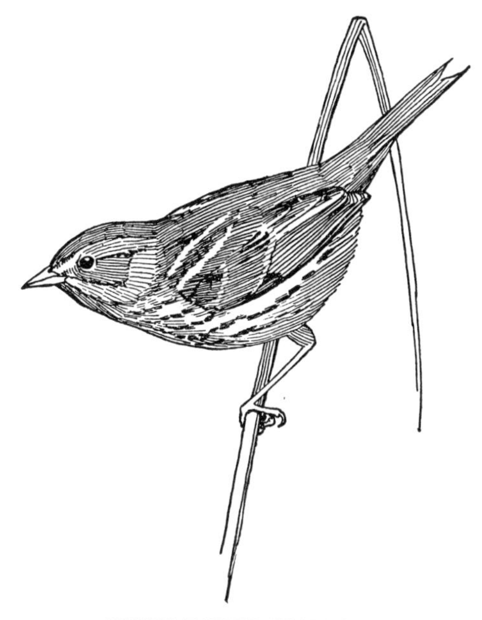

SHARP-TAILED SPARROW

State 14 continues south for 11 miles from the Inlet to the Maryland border, passing additional spots of interest such as Fresh Pond, Salt Pond, the Ocean View area, and Little Assawoman Bay.

SELBYVILLE
Cypress Swamp | Trap Pond State Park

West of Selbyville an extensive swamp sprawls across the Delaware-Maryland border, embracing the Pocomoke River and its northern tributaries and known as **'Cypress Swamp'** on Delaware maps and 'Pocomoke Swamp' in Maryland (*see under* **Pocomoke City, Maryland**), and often called 'Cedar Swamp' locally. Its vegetation is more distinctively southern than that in any other part of Delaware. In the approximately 6,000 acres owned by Delaware Wildlands, Inc., the cypresses, loblolly pines, water oaks, and red bays growing above the thick undergrowth form a suitable habitat for transient warblers and resident Prothonotary and Yellow-throated Warblers.

The trip from Selbyville includes some first-rate places for spotting birds. Drive west on State 54 toward Gumboro. Stop at the wooded area marked 'Delaware Wildlands' and watch or listen for resident warblers—Black-and-white, Prothonotary, Worm-eating, Northern Parula, and Yellow-throated—and possible transients. Continue west on State 54 until shortly beyond a sharp left curve, then turn left on a dirt road (Road 418), stopping frequently to check for cuckoos, warblers, and possibly Pileated Woodpeckers.

Road 418 eventually reaches an open, cultivated area, where Road 419 enters on the right. Blue Grosbeaks are fairly common summer residents on the edges of fields hereabout. Continue on Road 418 and cross the Maryland boundary at the edge of the forest. The Maryland stretch is good for warblers. Between the state line and the first dirt road to the left, inspect the thick undergrowth for Swainson's Warblers, which arrive in April to begin nesting.

The best way to return to State 54 is to continue straight through the Swamp, turn right on the first paved road into Delaware, then turn north on Road 419/420.

Another good place for spring bird finding is **Trap Pond State Park** (965 acres). Drive west from Selbyville on State 54 to Gumboro, turn right on State 26 and go for about 5 miles to Shaft Ox Corner, then continue north on State 25 for 0.9 mile, turn left on State 24, and clock 7 miles to a sign marking a left turn into the Park on Road 449. Within the Park, walk along the short dirt road opposite the turn to the campground, and examine the area around the sewage treatment plant. Inspect the bushes between the road and the dam (which forms the 90-acre Pond) and around the Pond itself.

Walking along another dirt road between the campground and Cypress Point in the Park is probably one of the best ways to look and listen for Pileated Woodpeckers, Brown-headed Nuthatches, Yellow-throated Warblers, and Summer Tanagers. Any open fields in the area should yield, in late spring or early summer, breeding Vesper Sparrows—scarce birds that move from year to year and require some effort to find.

SMYRNA
Woodland Beach Wildlife Area | Bombay Hook National Wildlife Refuge

The **Woodland Beach Wildlife Area** (3,664 acres), interspersed among private lands, extends along Delaware Bay south from Road 321 to about a mile south of State 6 and, in places, as far west as State 9. For an idea of bird-finding possibilities, go east from Smyrna on State 6 for 5.2 miles, left on State 9 for 1.0 mile, and right on Road 79 for 0.5 mile. Near the end of this short road an observation tower stands on the left and a farm with several ponds on the right—a fine setting for herons and egrets, geese, surface-feeding ducks, and shorebirds, especially during migration.

Return to State 9, turn left (south) to State 6, and then go east across a section of the Area toward Woodland Beach.

Perhaps Delaware's finest and best-known place for bird finding is the **Bombay Hook National Wildlife Refuge** (16,280 acres), extending 8 miles along Delaware Bay and west to State 9 in the northern section. To reach Refuge headquarters from Smyrna, drive south on US 13 to just outside town, turn left at the Refuge sign on State 12 and drive on for 5 scant miles, then right on State 9 and, almost immediately, left on State 85, continuing for about 2 miles.

The Refuge includes 10,500 acres of brackish tidal marsh, 1,200 acres of impounded fresh water, 1,000 acres of croplands, and the remainder, of brushy, timbered swamps and grassy, timbered uplands. The swamp forests consist of red maple, black gum, and willow; the timbered uplands, of hickory, tulip tree, and oak. Dike roads permit driving around Raymond Pool (the impoundment closest to headquarters) and Bear Swamp, along one side of Shearness Pool, and across one end of Finis Pool. Observation towers overlook all pools except Finis.

Among birds nesting in the Refuge are the Pied-billed Grebe, Least and American Bitterns, Canada Goose, Mallard, American Black Duck, Gadwall, Blue-winged Teal, Northern Shoveler, Wood Duck, Osprey, Common Gallinule, Purple Gallinule, and Willet, as well as King, Clapper, and Virginia Rails.

Waterfowl migrations—from February to April and from September to December—bring Canada Geese, Snow Geese, and great numbers of surface-feeding and diving ducks. The population of Canada Geese peaks in late November. During shorebird migration in May and from July to October, a great variety of species use the mud flats, muddy edges of tidal creeks, and grassy fields, most regularly expected being the Semipalmated Plover, Piping Plover, Black-bellied Plover, Greater and Lesser Yellowlegs, Red Knot, Pectoral Sandpiper, White-rumped Sandpiper, Least Sandpiper, Dunlin, Short-billed Dowitcher, Semipalmated Sandpiper, and Sanderling. Other shorebirds that may appear include the Lesser Golden Plover, Upland Sandpiper, Hudsonian Godwit, American Avocet, and Black-necked Stilt.

The Refuge is excellent for sighting hawks at all times, and, during the cold months, for Rough-legged Hawks and occasionally Bald Eagles and Northern Goshawks. Late in summer come many wading birds—Little Blue Herons, Great Egrets, Snowy Egrets, Glossy Ibises, and others.

During spring and fall migrations one can see large flocks of swallows on the Refuge and along State 9, with Tree and Bank Swallows predominating. Finis Pool is one of the better Refuge areas for warblers and other passerines during migration. The Christmas bird counts, covering the Refuge and adjacent areas, average some 140 species.

In winter, when leaving the Refuge, take State 12 back toward Smyrna, turn left on Road 326 (second crossroad), and follow it to the end, watching for (Northern) Horned Larks, Water Pipits, and Lapland Longspurs in the fields on both sides of the road and around farm buildings. Carefully look over all flocks of blackbirds for a few Brewer's Blackbirds.

WILMINGTON

Alapocas Woods | Rockford Park | Brandywine Creek State Park | Hoopes Reservoir

Alapocas Woods is a county park of 109.6 acres of mature, mixed deciduous woods with a brushy understory. Several valleys with small streams sloping toward Brandywine Creek make this park exceptionally good in the spring for observing migrating thrushes, vireos, warblers, and other passerines. To reach the Woods from city center, go northwest on State 52 (Pennsylvania Avenue) to State 141 and turn right. After 0.7 mile, State 141 passes the Dupont Experimental Station, turns left and follows the Dupont property fence for 0.6 mile to Alapocas Drive (Road 268), which runs along the northeast edge of the Woods.

Across Brandywine Creek from Alapocas Woods is **Rockford Park** (103.7 acres), similar in terrain but more fully landscaped, and fine for migrating warblers in spring and finches in winter. To reach the main entrance, go northwest from city center on State 52 to Rising Sun Lane, proceed right for 0.2 mile to 19th Street, and right again for 0.1 mile.

Brandywine Creek State Park (434 acres) is an excellent example of northern Delaware's diversified habitat. From the floodplain of Brandywine Creek, the land rises to 322 feet above sea level. Although a large part of the Park consists of old fields once under cultivation, some wooded portions have been undisturbed for at least two centuries, with tulip trees predominating. On the floodplain of the Creek, the woods consists largely of red maple and boxelder, with an understory of spicebush. A small cattail marsh borders a portion of the Creek.

A weekly, year-round census of the birds of the Park, conducted by members of the Delmarva Ornithological Society during 1969, 1970, and 1971, produced a total of 185 species plus a hybrid, thus indicating that the area supports a varied bird population. The list contains all eastern thrushes and vireos, thirty-three warblers, and twelve sparrows. During migration the Park entertains such species as the Rough-legged Hawk, Bald Eagle, Peregrine Falcon, Whip-poor-will, Least Flycatcher, Cliff Swallow, Prothonotary Warbler, Mourning Warbler, Brewster's Warbler (hybrid), and Yellow-throated Warbler.

To reach the Park, drive northwest from city center on State 52 to State 100, right for 2.3 miles, and then right again on Road 232 (Adams Dam Road) for 0.3 mile.

Hoopes Reservoir, northwest of Wilmington, is about 1.5 miles long and 0.25 mile wide, with many coves and points giving the illusion of a natural body of water. Surrounded by deciduous and coniferous woods and widely separated homes, it attracts in the winter a variety of waterbirds and waterfowl, including the Common Loon, Red-necked and Pied-billed Grebes, Canada Goose, Mallard, American Black Duck, Redhead, Ring-necked Duck, Canvasback, Lesser Scaup, and Common Merganser. From Wilmington take State 52 northwest for about 3 miles and turn left on State 82, proceeding for 0.8 mile to a causeway over the Reservoir. Since the dam and surrounding lands are private, the causeway and one spot on each road along the east and the west sides are the only vantage points for looking at birds on the Reservoir.

Florida

BROWN PELICANS

Some states may claim species and subspecies of birds not found elsewhere in the United States; others may claim enormous communal gatherings, or attract great winter populations, or lie along important lanes of migration; but no one state has all of these distinctive features within its boundaries—except Florida. Hence, for bird finding, no other area in the country is quite its equal.

Four peculiarities of the Florida peninsula account in large measure for the state's ornithological features:

1. Lying in close proximity to the West Indies, the peninsula draws from these islands certain species that do not appear anywhere else on the continent. The majority of these, often assisted by hurricane winds, appear as visitants, but a few—e.g., the

Black-whiskered Vireo—have established themselves as regular breeding residents.

2. The peninsula's somewhat remote position has provided the isolation favorable for the development of more than twenty races or subspecies of permanent-resident birds.

3. The peninsula's vast marshes and wooded swamps, its countless lagoons and other waterways, its hammocks, sand scrub, and prairies, in combination with a consistently mild climate and year-round abundance of food, have formed so-called preserving influences. These factors serve, first, to prolong the existence of species extirpated in most other parts of the country (e.g., the extinct Carolina Parakeet continued to survive in peninsular Florida after it had disappeared elsewhere, and the nearly extinct Ivory-billed Woodpecker retained—possibly still maintains—a foothold on the peninsula); second, to hold populations of species such as the Scrub Jay and the Burrowing Owl long since separated by unknown factors from their representatives in western United States; third, to support exceptionally large nesting colonies of pelicans, herons, egrets, and ibises; and fourth, to harbor in winter months an unusually rich variety of birds from northern and western regions.

4. Because it projects toward South America and the intervening West Indian islands, the peninsula is followed by a number of species passing to and from their wintering grounds on the southern continent or the islands. Thus in spring and fall Florida has many transients.

An ornithological feature of the southeastern peninsula, centered in metropolitan Miami, is the prevalence of many introduced exotic species. Owing to the aforementioned preserving influences, some of these species have become firmly established, among them the Black-whiskered Bulbul, Spotted Oriole, and several species of parrots and parakeets. The chances are that these and other exotic species, as they become established, will gradually move northward, to become a permanent element in the state's avifauna.

The Florida coast—excluding the Florida Keys to the south but including all other islands and the many indentations reached by the tides—has a shoreline extending about 4,000 miles from the Georgia boundary on the Atlantic to the Alabama boundary on the Gulf of Mexico. The shore on the Atlantic (east) coast consists almost entirely of sandy beaches and dunes on ribbon-like reefs and slender islands that parallel the mainland and shelter the narrow intervening salt-water lagoons sometimes called rivers. A considerable part of the Gulf (west) shoreline also has beaches and dunes, many of them on barrier reefs and islands, especially from Apalachee Bay west to Pensacola Bay.

On the eastern coast, salt marshes fringe the lagoons and river mouths from the Georgia line to Matanzas Inlet, and the borders of Mosquito Lagoon and Indian River farther south. On the western coast the marshes are much more numerous and extensive, bordering some part of all northwestern bays and stretching southward on the peninsula in almost continuous succession from Apalachee Bay to Charlotte Harbor.

Glassworts, switch grasses, and sharp-pointed rushes constitute the principal vegetation in the east-coast marshes, and rushes, in the west-coast marshes. Mangroves grow in scattered clumps in the marshes of Mosquito Lagoon as well as in Tampa Bay; farther south they become increasingly abundant, and the salt marshes disappear. In the extreme south, below Biscayne Bay on the east and Charlotte Bay on the west, mangroves (both red and white) and buttonbushes are the dominant growth on most of the low, wet coastal margins and on the innumerable islets immediately offshore—thus forming mangrove swamps.

Birds breeding regularly on the Florida coast include a wide variety of species. On a few small islands in bays and lagoons are colonies of Brown Pelicans whose nests are usually in the mangroves. Black Rails probably breed in many salt marshes as well as in fresh-water marshes in the interior, although very few nests have been reported. The interior marshes west from Titusville on the east-central coast constitute the most promising areas for nests of these elusive birds.

On both eastern and western coasts, Gray Kingbirds are familiar summer residents, placing their nests in trees and shrubs, sometimes near bays and lagoons, sometimes in or near marshes. Boat-tailed Grackles are common, often abundant, along the coasts, where they nest in salt marshes. In northwestern Florida they are salt-marsh inhabitants exclusively, as they are in other states; but in other sections of Florida they inhabit also the fresh-water marshes of the interior. Seaside Sparrows reside permanently in the salt marshes of the northeast coast south to the Titusville area and of the northwest coast south to Tampa; farther south they occupy interior marshes as well. Other species that nest regularly on the beaches, on small islands, or in salt marshes are the following:

Clapper Rail
American Oystercatcher (*rare southward*)
Snowy Plover
Wilson's Plover
Willet
Laughing Gull
Little Tern
Black Skimmer
Marsh Wren (*northern Florida only*)

The Florida interior is principally a very low plain of sand and clay, its surface cut by countless streams and pitted by some thirty thousand lakes varying in size from small ponds to Lake Okeechobee, one of the largest natural fresh-water bodies within United States boundaries. Northwestern Florida, like neighboring sections of Georgia and Alabama, has a rolling terrain. This same terrain forms a ridge or 'backbone' running south down the center of the peninsula almost to Lake Okeechobee. Along this ridge, near Lake Wales, about halfway down the peninsula, rises Iron Mountain (295 feet), highest point in the state. Although the interior has many natural regions of differing physiography, the following are considered of major importance with respect to birdlife: pinelands, prairies, hammocks, scrublands, marshes, and swamps.

The *pinelands*, typically open forests predominantly of pine, cover more territory than any of Florida's other natural regions. They are of two kinds: pine flats or flatwoods, and high pinelands; but the two are not always sharply distinguished. The flatwoods are widely distributed from the northwestern part of the state to the Miami region. In northern and central Florida they are comparatively level areas supporting both longleaf and slash pines, together with dense thickets of saw palmetto, myrtle, rosemary, and gallberry. Since the flatwoods are poorly drained, they occasionally have wet places where cypresses and various hardwoods thrive, sometimes in swamps of considerable size. Good examples of flatwoods are south of Tallahassee in the St. Marks National Wildlife Refuge and along US 92 west of Daytona Beach. In southern Florida, near Miami, the flatwoods are much the same in general appearance, but the slash pine is the principal tree growth, and the undergrowth contains many plants of tropical origin. The high pinelands are characteristic of the rolling northwestern country and of the 'backbone' of the peninsula. Most of the trees are longleaf pines, usually widely spaced; the undergrowth is sparse. Of birds occupying the pinelands, the Red-cockaded Woodpecker, Brown-headed Nuthatch, Pine Warbler, and Bachman's Sparrow are the most typical; they are seldom found in any other natural region.

The *prairies* are open, flat, seemingly endless areas mainly in central Florida. Kissimmee Prairie, the most extensive, in south-central Florida (reached from both Okeechobee and Kenansville), is relatively dry and grass-covered, with a scanty growth of saw palmettos and scrub oaks interspersed with clusters of cabbage palmettos. There are also many marshes of varying sizes, irregularly shaped tracts of flatwoods, and hammocks (*see below*). The Crested Caracara and Burrowing Owl are two Florida species limited largely to the prairies; another, the Grasshopper Sparrow, is still more limited, apparently existing only in certain sections of the Kissimmee Prairie, notably near Kenansville. The Sandhill Crane is found on the Kissimmee Prairie more commonly than elsewhere in Florida, its nesting habitat being the isolated, undisturbed wet places, usually marshes.

Hammocks are island-like wooded tracts, their character and size varying according to their location. As a rule, they are high enough above their surroundings to have a dry soil that supports a dense growth of hardwoods and shrubs. Some hammocks are on prairies, others are in marshes, flatwoods, and swamps. In most cases they are small, and yet their birdlife is invariably richer than that of the larger natural regions surrounding them. Two particularly fine hammocks are the Royal Palm and the Mahogany in Everglades National Park (*see under* **Homestead**); another is in Highlands Hammock State Park (*see under* **Sebring**).

Scrublands are tracts characterized by whitish sand and supporting sand pines and almost impenetrable thickets of saw palmetto, rosemary, and various scrub oaks. Typical scrublands occur along both eastern and western coasts of the peninsula and at widely scattered points in the interior. The Scrub Jay is the only species restricted to this natural region. Good examples of scrub country occupied by this species may be observed along State A1A south from Marineland (*see under* **St. Augustine**).

Fresh-water marshes are scattered here and there throughout the state. Dominating all such marshes—if not the topography of the state—are the famous Everglades. No words can adequately picture their vastness and their importance as wildlife reservoirs. At least 50 miles across from east to west at their widest point, they extend for more than 100 miles from the southern and eastern shores of Lake Okeechobee southward to Whitewater Bay on the southern end of the peninsula. Much of the northeastern part of the Everglades lies within the Loxahatchee National Wildlife Refuge (*see under* **Delray Beach**), and here the bird finder may see some of the few Snail Kites in the United States. The southern parts of the Everglades are in the Everglades National Park.

Swamps of varying size, many of them immense, with a heavy growth of cypresses and hardwoods, are distributed throughout Florida, but occur frequently along the margins of rivers nearing the western coast. Of these innumerable swamps, two are strongly recommended to the bird finder because of their accessibility, vastness, natural charm, and variety of birdlife: the swamp along the Wakulla River in northern Florida (*see under* **Tallahassee**) and the cypress swamp in Corkscrew Sanctuary in southern Florida (*see under* **Immokalee**). Both swamps are fine habitat for Limpkins. The mangrove swamps on the extreme southern coast have a small assortment of landbirds, including the White-crowned Pigeon, Mangrove Cuckoo, Black-whiskered Vireo, and Prairie Warbler—species not generally present in other Florida habitats. Specialties though these birds may be, they are overshadowed in popular interest by the tremendous waterbird colonies in the mangrove swamps in Florida Bay (*see under* **Homestead**).

An unusually large number of birds breed in Florida's fresh-water marshes and swamps, and in the shrubs and low trees along their edges. Among species nesting regularly in these environments are the following:

Pied-billed Grebe
Double-crested Cormorant
American Anhinga
Great Blue Heron
Green Heron
Little Blue Heron
Cattle Egret
Great Egret
Snowy Egret
Louisiana Heron
Black-crowned Night Heron
Least Bittern
American Bittern (*rare*)
Wood Stork
Glossy Ibis
White Ibis
Mottled Duck (*peninsula only*)
Wood Duck (*except tip of peninsula*)
Sandhill Crane (*peninsula only*)
Limpkin
King Rail
Purple Gallinule
Common Gallinule
Common Yellowthroat
Boat-tailed Grackle

The Bald Eagle nests in a few widely separated places in Florida and more frequently in Everglades National Park than elsewhere in eastern United States. The Osprey nests throughout the state; in a few places where it is common, its bulky aerie is conspicuous in a high tree or on a utility pole overlooking a river or lake.

A species to look for between February and November in central and southern Florida is the Swallow-tailed Kite. Since it is fairly common, and customarily searches for food in open country, especially near marshes, it is not easily missed. Other species to watch for are the Short-tailed Hawk—a rare permanent resident in southern Florida, and the Black-necked Stilt—a spring and summer resident of central and southern Florida, on the shores and mud flats of certain coastal lagoons and marshes and on lakes in the interior.

Throughout Florida, the following species breed regularly in wooded areas (hammocks, pinelands, and swamps containing cypress and hardwoods) or in dry, open, and semi-open areas (grassy parts of pinelands and prairies, shrubby thickets, fields, orchards, and dooryards):

WOODED AREAS

Turkey Vulture
Black Vulture
Mississippi Kite (*northwestern Florida only*)
Red-shouldered Hawk
Broad-winged Hawk
Yellow-billed Cuckoo
Smooth-billed Ani
Common Screech Owl
Barred Owl
Chuck-will's-widow
Yellow-shafted Flicker
Pileated Woodpecker
Red-bellied Woodpecker
Hairy Woodpecker
Downy Woodpecker
Great Crested Flycatcher
Acadian Flycatcher (*except southern Florida*)
Eastern Pewee (*northern Florida only*)
Fish Crow
Carolina Chickadee (*except southern Florida*)
Tufted Titmouse (*except southern Florida*)
White-breasted Nuthatch (*except southern Florida*)
Blue-gray Gnatcatcher (*except southern Florida*)
Yellow-throated Vireo (*mainly northern Florida*)
Red-eyed Vireo (*except southern Florida*)
Prothonotary Warbler (*mainly northern and central Florida*)
Swainson's Warbler (*mainly northwestern Florida*)
Northern Parula Warbler (*except southern Florida*)
Yellow-throated Warbler (*except southern Florida*)
Hooded Warbler (*mainly northwestern Florida*)
Summer Tanager (*except southern Florida*)

OPEN AND SEMI-OPEN AREAS

Common Bobwhite
Mourning Dove
Ground Dove
Red-headed Woodpecker
Eastern Kingbird
Blue Jay
Carolina Wren
Northern Mockingbird
Brown Thrasher (*northern Florida only*)
Eastern Bluebird
Loggerhead Shrike
White-eyed Vireo
Eastern Meadowlark
Orchard Oriole (*mainly northern Florida*)

OPEN AND SEMI-OPEN AREAS (*Cont.*)

Common Grackle
Northern Cardinal
Painted Bunting (*except southern Florida*)
Rufous-sided Towhee

The Florida Keys are a chain of small, low-lying islands extending for more than a hundred miles south and west from the peninsula, and surrounded by a shallow sea noted for its green, sometimes purplish, color. Sandy beaches, strewn with fascinating varieties of shells, line the islands. Many have wet areas covered with mangroves, as well as uplands where coconut and silver palms, gumbo-limbos, wild tamarinds, poisonwoods, mahoganies, and other tropical trees grow in jungle profusion.

Birdlife, though limited in variety, is exciting by reason of the West Indian elements represented. With the exception of the southern part of the peninsula, the Keys constitute the only nesting place in the United States for the Great 'White' Heron (color phase of the Great Blue Heron), Reddish Egret, Roseate Spoonbill, White-crowned Pigeon, Mangrove Cuckoo, and Black-whiskered Vireo. The Keys, many of them, are readily accessible via the Overseas Highway (US 1), which runs from Miami to the island of Key West near the end of the chain. Bird finding is good in any season.

Although most of the landbirds breeding in Canada and eastern United States and wintering in South America migrate directly across the Gulf of Mexico, impressive numbers pass through peninsular Florida and the West Indies. Some species, such as the Blackpoll Warbler, use the Florida–West Indies route almost entirely, and others use both routes; in either case these species are common, often abundant, transients in Florida in spring and fall. Northwestern Florida lies in the path of birds using the trans-Gulf route. Here the spring migration shows the same remarkable irregularities observed on the Mississippi coast (*see also* **Mississippi** chapter). Northern shorebirds are prevalent in Florida in all seasons except early summer, but are especially abundant in spring and fall, when their populations are augmented by individuals passing to and from wintering grounds farther south than Florida.

WHITE-CROWNED PIGEON

The majority of transients passing through Florida may be expected within the following dates:

Shorebirds: 15 March–15 May; 15 August–20 October
Landbirds: 15 March–1 May; 15 September–10 November

Even though the total number of species breeding regularly in Florida is small compared with totals in most other states, the number in winter (counting permanent residents) is exceptionally large. Christmas bird counts taken in such widely separated localities as Jacksonville, St. Marks on Apalachee Bay, Cocoa, Sarasota, Fort Lauderdale, and Everglades National Park almost invariably exceed 150 species, at least half of them from regions north and west of Florida. The Lesser Scaup is the most common of the wintering waterfowl, but there is a rich variety of species. The waterfowl concentrations in the St. Marks, Merritt Island, Loxahatchee, and 'Ding' Darling National Wildlife Refuges are most impressive. Northern Gannets appear on both eastern and western coasts, and gulls and terns frequent all the larger watercourses throughout the state. Painted Buntings—the male in bright plumage—often show up in dooryards with feeding stations. American Kestrels are a familiar sight on utility wires, and such other northern landbirds as Myrtle Warblers and Palm Warblers are common, if not abundant. A marked peculiarity of winter birdlife is a western element, represented by American White Pelicans, Long-billed Curlews, Marbled Godwits, American Avocets, Western Kingbirds, and Scissor-tailed Flycatchers among others.

A final comment. . . . The bird finder driving the Florida highways in any season notices the conspicuous abundance of several species: Cattle Egrets in ditches and in pastures with livestock; both Turkey and Black Vultures soaring on thermals; and Fish Crows, Northern Mockingbirds, and Boat-tailed Grackles almost anywhere, flying, perching, or foraging on the ground.

Authorities

Clara Bates, C. Wesley Biggs, Charles M. Brookfield, Allan D. Cruickshank, Whitney H. Eastman, William G. Fargo, Merritt C. Farrar, Samuel A. Grimes, Frances Hames, John William Hardy, Joseph C. Howell, Margaret H. Hundley, Paul T. Kreager, Rupert J. Longstreet, Reginald R. McKay, Herbert R. Mills, Hilda Morris, Donald J. Nicholson, W. H. Nicholson, Clifford Pangborn, Alexander Sprunt, Jr., Alexander Sprunt IV, Henry M. Stevenson, Louis A. Stimson, Herbert L. Stoddard, Francis M. Weston.

Reference

Where To Find Birds in Florida. Edited by Margaret C. Bowman and Herbert W. Kale II. Florida Audubon Society, 1976. Available from F.A.S. Headquarters, 921 Lake Sybelia Drive, Maitland, Fla. 32751.

ALLIGATOR ALLEY

This superhighway, an 84-mile stretch of State 84 straight across southern Florida between Fort Lauderdale and Naples, traverses unspoiled parts of the eastern Everglades and western Big Cypress Swamp. The vastness of both natural regions is nowhere more impressive at ground level than from this road. Turnouts for parking are frequent, often near sloughs or paralleling canals attractive to grebes, herons, egrets, and gallinules throughout the year. During winter, Northern Harriers often fly low over the Everglades; during any season, Red-shouldered Hawks perch on stubs in the Cypress Swamp.

DAYTONA BEACH

This city beside the Atlantic has an eastern division built up along the ocean front and separated from a western division by a tidewater lagoon called the Halifax River.

The ocean-front beach, a magnificent stretch of white sand 500 feet wide, is so smooth and hard that at low tide it is used as a speedway. From the center of the eastern division, the bird finder may drive 6 miles north and 10 miles south on the beach, searching for birds in a way that can hardly be excelled for ease and convenience. In winter the same species of gulls and terns in evidence at Matanzas Inlet (*see under* **St. Augustine**) are present here, and Northern Gannets often appear offshore. Shorebirds present in winter include the following, in decreasing order of abundance: Sanderling, Ruddy Turnstone, Piping Plover, Semipalmated Sandpiper, and Black-bellied Plover. Beginning in April, the variety of shorebirds increases with the Semipalmated Plover, Wilson's Plover, Willet, Least Sandpiper, sometimes the Spotted Sandpiper and Dunlin, and perhaps a dowitcher. In May, Red Knots are usually numerous, sometimes appearing in flocks of a hundred or more individuals.

If the bird finder visits Daytona Beach in winter, he should stop at the Halifax River sea wall near the Yacht Club on South Beach Street; here Herring Gulls, Ring-billed Gulls, Laughing Gulls, Bonaparte's Gulls, and Lesser Scaups are so accustomed to being fed by human beings that they have become remarkably tame. If the bird finder holds up food, very likely one of the gulls hovering overhead will swoop down to snatch it.

West of Daytona Beach on US 92, one passes a great expanse of typical Florida flatwoods, among which are numerous cypress ponds. The edges of the ponds are especially good bird-finding spots because they attract species characteristic of both environments. In one such spot—7 miles west of Daytona Beach on State 92—on any spring day, preferably in April, it is possible to draw up a long list that will include northbound transients together with permanent and summer residents.

Among resident species should be the Turkey Vulture, Black Vulture, Red-shouldered Hawk, Mourning Dove, Yellow-shafted Flicker, Pileated Woodpecker, Red-bellied Woodpecker, Hairy Woodpecker, Downy Woodpecker, Red-cockaded Woodpecker, Great Crested Flycatcher, Carolina Chickadee, Tufted Titmouse, Brown-headed Nuthatch, Carolina Wren, Northern Mockingbird, Eastern Bluebird, Blue-gray Gnatcatcher, Loggerhead Shrike, White-eyed Vireo, Prothonotary Warbler, Northern Parula Warbler, Yellow-throated Warbler, Pine Warbler, Common Yellowthroat, Eastern Meadowlark, Common Grackle, Northern Cardinal, and Bachman's Sparrow.

DELRAY BEACH
Loxahatchee National Wildlife Refuge

As many as 145,635 acres of the northeastern reaches of the Everglades comprise the **Loxahatchee National Wildlife Refuge.** Headquarters lies on the eastern side of the Refuge, west off US 441, 3 miles north of its intersection with State 806 coming from Interchange 32 on I 95 (2 miles distant) and Delray Beach (10 miles distant).

The Refuge, bounded entirely by levees, maintains the natural water level of the Everglades—so severely drained in other northern areas, with consequent reduction in plant and animal resources. Within the Refuge, therefore, is a fine sampling of northern Everglades features, such as sawgrass marshes, wet prairies, sloughs, hammocks, and associated wildlife as they existed years ago.

By virtue of its being one of the last undrained areas of the northern Everglades, protected from shooting and excessive human disturbances, the Refuge has some of the few Snail Kites in the United

States. Here they find sufficiently extensive habitat for nesting and, more importantly, an ample supply of fresh-water snails (*Pomacea paludosa*), on which they depend exclusively for food. Except while nesting, from December through May, the kites are widely dispersed and seldom seen on the Refuge.

Numerous and conspicuous at any time are the wading birds—Great Blue Herons, Little Blue Herons, Cattle Egrets, Great Egrets, Snowy Egrets, Louisiana Herons, and Black-crowned Night Herons. During the day they feed, widely scattered, along canals and sloughs, and toward evening converge in large flocks as they fly to roost for the night in distant hammocks. Also numerous the year round are American Anhingas, Mottled Ducks, Sandhill Cranes, Common Gallinules, and (though inconspicuous) Limpkins and King Rails. The most common and noticeable predator is the permanent-resident Red-shouldered Hawk.

Winter brings many waterfowl, chiefly Common Pintails, Green-winged and Blue-winged Teal, American Wigeons, Wood Ducks (some are permanent residents), Ring-necked Ducks, and Hooded Mergansers, usually a few Fulvous Tree Ducks, Northern Shovelers, Ruddy Ducks, and occasionally such rarities as an American Black Duck or a Masked Duck.

A good vantage point for seeing Snail Kites as well as many other birds on the Refuge is along the levee just west of headquarters. Also on the levees, where alligators sun themselves, are coveys of Common Bobwhites. Now and then, Smooth-billed Anis put in an appearance close to headquarters or elsewhere in wooded places.

FORT WALTON BEACH
Santa Rosa Island

From US 98 in this western Florida city, turn off and cross the bridge to **Santa Rosa Island,** and then proceed eastward as far as the bridge across East Pass. Park the car and look over the sand bars and outlying waters. From the first of November until April, many of the same species of transient and wintering waterbirds, ducks, and shorebirds that are observed at the western end of Santa Rosa Island (*see under* **Pensacola**) may be seen here as well. Walk south to the Gulf of Mexico and westward along the beach, where a few Snowy Plovers usually reside all year long.

GAINESVILLE
Morningside Park

For sighting common woodland birds in northern Florida, **Morningside Park** (278 acres), owned by this city, is ideal. Forested largely with longleaf pine, oak, and palmetto, with a small stand of bald cypress, the Park is prime habitat for permanent-resident Red-cockaded Woodpeckers, Brown-headed Nuthatches, Pine Warblers, and Bachman's Sparrows. The Park lies about 2 miles east of downtown Gainesville just north of State 26. A sign on the highway indicates the direction of the entrance road, which leads to a nature center and to nature trails through the west side of the Park.

HOMESTEAD
Everglades National Park

The southernmost region of Florida from Miami west and south constitutes the Everglades, which, contrary to popular belief, are neither desolate waste nor ghastly wilderness defying penetration; they are a great fresh-water marsh, exotic in aspect and teeming with wildlife.

Everglades National Park embraces 2,100 square miles of this vast area with its flat, low-lying terrain—nowhere higher than 10 feet—and countless fresh-water channels and sloughs. The interior is largely sawgrass country—open prairies or 'glades,' in some places hundreds of miles of sawgrass as far as the eye can see—flooded during the summer rains and often ravaged by fire as spring follows the winter drought.

Here and there, where the elevation is slight, loom many tree islands, most of them 'bayheads' comprised of low stands of bay, holly, and magnolia, overgrown with innumerable vines to form a thick tangle. They are often dome-shaped, since the trees tend to grow taller toward the center of the islands. The other islands are 'hammocks,' best described as dense, moist, jungle-like forests whose trees support a lush epiphytic flora consisting of orchids, bromeliads, ferns, and other air plants. The trees themselves—mostly large—represent many broad-leafed varieties of West Indian derivation.

The only other forested places in the Park are the pinelands, restricted to a few limestone elevations. From the sawgrass country southward and westward stretch the mangrove swamps—thickets of low mangroves, some deeply rooted in wet saline soils, others in tide-washed flats or in salt-water shallows.

To reach the Park's main entrance, drive 12 miles south and west from Homestead on State 27, following the directional signs. Just within the Park, stop at the visitor center for a map and information.

From the entrance a 38-mile paved road traverses the Park to Flamingo on Florida Bay. Along the road in winter, expect to see Red-shouldered Hawks, American Kestrels, Yellow-shafted Flickers as well as an abundance of American Crows, Eastern Meadowlarks, Red-winged Blackbirds, and Boat-tailed Grackles. After late February, look for Swallow-tailed Kites, which return from the southern wintering grounds to nest. In any season, watch the sky and horizon for one or two Bald Eagles in flight (always a possibility) and for Ospreys (almost a certainty).

Just after leaving the visitor center and before crossing the bridge over Taylor Slough, note the vast stretches of sawgrass on both sides of the road. (Cape Sable) Seaside Sparrows reside here year round but can be located readily only during breeding season, beginning in spring, when they reveal themselves by singing.

Royal Palm Hammock. After driving about 2 miles, watch for a sign indicating the Royal Palm Ranger Station on the left. Turn in and park the car. The Ranger Station occupies Royal Palm Hammock, with a luxuriant growth of vegetation including a few royal palms (growing rarely elsewhere in the Park), live oaks, and giant leather ferns. Pileated Woodpeckers frequent the dead live oaks, and Barred Owls call during the day. The Gumbo Limbo Trail leads through some of the best parts of the hammock, but the primary attraction is the Anhinga Trail, which offers exceptionally close views of alligators and many different species of waterbirds and marsh-loving birds—and unusual opportunities for photography.

SWALLOW-TAILED KITE

Anhinga Trail. From the Ranger Station the Anhinga Trail runs for about a half-mile into Taylor Slough, past a stretch of open water, along an old road grade paralleled by water-filled ditches, and onto an elevated boardwalk. During its course the Trail passes some habitats dominated by sawgrass, others by tall, dense shrubs, and still others by emergent herbaceous plants, among them waterlilies with huge pads. The variety of habitats is only one of the factors contributing to the Trail's uniqueness. Just as important are the great abundance of fish and aquatic invertebrates, which provide a generous supply of food for birds, and the daily, almost steady stream of human visitors. Lured to the area by food and gradually accustomed to human movements, the birds stay to feed and rest, ignoring the people.

Common Gallinules and American Coots are numerous on the open water. In the ditches are Green Herons, Little Blue Herons,

Snowy Egrets, Louisiana Herons, American Bitterns (winter only), and White Ibises. Sometimes a Limpkin or a King Rail appears momentarily in the background. American Anhingas are everywhere conspicuous, often sunning themselves on bushes or stubs. Purple Gallinules walk confidently on the lily pads below the boardwalk. Pied-billed Grebes, Belted Kingfishers, and sometimes Great Blue Herons feed in the big pool at the end of the boardwalk.

Returning to the Park road and continuing south, stop at Mahogany Hammock, 19.5 miles from the Park entrance, for a brief hike on the elevated boardwalk that circles through a fine stand of Madeira mahoganies. Be alert to the possibility of hearing or seeing the Pileated Woodpecker and the Barred Owl.

West Lake, in view from the road 30.5 miles from the Park entrance, has American Coots and a small variety of ducks, including the Ring-necked, during the winter months, and wading birds all year long. Shortly beyond West Lake comes Mrazek Pond, also in view from the road and with similar though somewhat smaller aggregations of birds.

Between West Lake and Flamingo, watch carefully for the rare Short-tailed Hawk, more likely to show up here than elsewhere along the road.

Flamingo. At Flamingo, which overlooks Florida Bay from east of Cape Sable—the southernmost tip of the Florida mainland—are available a restaurant, motor lodge, cottages, campground, grocery store, service station, marina, and National Park services such as an evening program by naturalists. Request for lodge and cottage reservations should be addressed to Everglades Park Company (18494 South Federal Highway, Miami, Florida 33157).

During winter, be sure to see the 'freeloaders'—Brown Pelicans, Ring-billed and Laughing Gulls—waiting at the Flamingo marina for returning fishermen. Every boat that comes into view they swarm over and tag after, responding instantly to any action that bears promise of discarded fish, unused bait, or anything else edible. Scan the mud flats, sand bars, and shallow water offshore in Florida Bay for large gatherings of Black Skimmers, and for Long-billed Curlews, Marbled Godwits, and other shorebirds. And do not overlook the tops of the piling close to shore for resting birds such as Double-crested Cormorants and Royal, Sandwich, and Caspian Terns.

North of the road between the cottages and campground, look over the sewage disposal bed for Great 'White' Herons, Great Egrets, Roseate Spoonbills, and other wading birds as well as for a few waterfowl that may include Fulvous Tree Ducks. Be at the campground within a half-hour before sundown to watch flocks of White Ibises flying low overhead on their way from inland swamps to pass the night on the mangrove islands out in the Bay.

Bear Lake. Worthwhile is a walk northwest from Flamingo to Bear Lake, accessible by 2 miles of wilderness road closed to cars. Ask a Park naturalist for specific directions and be prepared for mosquitoes. Shallow and surrounded by mangrove jungles with intervening channels and sloughs, Bear Lake usually features large numbers of American White Pelicans, Little Blue Herons, Great Egrets, Snowy Egrets, Louisiana Herons, and White Ibises, together with smaller numbers of American Anhingas, Great 'White' Herons, Wood Storks, and Roseate Spoonbills. During a spring and summer walk, watch or listen for White-crowned Pigeons, Mangrove Cuckoos, Black-whiskered Vireos, and Prairie Warblers.

Besides the entrance to Everglades National Park from Homestead, there is an approach south from US 41—the Tamiami Trail, about 35 miles west of Miami and 75 miles east of Naples. Access is by public transportation (private cars are prohibited) south along a canal for 7 miles—with stops to observe wading birds—to the Shark Valley area, which is overlooked by a 35-foot observation tower of modern design. From its top—reached by a stairless, circular ramp—one is treated to a panorama of the Everglades and, through binoculars or telescope, perhaps some good views of a few Sandhill Cranes.

Access to the Ten Thousand Islands on the northwestern fringe of the Park is by water from Everglades City. Inquire here at the Park's Gulf Coast Ranger Station about boat trips for viewing the rich assortment of birds along the channels and creeks that wind through this fine wilderness area.

IMMOKALEE
Corkscrew Swamp Sanctuary

A must for the bird finder in southwestern Florida is **Corkscrew Swamp Sanctuary,** owned and operated by the National Audubon Society, on State 846, 16 miles west of this farming center, in the heart of the Big Cypress Swamp.

The Sanctuary embraces 10,422 acres containing the last stand of virgin bald cypress in Florida, with bordering areas of pine flatwoods, wet prairie, and pond-cypress swamp. One section of the Sanctuary is open to the public, by way of a self-guiding nature trail—a mile-long boardwalk that loops through the major habitats.

Starting at the entrance gate, the trail first traverses flatwoods of slash pine and saw palmetto, where one should be alert for Red-cockaded Woodpeckers, Brown-headed Nuthatches, Pine Warblers, and Rufous-sided Towhees. Then, after crossing a strip of treeless wetland thickly covered with grasses and sedges—often sprinkled with showy wildflowers in the spring—the trail abruptly penetrates a subtropical forest of almost overwhelming density. Here tower the magnificent cypresses, centuries old, festooned with epiphytes (air plants) of astonishing variety. The trees one sees on entering the forest are 'pond cypresses.' Although truly enormous, they seem small when compared to the trees one soon encounters in the interior—the giant 'bald cypresses,' which drop their feathery foliage in winter and appear 'bald.' Some are probably seven centuries old.

The birds that first catch the eye in any season are the conspicuous wading birds, some flying over, others stalking food or resting. Wood Storks are common, understandably, since in the Sanctuary is found one of the largest nesting colonies of the species in the United States. Also common are Great Blue Herons, Great Egrets, Little Blue Herons, White Ibises, and, to a less extent, Green Herons, Black-crowned Night Herons, and American Bitterns.

Other birds whose presence one notices eventually are American Anhingas; Red-shouldered Hawks, quite tame, often perched quietly; Limpkins, characteristically skulking through the lower swamp cover in search of the large snail, *Pomacea,* their principal food; Pileated Woodpeckers, their oblong excavations apparent on old tree trunks and stubs, even though the birds may not appear

WOOD STORK

themselves; Barred Owls, sometimes calling in the late afternoon or throughout a heavily overcast day; and several passerine birds—Tufted Titmice, Carolina Wrens, Blue-gray Gnatcatchers, White-eyed Vireos, Red-winged Blackbirds, Common Grackles, and Northern Cardinals. In winter one can expect Yellow-bellied Sapsuckers and such warblers as the Black-and-white, Yellow-throated, and Palm; in spring and summer, Swallow-tailed Kites, frequently soaring high over the cypresses.

JACKSONVILLE

Fort George Island | Little Talbot Island State Park | North Jetty of the St. Johns River | Mayport

This large city in northeastern Florida, on the St. Johns River about 20 miles from the Atlantic, has in its vicinity a wide variety of permanent-resident species and also, since it lies along the Atlantic flyway, many transients.

Two short trips that will yield birds of particular interest are recommended.

1. Drive north from Jacksonville on US 17 or I 95. Turn right

(east) on State 105, a half-mile north of the Trout River Bridge, and proceed slowly, scanning the many creeks and rivers where birds invariably congregate. At Pilot Town, turn left on State A1A around **Fort George Island**. Visit the Kingsley Plantation, 3 miles to the left off State A1A. Here, among palms and live oaks planted long ago by slaves, and in the hammocks, one may find throughout the year such species as the Ground Dove, Pileated Woodpecker, Yellow-throated Warbler, and Rufous-sided Towhee; in spring and summer, the Chuck-will's-widow, Northern Parula Warbler, Summer Tanager, and Painted Bunting; in winter, the Purple Finch.

Continue north on State A1A and turn right into **Little Talbot Island State Park** (2,500 acres). In winter the wax-myrtle thickets teem with Myrtle Warblers. The Red-tailed and Red-shouldered Hawks are permanent residents, and the Osprey is a visitant from February through September. Waterbirds, ducks, and shorebirds are numerous on the ocean front. The American Oystercatcher feeds here, and large groups of Black Skimmers are especially impressive.

Return south on State A1A. At a point where State A1A turns right to Pilot Town, turn left (east) on a rough road to the **North Jetty of the St. Johns River.** Leave the car and walk over the flat-topped rocks of the Jetty. Brown Pelicans and Black Skimmers, as well as various plovers, sandpipers, gulls, and terns, gather here at all times of the year. Offshore, Northern Gannets may appear from September through April, Red-throated Loons from September through May, and rarities such as Common Goldeneyes from October through March.

2. From Jacksonville, drive east on US 90 across the Intracoastal Waterway and turn north on State A1A toward the St. Johns River and the little fishing village of **Mayport.** About a mile or so before reaching the village and just as the road curves to the River, turn left on a sand road, park the car, and proceed westward on foot. After passing salt marshes, fields, and sand spits, the road crosses two creeks. Look for King Rails, Clapper Rails, and Seaside Sparrows at any time of year, for Water Pipits in winter, and for Black-necked Stilts in spring and summer.

Return to State A1A and continue north to Mayport. Turn east on Wonderwood Road at the sign 'Kathryn Hanna Abbey Park,' and go on to this city Park and the ocean beach. Chances are always good, particularly in winter, of seeing American Anhingas and herons. Many shorebirds, gulls, and terns appear on the beach.

Return to Jacksonville on State A1A, by way of the beaches and US 90. Stop occasionally at one of the parking places and walk to the ocean to look for more shorebirds, gulls, and terns. Turn west on US 90 and, after crossing the Intracoastal Waterway, take a side trip into the pine flats by turning right on Pablo Road and going on for a short distance. Brown-headed Nuthatches, Pine Warblers, and Bachman's Sparrows reside in the pines the year round, and, during spring and summer, possibly Swallow-tailed Kites as well.

KENANSVILLE
Kissimmee Prairie

A section of **Kissimmee Prairie** can be reached from Kenansville, on US 441 between St. Cloud and Okeechobee, by driving 6 to 8 miles southwest on a dirt road. This area, with its mixture of pinelands, cypress swamps, and prairies, holds a remarkable variety of birds.

Because the habitats are often intermixed, it is impossible to state precisely where a particular species may be found. In general, however, one may expect to find in certain habitats the following breeding birds, usually after April: *on the prairies where there are thin grasses, palmettos, and St. John's worts,* Sandhill Crane, Burrowing Owl, Common Nighthawk, Grasshopper Sparrow; *in the pinelands,* Wild Turkey, Red-cockaded Woodpecker, Brown-headed Nuthatch, Pine Warbler, Summer Tanager, Bachman's Sparrow; *in the cypress swamps,* American Anhinga, Wood Duck, Pileated Woodpecker, Prothonotary Warbler, Northern Parula Warbler.

KEY WEST
Plantation Key | Cowpens Key | Upper Metecumbe Key
Lower Metecumbe Key | Fiesta Key | Long Key
Grassy Key | Vaca Key | Little Duck Key | Missouri Key
Bahia Honda Key | Big Pine Key | Sugarloaf Key
Stock Island | Key West | Dry Tortugas

For bird finding, a trip to the Lower Florida Keys on the Overseas Highway (US 1) from Miami to Key West, a distance of 165 miles, is unlike any other anywhere. Following an old railroad bed from key to key over bridges, trestles, and causeways, this remarkable road permits a leisurely drive over sea and land toward the Tropic of Cancer, only 75 miles south of Key West, the southernmost city in the United States. At a few points on the route, notably on the Seven-mile Bridge, land is almost beyond view. One may turn off almost anywhere to watch the birds flying overhead, resting on the water, or perching on the wires and guard rails. One may park on the keys for exploration on foot along the beaches and through the wooded areas.

Some of the birds in any season are the Magnificent Frigatebirds, Great 'White' Herons, Reddish Egrets, Roseate Spoonbills, Red-shouldered Hawks, Ospreys, White-crowned Pigeons (often flying across the roads), and Smooth-billed Anis; in spring and summer, Mangrove Cuckoos, Gray Kingbirds (conspicuous on wires), Black-whiskered Vireos, (Cuban) Yellow Warblers, and Prairie Warblers; in winter, Western Kingbirds and Scissor-tailed Flycatchers (conspicuous on wires).

Below are some suggestions for the bird finder during a trip south on the Overseas Highway.

Plantation Key. At Mileage Marker 89, turn right toward Florida Bay. The second house on the left is an office of the National Audubon Society, with information available on bird finding in the Florida Keys, including Plantation Key itself.

West from Plantation Key in Florida Bay is 10-acre **Cowpens Key,** a sanctuary of the National Audubon Society. Low-lying, mangrove-covered, and surrounded by shallow water, it has a large nesting colony of Roseate Spoonbills, as well as many nesting Brown Pelicans, Great 'White' Herons, Reddish Egrets, and other wading birds. Landing is prohibited during nesting season, but one may readily view the aggregation from offshore. Make arrangements through the National Audubon Society office for a boat trip around the island.

Upper Metecumbe Key. In Islamorada, stop at the Chamber of Commerce and obtain a map of the Florida Keys and Key West. South of town, take the nature trail, starting near Green Turtle Inn on the left and going through a hammock to the ocean beach. In an

ROSEATE SPOONBILL

open area along the way, back from the shelter on the beach at right, look especially for White-crowned Pigeons, Smooth-billed Anis, Western Kingbirds, and Scissor-tailed Flycatchers.

Lower Metecumbe Key. South of Mileage Marker 74 at the southern end of the island on the ocean side are a sand spit and mud flats where shorebirds, gulls, and terns congregate in large numbers (except during summer). Common shorebirds include Semipalmated, Piping, and Black-bellied Plovers, Ruddy Turnstones, Greater and Lesser Yellowlegs, Least Sandpipers, Short-billed Dowitchers, Semipalmated and Western Sandpipers, and Sanderlings. Ring-billed and Laughing Gulls are usually numerous all year long, as are Royal Terns and Black Skimmers. Along Caloosa Cove Beach, on the oceanside in the same area, Wilson's Plovers reside all year, and a few Sandwich and Caspian Terns appear in winter.

Fiesta Key. The shores on both sides of the Overseas Highway are worth watching for shorebirds, gulls, and terns.

Long Key. Keep an eye on the shore and all shallow ponds on both sides of the Highway for feeding and resting Roseate Spoonbills, Reddish Egrets, and other wading birds, as well as shorebirds, gulls, and terns. Near the southern end of the island, enter the Long Key State Recreational Area and drive to the beach; then walk north on the beach—a good place for shorebirds.

Grassy Key. Just beyond Mileage Marker 59, Guava Avenue, and then farther south, Tropical, Peachtree, and Kyle Avenues in succession, leave the Overseas Highway from the right and run parallel across shallow mangrove ponds, collectively called Lake Edna. The ponds are excellent for Wilson's Plovers and transient and wintering shorebirds, all readily viewed from the roads.

Vaca Key. At Mileage Marker 50, opposite Marathon State Bank in Marathon, turn left from the Highway, go on to Marathon High School, leave the car, and walk back of the School to the ocean, where the mud flats are attractive to wading birds and shorebirds. On the first road beyond Marker 50, turn left to the Sombrero Beach Club and take the drive circling the golf course, meanwhile looking for Burrowing Owls, which nest in the rough adjacent to the fairways during late spring and summer, and, at other seasons, perch in the casuarina trees ('Australian pines') surrounding the course.

Little Duck Key. On arriving from the Seven Mile Bridge, stop to search for Mangrove Cuckoos in the mangroves near the highway, and scan the shallow pond and the sand bars for shorebirds.

Missouri Key. The mud flats on the bayside are excellent for sighting shorebirds.

Bahia Honda Key. Toward the northern end, an ocean inlet left of the Highway attracts many wading birds, among them a few Yellow-crowned Night Herons and Reddish Egrets, and is often frequented by Sandwich Terns. Near the southern end, on the left just beyond Mileage Marker 37, is the entrance to Bahia Honda State Park. Drive in here, park the car, and walk along the beach for shorebirds and Black Skimmers.

Big Pine Key. Near Mileage Marker 33, beyond a paved road very soon after reaching this large island, is an unpaved road on the left. Take this road through fine wooded areas for many landbirds, among them the Red-shouldered Hawk, White-crowned Pigeon, Smooth-billed Ani, Red-bellied Woodpecker, Gray Kingbird, Black-whiskered Vireo, and (Cuban) Yellow Warbler. Return to the Highway, continue south. Beyond Marker 31, turn right on State 940 and follow signs to headquarters of Key Deer National Wildlife Refuge. Inquire here about opportunities for seeing birds on Big Pine Key, and observe several key deer in pens. Ground Doves commonly feed in the pens. Go back to State 940 and continue for 2 miles, taking the first paved road to the left for a half a mile, then the first left on an unpaved road. As soon as this road becomes impassable, park the car and proceed on foot into a big woods. Along the way, watch especially for Short-tailed Hawks and Bald Eagles and be alert for wild key deer. This woods, comprised of many kinds of tropical trees, is a haven for many transient and wintering passerines.

Sugarloaf Key. Turn left off the Highway on State 939, which circles a large lake on the right and returns to the Overseas Highway farther south. The mangroves bordering parts of the lake are excellent for Mangrove Cuckoos.

Stock Island. Beyond Mileage Marker 5, just before the Highway leaves Stock Island over Cow Key Channel to Key West, take the last road on the right, passing some county service buildings, to the Key West Botanical Gardens. Here thousands of exotic trees and shrubs, unattended and growing wild, provide the best cover for landbirds in the Lower Florida Keys. Time spent along the trails through the Gardens is certain to be rewarding.

Key West. This island, about 4 miles long and 2 miles wide, makes a gratifying climax to the trip. In any season the waterfront is exceptionally good for seeing Magnificent Frigatebirds. During winter, Painted Buntings are quite numerous around dooryard feeding stations. On Truman Avenue near its intersection with Simonton Street, either in the adjacent trailer parks or at a feeding station back of the Catholic church and convent, Inca Doves are permanent residents. West Beach, on the south shore between White and Bertha Streets, not only yields transient and wintering shorebirds but also becomes a vantage point for Sandwich Terns in winter. The Salt Ponds north of South Roosevelt Boulevard, between Bertha Street and Key West International Airport, are worth inspecting for transient and wintering shorebirds, including such possibilities as American Avocets; for Black-necked Stilts in spring and summer; and for occasional Fulvous Tree Ducks. Near the Airport terminal, entered from South Roosevelt Boulevard, the (Bahaman) Common Nighthawk may be seen or heard performing just before dark in late spring and summer.

In the Gulf of Mexico, 70 miles west of Key West, are the remote **Dry Tortugas,** seven small, low-lying keys—an archipelago replete with history and ornithological attractions.

On Garden Key looms old Fort Jefferson, an incredible citadel of brick masonry erected in 1846 to serve as a Gibraltar for control of the Florida Straits. The largest fortification of its type in the New World, surrounded by moats and equipped with mammoth guns, it figured in four wars but never fired a shot. Neglected for years, stripped by vandals, and ravaged by hurricanes, Fort Jefferson was a ghost of its former self when, in 1937, it was placed under the National Park Service as a National Monument.

On nearby Bush Key, a tiny isle almost under the shadow of the ghost fortress, vast numbers of Sooty Terns nest on an open, sandy area no larger than 7 acres. In addition, smaller numbers of Brown Noddies nest in low shrubs, and a few Roseate Terns, on an adjoining sand spit. Bush Key is the only nesting locality of Sooty Terns and Brown Noddies in North America. The nesting season is from mid-May through August.

Frequenting the Tortugas during the terns' nesting season are Brown Pelicans and Magnificent Frigatebirds, these latter often observed relieving the parent terns of their hard-earned catch of fish. Masked and Brown Boobies are regular visitants, perching on the buoys and channel markers. A few Smooth-billed Anis reside on Loggerhead Key.

Ample anchorage is available at Fort Jefferson for yachts and chartered boats, which may be procured at Key West. Flights may be chartered at either Key West or Marathon. There are no accommodations; all visitors must bring their own food, water, and camping gear. For permission to remain overnight, write Superintendent, Everglades National Park (Box 279, Homestead, Florida 33030).

LAKE WALES
Mountain Lake Sanctuary

Three miles north of this small resort is the **Mountain Lake Sanctuary** (117 acres) on the slope of Iron Mountain (elevation 295 feet), highest point in Florida. Within the Sanctuary's borders are a pond and many trees and shrubs, predominantly native, with the addition of some exotics to give variety and color. From the summit of Iron Mountain rises the 205-foot Singing Tower, the gift of Edward William Bok. The stellar ornithological attraction is a year-round, free-flying population of Wood Ducks, some of which nest in boxes installed especially for them. Other attractions are: Barn Owls nesting in a box high up on the Tower; Common Screech Owls occupying a box on a tree; and Great Crested Flycatchers nesting in a hanging bird house.

Set up at close range to feeders is a blind from which one may observe and photograph Common Bobwhites, Red-bellied Woodpeckers, Tufted Titmice, Carolina Wrens, Northern Mockingbirds, Brown Thrashers, Yellow-throated Warblers, Northern Cardinals, Rufous-sided Towhees, White-throated Sparrows, and other species that are either winterers or permanent residents. Meandering through the Sanctuary is a nature trail with numbered markers keyed to a booklet available at the information booth near the Tower.

MIAMI
Rickenbacker Causeway | Virginia Key | Virginia Beach | Key Biscayne | Matheson Hammock County Park | Fairchild Tropical Garden | Parrot Jungle | Hialeah Park and Race Track | Greynolds Park

Despite its burgeoning size, Miami still has several areas that are ornithologically rewarding.

From US 1 in Miami, turn east at S.E. 26th Road across the **Rickenbacker Causeway** to **Virginia Key, Virginia Beach,** and **Key Biscayne.** Watch for birds along the shores on both sides of the Causeway. Certain to be sighted are Laughing Gulls and Royal

Terns at any time of year, Little Terns in summer, and Herring and Ring-billed Gulls in winter, along with such possibilities as Glaucous, Greater Black-backed, and Bonaparte's Gulls, and Caspian Terns. Look also for shorebirds: Willets at any time of year, Wilson's Plovers in summer, and a variety in fall, winter, and spring, among them Semipalmated, Piping, and Black-bellied Plovers, Ruddy Turnstones, Red Knots, Least Sandpipers, and Sanderlings.

On Virginia Key, opposite the western end of the Seaquarium parking lot, a road leads north to the Miami Sewage Disposal Plant. Outside the gate, drive right and follow the fence as closely as possible to a high dike, park the car, and walk along the dike for about a mile toward the Miami skyline (in view across Biscayne Bay) to extensive mud flats attractive to herons and egrets at any time of year, and wintering shorebirds such as Greater and Lesser Yellowlegs, Dunlins, dowitchers, and Western Sandpipers.

Slightly east of the Seaquarium parking lot, a road heads north between two large government laboratories to Virginia Beach. Drive beyond the stop sign near the center of the parking lot and through a lane near the picnic pavilion. Park the car and walk straight past a no-entry sign on a road to a curve; continue about 10 yards and bear right through a woods to a large fresh-water pond. Fulvous Tree Ducks rest here in winter. Listen for the Smooth-billed Anis. Return to the parking lot and walk northward to a restored beach. Short paths lead from the beach into mangroves, where there are Gray Kingbirds and Black-whiskered Vireos in the summer. On Key Biscayne, stop at the third and northernmost parking lot and walk north on paths through woods that sometimes abound with transient and wintering landbirds. Return on the beach for shorebirds.

From downtown Miami, drive south on US 1, left on Le Jeune Road (S.W. 42nd Avenue) to the traffic circle, and continue south on Old Cutler Road, which bisects **Matheson Hammock County Park** (583 acres). Turn left into the Park and leave the car. The section of the Park west of Old Cutler Road has a county nursery, with dense thickets and many native and exotic trees and shrubs planted in groups separated by wide, open areas. The section east of Old Cutler Road has a small area of original hammock, an open grove of live oaks—now a picnic area—and a small stand of slash pine, a lake, a series of small ponds, and a partly drained marsh extending to mangroves bordering Biscayne Bay. Adjacent to the Park on the south is **Fairchild Tropical Garden** (83 acres), also reached from Old Cutler Road.

Both Park and Garden are recommended especially for transient and wintering landbirds. Such species as the Black-and-white, Cape May, and Black-throated Blue Warblers, American Redstarts, and Bobolinks are common in migration; other species such as Blue-gray Gnatcatchers and Palm Warblers are common throughout winter. Black-whiskered Bulbuls and Spotted Orioles reside here, as they do in many others areas of metropolitan Miami. Among other permanent residents are both Mourning and Ground Doves, Common Screech Owl, Red-bellied Woodpecker, Great Crested Flycatcher, White-eyed Vireo, Prairie Warbler, Eastern Meadowlark, and Rufous-sided Towhee. Black-whiskered Vireos appear in summer.

Parrot Jungle (10 acres), south from downtown Miami on US 1, then left on Red Road (S.W. 57th Avenue), is an attractive spot, with its huge cypresses, live oaks, strangler figs, and other tropical and semitropical vegetation. On exhibit are many exotic birds—macaws, cockatoos, parakeets, peafowl, cranes, flamingos, and pheasants. The macaws, which fly freely, are the outstanding feature of the exhibit. During winter, both wild Indigo and Painted Buntings are common, apparently being attracted by the prevalance of food for the exotics.

Hialeah Park and Race Track, north from downtown Miami on I 95, then left on State 828, has within the race track oval beautifully landscaped ponds where a colony of free-flying, permanent-resident American Flamingos has become completely established and produces young every year. In winter, the same ponds attract small flocks of ducks, including Blue-winged Teal, Northern Shovelers, and Ring-necked Ducks, and the plantings of trees and shrubs around the ponds provide habitat for many small permanent-resident landbirds as well as for transient and wintering species.

Greynolds Park (232 acres) north from downtown Miami on US 1, then left on N.E. 171st Street, has shallow ponds and canals with wooded islands supporting large, readily observed colonies of wading birds: Great Blue, Green, Little Blue, and Louisiana Herons, Cattle, Great, and Snowy Egrets, White Ibises, and hybrid Scarlet × White Ibises. Other birds either nesting or resting here at the

same time are Double-crested Cormorants, American Anhingas, Black-bellied Tree Ducks, Common Gallinules, American Coots, and Fish Crows. Although the nesting season extends from March to September, almost all the above species turn up at other times of year. In addition to its ponds and canals, Greynolds Park has fine wooded areas worth exploring for small landbirds such as the Spotted Oriole and other introduced species.

The bird finder interested in the great array of exotic species now established in metropolitan Miami is advised to contact the Department of Zoology at the University of Miami, Coral Gables, for information on their specific localities. Some species, such as the Canary-winged Parakeet and Hill Mynah, nest on the campus itself.

OKEECHOBEE

Kissimmee Prairie | Lake Okeechobee

The most famous of Florida prairies is **Kissimmee Prairie**, which extends roughly from St. Cloud, not far from Orlando, southward to **Lake Okeechobee.** Probably its most productive section for bird finding is the area immediately northwest of Lake Okeechobee. Here are prairie birds existing near species attracted by open water, shores, marshes, and streams. Two routes are recommended for reaching the habitats of these birds, both leading from the town of Okeechobee on the north side of the Lake.

The Lake Route. Drive south from Okeechobee on US 441, then right on State 78 for about 4 miles to where the road crosses the dike. Walk along the dike for Burrowing Owls. Eight miles from Okeechobee, State 78 crosses the Kissimmee River at the point where it empties into the Lake. This is a particularly good spot for birds. Continue on State 78 to the Indian Prairie Canal, 17 miles from Okeechobee—bird finding is good all the way, and the Canal banks are worth exploring for birds.

Retrace State 78 about half a mile to Harney Pond Canal, then turn left onto an unnumbered road and drive for about 3 miles to the entrance of the Brighton Seminole Indian Reservation in palm-hammock country. Continue to Reservation headquarters, where visitors are welcome. The oaks and palms surrounding headquarters attract a surprising assortment of landbirds. From headquarters, stay on the road going north, leaving the Reservation a few hundred yards beyond. At this point begin watching particularly for Crested Caracaras and Burrowing Owls. The road continues straight north—a distance of 7 miles from Reservation headquarters—and meets State 70. Turn east for Okeechobee.

In any season the Lake Route should yield a long list of species, including American White Pelicans (except summer), Wood Storks, Glossy Ibises, Mottled Ducks, Swallow-tailed Kites (except winter), Crested Caracaras, Greater and Lesser Yellowlegs (except early summer), Black-necked Stilts (except in winter), Gull-billed Terns (in summer), Loggerhead Shrikes, Eastern Meadowlarks (singing in both summer and winter), and Boat-tailed Grackles.

The Prairie Route. Drive west from Okeechobee for 8 miles on State 70 to the bridge across the Kissimmee River. Park the car and walk along the River bank to the left. Here Limpkins are numerous, and there is a good chance of observing American Anhingas, White and Glossy Ibises, and other waterbirds. Except in summer, various kinds of warblers can be observed in the nearby live oaks.

Return to the car, cross the bridge, and continue west on State 70 to a point about 14 miles from Okeechobee, where an unnumbered road leads off to the right. Take this road, 9 miles of which crosses a typical prairie. Burrowing Owls often have their burrows in the road shoulder. Watch for Sandhill Cranes and Grasshopper Sparrows. About 8.5 miles from State 70, the road crosses a railroad track, and, 0.5 mile beyond, joins another road at right angles. Turn right here and continue past the Pearce Ranch House (the sign at the gate reads 'P-4 Ranch'), then make a sharp left turn onto a long wooden bridge across the Kissimmee River. Look for different kinds of waterbirds here.

Beyond the bridge, drive along a causeway to its end, where a house will be seen at the left; then bear right. This is the hamlet of Basinger. Continue east to another causeway and concrete bridges across Chandler Slough, a beautiful cypress swamp. Park the car and walk over the causeway and the bridges, which make excellent points for observing American Anhingas and Limpkins, together with egrets, herons, ibises, and many small landbirds.

When a quarter-mile beyond the last concrete bridge, take the right fork onto a road that passes through prairie country, dotted with palm hammocks. Again, watch for Sandhill Cranes, Burrowing Owls, and Grasshopper Sparrows. Eventually this road meets State 70 on the outskirts of Okeechobee. Turn left for the center of town.

PENSACOLA
Santa Rosa Island

At the west extremity of western Florida are good opportunities for finding transient and wintering waterbirds, ducks, and shorebirds. Any time between November first and April first, take the following trip.

From downtown Pensacola, drive south on US 98 across Pensacola Bay Bridge and, 2 miles beyond, turn off right to Pensacola Beach on **Santa Rosa Island,** crossing the bridge over Santa Rosa Sound. From both sides of the bridge and the road to the Pensacola Beach Casino, watch for Common Loons, Horned Grebes, Brown Pelicans, Double-crested Cormorants, Lesser Scaups, Common Goldeneyes, Buffleheads, Red-breasted Mergansers, Herring Gulls, Ring-billed Gulls, Laughing Gulls, Forster's Terns, Common Terns, and Royal Terns. At the water's edge before and beyond the bridge, look for shorebirds—which may include Black-bellied Plovers, Dunlins, and Sanderlings. Park the car at the Casino and, with binoculars or telescope, scan the offshore waters of the Gulf of Mexico for Northern Gannets, which sometimes appear in December and remain until April.

For another bird-finding area on Santa Rosa Island, *see under* **Fort Walton Beach.**

ST. AUGUSTINE
Anastasia State Recreation Area | St. Augustine Beach | Matanzas Inlet | Washington Oaks State Gardens

A trip south from St. Augustine just inland from the Atlantic coast is rewarding ornithologically in any season.

Leave St. Augustine on State A1A across the Matanzas River on the Bridge of the Lions to Anastasia Island; continue south on State A1A for 2.5 miles and bear left into **Anastasia State Recreation Area** (1,035 acres). Here are sand flats and beach attractive to a variety of wading birds, including Great Blue Herons, Cattle Egrets, Snowy Egrets, and Louisiana Herons. Shorebirds such as Semipalmated Plovers, Piping Plovers, Black-bellied Plovers, Ruddy Turnstones, Least Sandpipers, Dunlins, and Semipalmated Sandpipers are present from fall to spring, and Black-necked Stilts after April.

Return to State A1A and continue south 2 miles to **St. Augustine Beach.** At low tide, walk north on the Beach for about a mile, then inland over the sand dunes to Salt Run, where there are mud flats. In late spring and summer a few American Oystercatchers stop here.

Return to State A1A and continue south, crossing the bridge over **Matanzas Inlet,** which separates the southern end of Anastasia Island from the mainland. Overlooking a great expanse of marshes, lagoons, sand spits, and open sea, the bridge (no parking allowed on it) provides an excellent vantage point for Brown Pelicans any time of year and, in fall, winter, and early spring, for a variety of shorebirds, also Greater Black-backed Gulls, Herring Gulls, Ring-billed Gulls, Laughing Gulls, Bonaparte's Gulls, Forster's Terns, Common Terns, Royal Terns, Caspian Terns, and Black Skimmers. Occasionally on stormy days, Northern Gannets may sail by, sometimes diving far offshore. Immediately south of the bridge, State A1A passes an area of scrub oaks where Painted Buntings reside from April through July.

After passing Marineland, about 2 miles south of the bridge, State A1A traverses scrubby country. Watch for Scrub Jays on low shrubs.

Three miles south of Marineland, State A1A passes through **Washington Oaks State Gardens** (340 acres), well worth a visit. First turn left to the picnic area near the beach and look for Scrub Jays in the low shrubs along the way. Then backtrack, cross the highway, and enter the garden area, which features both native and exotic trees and shrubs. Stop now and then along the drives and search for birds: in any season, Carolina Chickadees, Carolina Wrens, White-eyed Vireos, and Rufous-sided Towhees; in early spring, throngs of transient warblers and other passerines; in breeding season (under way by April), Yellow-billed Cuckoos, Gray Kingbirds, Great Crested Flycatchers, Yellow-throated Vireos, and Summer Tanagers.

ST. PETERSBURG
Lake Maggiore Park | Fort De Soto Park | Sunshine Skyway

This city, on the west-central coast of Florida and on a peninsula jutting southward between Tampa Bay and the Gulf of Mexico, has two parks that are highly recommended for bird finding.

Lake Maggiore Park. Drive east from US 19 on 22nd Avenue South, south on 9th Street South, which soon passes along the eastern boundary of the Park, then westward on Country Club Way around the southern boundary. Shortly after entering Country Club Way, stop on the right and enter the Park on foot via the nature trail. This winds through a wooded area, good habitat for transient and wintering passerine birds, to Lake Maggiore, where many ducks may be viewed during winter months, and many waterbirds—American Anhingas, Double-crested Cormorants, herons, and egrets—in any season. A small island near the northern shore of Lake Maggiore is a favorite roost for waterbirds.

Fort De Soto Park. Embracing long, slender Mullet Key south of the city between Tampa Bay and the Gulf, the Park may be reached by driving west from US 19 on 54th Avenue South across the Pinellas Bayway, then south on State 693 to the entrance. Continue south into the Park on a paved road to its intersection with a road that runs left and right for the entire length of the Key and thus provides ready access to the beaches on both Bay and Gulf sides as well as to interior fields, pine woods, and brushlands. On these beaches foregather many hundreds of transient and wintering shorebirds, frequently including a few Long-billed Curlews and Whimbrels. Ospreys reside the year round, and, in winter, Forster's and Caspian Terns patrol the shores. American White Pelicans show up in small numbers offshore on the Bay side; and Ruby-crowned Kinglets, Palm Warblers, American Goldfinches, and Savannah Sparrows are among the many passerine birds occupying the interior habitats.

The causeways of the **Sunshine Skyway,** over which US 19 crosses the entrance to Tampa Bay south of St. Petersburg, provide places to park the car and view shorebirds on the bordering beaches. Look especially for American Oystercatchers (possibly summer residents), and Marbled Godwits during winter. Many wading birds and occasionally a few Roseate Spoonbills may be expected at any time of year.

SANIBEL ISLAND

J. N. 'Ding' Darling National Wildlife Refuge | Shelling Beaches

This semitropical island in the Gulf of Mexico, 3 miles off the southwest coast of Florida, is as rewarding for bird finders as it is for vacationers in search of an uncrowded place for quiet and relaxation. Despite its ready access over a causeway from Punta Rassa via State 867 from Fort Myers—and the consequent invasion of 'developers'—Sanibel retains much of its charm and wildlife values, thanks to an enlightened local citizenry and conservation groups working with the United States Fish and Wildlife Service. A major result of this collaboration was the establishment of the **J. N. 'Ding' Darling National Wildlife Refuge** (2,500 acres), named in memory of the distinguished cartoonist-conservationist, a Sanibel Island resident and a founder of the National Wildlife Refuge System.

A 12-mile-long barrier island of sand and shell, Sanibel has low wooded ridges along its middle course, flanked by broad beaches on the Gulf side and broken into a labyrinth of mangrove-bordered sloughs and salt-water lagoons on the north side toward the mainland. The Refuge proper embraces most of the sloughs and sheltered lagoons of Tarpon Bay; disjoined parts include the Bailey Tract south of the Refuge proper and Point Ybel, site of Refuge headquarters, at Sanibel's eastern extremity.

After stopping in at headquarters for directional information and a list of birds, go to the Refuge proper and take the loop drive on dikes with salt flats at low tide on one side, and canals and impounded pools on the other. Stay in the car so as to watch at close range a variety of wading birds and ducks that would otherwise be disturbed. In any season, using a camera with telephoto lens, one can often take frame-filling pictures from the car of Double-crested Cormorants, American Anhingas, Great Blue Herons, Little Blue

Herons, Cattle Egrets, Great Egrets, Snowy Egrets, Louisiana Herons, White Ibises, and, from September through February, ducks such as Green-winged and Blue-winged Teal, American Wigeons, Northern Shovelers, Lesser Scaups, and Red-breasted Mergansers. During this same September/February interval, Black Skimmers frequent the pools, and such shorebirds as Greater and Lesser Yellowlegs and Short-billed Dowitchers tarry and feed on the wet flats. The sight of a few Roseate Spoonbills is always a possibility.

The Refuge is a breeding area for Mottled Ducks, Common Gallinules, Willets, and Black-necked Stilts. Red-shouldered Hawks are permanent residents. Usually indifferent to the presence of cars, they are easy to spot. Several pairs of Ospreys nest, their aeries sometimes in view from the loop drive. Ground Doves are common on the dikes and often fly up ahead of the car. White-eyed Vireos and Prairie Warblers are among the passerine birds heard or seen in the mangroves flanking the canals and pools.

In Tarpon Bay and its sheltered lagoons, rafts of Lesser Scaups pass the winter, as do a few American White Pelicans. Inquire at headquarters about the best vantage points for observing them.

The Shelling Beaches—beaches on the Gulf side of Sanibel—are famed for some four hundred varieties of shells continually washed up by the tides; thus beachcombing is a favorite pastime for many vacationers. For bird finders the beaches are excellent for transient and wintering shorebirds: vast numbers of Black-bellied Plovers, Ruddy Turnstones, and Sanderlings, somewhat smaller numbers of Semipalmated and Piping Plovers, Red Knots, and Least and Semipalmated Sandpipers. Gulls and terns are also numerous. Wilson's Plovers, Sandwich Terns, and Little Terns are summer residents. Laughing Gulls and Royal Terns are present the year round, Ring-billed Gulls and Forster's Terns, in winter only.

SARASOTA
Myakka River State Park

From Sarasota, drive south 3 miles on US 41, then west on State 72 for 14 miles to **Myakka River State Park,** Florida's largest, with

28,875 acres. From the entrance and camping area a scenic road twists through a densely forested area bordering the Myakka River and shortly follows the south side of Upper Myakka Lake into which the River flows. The Park is probably the best place in Florida for seeing Wild Turkeys. Look for them in early morning and late afternoon, when they often cross the road and sometimes show up in the picnic areas.

While these birds are the Park's principal ornithological feature, there are many wintering ducks and permanent-resident wading birds, easily viewed from a boardwalk into Upper Myakka Lake. The Park operates a trackless train for 7 miles into the marshes to a 40-foot tower overlooking a spectacular nesting colony of White Ibises. Red-shouldered Hawks and Pileated Woodpeckers are residents in the dense woods; 'freeloading' Northern Mockingbirds and Northern Cardinals are constantly alert for handouts around the campground and picnic tables. Near the entrance, Barred Owls are noisy at night. Although Sandhill Cranes are present in remoter parts of the Park, they are more readily seen on open ranch lands south of State 72 as it approaches the Park from US 41.

SEBRING
Highlands Hammock State Park

Sebring, on the west edge of the Kissimmee Prairie in south-central Florida, has large lakes and parks, lawns and shade trees—all suitable places for a variety of birds.

Highlands Hammock State Park (3,800 acres), 3 miles west of town just off US 27 on State 634, has a particularly wide variety of birds because of a correspondingly wide variety of habitats: a typical hammock heavily grown to live oaks, gums, bays, and magnolias; pine woods and pine scrub; lagoons and a large cypress swamp through which a footbridge circles. Almost every section of the Park is accessible by fine trails without marring much of its primeval state.

Birds present all year include the American Anhinga, herons, egrets, ibises, Pileated and Red-cockaded Woodpeckers, Scrub Jay, and Brown-headed Nuthatch. Yellow-throated Warblers are conspicuous in the camping area, where they habitually inspect picnic tables for bits of food. In spring and early summer, listen for Chuck-will's-widows during early evening; look for Prothonotary Warblers in the cypress swamp; and do not be surprised to see several Swallow-tailed Kites at one time soaring high over forested parts of the Park. Jackson Lake, passed by the entrance road, attracts many waterbirds.

TALLAHASSEE
Wakulla River | St. Marks National Wildlife Refuge

South of Florida's capital city are two opportunities no bird finder should miss. One is to go to Wakulla Springs and take the 'jungle cruise' on the **Wakulla River,** which flows through some of the most beautiful stretches of wilderness in northern Florida. The other is to visit the St. Marks National Wildlife Refuge.

The Wakulla River rises at the Springs—actually one huge spring, deep and crystal clear—and takes a leisurely course toward the Gulf of Mexico; it is flanked by dense aquatic shrubs and various moss-draped hardwoods, including giant cypresses. Leading away from the River on both sides, and sometimes arched over by lush vegetation, are many channels connecting with nearby sloughs often obscured from view. Birdlife, exceptional in diversity and abundance in any season, can be observed with remarkable ease and convenience.

From Tallahassee, drive south on State 363 through the town of Wakulla and across the Wakulla River. About 4 miles beyond Wakulla and 21 miles from Tallahassee, turn right on a well-marked road and continue for about 2.5 miles to Wakulla Springs, where there is a hotel on landscaped grounds. First, take time to inspect the great spring from a glass-bottomed boat. The water is so clear that objects are recognizable on the bottom, 100 feet down. Then go on the jungle cruise, a guided trip, for a mile down the Wakulla River. If one wishes to explore the River beyond that point, he can hire a boat at the Upper Bridge, crossed by State 363, and row up the River. The section of river between Upper Bridge and Wakulla Springs is about 3 miles long.

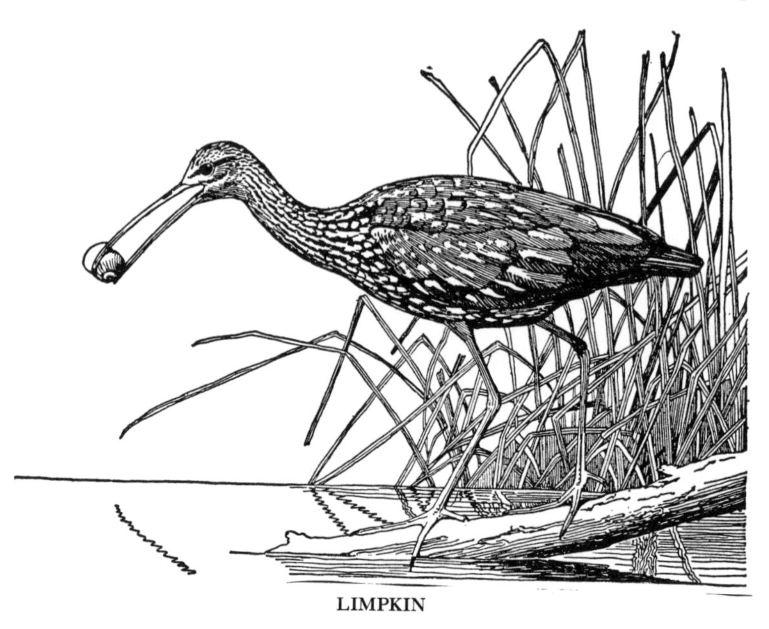

LIMPKIN

If possible, take the cruise in spring, preferably in April, when the breeding season is well under way. As soon as the trip starts, one will hear the wailing, endlessly repetitious cries of Limpkins from the dense borders of the stream. Eventually the large, rail-like birds will come into view. Some are surprisingly tame, thus permitting very close observation.

As the trip continues, one will hear Prothonotary and Northern Parula Warblers, and catch sight of broods of Wood Ducks hurrying along ahead of the boat, as well as Purple Gallinules searching for food among the aquatic plants edging the shore. Very likely there will also be views of Swallow-tailed and Mississippi Kites soaring overhead, alarmed American Anhingas diving into the water and later reappearing with only their heads and snake-like necks showing above the surface, and Ospreys circling about the aeries high in the cypresses.

In any season, cormorants, egrets, and herons are present, as are Limpkins, Wood Ducks, American Anhingas, and Ospreys. In winter, Pied-billed Grebes, Common Gallinules, and American Coots, together with several kinds of ducks, appear in impressive numbers.

St. Marks National Wildlife Refuge, comprising 65,136 acres of land along Apalachee Bay in the Gulf of Mexico, is not far from Wakulla Springs, yet offers quite different bird-finding possibilities. From the open water of the Bay to the interior of the Refuge, the succession of natural areas, together with characteristic birdlife, is as follows:

1. *Coastal marsh,* a mile or more in width, consisting principally of sharp-pointed rushes. Nesting birds include Clapper Rails, Willets, Marsh Wrens, and Seaside Sparrows. The marsh is frequently interrupted by tidal streams and by 'islands.' The latter, timbered with southern red cedar and slash pine, are breeding localities for Ground Doves and Gray Kingbirds.

2. *Sand flats,* frequented by shorebirds in all seasons and by Canada Geese in winter.

3. *Flatwoods,* the largest natural area of the Refuge, supporting slash pine and saw palmetto, with sand ridges, supporting scrub oaks and longleaf pine. Nesting birds include Wild Turkeys, Common Nighthawks, Red-bellied Woodpeckers, Red-cockaded Woodpeckers, Brown-headed Nuthatches, Eastern Bluebirds, Yellow-throated Warblers, Pine Warblers, and Bachman's Sparrows. Here and there among these flatwoods and sand ridges are hardwood swamps, attractive to Wood Ducks, Prothonotary Warblers, and Northern Parula Warblers. And there are also many ponds and artificial pools fringed by cattails, bulrushes, and other aquatic plants. These serve as nesting habitats for King Rails, Purple Gallinules, Common Gallinules, Least Bitterns, Marsh Wrens, and Red-winged Blackbirds in late spring and summer, and as feeding and resting grounds for waterfowl in the winter.

Probably the chief ornithological attraction of the Refuge is the wintering concentration of waterfowl between mid-November and mid-January. Usually the most abundant are Canada Geese, Mallards, Common Pintails, American Wigeons, Redheads, Ring-necked Ducks, Lesser Scaups, Buffleheads, and Red-breasted Mergansers. Other species well represented are American Black

Ducks, Gadwalls, Green-winged and Blue-winged Teal, Northern Shovelers, and Ruddy Ducks. Additional ornithological features are several pairs of Bald Eagles, which begin nesting in early winter, and a few small-sized nesting colonies of Double-crested Cormorants, Great Blue Herons, Great and Snowy Egrets, and Louisiana Herons, which become established in April, and sometimes earlier.

Refuge headquarters is 4 miles south of Wakulla on US 98. Inquire here about foot trails, observation points, and nature drives on the Refuge dikes.

TAMPA

Green Key | Whiskey Stump Key | Alafia River Spoil Banks | Courtney Campbell Causeway

Among the finest waterbird colonies along the Florida coast are those occupying small islands and banks on the east side of Hillsboro Bay, a northeasterly extension of Tampa Bay. The following sites—all sanctuaries of the National Audubon Society—are worth visiting.

Green Key (6 acres), a half-mile offshore and 10 miles south of Tampa, receives its name from being completely covered with mangroves. The majority of its many hundred nesting waterbirds are Little Blue Herons, Great Egrets, Snowy Egrets, and Louisiana Herons. Others are Brown Pelicans, Double-crested Cormorants, American Anhingas, Great Blue Herons, Green Herons, Cattle Egrets, Black-crowned Night Herons, Yellow-crowned Night Herons, and White Ibises. Almost overshadowed by these large birds are a few pairs of Gray Kingbirds, Black-whiskered Vireos, Prairie Warblers, and Boat-tailed Grackles.

Whiskey Stump Key (3 acres), a quarter-mile from Green Key, is comparatively high, with a cover of cabbage palms and mangroves. Species known to be occupants are the Green Heron, Yellow-crowned Night Heron, Barn Owl, Chuck-will's-widow, Gray Kingbird, Fish Crow, Black-whiskered Vireo, and Prairie Warbler.

The **Alafia River Spoil Banks** (50 acres), at the mouth of the Alafia River and closer to shore than Green Key, were formed by dredging. Much of the surface is covered by mangroves in which many hundreds of herons and egrets nest, together with smaller numbers of White Ibises. The surface elsewhere, either covered with grasses or largely devoid of vegetation, attracts a few nesting Willets, Ground Doves, Common Nighthawks, occasionally Clapper Rails, and, in some years, small colonies of Laughing Gulls, Little Terns, and Black Skimmers.

Nesting activities in all colonies are usually under way by late February, but bird-finding visits are more rewarding from late May through July, when the young are well developed and readily observed.

When planning to visit the above preserves, first obtain permission from the Sanctuary Department of the National Audubon Society (950 Third Avenue, New York, New York 10022), and then arrange for the trip with the local Audubon warden who patrols the sanctuaries during nesting season.

A large variety of waterbirds, ducks, and shorebirds can be observed at any time of year from the **Courtney Campbell Causeway,** traversed by State 60 between Tampa and Clearwater. Along the Causeway, which spans the northern part of Old Tampa Bay for 9.5 miles, are beaches, mud flats, and shallows attractive in any season to all species of wading birds that nest on the above-mentioned islands in Hillsboro Bay. In addition, during fall, winter, and early spring one may expect to see such species as the Common Loon, Horned Grebe, Common Pintail, Lesser Scaup, Red-breasted Merganser, Herring Gull, Ring-billed Gull, Forster's Tern, Black-bellied Plover, Ruddy Turnstone, Greater and Lesser Yellowlegs, Red Knot, Dunlin, Short-billed Dowitcher, and Sanderling.

TITUSVILLE

Merritt Island National Wildlife Refuge | St. Johns National Wildlife Refuge | Ulumay Wildlife Refuge | Port Canaveral Turn Basins | Jetty Park

Off the east-central coast a complex water-bound area comprising the Canaveral Peninsula and Merritt Island offers some of the best bird finding in the state. The Peninsula, across Indian River (Intracoastal Waterway) from Titusville, parallels the coast north and south for many miles. Although ornithologists often refer to the whole Peninsula as a part of Merritt Island, the Island itself lies inland from the Peninsula, separated by Banana Creek on the north, Banana River on the east, and Indian River on the west. Larger than the Peninsula, Merritt Island is about 7 miles wide and 30 miles long.

Occupying a part of both the Peninsula and the Island is 140,393-acre **Merritt Island National Wildlife Refuge.** It consists largely of marshes, impoundments of either fresh or brackish water, and outlying salt-water lagoons and mangrove swamps. Elsewhere on dry ground, barely above water level, are citrus groves, scattered stands of slash pines, and hammocks supporting oaks, saw palmettos, and cabbage palms. The John F. Kennedy Space Center occupies the southeastern parts of the Refuge. Because of the Space Center's security requirements, Refuge areas adjacent to the Center are often closed to public access.

To see the widest variety of birds in the Canaveral Peninsula–Merritt Island complex, two separate trips from the mainland are recommended.

1. From US 1 at Titusville, drive east on State 406 across Indian River by bridge and causeway to a fork where State 402 branches right to the Refuge visitor center. Bear left at this fork on State 406 and, immediately after crossing railroad tracks, turn abruptly left onto an unimproved road following the tracks and overlooking open water on both sides. Take the first right onto an impoundment road and each right turn thereafter on obviously traveled roads back to State 406; then turn north to Dummit Cove Campground and walk a short distance east on various unimproved roads and trails to Mosquito Lagoon. Returning to the Campground, drive back on State 3, passing the intersection with State 406, to State 402. Turn west here to the visitor center and Titusville. This entire trip, east of the bridge, is within the Refuge.

During late fall and winter, hordes of waterfowl are in view from the causeway as well as on all the water impoundments and Mosquito Lagoon. Lesser Scaups are the most abundant; also impressive in numbers are Gadwalls, Common Pintails, Green-winged and Blue-winged Teal, American Wigeons, Northern Shovelers, Redheads, Ring-necked Ducks, Hooded Mergansers, and Red-breasted

Mergansers. Mottled Ducks are present but never abundant. Look for a few Fulvous Tree Ducks on Mosquito Lagoon.

Other birds present during the same season in suitable habitats include Common Loons, Horned Grebes, American White Pelicans, Northern Harriers, and Caspian Terns. In spring and summer, Black-necked Stilts are breeding residents, and a few Roseate Spoonbills may be sighted, even though they do not nest. In any season there are Pied-billed Grebes, Brown Pelicans, Double-crested Cormorants, American Anhingas, an array of herons and egrets, Least Bitterns, Wood Storks, both Glossy and White Ibises, Mottled Ducks, King and Clapper Rails, Common Gallinules, American Coots (very abundant in winter), Willets, Royal Terns, and Black Skimmers. There is always a likelihood of seeing Red-tailed Hawks, Bald Eagles, and Ospreys, all of which breed in the area.

Although a few Black Rails and (Dusky) Seaside Sparrows probably reside the year round in marshy stretches of glassworts, grasses, and rushes traversed by State 406 and 402, a more productive area lies on the mainland in the **St. Johns National Wildlife Refuge** (4,000 acres) southwest of Titusville. Thus, on returning to Titusville, take US 1 south and turn west on State 50. Exactly 2.8 miles west of State 50's intersection with I 95, turn north on Hacienda Drive, park, and continue on foot. For the next 3 miles inspect the switch-grass marsh west of the road for rails and sparrows. Both species are easier to find in spring, when they are more vociferous.

2. For a second trip drive south from Titusville on either I 95 or US 1 to Cocoa and turn east on State 520 across Indian River by bridge to Merritt Island. At the Merritt Island Shopping Center, turn northeast on Sykes Creek Parkway (Palmetto Avenue and Audubon Road on some maps). Shortly after crossing Sykes Creek, stop at the **Ulumay Wildlife Refuge** (461 acres), owned by Brevard County and embracing a salt marsh with bordering mangroves and brushland.

Along a nature trail on a brushy dike one may observe, in any season, such species as the Green Heron, Glossy and White Ibises, Mottled Duck, King and Clapper Rails, White-eyed Vireo, Prairie Warbler, and Red-winged Blackbird; during winter, the Northern Harrier, Sora, Common Snipe, Myrtle Warbler, Palm Warbler,

Common Yellowthroat, Swamp Sparrow, and Song Sparrow, together with great numbers of wading birds and more than a dozen species of ducks.

Return to State 520 and continue eastward across the Banana River by bridge and causeway. From late fall through winter the River is generally stippled with Common Loons, Horned Grebes, and from twelve to fifteen species of ducks. A good place for observing them is the parking area behind Cape Canaveral Hospital. At Cocoa Beach on the Peninsula, turn north on State A1A. On approaching Port Canaveral, leave the highway eastward on roads skirting the **Port Canaveral Turn Basins.**

At low tide the shallow pools and mud flats around these basins attract many transient and wintering shorebirds: Semipalmated Plover, Piping Plover, Black-bellied Plover, Ruddy Turnstone, Whimbrel, Spotted Sandpiper, Greater Yellowlegs, Lesser Yellowlegs, Red Knot, Least Sandpiper, Dunlin, Short-billed Dowitcher, Semipalmated Sandpiper, Western Sandpiper, Marbled Godwit, Sanderling, and American Avocet. During winter a Short-eared Owl or a Water Pipit occasionally appears along the grassy borders of the basins.

Continue eastward along the south side of Port Canaveral to **Jetty Park** overlooking the Atlantic and the entrance to the Port. From here, during late afternoons in winter, one may see scores of gulls and terns, of several species, following the shrimp boats on their return from the sea; and on stormy winter days one can sometimes spot a Northern Gannet, a Pomarine Jaeger, or a Parasitic Jaeger in flight not far offshore. Scrub Jays are numerous in brushy areas of the Park.

VERO BEACH

Pelican Island | Pelican Island National Wildlife Refuge

In Indian River along the east-central coast of peninsular Florida, lies tiny **Pelican Island,** about 10 miles north of Vero Beach near the town of Sebastian. Its small size—about 3 acres—belies its fame and importance.

During the last century—and no one knows how long before—

Pelican Island was the only nesting site of Brown Pelicans on the Atlantic coast of Florida. As the century waned, the Island's uniqueness became increasingly well known and a prime attraction for many people. Some of these were naturalists relishing the chance to observe pelicans at their nests; others were gunners mindlessly shooting the birds from boats. The naturalists, though angered by the senseless killing, made no concerted effort to stop it. Then at the turn of the century, the flourishing millinery trade, demanding among other plumes the big wing-feathers of pelicans, threatened the annihilation of the entire colony. By then thoroughly alarmed and organized, the naturalists urged President Theodore Roosevelt to take strong action at once. This he did by executive order in 1903, declaring Pelican Island a federal reserve—the first of what is now well over 330 units in the present National Wildlife Refuge System.

In later years other islands in Indian River adjacent to Pelican Island were acquired by the Federal Government and incorporated in what is now the **Pelican Island National Wildlife Refuge** (4,359 acres). Pelican Island proper remains an important nesting site of Brown Pelicans and, in addition, of Double-crested Cormorants, American Anhingas, Cattle and Great Egrets, Louisiana Herons, Wood Storks, and White Ibises. The nesting season of all these species begins in late February, reaching its peak in April and May and ending by October.

Inquiries about boat transportation to Pelican Island and permission to land during nesting season should be directed to Refuge Manager, Merritt Island National Wildlife Refuge (P.O. Box 6504, Titusville, Florida 32780). Visitors are usually requested to stay in their boats just offshore, where they can readily view the birds without disturbing them.

Georgia

WILD TURKEYS

Georgia, the largest state east of the Mississippi, offers many birds of wide variety: Wild Turkeys in the deep forests and swamps; Sandhill Cranes on the peculiar 'prairies' of the great Okefenokee Swamp; Winter Wrens, Veeries, and Rose-breasted Grosbeaks on the wooded slopes of Brasstown Bald Mountain; American Oyster-catchers on undisturbed beaches; Clapper Rails and Seaside Sparrows in salt marshes; Painted Buntings in the shrubby pastures and woodland edges of the coastal lowlands; and throughout the state, in suitable habitats, such birds as King Rails, Yellow-billed Cuckoos, Pileated Woodpeckers, Acadian Flycatchers, Carolina Wrens, Yellow-throated Vireos, Summer Tanagers, and Blue Grosbeaks.

The entire southern half of Georgia is on the Coastal Plain—low sandy country, almost flat near the coast but increasingly hilly northward. In the longleaf pines—the dominant forest growth—Red-cockaded Woodpeckers are fairly common permanent residents. In the southeastern Coastal Plain near the Florida line lies famous Okefenokee Swamp, where, in the parts that are always wet, trees such as cypress and tupelo attain gigantic size. Though the primeval charm of the Okefenokee has been somewhat marred by commercial enterprises, it is still one of the great reservoirs of wildlife in southeastern United States. American Anhingas, Great Egrets, Little Blue Herons, White Ibises, and Wood Ducks are among the waterbirds and waterfowl enjoying the solitude of its waterways.

The Coastal Plain's eastern boundary is a series of irregularly shaped, sandy islands known as 'sea islands,' which serve as a barrier against the Atlantic. Wide, gently sloping beaches meet the ocean itself. Behind them, above the high-tide mark, appear parallel ridges of sand dunes and then either forested stretches with magnolias, pines, and moss-bannered live oaks, or open areas grown to scrub palmetto and myrtle. Separating the sea islands from the mainland is a labyrinth of tidal bays, channels, and lagoons flanked by marshes and mud flats. Tropical in aspect, favored by a mild climate, and abounding in birdlife, the sea islands provide a delightful experience for the northern bird finder. Transient and wintering waterbirds and waterfowl congregate on the outlying waters; heron and egret colonies swell the population of birds breeding in the woods and thickets. On undisturbed beaches or in the nearby salt or brackish marshes, the following birds nest regularly:

Clapper Rail
American Oystercatcher (*not commonly*)
Wilson's Plover
Willet
Little Tern (*not commonly*)
Marsh Wren
Boat-tailed Grackle
Seaside Sparrow

The northern half of Georgia is marked off from the Coastal Plain by the Fall Line, which extends southwestward across the state from Augusta through Milledgeville and Macon to Columbus. The southern two-thirds of this northern portion occupies the rolling Piedmont Plateau and has elevations ranging from 400 feet in the south to 1,200 feet in the north. The forests consist for the most part of a coniferous and hardwood growth, with loblolly pines, shortleaf pines, red oaks, white oaks, and hickories among the predominating stands.

The northern third, the Mountain Region, embraces a section of the Blue Ridge Mountains and various other ridges. Near the North Carolina line is the state's highest peak, Brasstown Bald Mountain (4,784 feet). The forests of the Mountain Region are chiefly deciduous, containing chestnut oaks, red oaks, and maples, with an understory of mountain laurel and rhododendron. Northern conifers are represented by white pines and hemlocks, usually in cool ravines and on shaded slopes. The highest summits are either 'bald'—covered with grasses and shrubs—or capped by deciduous trees.

The transition from Coastal Plain to Mountain Region is gradual. This is true of elevation and vegetation—and birdlife. Species such as the Purple Gallinule, Ground Dove, Fish Crow, Prothonotary Warbler, and Swainson's Warbler, sometimes quite common in certain localities on the Coastal Plain, are rare on the Piedmont Plateau. On the other hand, the Eastern Phoebe, American Robin, Black-and-white Warbler, Yellow Warbler, and American Goldfinch are seldom, if ever, found nesting on the Coastal Plain; yet through the Piedmont northward to the Mountain Region their numbers steadily increase.

Birds breeding regularly in the Mountain Region are listed below. Those marked with an asterisk are confined to the higher mountains.

Ruffed Grouse
Whip-poor-will
*Northern Raven
*Winter Wren
Bewick's Wren
Solitary Vireo
Golden-winged Warbler
Black-throated Blue Warbler
Black-throated Green Warbler
Blackburnian Warbler
Chestnut-sided Warbler
Ovenbird (*also locally on Piedmont Plateau*)
*Canada Warbler
American Redstart (*also western part of state*)
Scarlet Tanager
*Slate-colored Junco
*Rose-breasted Grosbeak
Song Sparrow

Throughout Georgia, in the forests and on the farmlands (fields, wet meadows, brushy lands, orchards, and dooryards), the following species breed regularly:

FORESTS

Turkey Vulture
Black Vulture (*except in mountains*)
Red-shouldered Hawk
Broad-winged Hawk (*except on Coastal Plain*)
Yellow-billed Cuckoo
Common Screech Owl
Barred Owl
Chuck-will's-widow (*except in mountains*)
Yellow-shafted Flicker
Pileated Woodpecker
Red-bellied Woodpecker
Hairy Woodpecker
Downy Woodpecker
Great Crested Flycatcher
Acadian Flycatcher
Eastern Pewee
Blue Jay
Carolina Chickadee
Tufted Titmouse
White-breasted Nuthatch
Brown-headed Nuthatch (*except in mountains*)
Wood Thrush
Blue-gray Gnatcatcher
Yellow-throated Vireo
Red-eyed Vireo
Northern Parula Warbler (*except on Piedmont Plateau*)
Yellow-throated Warbler (*except on Piedmont Plateau*)
Pine Warbler
Louisiana Waterthrush
Kentucky Warbler
Summer Tanager

FARMLANDS

Common Bobwhite
Mourning Dove
Red-headed Woodpecker
Eastern Kingbird
Carolina Wren
Northern Mockingbird
Gray Catbird (*except in extreme south*)
Brown Thrasher
Eastern Bluebird
Loggerhead Shrike (*except in mountains*)
White-eyed Vireo
Prairie Warbler (*except southern part*)
Common Yellowthroat
Yellow-breasted Chat
Eastern Meadowlark
Orchard Oriole
Common Grackle
Northern Cardinal
Blue Grosbeak
Indigo Bunting (*except near coast*)
Rufous-sided Towhee
Grasshopper Sparrow (*northern part only*)
Bachman's Sparrow (*except in mountains*)
Chipping Sparrow
Field Sparrow

Valleys of the larger rivers flowing southeastward through the state serve as migration routes for many birds, but none can compare with the coast, which is on the Atlantic flyway. In such places as Tybee Island (reached from Savannah) and Jekyll, St. Simons, and Sea Island (reached from Brunswick) one may see impressive transient aggregations. The main flights may be expected within the following dates:

Waterfowl: 15 February–1 April; 25 October–15 December
Shorebirds: 20 April–20 May; 10 August–15 October
Landbirds: 25 March–5 May; 10 September–5 November

Although bird finding in inland Georgia during winter rarely yields heavy concentrations, one may see many different kinds. At Augusta and Atlanta, for example, one may easily observe eighty or more species on a January or February day. On the coast one may find a much larger number, owing to the presence of such birds as loons, cormorants, herons, ducks, shorebirds, gulls, and terns. Hence, the coastal areas from Savannah to Brunswick are considered the best places in Georgia for winter bird finding.

Authorities

Thomas D. Burleigh, Margaret Davis Cate, J. Fred Denton, Raymond J. Fleetwood, William W. Griffin, Frederick V. Hebard, Edgar S. Jaycocks, Michael B. Lilly, Sara Menaboni, Dorothy P. Neal, Robert A. Norris, Eugene P. Odum, Mabel T. Rogers, Herbert L. Stoddard, Ivan R. Tomkins, Barbara Woodward.

References

Georgia Birds. By Thomas D. Burleigh. Norman: University of Oklahoma Press, 1958.

A Birder's Guide to Georgia. Edited by Daniel W. Hans. Georgia Ornithological Society (P.O. Box 362, Atlanta, Ga. 30301), 1975.

ATLANTA

Piedmont Park | Chattahoochee River Bottomlands | Stone Mountain Park

Situated on the Piedmont Plateau, a region of rolling hills, Atlanta and its environs offer opportunities for observing landbirds at all seasons, but the absence of large lakes and marshes limits the number of waterbirds and waterfowl.

Fine for birds the year round is **Piedmont Park,** a 196-acre tract 2.5 miles northeast of center city, on the east side of Piedmont Avenue at 14th Street. Among the oaks, hickories, magnolias, and willows, in the shrubbery, and on the lawns, are permanent residents such as the Red-bellied Woodpecker, Carolina Chickadee, Tufted Titmouse, White-breasted Nuthatch, Carolina Wren, Northern Mockingbird, Brown Thrasher, Eastern Meadowlark, Northern Cardinal, and Rufous-sided Towhee. A small lake attracts an occasional Pied-billed Grebe or Green Heron and, except in late spring and summer, sometimes a Blue-winged Teal, Lesser Scaup, or American Coot. Summer residents in Piedmont Park include the Great Crested Flycatcher, Eastern Pewee, Wood Thrush, Blue-gray Gnatcatcher, White-eyed Vireo, Red-eyed Vireo, Common Yellowthroat, Yellow-breasted Chat, Hooded Warbler, Summer Tanager, and Indigo Bunting.

For bird-finding opportunities outside Atlanta, drive west from I 285, which circles the city, on US 78 (Bankhead Highway). At the Bankhead Bridge over the Chattahoochee River, leave the car in the parking area just north of the Highway on the east side of the River; follow fishermen's trails upstream along the River bank and the brushy edges of sloughs and through the deciduous woods—cottonwoods, birches, oaks, and hickories, with a scattering of beeches and maples—of the **Chattahoochee River Bottomlands.** During migration—in April/May and in September/October—this is an excellent place for such transients as the Black-billed Cuckoo, Hermit Thrush, Swainson's Thrush, Gray-cheeked Thrush, Veery, Solitary Vireo, Black-and-white Warbler, Worm-eating Warbler, Golden-winged Warbler, Tennessee Warbler (September and October only), Northern Parula Warbler, Magnolia Warbler (September and October only), Cape May Warbler (April and May only), Black-throated Blue Warbler, Myrtle Warbler, Black-throated Green Warbler, Blackburnian Warbler, Chestnut-sided Warbler, Bay-breasted Warbler, Blackpoll Warbler (April and May only), Prairie Warbler, Palm Warbler, Ovenbird, Northern Waterthrush, Canada Warbler, and Scarlet Tanager.

Among regular breeding species in May and June are the Yellow-billed Cuckoo, Acadian Flycatcher, Gray Catbird, Yellow-throated Vireo, Prothonotary Warbler, Blue-winged Warbler, Yellow-throated Warbler, Louisiana Waterthrush, Kentucky Warbler, American Redstart, and Blue Grosbeak. Some of the winter residents are the Eastern Phoebe, Brown Creeper, Winter Wren, Golden-crowned Kinglet, Ruby-crowned Kinglet, Orange-crowned Warbler, Rusty Blackbird, American Goldfinch, Slate-colored Junco, White-throated Sparrow, Fox Sparrow, Swamp Sparrow, and Song Sparrow. Occasionally in winter (November through February), Great Blue Herons, American Black Ducks, and Wood Ducks show up along the River or in the sloughs.

From I 285, drive east on US 78, Stone Mountain Parkway, to **Stone Mountain Park** (3,200 acres). From the East Gate proceed along Jefferson Davis Drive to Robert E. Lee Boulevard, which circles Stone Mountain. Rising 1,683 feet above sea level and 825 feet above a lake at the base, Stone Mountain is a granite monolith, said to be the largest body of exposed granite in the world. The western side of the Mountain slopes gently; a 1.3-mile trail starting at Confederate Hall leads to the top. Pine, oak, hickory, beech, and maple woods cover its base; scattered pines, cedars, and clumps of bushes dot the exposed and exfoliated surface of the upper slopes.

Some permanent residents in the Park are the Turkey Vulture, Black Vulture, American Kestrel, Yellow-shafted Flicker, Pileated Woodpecker, and Brown-headed Nuthatch. Nesting in late spring and summer are the Common Nighthawk, Blue-gray Gnatcatcher, Yellow-throated Warbler, Orchard Oriole, Summer Tanager, Blue Grosbeak, and Bachman's Sparrow; in winter the scattered clumps of bushes on the mountainside are favorite haunts of the Bewick's Wren.

AUGUSTA

Savannah River Bottoms | Merry Brothers Brickyard Ponds | Municipal Airport

For observing birds in the vicinity of Augusta, the **Savannah River Bottoms** southeast of the city are the best and most accessible. To reach the area, drive east to the end of Broad Street, continue on

State 28 for 0.75 mile, then turn right on Lover's Lane. This road traverses several habitats for common breeding birds: hayfields for Eastern Meadowlarks and Grasshopper Sparrows; hedgerows dividing the hayfields and fringing the road for Indigo and Painted Buntings; lagoons, bordered by moss-hung cypresses and tupelos, for Prothonotary, Northern Parula, and Yellow-throated Warblers; drier woods for Kentucky and Hooded Warblers.

Lover's Lane continues for 3 miles, ending on the levee, in the vicinity of which are many of the same kinds of birds. One may follow the levee southward for 4 miles to the New Savannah Lock and Dam, a recreational area. In late spring and summer a few Dickcissels occasionally nest in fields along the levee, and now and then a Mississippi Kite searches for food over the fields. On any one of the short side roads flanked by cane, listen and search for Swainson's Warblers. The hardwoods in the Lock and Dam area are attractive to Pileated Woodpeckers and a variety of other breeding birds as well as to many transient and wintering passerines.

The **Merry Brothers Brickyard Ponds,** where clay has been mined for a century—thus forming shallow bodies of water and marshes covering nearly a square mile—offer the chance to see herons and egrets in summer and winter. Drive east to the end of Broad Street, then turn right on East Boundary Street and follow it to its end at the Brickyard Ponds. Obtain permission to explore the area from the person on duty at the entrance. Among birds nesting in the marshes are Pied-billed Grebes, Least Bitterns, King Rails, Common Gallinules, and American Coots.

For breeding (Prairie) Horned Larks, scan the runway edges in early spring at the **Municipal Airport,** which is reached from the city by driving west on US 78 and 278 (Gordon Highway), following directional signs.

BRUNSWICK

Jekyll Island State Park | St. Simons Island | Sea Island

Excellent for observing spring migration on the Atlantic coast is semitropical **Jekyll Island State Park** (5,760 acreas), with its wide, 9-mile beach. Although a public resort, Jekyll Island has extensive forests of longleaf pine in combination with oak, tulip tree, and other hardwoods, and undisturbed open areas where scrubby growth predominates. Jekyll Island, only 9 miles from Brunswick, is easily reached by driving south on US 17 (State 25) and east on State 50.

The time to visit Jekyll Island is during the last week of April and the first week of May, when it is possible to find at least ninety waterbird, waterfowl, shorebird, and landbird species by walking along the beach—preferably at low tide—and along bridle trails and shell roads leading through the woods. A few Whimbrels and large numbers of Painted Buntings are invariably among the rewards of a trip over the Island.

St. Simons Island and adjacent **Sea Island,** north of Jekyll Island, are excellent for birds the year round. Habitats to investigate are the open sea and the tidewater rivers and channels; gently sloping sand beaches; dunes, covered in many places with scrubby growth; extensive salt marshes; woods, consisting of live oaks, pines, palms, and other trees, with thick undergrowth and brushy edges; fresh-water ponds, frequently choked with aquatic plants and surrounded by dense thickets; and scattered fields, including a golf course.The following tour—best made in late March or April—includes many of these habitats.

From US 17 north of Brunswick take the Torras Causeway to St. Simons Island over the Frederica River, the Intracoastal Waterway, and the neighboring marshes and tidal streams. In any season these marshes should yield Willets, Clapper Rails, and Seaside Sparrows. In fall and early spring, and sometimes in winter, they should yield Sharp-tailed Sparrows as well.

Still on the Island, proceed directly southeast on Kings Way to the intersection with Retreat Avenue. Turn right onto Retreat Avenue and go to the Sea Island Golf Club on the southern edge of land. The open fairways and water hazards, the salt marshes on the side toward St. Simons Sound, and the massive pines and live oaks bordering the course elsewhere, will produce many birds, particularly if visited early in the morning. Look for Pileated Woodpeckers in the heavier woods.

From the Golf Club, drive eastward on Ocean Boulevard to St. Simons Village, where one may see Jekyll Island from a pier extending into St. Simons Sound. The pier and its surroundings are good vantage points for observing waterbirds, waterfowl, and shorebirds, though an even better one is on Sea Island, mentioned below.

From St. Simons Village drive northwest on Kings Way to the intersection with Retreat Avenue; turn right on Frederica Road and go north and then west to Christ Episcopal Church, a low-gabled frame structure in a grove of moss-hung live oaks. The district is good for finding such permanent- or summer-resident birds as Blue-gray Gnatcatchers, Northern Parula Warblers, Yellow-throated Warblers, Northern Cardinals, and Painted Buntings. An equally good bird-finding area is that surrounding Fort Frederica in the Fort Frederica National Monument (75 acres), just beyond the church on the same road.

Return on Frederica Road and turn east on Sea Island Road to Sea Island, about 2 miles distant, over the Black Banks River and its adjacent salt marshes. These marshes are as good for birds as those crossed by the Torras Causeway. The Sea Island beach is of great ornithological interest, but cannot be reached by car. Hence one should either leave the car at Cloister Inn near the swimming pool, walk to the beach, and then south to the southern point of Sea Island, or drive to the Sea Island Fishing Camp at the north end of the beach and walk south along the beach from there. The best bird-finding spots are the beach and the sand spits near the Fishing Camp, where the Hampton River enters the ocean, and near the southern point of the Island, where the Black Banks River enters the ocean.

From December through March, birds to be expected are the Brown Pelican, American Oystercatcher, Piping Plover, Semipalmated Plover, Killdeer, Black-bellied Plover, Ruddy Turnstone, Greater Yellowlegs, Least Sandpiper, Semipalmated Sandpiper, Western Sandpiper, Sanderling (abundant), Herring Gull, Ring-billed Gull (abundant), Laughing Gull, Bonaparte's Gull, Forster's Tern, Little Tern, Royal Tern, Caspian Tern, and Black Skimmer. In October, late March, and April, shorebirds are more abundant, and a list will probably include the Whimbrel, Red Knot, and Short-billed Dowitcher. Offshore, Common Loons, Double-crested Cormorants, Lesser Scaups (great numbers), and Red-breasted Mergansers may be viewed throughout the winter.

CLAYTON
Rabun Lake | Seed Lake | Burton Lake

In the extreme northeastern corner of Georgia are three bodies of water, **Rabun Lake, Seed Lake,** and **Burton Lake,** created by the impounding of the Tallulah River for power development. All three Lakes, whose surfaces reflect the surrounding mountains, are almost completely encircled by forests of mixed hardwoods, pines, and hemlocks, though there are a few open areas along streams and roadsides where shrubs grow extensively. Some of the birds breeding in suitable habitats are the Pileated Woodpecker, Least Flycatcher, Carolina Chickadee, Wood Thrush, White-eyed Vireo, Yellow-throated Vireo, Yellow Warbler, Pine Warbler, Prairie Warbler, Scarlet Tanager, Summer Tanager, Blue Grosbeak, and Indigo Bunting.

One can reach the bird-finding areas by driving south from Clayton on US 441 for 9 miles to Lakemont, then west on a road that soon winds along the lake shores first of Rabun, then of Seed and Burton.

CLEVELAND
Brasstown Bald Mountain

There are in extreme northeastern Georgia many majestic mountains and wooded valleys where bird finding can be productive amid picturesque surroundings. The following trip should be worthwhile at any time from mid-April to July, though perhaps especially so between mid-May and mid-June, when most of the birds are in full song and when the rhododendrons, mountain laurels, and azaleas are blooming. From Cleveland—on US 129 about 24 miles north of Gainesville—drive north on State 75, west on State 66, then north on an all-weather road up **Brasstown Bald Mountain,** Georgia's highest peak at 4,784 feet. Within a half-mile of the summit, where the road is closed to private vehicles, park and either take the shuttle bus or walk to the fire tower and visitor center at the top.

Brasstown Bald Mountain lies in the Chattahoochee National Forest, where the principal trees are red and white oaks, red maples, hickories, white pines, and hemlocks. Much of this tree growth extends up Brasstown Bald, with hemlocks predominant on the cooler north slopes. The summit, however, is 'bald' and, except for a few scrub oaks and small firs, covered with grasses and thickets of rhododendron and other shrubs.

The following warblers are summer residents on Brasstown Bald: Black-and-white, Swainson's, Worm-eating, Northern Parula, Black-throated Blue, Black-throated Green, Blackburnian, Chestnut-sided, Ovenbird, Louisiana Waterthrush, Kentucky, Hooded, Canada, and American Redstart. Though most of the warblers appear—where habitats are favorable—at any altitude up to 4,000 feet, generally the Blackburnian stays above 2,000 feet, the Black-throated Blue and Chestnut-sided above 2,600, the Canada above 4,000 and the Swainson's, Worm-eating, Louisiana Waterthrush, Kentucky, and Hooded below 3,000.

Besides warblers, birds for special note are the Winter Wren above 4,000 feet, the Veery and Slate-colored Junco above 3,500, the Rose-breasted Grosbeak above 2,700, and the Broad-winged Hawk, Ruffed Grouse, Solitary Vireo, and Scarlet Tanager on all wooded slopes. Most of the birds are readily heard or seen from the road. From the summit, look for permanent-resident Northern Ravens passing in flight below or, if the day is sunny, soaring on the thermals.

MACON
Piedmont National Wildlife Refuge

For a good representation of birdlife characteristic of the Piedmont Plateau, visit the **Piedmont National Wildlife Refuge** (34,673 acres), north of Macon. Extending for 18 miles along the east side of the Ocmulgee River, the terrain has red-clay hills with gentle, sometimes steep, slopes. Two-thirds of the area is covered by loblolly and shortleaf pines interspersed with hardwoods on a few upland tracts and on the bottomlands along the many small streams that flow into the Ocmulgee. The remaining third is abandoned farmland reverting to forest. Although there are no natural lakes, pools, or marshes, water, impounded in several areas, provides habitat for waterfowl.

The Common Bobwhite and Wild Turkey are permanent residents. Clans of Red-cockaded Woodpecker occupy one particular tract of pine. Other species representing the breeding population are the Mallard, Wood Duck, Red-tailed Hawk, Yellow-billed Cuckoo, Chuck-will's-widow, Pileated Woodpecker, Acadian Flycatcher, Carolina Chickadee, Tufted Titmouse, Brown-headed Nuthatch, Brown Thrasher, Wood Thrush, Blue-gray Gnatcatcher, Yellow-throated Vireo, Pine Warbler, Prairie Warbler, Hooded Warbler, Summer Tanager, Blue Grosbeak, Indigo Bunting, Rufous-sided Towhee, Bachman's Sparrow, and Field Sparrow. Most of the transients are landbirds, particularly passerine species. The two peaks of migration are usually in mid-April and late October; the April peak is invariably the larger in number of species.

Wintering birds include the Ring-necked Duck, Yellow-bellied Sapsucker, Winter Wren, Bewick's Wren, Hermit Thrush, Golden-

BROWN-HEADED NUTHATCH

crowned Kinglet, Ruby-crowned Kinglet, Myrtle Warbler, Slate-colored Junco, White-throated Sparrow, Fox Sparrow, Swamp Sparrow, and Song Sparrow.

To reach Refuge headquarters from Macon, drive northward on US 129 to Gray, then northwestward on State 11 through Round Oak. One mile north of Round Oak, turn left on a road marked by a Refuge sign, to headquarters 3.5 miles beyond.

SAVANNAH

Bonaventure | Whitemarsh Island | McQueen Island | Tybee Island | Harris Neck National Wildlife Refuge | Blackbeard Island National Wildlife Refuge

On a bluff above the Savannah River and a short 14 miles from the Atlantic this imposing city is within easy reach of many woods, extensive salt- and fresh-water marshes, and broad stretches of sand beach—areas that together attract a varied and abundant birdlife. US 80 east from the city leads to some of the best of these areas.

In Thunderbolt, on US 80 about 6 miles east from downtown Savannah, turn left and proceed about a mile on Bonaventure Road to **Bonaventure,** a historic old plantation, now a cemetery. In spring and summer a stroll under the ancient, moss-hung live oaks and among the shrubbery should yield many breeding birds, including Carolina Wrens and Northern Parula, Yellow-throated, and Prairie Warblers; along the marshy edges of the bluff near the Wilmington River, Marsh Wrens and Common Yellowthroats. Bonaventure is one of the best spots for transient and wintering birds of many kinds.

Return to US 80 and continue east along a palmetto- and oleander-lined causeway over a succession of tidal rivers and islands to the coast. First, the highway crosses (at 6.5 miles from downtown Savannah) the Wilmington River—part of the Intracoastal Waterway—to **Whitemarsh Island,** consisting of strips of salt marsh alternating with higher wooded land. Next, the highway crosses Turner Creek to Wilmington Island, where there is more extensive high ground with forests of pine and live oak. From here US 80 crosses (at 13 miles) the Bull River to **McQueen Island,** where salt marshes stretch eastward for more than 4 miles, and several miles farther to the north and south.

One mile beyond the bridge over the Bull River, in the marshes to the left of the causeway, look for both Sharp-tailed Sparrows and Seaside Sparrows in winter and an occasional wintering Sedge Wren. A little beyond, on the right side, are the mud flats of Oyster Creek, excellent for shorebirds from the first of October until May, but particularly during winter. Species that may be expected are the American Oystercatcher, Semipalmated Plover, Killdeer, Black-bellied Plover, Whimbrel, Greater and Lesser Yellowlegs, Least Sandpiper, Dunlin, Semipalmated Sandpiper, and Western Sandpiper. In summer, Willets nest in the grassy edges of the road, Marsh Wrens and Boat-tailed Grackles at the edges of the marsh creeks, and Clapper Rails in the marsh itself. Farther on, turn off State 80 to the Fort Pulaski Bridge over the South Channel to Cockspur Island. In winter this waterway is worth scanning for waterfowl—especially for Canvasbacks—and for various waterbirds.

Return to US 80 and, continuing east, cross Lazaretto Creek (at 18 miles) to **Tybee Island.** At about 100 yards beyond the bridge, turn left, and follow this road a short distance until a thicket of live oaks and cassias appears on the right. Deep in this thicket is a nesting aggregation of Great Blue Herons, Cattle Egrets, Great Egrets, Snowy Egrets, Louisiana Herons, Black-crowned Night Herons, and a few Green Herons. Though nesting season is from April through July, the birds remain in the vicinity well into fall. In the evening they gather from far and near and settle into the thicket for the night, providing a memorable spectacle.

Again, return to US 80 and go east to Fort Screver, an old army post (21.5 miles). Turn left and continue to an old dock at the north end of the Island. Then bear right, drive up on the old battery near Tybee Lighthouse and follow the beach around to US 80.

The entire beach is good for waterfowl, waterbirds, and shorebirds in almost any season. Little Terns and Wilson's Plovers nest from April to July on the sand above the high-tide mark west of the old dock toward Lazaretto Creek. Ruddy Turnstones, Red Knots, Sanderlings, and other shorebirds frequent the beach in any season except early summer; Purple Sandpipers, the jetties and groins in winter. Common Loons, occasionally Northern Gannets, Double-

WILSON'S PLOVERS

crested Cormorants, and Red-breasted Mergansers show up offshore in winter, Brown Pelicans in all seasons; Herring Gulls, Ring-billed Gulls, Bonaparte's Gulls, and Common Terns appear commonly on or near the shore from November to April; Laughing Gulls, Forster's Terns, Royal Terns, Caspian Terns, and Black Skimmers, at any time of year.

South from Savannah along the Atlantic coast to Brunswick, are several wildlife refuges, two of which—**Harris Neck National Wildlife Refuge** and **Blackbeard Island National Wildlife Refuge**—offer especially good bird finding. To reach both, drive south on I 95 for 31 miles beyond its intersection with I 16, then exit south on US 17. For the Harris Neck Refuge, turn east on State 131 and go 7 miles to the entrance. For the Blackbeard, turn east from US 17, 7 miles south of its intersection with State 131, on an unpaved road, and drive 6 miles to the community of Shellman Bluff. Hire a boat here for a trip to Refuge headquarters, 18 miles away.

The Harris Neck Refuge, formerly a military airfield and occupying only 2,687 acres, has nonetheless a variety of habitats, all readily accessible by a network of paved roads. Its small, man-made ponds, which support a resident flock of Canada Geese, are frequented by Pied-billed Grebes and Wood Ducks the year round and over a dozen species of ducks from September to April. Wading birds—

especially Cattle Egrets along with smaller numbers of Great Blue Herons, Green Herons, Little Blue Herons, Great Egrets, Snowy Egrets, Louisiana Herons, and White Ibises—are present in any season.

In the extensive salt marshes flanking the tidal streams that form much of the Refuge boundary, Clapper Rails, Willets, Marsh Wrens, Boat-tailed Grackles, and Seaside Sparrows are among the commonest species. At low tide the mud flats at the north end of the Refuge draw impressive numbers of transient shorebirds. Inland from the streams and marshes the stands of live oak and other wooded tracts—often with brushy edges—prove attractive in winter to an abundance of Northern Parula Warblers and Painted Buntings, and somewhat fewer Great Crested Flycatchers, Eastern Pewees, Yellow-throated Vireos, Prairie Warblers, Hooded Warblers, Orchard Orioles, and Summer Tanagers.

The Blackbeard Island Refuge consists of one 5,618-acre island with a broad, sandy beach, backed by low dunes, fronting on the Atlantic, and salt marshes on the opposite side extending toward the mainland. The interior features low-lying, parallel ridges covered by live oaks and separated by fresh-water marshes and numerous ponds. Large parts of the Island still support virgin live oaks, slash pines, hollies, and magnolias.

In fall and winter, large numbers of ducks stay on the ponds, principal species, in order of decreasing abundance, being the Ring-necked Duck, Canvasback, American Wigeon, Wood Duck, Mallard, Gadwall, Northern Shoveler, Green-winged Teal, American Black Duck, Ruddy Duck, and Common Pintail. Lesser Scaups frequent the surf off the beach. American Coots also are present in large numbers. Waterbirds breeding commonly are the Pied-billed Grebe, American Anhinga, Great Blue Heron, Green Heron, Little Blue Heron, Great Egret, Snowy Egret, Louisiana Heron, Black-crowned Night Heron, Least Bittern, White Ibis, King Rail, and Purple Gallinule.

In late spring and early summer, Wilson's Plovers and occasionally American Oystercatchers nest on the beach and here, at other seasons, many transient shorebirds feed and rest. Northern small landbirds, particularly fringillids, abound in the winter. In general, breeding landbirds are fewer in variety and number than in the Harris Neck Refuge, although such species as the Northern Parula Warbler, Yellow-throated Warbler, Pine Warbler, and Painted Bunting are nevertheless common. The salt marshes are equally as productive ornithologically as in the Harris Neck Refuge.

WAYCROSS

Okefenokee Swamp | Okefenokee National Wildlife Refuge

Okefenokee Swamp, in southeastern Georgia near the Florida line, is a trackless wilderness of marsh and inundated woodland covering nearly 700 square miles. Despite efforts to drain the area with a view to removing the timber, the quiet amber waters overspread with flowering plants remain, along with the giant trees decked with skeins of moss.

The eastern part of the Swamp has extensive marshes, known locally as 'prairies.' In water that is relatively shallow, pickerel weeds, bladder worts, floating hearts, waterlilies, and many other water-loving species thrive in rank profusion. Here and there emerges an island (or 'hammock') of cypresses and tupelos, and sometimes of longleaf and slash pines. The western part is more typically swampland, having even shallower water, with cypresses, tupelos, bays, and magnolias dominant, except where timber-cutting operations have taken their toll.

Okefenokee Swamp is a great saucer-shaped depression filled from prairie springs and drained by two small rivers: the St. Marys, which wends its way to the Atlantic, forming part of the Georgia–Florida boundary in its course; and the famed, idyllic Suwannee, the principal outlet, which arises deep in the Swamp and meanders westward and then southward through miles of eerie stillness broken only by such wilderness sounds as the songs of birds and the bellowing of alligators, and finally reaches the Gulf of Mexico through Florida. As the water drains through the cypress forests, it is stained the color of tea.

Different parts of the Okefenokee, particularly to the west, deepen into large ponds, or lakes, which are probably swollen sections of the Suwannee. Seemingly everywhere on the prairies and swampland small watercourses twist, turn back upon themselves, and come to a dead end. Many are open to motor crafts, but others, especially in the prairies, are so clogged by vegetation that only pole boats can be used. Whether or not navigation is possible, only one familiar with the intricate geography of the Okefenokee can run the maze without becoming hopelessly confused.

The breeding bird population is outstanding more for abundance than for variety of species. A more notable permanent resident, considered fairly common in the prairie habitat, is the Sandhill Crane. Its nests contain eggs as late as the first week of April. Other species nesting commonly, usually by late April, include the Great Blue Heron, Green Heron, Little Blue Heron, Cattle Egret, Great Egret, Wood Stork, White Ibis, Wood Duck, Red-shouldered Hawk, King Rail, Ground Dove, Barred Owl, Chuck-will's-widow, Pileated Woodpecker, Great Crested Flycatcher, Acadian Flycatcher, Eastern Pewee, Fish Crow, Brown-headed Nuthatch, White-eyed Vireo, Prothonotary Warbler (the commonest parulid), Northern Parula Warbler, Yellow-throated Warbler, Pine Warbler, Red-winged Blackbird, Rufous-sided Towhee, and Bachman's Sparrow. Mallards, American Black Ducks, Ring-necked Ducks, Hooded Mergansers, and smaller numbers of other ducks find the open waters a winter haven.

The **Okefenokee National Wildlife Refuge,** with headquarters at Waycross on US 1, embraces much of the Okefenokee—340,788 acres in all. The Refuge has three entrances, each separated by many miles from the others and each providing access to a distinctive part: the northern, at Okefenokee Swamp Park; the eastern, at Camp Cornelia; and the western, at Stephen Collins Foster State Park.

Okefenokee Swamp Park, reached from Waycross by driving southeast on US 1 for 8 miles, then bearing right on State 177, lies largely outside the Refuge boundary. Owned and managed by a private organization, the Park has an interpretive center, a boardwalk, an observation tower, and exhibits of captive wildlife. Boat rides are available into the Refuge, following watercourses that meander through a cypress forest and typical Okefenokee prairie.

Camp Cornelia, reached from Waycross by driving southeast on US 1 for 35 miles to Folkston and then west for about 7 miles, is administered by the Refuge. It has a visitor center and a loop drive leading to nature trails, as well as a boardwalk to an observation tower overlooking extensive prairies. One may hire boats for cruises on the Suwannee Canal to a series of prairies—Chesser, Mizell, Grand, and others—for marsh birds the most rewarding in the Okefenokee. If possible, hire a guide for any trip, since the watercourses to the prairies, though marked, are commonly lily-choked and thus difficult to follow.

Foster State Park (80 acres), reached from Waycross by driving southwest on US 84 for 26 miles to Homerville, turning left on US 441 to Fargo, and then northeast on State 177 for about 15 miles, is deep in the Refuge. There are boats, with a guide if desired, for a trip east from the Park on Billy's Lake, northward through Minnie's Run—a narrow course of dark water that winds peacefully for 2.5 miles under towering cypresses and tupelos—to Minnie's Lake, and then along another tree-shaded watercourse northward to Big Water Lake. For wading and forest birds this is the most rewarding trip in the Okefenokee.

Illinois

DICKCISSEL

The tourist who has driven across the plains of central Illinois in June remembers the verdant farmlands, the grainfields that alternate with livestock pastures, the thin, straight lines of fences, or hedgerows of Osage orange, that separate them. The bird finder who has made the same trip remembers still more—the Dickcissels, the black-throated males on fences, telephone wires, weed stalks, and hedges, and the songs, lusty and unmusical, that pour incessantly from their uplifted mouths. Both tourist and bird finder gain the impression from the trip that Illinois is a flat to gently-rolling agricultural country, its terrain unbroken, its birdlife the same from border to border. The truth is that the farmlands constitute but one of several quite diversified areas for birds within the state.

Illinois is remarkably level, the variation in elevation being less than a thousand feet. The lowest point (279 feet) is at the southernmost extremity of the state, near Cairo, on the river bed where the Ohio and the Mississippi merge; its highest point (1,241 feet) is in the extreme northwestern corner of the state, on Charles Mound. Other sections that can be considered decidedly hilly are in the southern tip, where there is an eastward extension of the Ozarks (sometimes called Little Ozarks), and the blufflands along the big rivers—chiefly the Ohio along the southeast boundary, and the Illinois, which flows southwestward through the state to the Mississippi at Grafton.

Roughly 40 per cent of the state was originally forested; the rest was grassland or prairie. The southern third supported an almost continuous forest cover; elsewhere, forests stood along the river bluffs and bottomlands, or occasionally in the grasslands as prairie groves. Trees predominant in most of the forests were maples, oaks, elms, basswood, ashes, hickories, and, sometimes, beeches; willows, sycamores, and cottonwood were prevalent on the bottomlands. The meeting place of forest and grassland was an area of low trees, shrubs, and rank weeds.

With the development of agriculture, the forests were for the most part cleared away and replaced, as was the grassland, by farmlands. Most of the bird species occupying the original grassland survive today in grassy pastures or in grassy fields allowed to remain fallow. Despite the great reduction of forests, there are still everywhere in the state timbered tracts and shrubby places that attract a wide variety of forest and forest-edge birds. At the present time the following species breed regularly in the forests (including remnants of prairie groves) and farmlands (fields, wet meadows, shrubby lands, orchards, and dooryards).

FORESTS

Red-shouldered Hawk
Yellow-billed Cuckoo
Black-billed Cuckoo (*except in southern Illinois*)
Common Screech Owl
Barred Owl
Whip-poor-will
Yellow-shafted Flicker
Red-bellied Woodpecker
Hairy Woodpecker

FORESTS (*Cont.*)

Downy Woodpecker
Great Crested Flycatcher
Acadian Flycatcher
Eastern Pewee
Blue Jay
Black-capped Chickadee (*northern Illinois*)
Tufted Titmouse
White-breasted Nuthatch
Wood Thrush
Blue-gray Gnatcatcher
Red-eyed Vireo
Warbling Vireo
Prothonotary Warbler
Cerulean Warbler
Ovenbird
Louisiana Waterthrush
American Redstart
Baltimore Oriole
Scarlet Tanager
Rose-breasted Grosbeak (*chiefly northern Illinois*)

FARMLANDS

Common Bobwhite
Mourning Dove
Red-headed Woodpecker
Eastern Kingbird
Eastern Phoebe
(Prairie) Horned Lark
Tree Swallow
Barn Swallow
House Wren
Carolina Wren
Northern Mockingbird
Gray Catbird
Brown Thrasher
Eastern Bluebird
Loggerhead Shrike (*southern Illinois*)
Yellow Warbler
Common Yellowthroat
Bobolink (*northern Illinois*)
Eastern Meadowlark
Western Meadowlark
Common Grackle
Northern Cardinal
Indigo Bunting
Dickcissel
American Goldfinch
Rufous-sided Towhee
Savannah Sparrow (*northern Illinois*)
Grasshopper Sparrow
Henslow's Sparrow
Vesper Sparrow (*except southern Illinois*)
Chipping Sparrow
Field Sparrow
Swamp Sparrow (*northern Illinois*)
Song Sparrow

Two upland species not on these lists require special comment. One is the Greater Prairie Chicken, which formerly ranged over all the grassland of the state but declined sharply with the onset of farming. Today this fine bird occupies only territories where farming has ceased, permitting a sizable expanse of land to lie idle during nesting season. Southeastern Illinois (*see under* **Newton**) has the largest remaining population in the state. The other is Bell's Vireo, a summer resident of central and southern Illinois that breeds in shrubby vegetation, often sharing that habitat with the White-eyed Vireo.

Because the north/south length of Illinois is 378 miles, the birdlife in the southern part includes a number of species characteristic of southern environments. The following nest regularly in central Illinois and increase in abundance southward.

Turkey Vulture
Black Vulture (*southern Illinois*)
Chuck-will's-widow (*southern Illinois*)
Carolina Chickadee
Bewick's Wren
White-eyed Vireo
Worm-eating Warbler
Yellow-throated Warbler
Kentucky Warbler
Yellow-breasted Chat
Hooded Warbler
Summer Tanager

An area quite unlike any other in the state is Illinois State Beach Park, north of Waukegan, on the Lake Michigan shore. Here are sand beach, dunes, and, farther inland, tree-covered ridges and intervening marshes. The birdlife, while rich in variety, possesses no species unusual for the state. A very different area is the Sand Ridge State Forest (*see under* **Havana**), where there are sand hills and sand prairie, with birdlife distinguished by numerous Lark Sparrows.

Marshes and swamps are numerous in all parts of Illinois. In the Chicago area there are a few marshes where one may observe a typical bittern-rail association. McGinnis Slough Wildlife Refuge is one. Along the big rivers there are swampy bottomland woods where Wood Ducks and Prothonotary Warblers are common. A good example of this habitat in central Illinois is the Illinois River bottomlands in Chautauqua National Wildlife Refuge (*see under* **Havana**).

In the wet bottomlands of the Ohio and the Mississippi in the extreme south, there are cypresses, sweet gums, and other southern trees, giving a semitropical aspect. Typical are the bottomland forests in Crab Orchard National Wildlife Refuge (*see under* **Carbondale**). The Fish Crow, Swainson's Warbler, and other species of southern affinities are regular nesting birds in some of these southernmost Illinois areas.

Because of its relation to the Mississippi River Valley and the converging Ohio, Kaskaskia, and Illinois Rivers among others, Illinois occupies a unique position with respect to the great Mississippi flyway. Thus, in fall, birds pass through, and on both sides of, the state from their breeding grounds in north-central North America, which extends as far east as western Quebec and the east coast of Hudson Bay and as far west as Montana, Alberta, the Mackenzie Delta, and northwestern Alaska. All kinds of birds—waterbirds, waterfowl, shorebirds, and landbirds—comprise the traffic, but the group whose movements arouse greatest popular interest are the waterfowl.

Enormous numbers of ducks and geese migrate through the state, following the river valleys south to their confluence with the Mississippi. Along the way the birds feed and rest on the larger bodies of water, many of them impoundments such as Lake Shelbyville, the Carlyle Reservoir, Rend Lake between Mt. Vernon and Benton, and Lake Kinkaid east of Murphysboro. Tremendous concentrations of Mallards are seen in the Chautauqua National Wildlife Refuge. Equally spectacular are the huge aggregations of Canada Geese in the Crab Orchard National Wildlife Refuge and the Horseshoe Lake Wildlife Refuge (reached from Olive Branch; not to be confused with the Horseshoe Lake near East St. Louis).

In spring the passage of birds along the Mississippi flyway is no less heavy than in fall, and proceeds in the same manner, but in reverse. The main flights in spring and fall may be expected within the following dates:

Waterfowl: 1 March–10 April; 15 October–1 December
Shorebirds: 1 May–1 June; 1 August–5 October
Landbirds: 15 April–15 May; 1 September–25 October

Winters in northern Illinois, including the Chicago area, usually bring Evening Grosbeaks and Common Redpolls, occasionally Pine

Grosbeaks, Red Crossbills, White-winged Crossbills, and sometimes a few Bohemian Waxwings. A good place to find them is the Morton Arboretum at Lisle. Where the water remains open along the Chicago lakefront, Oldsquaws and a few White-winged Scoters appear. Snow Buntings frequently show up on the beaches north of Chicago, and Lapland Longspurs abound on open fields as far south as central Illinois.

Bald Eagles began wintering along the Mississippi River after the 1930s when locks and dams were constructed, keeping the water above and below them free of ice for the eagles' favorite food—fish, principally the weak-swimming gizzard shad 3 to 4 inches in length. The great birds sometimes capture them by swooping low over the water and grasping them in talons or bill; but the most common and successful method is to wade into shallow water and seize them in the bill. Between feeding sessions the birds loaf in big trees near the open water and usually roost at night in the same trees or those not far away. Bald Eagles begin showing up by mid-November, reach peak abundance between mid-January and mid-February, and gradually disappear from that time on through March. During midwinter the bird finder may observe at one place, at one time, more Bald Eagles than anywhere else south of Alaska. The Official Highway Map of Illinois, issued annually by the Department of Transportation (2300 South 31st Street, Springfield, Illinois 62706), shows the location of each lock and dam by number. Some locks and dams attract more Bald Eagles than others—e.g., the northernmost, Number 12 (*see under* **Savanna**), and Number 19 (*see under* **Hamilton**). Numbers 15 (at Rock Island) and 17 (near New Boston) are very attractive. The open water usually brings a good representation of Common Goldeneyes and Common Mergansers as well.

Authorities

Richard A. Anderson, William J. Beecher, Frank C. Bellrose, Jr., Dale E. Birkenholtz, Marilyn Campbell, James Earl Comfort, Eugene E. Crawford, Virginia S. Eifert, Hattie Ettinger, Elton Fawks, William G. George, Jean W. Graber, Richard R. Graber, William N. Kelly, S. Charles Kendeigh, Vernon M. Kleen, Robert B. Lea, George M. Link, Peggy Muirhead MacQueen, Ernestine Magner, Marshall Magner, Russell L. Mixter, Vernal S. Nave, W. D. Petzel, Donald T. Reis, William A. Sausaman, Preston W. Shellbach, Robert L. Smith, William E. Southern, Patrick M. Ward, Mrs. Theron Wasson, Ralph E. Yeatter.

References

Bird Finding in Illinois. Compiled by Elton Fawks; edited by Paul H. Lobik. Illinois Audubon Society (1017 Burlington Avenue, Downers Grove, Ill. 60515), 1975. Compiled before 1966.

Top Birding Spots near Chicago. By Jeffrey Sanders and Lynn Yaskot. Privately printed, 1975. Available from Top Birding Spots, 1093 Elm Street, Winnetka, Ill. 60093.

BARRINGTON
Crabtree Nature Center

Northwest of Chicago, some 1,100 acres of former farmlands, marshes, ponds, two lakes, and small forested tracts are embraced by the **Crabtree Nature Center,** operated by the Forest Preserve District of Cook County. The center is easily reached from I 90 (the Northwest Tollway) by exiting on Barrington Road and proceeding north 2.5 miles, then continuing left on Palatine Road for 1.0 mile to the entrance on the right.

From the interpretive building a self-guiding nature trail winds nearly 1.5 miles through woods of oak and hickory, around two ponds and a marsh—all good habitats for some of the commoner birds of northern Illinois. Along the way a side trail leads to an observation tower overlooking Crabtree Lake, where many Canada Geese rest and numerous ducks such as Mallards, Gadwalls, Common Pintails, Blue-winged Teal, American Wigeons, and Northern Shovelers feed and tarry in early spring and late fall.

CARBONDALE
Crab Orchard National Wildlife Refuge

East of this community in the Little Ozarks of southern Illinois is the **Crab Orchard National Wildlife Refuge** (43,017 acres), noted for the great numbers of waterfowl that pass the winter on its three artificial lakes—by name Crab Orchard (7,000 acres), with its edging of cattail marshes; Little Grassy (1,000 acres), with its steep, rocky shores covered with mixed hardwoods; and Devil's Kitchen (800 acres). In addition the gently rolling terrain of the Refuge has open farmlands; upland forests of oak, hickory, and maple; bottomland forests of sycamore, sweet gum, and cypress; and groves of introduced shortleaf and loblolly pines.

Waterfowl concentrations begin in October and extend through March. The huge wintering population of Canada Geese is rivaled elsewhere in Illinois only at Horseshoe Lake Conservation Area farther south (*see under* **Olive Branch**). Snow Geese are common from late October to mid-December. Of the twenty-two species of ducks recorded on the Refuge, most common are the Mallard, American Black Duck, Common Pintail, Green-winged Teal, American Wigeon, Northern Shoveler, Ring-necked Duck, Common Goldeneye, Bufflehead, Ruddy Duck, Hooded Merganser, and Common Merganser. Wood Ducks nest on the Refuge, as does a resident flock of Canada Geese.

As many as a hundred species are known to breed on the Refuge, among them the Green Heron, Black-crowned and Yellow-crowned Night Herons, Red-tailed and Red-shouldered Hawks, Northern Harrier, American Woodcock, Upland Sandpiper, Great Horned

CANADA GEESE

and Barred Owls, Whip-poor-will, Red-bellied and Red-headed Woodpeckers, (Prairie) Horned Lark, Carolina Chickadee, Bewick's and Carolina Wrens, Blue-gray Gnatcatcher, White-eyed Vireo, Prothonotary and Prairie Warblers, Louisiana Waterthrush, Orchard Oriole, Summer Tanager, Indigo Bunting, and Dickcissel. Great waves of flycatchers, vireos, warblers, fringillids, and other passerines go through the Refuge from March to mid-May

To reach Refuge headquarters, drive 11 miles east from Carbondale on State 13, then go south on State 148 to the entrance. On the way, State 13 crosses the northern part of Crab Orchard Lake and affords excellent vantage points for seeing waterfowl. Good roads provide access to much of the Refuge, and there are boats for rent at all three lakes.

CHAMPAIGN-URBANA
Busey Woods

Busey Woods, a city-owned tract of 40 acres a mile north of US 45 in the Urbana business district, is largely overgrown with hawthorn, willow, and other small trees and shrubs. Stands of oak cover portions of the tract, and there are water areas, including a small oxbow lake and a small stream. The Woods may be reached from US 45 by driving north on Broadway Avenue along Crystal Park and turning left into the swimming pool entrance. Leave the car in the parking lot north of the pool and walk north over a bridge.

Birds appearing regularly during migration include the American Woodcock, both Swainson's and Gray-cheeked Thrushes, and many species of warblers. Among species nesting in summer are the Yellow-billed Cuckoo, Red-bellied Woodpecker, Great Crested Flycatcher, Eastern Pewee, Tufted Titmouse, Wood Thrush, White-eyed Vireo, Yellow-breasted Chat, Rose-breasted Grosbeak, and Indigo Bunting.

CHICAGO
Grant Park | Jackson Park | Lincoln Park | McGinnis Slough Wildlife Refuge | Palos Park Forest Preserve | Morton Arboretum

On the southwest shore of Lake Michigan from the Indiana Line northward for some 25 miles, extends the impressive Chicago lakefront, its beaches, breakwaters, docks, and parkways backed by a serrated wall of buildings including, about halfway, the tall towers and massive blocks that mark the eastern edge of the Loop, the city's commercial center. To the west, north, and south, covering more than 200 square miles of what was once flat marshland, are the great city itself and the suburban settlements. In both the city and its immediate vicinity are numerous parks and preserves where wild birds abound.

Stretching along the lakefront is a chain of parks. The central one, **Grant Park** (303 acres), lies east of the Loop between Michigan Avenue and Chicago Harbor, and extends south as far as Roosevelt Road, and north to Randolph Street. Liberally landscaped with trees and shrubs, Grant Park is good for transient kinglets, vireos, warblers, and fringillids in April and early May, and again in September. From October to April, ducks and gulls enjoy the Harbor's open water, the gulls sometimes resting on the shore and breakwaters or scavenging in the Park. A walk along the shore after a winter storm may yield such uncommon species as a Glaucous Gull, an Iceland Gull, or a Black-legged Kittiwake.

An inlet just north of Grant Park, on the north side of the Naval Reserve Armory at the east end of Randolph Street, is particularly attractive to waterbirds and waterfowl. Horned Grebes stop here in fall and spring; Lesser Scaups, Common Goldeneyes, Oldsquaws, a few White-winged Scoters, Common Mergansers, and Red-breasted Mergansers arrive late in fall and stay through winter if the water remains open.

Fronting on the Lake 5 miles south of the Loop between 56th and 67th Streets, and bounded on the west by Stony Island Avenue, is **Jackson Park** (543 acres), with lagoons dotted by tiny islands and bordered by cattails and other aquatic plants. The lagoon containing Wooded Island, with its plantings of deciduous trees and fruit-bearing shrubs, is the focal point for bird finding. Mallards are year-round residents on the lagoon, and are joined during migration and mild winters by more of their kind and other species of ducks. When migration is under way in spring, many small landbirds find haven in the shrubbery on Wooded Island. Warblers are especially numerous here at the height of their migration, between 5 and 10

May. Wooded Island has trails accessible by two footbridges reached from the 63rd Street entrance, two blocks east of Stony Island Avenue.

Paralleling the Lake north of the Loop, from North Avenue to Foster Avenue and accessible from Lake Shore Drive (US 41), is **Lincoln Park** (1,100 acres), where a great variety of trees and fruit-bearing shrubs surrounds harbors, ponds, and lagoons. Probably the best area for bird finding is a partly wooded peninsula extending into the Lake and reached by Montrose Drive from the eastern end of Montrose Avenue (Montrose Avenue Exit from Lake Shore Drive). From the peninsula, which encloses tiny Montrose Harbor and affords a fine view of the Chicago skyline, one may see gulls at almost any time of year, a few diving ducks in spring and fall and occasionally in winter.

Beginning in mid-July and continuing through September, the peninsula shore and Montrose–Wilson Beach north of the peninsula become gathering places for shorebirds in the early morning, when there is little human disturbance. Among species appearing fairly regularly are the Semipalmated Plover, Black-bellied Plover, Ruddy Turnstone, Greater and Lesser Yellowlegs, Least Sandpiper, Dunlin, and Semipalmated Sandpiper. Many transient flycatchers, warblers, fringillids, and other passerines linger in spring (April and early May) and fall (September and early October) in the wooded sections of the peninsula. Another good area is the wooded section north from Belmont Harbor, reached via Belmont Drive (Irving Park Boulevard Exit from Lake Shore Drive).

The Forest Preserve District of Cook County (an organization distinct from the Chicago Park District, which manages the city parks described above) embraces 65,000 acres of sanctuaries or reservations of native landscape, bordering the Des Plaines River, the north branch of the Chicago River, and other waterways—in most cases outside Chicago city limits. Much of the Forest Preserve lies along the Des Plaines River, which flows from north to south through the western suburbs.

Through the woods—chiefly of oak and maple—wind many miles of well-marked trails for hiking, bicycling, and horseback riding. Fireplaces, shelters, and picnic grounds are in convenient locations. Folders containing trail maps and information on trail facilities are available at headquarters of the Forest Preserve District of Cook County (536 North Harlem Avenue, River Forest, Illinois 60305). The folders provide directions for reaching the trails from the Loop by rapid transit lines, Chicago surface lines, bus, and connecting local transportation. With relatively little effort the bird finder can enter a woodland environment where birds may be seen and enjoyed in any season.

Of the many natural areas controlled by the Forest Preserve District, the **McGinnis Slough Wildlife Refuge** and **Palos Park Forest Preserve** are among the best for variety and abundance of birds. The McGinnis Slough Refuge, about 20 miles southwest of the Loop, has 975 acres, 314 of which comprise shallow McGinnis Slough. Bordering this body of water are marshes containing cattails, arrowheads, giant bur reeds, and various marsh grasses. Beyond the marshes, on higher, rolling terrain, are oak forests interspersed with dense thickets and abandoned fields producing shrubs and tall grasses. Except that the natural food supply and nesting habitats for the ducks and other birds frequenting McGinnis Slough and the adjoining marshes have been artificially augmented, the Refuge remains in its original condition.

Because of different habitats, birds of a wide variety nest on the Refuge, beginning usually in late May. These include: *in marshes,* Pied-billed Grebe, Green Heron (nests in nearby willows and other trees), Black-crowned Night Heron, American Bittern, Least Bittern, Mallard, Blue-winged Teal, Wood Duck (nests in adjoining woods), King Rail, Virginia Rail, Sora, Common Gallinule, American Coot, Black Tern, Marsh Wren, Red-winged Blackbird, and Swamp Sparrow; *in woods,* Red-shouldered Hawk (occasionally), American Kestrel, Mourning Dove, Yellow-billed and Black-billed Cuckoos, Common Screech and Long-eared Owls, Whip-poor-will, Yellow-shafted Flicker, Red-headed Woodpecker, Great Crested Flycatcher, Eastern Pewee, Blue Jay, Wood Thrush, Cedar Waxwing, Red-eyed and Warbling Vireos, Ovenbird, American Redstart, Baltimore Oriole, Scarlet Tanager, and Rose-breasted Grosbeak; *in fields, forest edges, and brushy lands,* Ring-necked Pheasant, Killdeer, Traill's Flycatcher, (Prairie) Horned Lark, House Wren, Gray Catbird, Brown Thrasher, Eastern Meadowlark, Common Grackle, Northern Cardinal, Indigo Bunting, Dickcissel,

American Goldfinch, Rufous-sided Towhee, Savannah Sparrow, Henslow's Sparrow, Vesper Sparrow, Field Sparrow, and Song Sparrow.

Transient waterbirds and waterfowl appear, sometimes in impressive numbers, from early March to mid-April and from September to mid-December. They include the Common Loon, Horned Grebe, Canada Goose, American Black Duck, American Wigeon, Green-winged Teal, Northern Shoveler, Redhead, Ring-necked Duck, Lesser Scaup, Common Goldeneye, Bufflehead, Ruddy Duck, Hooded Merganser, Common Merganser, Herring Gull, and Ring-billed Gull. Bonaparte's Gulls, Forster's Terns, and Common Terns are spring and fall transients, most numerous in May. Little Blue Herons and Great Egrets are late-summer visitants; Great Blue Herons, though present throughout summer, are more common in late summer and early fall. The American Woodcock, a common transient and uncommon summer resident, arrives in March and departs in October or early November.

The McGinnis Slough Refuge may be reached from the Loop by highways connecting either with State 7 (Southwest Highway), which passes the Refuge on the southeast, or with US 45, which passes the eastern boundary. After passing the Refuge, turn right (off either route) onto 143rd Street and proceed to the entrance on the right. Although the Refuge has no roads open to the public, there are several observation points offering good views of its birdlife, along the entrance road as well as from roads surrounding the Refuge.

The Palos Park Forest Preserve, of about 10,000 acres north of the McGinnis Refuge, is a large tract of hardwood forest breached only by small lakes, marshy sloughs, a few upland fields with shrubs and small trees, and several major highways. Though the larger species of waterbirds and waterfowl are less common here, the birdlife is almost identical with that of the McGinnis Slough Refuge, and can be seen to better advantage, since excellent trails, open to the public, lead to the best spots for bird finding. Transient passerine birds are abundant in April, early May, September, and early October.

The Palos Preserve may be reached from the Loop by highways connecting with US 45, which passes south through the eastern side of the tract. Before undertaking a trip along the trails, obtain from the Forest Preserve District a folder containing a map clearly indicating all the trails. Longjohn Slough in the northern section of the Palos Park Preserve is particularly good for waterbirds and waterfowl. The Slough is a shallow pond about a half-mile across, with sedges and low bushes around the borders. To reach it, turn west off US 45 onto 95th Street and proceed to the intersection with Willow Springs Road. The Slough lies along the east side of the Road and is so close to it that one can see the birds at close range without leaving the car. Overlooking the Slough from the south on 104th Avenue is the Little Red Schoolhouse Nature Center, with self-guiding nature trails.

Morton Arboretum, about 25 miles west of the Loop, is a partly wooded, rolling 1,500-acre tract, privately endowed and administered. With more than four thousand species, varieties, cultivars, and hybrids of the woody plants of the North Temperate Zone now included in the living plant collection, it becomes a wonderland of color in both spring and fall. Within the Arboretum are fine plantings of spruce, fir, and pine, forests of oak and sugar maple, four ponds, and a reconstituted Illinois prairie. Many of the trees and shrubs along the roads and paths are labeled.

The Arboretum is the best spot in the Chicago region for bird finding in winter. Long-eared Owls and Saw-whet Owls are visitants. Hairy and Downy Woodpeckers, Blue Jays, Black-capped Chickadees, Tufted Titmice, White-breasted and Red-breasted Nuthatches, Brown Creepers, Cedar Waxwings, Northern Cardinals, Purple Finches, Pine Siskins, American Goldfinches, Slate-colored Juncos, and American Tree Sparrows can be seen regularly, and there is a good chance of observing Evening Grosbeaks and Common Redpolls. Bohemian Waxwings, Pine Grosbeaks, Red Crossbills, and White-winged Crossbills are visitants in certain winters.

In May, when the crab apples and hawthorns are blooming, waves of warblers and other passerine birds pass through the Arboretum, among them the Yellow-bellied Flycatcher, Red-breasted Nuthatch, Winter Wren, Hermit Thrush, Swainson's Thrush, Gray-cheeked Thrush, Veery, Golden-crowned and Ruby-crowned Kinglets, Solitary and Philadelphia Vireos, Black-and-white Warbler,

Golden-winged Warbler, Tennessee Warbler, Nashville Warbler, Northern Parula Warbler, Magnolia Warbler, Cape May Warbler, Black-throated Blue Warbler, Myrtle Warbler, Black-throated Green Warbler, Blackburnian Warbler, Chestnut-sided Warbler, Bay-breasted Warbler, Blackpoll Warbler, Palm Warbler, Northern Waterthrush, Wilson's Warbler, Canada Warbler, Rusty Blackbird, White-crowned Sparrow, White-throated Sparrow, and Fox Sparrow. The Connecticut Warbler is always a possibility; look for it in shrubby thickets.

To reach the Arboretum (address: Lisle, Illinois 60532), go west from the Loop on the Eisenhower Expressway to State 5 (East-West Tollway), and proceed west to State 53, which passes through the Arboretum. Turn north on State 53 and almost immediately enter the east entrance. Maps of the Arboretum are available at the east gatehouse.

DANVILLE
Forest Glen County Preserve

For bird finding in east-central Illinois, the **Forest Glen County Preserve** is unexcelled. Within its 1,800 acres on the west side of the Vermilion River south of Danville are beech-maple and oak-hickory woodlands, some stands of pine, meadows, old fields, brushy pastures, a small stretch of restored tall-grass prairie, and several small ponds. To reach the Preserve, take the Westville Exit from I 74, go south on State 1 to Georgetown, then turn east on Mill Street and follow directional signs to the entrance 6 miles beyond. After entering, stop at the visitor center for a map, and at the Willow Shores Nature Center for information on spots for particular birds.

Permanent-resident species include the Common Bobwhite, Common Screech, Great Horned, and Barred Owls, and Pileated, Red-headed, and Red-bellied Woodpeckers. Among the great array of summer-resident species are the following: *in wooded or brushy edges:* Red-tailed Hawk, Whip-poor-will (abundant), Great Crested, Acadian, and Traill's Flycatchers, Carolina Chickadee, Carolina Wren, White-eyed and Bell's Vireos, Blue-winged, Northern Parula, Cerulean, Yellow-throated, and Prairie Warblers, Louisiana Waterthrush, Kentucky Warbler, Yellow-breasted Chat, Orchard and Baltimore Orioles, Scarlet and Summer Tanagers, Rose-breasted Grosbeak, and Indigo Bunting; *in open areas:* (Prairie) Horned Lark, Eastern Bluebird (common; fifty nesting boxes available), Dickcissel, and Grasshopper, Henslow's, and Field Sparrows.

During winter, Rough-legged Hawks, Northern Harriers, and Short-eared Owls are usually regular visitants. In March and April, American Woodcock, a common nesting species, give their courtship displays above the old fields and brushy pastures at twilight and dawn. In May, small flocks of north-bound Lesser Golden Plovers are a common sight as they stop briefly on fields and croplands west of the Preserve.

DECATUR
Rock Springs Center for Environmental Discovery

Bordering the winding Sangamon River for 2 miles just southwest of this central-Illinois city is the **Rock Springs Center for Environmental Discovery.** The entrance, leading to the visitor center, is reached by driving south from Decatur on State 48, west on Rock Springs Road, then 0.5 mile north on Brazio Lane and turning left.

Within the Center's 1,000 acres, mostly forested, are bluffs with oak and hickory, ravines and stream margins with maple and basswood. Back from the wooded bluffs are grassy and shrubby areas that were formerly pastures and croplands. All habitats are accessible by 12 miles of foot trails starting from the visitor center.

The Pileated Woodpecker is a year-round resident. Among the sixty-some bird species breeding regularly are the Red-tailed Hawk, Whip-poor-will, Red-bellied and Red-headed Woodpeckers, Great Crested Flycatcher, Eastern Pewee, Carolina Wren, Eastern Bluebird, Loggerhead Shrike, Bell's and Warbling Vireos, Prothonotary Warbler, Orchard Oriole, Scarlet and Summer Tanagers, Rose-breasted Grosbeak, Indigo Bunting, Dickcissel, Grasshopper Sparrow, and Field Sparrow.

EAST ST. LOUIS

Frank Holten State Park | Cahokia Mounds State Park
Horseshoe Lake | Moredock Lake

This industrial city, across the Mississippi River from St. Louis, Missouri, lies on the broad floodplain. Rich farmlands near the city and on the steep bluffs farther east produce abundant crops. Despite the fact that the region is heavily settled, it has good localities for bird finding.

Frank Holten State Park (1,125 acres), a recreational area that was formerly swampland, is landscaped with deciduous trees and shrubs. Within its boundaries are three lakes of which the largest, Pittsburgh Lake, is now divided into two sections because of silting. In the tall weeds at its east end, permanent-resident Eurasian Tree Sparrows can always be found with a minimum of searching. Introduced in St. Louis in 1870, these Old World birds established themselves in the outskirts of that city and East St. Louis, and for a long time failed to spread elsewhere.

All three lakes attract transient waterbirds, waterfowl, and shorebirds. In fall (November) and spring (March), fifteen or more species of ducks may be seen in a single day. Mallards are the most abundant, and many remain throughout winter. Others to be expected are the Gadwalls, Common Pintails, American Wigeons, Lesser Scaups, and (throughout winter) Common Goldeneyes. Along with ducks, Pittsburgh Lake attracts also a few transient Common Loons, Horned Grebes, and Pied-billed Grebes. Trees and shrubs near the lakes are fine for transient passerines in April, early May, and September.

To reach Holten Park from I 70, take the Illinois Avenue Exit and proceed southeast on Illinois Avenue, bear left on State Street for 3.5 miles, and turn right on Kingshighway Avenue (State 111). This immediately intersects Lake Drive, which parallels the Park and provides access via several routes.

Four miles east of East St. Louis along Business US 40, the monotony of the flat country is unexpectedly broken by dozens of flat-topped Cahokia Mounds, relics of an ancient Indian race that existed a century or more before Columbus. A number of these mounds, including the largest, Monk's Mound, are in **Cahokia Mounds State**

Park (649 acres). Monk's Mound and most of the others are partly obscured by trees and shrubbery that have grown over them. Bird finding in the Park's unusual setting is especially good from 20 April to 10 May, when warblers are migrating. The Black-and-white, Tennessee, Nashville, Magnolia, Black-throated Green, Chestnut-sided, Bay-breasted, Blackpoll, and Palm are among the species often abundant.

To reach Cakohia Mounds Park from I 70, take the State 111 Exit and go south for a short distance, then east on Business US 40, which, after 2.5 miles, crosses the northern part of the Park and passes the entrance at the base of Monk's Mound.

Horseshoe Lake, northeast of the city, is a natural body of water owned by the state. Five miles in greatest length, it attracts a few waterfowl despite industrial development in its vicinity. East and north of the Lake are still-undisturbed marshes where there are a few nesting King Rails, Common Gallinules, American Coots, and Marsh Wrens. From I 70, take State 111 Exit, go north for 2.5 miles, then turn left on an unnumbered road bordering the north shore of the Lake. After scanning the open water and shoreline for birds, return to State 111 and continue north toward Roxana. Along the next 10 miles the highway traverses some marshes not yet drained, which should yield nesting birds as well as transients in spring and fall.

During wet seasons the low Mississippi bottomlands along the east side of the River, south of East St. Louis for over 50 miles to Prairie du Rocher, have many small ponds and sloughs drained by creeks edged with trees and shrubs. Attractive to ducks in spring and fall, the bottomlands (known locally as 'the levee area') are equally attractive to shorebirds—Semipalmated Plovers, Solitary Sandpipers, Greater and Lesser Yellowlegs, Pectoral Sandpipers, Least Sandpipers, Semipalmated Sandpipers, and others—in May, late August, and September; and to wading birds—Great Blue Herons, Green Herons, Little Blue Herons, Great Egrets, and Black-crowned and Yellow-crowned Night Herons—from April to October but with greater concentrations in August and September.

For a sampling of bird finding in the levee area, drive south from East St. Louis on State 3 to Waterloo, and west on State 156 for 10 miles to Valmeyer; then turn off and go north on a gravel road for

2.4 miles to marshy **Moredock Lake.** This is excellent for ducks, shorebirds, and herons. Park near the railroad and walk north along the tracks overlooking the Lake marshes from both sides. Watch particularly for Least Bitterns and Common Gallinules. Anyone wishing to explore the levee area further may return to Valmeyer and work his way south on the levee roads as far as Fort Chartres.

For Mississippi Kites, backtrack to State 3, turn south and continue for 31 miles to Ellis Grove, then turn right into Fort Kaskaskia State Park. The kites are known to nest here and can usually be seen, between late May and mid-August, along the Park road atop the bluffs overlooking the Mississippi River Valley or on the road along the River just below the Park.

GOREVILLE
Ferne Clyffe State Park

For bird finding in far-southern Illinois, one of the best of several parks is **Ferne Clyffe State Park** (1,073 acres), reached from I 57 by exiting at the intersection of State 148 and State 37, then turning south on State 37 to 1 mile beyond Goreville and the entrance on the left. Here a number of gorges and canyons radiate out from a central valley forested with hardwoods. Elsewhere in the Park are abandoned farmlands gradually acquiring a natural cover of shrubby thickets. More than 7 miles of foot trails lead through the Park's diversified habitats.

On any evening in late spring and early summer one may hear both Chuck-will's-widows and Whip-poor-wills. Among summer residents are Bewick's Wrens, Loggerhead Shrikes, eight species of warblers including the Worm-eating and Prairie, Orchard Orioles, both Scarlet and Summer Tanagers, Blue Grosbeaks, Indigo Buntings, and Chipping Sparrows. During migrations and throughout winter, the central valley is a haven for small landbirds.

GRAFTON
Pere Marquette State Park | Calhoun Division of the Mark Twain National Wildlife Refuge

At the confluence of the Illinois and Mississippi Rivers north of St. Louis, Father Jacques Marquette and his companions entered Illinois in 1673—the first white men to set foot on its soil. On State 100, 2 miles west of Grafton, a simple monument consisting of a rough-hewn stone cross marks the site. On the same road, 4.5 miles farther west, is the main entrance (right) to beautiful **Père Marquette State Park** (8,000 acres), largest of all Illinois state parks.

The terrain of the Park is hilly and, where limestone outcroppings form steep cliffs, even rugged. There are grassy upland meadows, richly forested bluffs, areas of thick brush, and, at the foot of the bluffs, flat, often flooded stretches of marshland. Among the sixty or more kinds of trees growing in the Park are silver maples, pecans, butternuts, and red cedars. Shadbush, redbud, wild plum, and dogwood, when flowering in spring, create a veritable flower garden.

An automobile road leads through the Park over successive hills that give splendid views of the Illinois River Valley. There are also miles of foot trails and bridle paths, along which, conveniently spaced, are facilities for picnicking and camping.

Four localities adjacent to or in the Park are especially recommended for bird finding: (1) *Graham Hollow.* From State 100, 2.4 miles west of Grafton and past the Marquette monument, turn right (north), proceed 1.3 miles, and park where the road forks. Explore here the brushy areas adjacent to Graham Creek and the woods on the nearby slopes for landbirds. (2) *Boat Harbor,* just off State 100, near Marquette Lodge and the Park's main entrance. From the parking areas here, scan the open water and marshy shoreline for waterbirds and waterfowl. (3). *Upper Picnic Area.* From Marquette Lodge, drive into the Park and uphill for 1.3 miles and park. Look for woodland birds along the bridle paths leading from this recreational area. (4) *Stump Lake.* Drive north on State 100 for 2 miles beyond the Park's main entrance; turn left on Dabbs Road, go by the pumping station, and park. Walk out on the levee, which provides a fine vantage point for wetland birds as it traverses the marshes and wooded bottomlands in this part of Stump Lake.

Birds nesting regularly in the Park and vicinity include the Red-tailed Hawk, Common Bobwhite, Yellow-billed Cuckoo, Barred Owl, Whip-poor-will, Pileated Woodpecker (unusually numerous),

Red-bellied Woodpecker, Red-headed Woodpecker, Great Crested Flycatcher, Carolina Wren, Wood Thrush, White-eyed Vireo, Warbling Vireo, Worm-eating Warbler (near Upper Picnic Area), Blue-winged and Prairie Warblers (Graham Hollow), Ovenbird, Kentucky Warbler, Yellow-breasted Chat, Orchard Oriole, both Scarlet and Summer Tanagers (on wooded slopes of Graham Hollow), Indigo Bunting, Rufous-sided Towhee, and Field Sparrow.

Green Herons, Wood Ducks, and Prothonotary Warblers commonly nest at Stump Lake, where Pied-billed Grebes, American and Least Bitterns, King Rails, Soras, Common Gallinules, and American Coots also breed. In August and September, Great Blue Herons and Yellow-crowned Night Herons, as well as a few Little Blue Herons, Great Egrets, Snowy Egrets, and Black-crowned Night Herons, visit the shallow waters around Boat Harbor. Except in mid-winter and summer months, there are hordes of Herring and Ring-billed Gulls and frequently Double-crested Cormorants and Ospreys. Only the Herring Gulls remain all winter.

South of Marquette State Park and across the Illinois River is Calhoun Point, formed by the junction of the Illinois and Mississippi Rivers. This area was formerly characterized by low ridges interspersed with flats, small marshes, and lakes. Bottomland forest covered most of the land above water level. Then in 1938, with the closing of the gates of the Alton Dam, downstream on the Mississippi, hundreds of acres were permanently flooded; thousands of trees were killed, new marshes were created, and the extent of the lakes increased. Today **Calhoun Division of the Mark Twain National Wildlife Refuge** embraces two of the re-formed lakes—Swan Lake (3,000 acres) on the south and west sides of the Illinois River, and Gilbert Lake (225 acres) on the east side opposite Marquette State Park—together with 1,700 acres of adjoining marshes, bottomland forest, and agricultural land.

The Refuge is notable ornithologically for its huge congregations of transient and wintering waterfowl, particularly Snow Geese, which are in greatest number in October and March and in somewhat lesser number through winter. A few wintering Bald Eagles are always a probable find.

Anyone wishing to visit the Refuge may cross the Illinois River on the free ferry (directional sign to the ferry on State 100 west of Grafton) and drive 2.8 miles to Refuge headquarters at Brussels, where he must obtain permission to enter the area.

HAMILTON
Montebello Conservation Park | Victory Park (Keokuk, Iowa)

One of the best vantage points along the Mississippi River for seeing an impressive number of wintering Bald Eagles is **Montebello Conservation Park** (3 acres), on the riverside north of Lock and Dam Number 19 and just north of Hamilton off State 96. An equally good vantage point is **Victory Park,** across the River in Keokuk, Iowa. This is reached by crossing the bridge to the center of town, then following Main Street east for 8 blocks and turning right into the Park, which overlooks the River. Look for Bald Eagles along the edges of open water or perched on nearby trees.

HAVANA
Chautauqua National Wildlife Refuge | Sand Ridge State Forest

The **Chautauqua National Wildlife Refuge** (5,124 acres) lies north of Havana in a natural overflow basin of the Illinois River. The main pool, Lake Chautauqua, which is impounded by 9.5 miles of dikes, covers about 3,500 acres. Outside the dikes are river bottoms, taking up nearly 500 acres, and sand-dune bluffs.

In spring and fall, enormous numbers of transient ducks gather on Lake Chautauqua, reaching peaks of abundance from mid-March to April and from mid-November to December. Among commonest ducks are Mallards (forming one of the greatest concentrations of the species in the state), American Black Ducks, Lesser Scaups, Common Goldeneyes, Ruddy Ducks, and Common Mergansers. Impressive numbers of Canada Geese and Snow Geese also stop off to feed and rest during their migratory passage in spring and fall.

In the river bottoms, which are forested with black willow, silver maple, sycamore, and cottonwood, breeding birds include Wood Ducks, Barred Owls, and Prothonotary Warblers. Either in these woods or on the adjacent sand-dune bluffs, where black oak and blackjack oak are the predominant tree growth, are other breeding birds such as Whip-poor-wills, Red-headed Woodpeckers, Carolina Wrens, Red-eyed and Warbling Vireos, American Redstarts, Orchard Orioles, Rose-breasted Grosbeaks, and Rufous-sided Towhees. During winter a few Bald Eagles stay on the Refuge or in its vicinity. In late afternoon they often come to roost in the big trees on the river bottoms.

Undoubtedly the top ornithological attraction is the huge concentration of shorebirds around Lake Chautauqua during the last of August and the first of September, when the water level is normally very low. Among some ten thousand shorebirds present one day in early September, as many as twenty species were noted. The Pectoral Sandpiper is usually the most abundant. Nearly as abundant are the Killdeer, Common Snipe, Spotted Sandpiper, Solitary Sandpiper, Lesser Yellowlegs, Least Sandpiper, and Semipalmated Sandpiper. Fairly common are the Semipalmated Plover, Greater Yellowlegs, Dunlin, Short-billed Dowitcher, Stilt Sandpiper, Buff-breasted Sandpiper, and Wilson's Phalarope. Other species often present include the Black-bellied Plover, White-rumped Sandpiper, Baird's Sandpiper, Long-billed Dowitcher, Western Sandpiper, and Sanderling.

To reach the Refuge from US 136, drive north for 9 miles on the eastern edge of Havana on Promenade Street (which soon becomes Manito Blacktop Road), and follow directional signs to headquarters. The best vantage points in the Refuge are accessible by roads and trails.

For a different area with a correspondingly different association of birds, visit **Sand Ridge State Forest,** not far northeast of the Chautauqua Refuge. Because roads leading to the area are unmarked and confusing to follow, the stranger is advised to return to Havana, drive east on US 136 for 12 miles, turn north for 5 miles to Forest City, then left on Forest City Road and left again on Sand Ridge Road to State Forest headquarters on the left. Obtain a map and advice on the most productive spots for bird finding.

Sand Ridge Forest embraces 6,389 acres of sandy hills covered by forests of native oaks and hickory. Here and there are abandoned farmlands, mostly planted with pines of various kinds, and small stretches of sand prairie. In late spring and summer, inspect the

LARK SPARROW

sand prairies, where Lark Sparrows are numerous. Watch all hedgerows for Dickcissels, Indigo Buntings, and an occasional pair of Blue Grosbeaks. Around buildings, look for Bewick's Wrens as well as House Wrens; both are common. In the forests, search for the Scarlet Tanager and Summer Tanager, as well as the Yellow-billed Cuckoo and Black-capped Chickadee. At dusk, Common Nighthawks in flight over open places are a familiar sight. During winter the pine plantations often attract Red Crossbills and sometimes White-winged Crossbills, too—in years when the trees bear a good crop of cones.

JACKSONVILLE
Areas for Eurasian Tree Sparrows

Eurasian Tree Sparrows are common on the outskirts of this city in west-central Illinois, as they are in the vicinity of St. Louis, Missouri, where they were introduced in North America over a hundred years ago (*see under* **East St. Louis**). To find the sparrows in spring and fall, take the following 5-mile loop trip:

From the intersection of US 36 with US 67 in Jacksonville, drive east on US 36 for about a mile; after a railroad underpass, take the first road south and follow it past the Jacksonville Country Club on the left and Mauvais Terre Lake on the right. At the next three intersections, turn right, circling the Lake and returning to Jacksonville via Vandalia Road. During this trip, watch all hedges and wires on which the sparrows may be sitting, also all dead stubs and fence posts in which they may be nesting. Like House Sparrows, they are easy to spot.

In fall and winter, the tree sparrows gather in flocks, sometimes numbering fifty to a hundred individuals, and spend much of their time in the hedges of multiflora rose, where they feed on the fruits, or center their activities elsewhere in brush piles and thickets. After corn is harvested in September, a reliable place to see them is around a grain elevator beside the railroad in the little town of Woodson, reached from Jacksonville by driving 8 miles south on State 67. The elevator rises east of the highway, where it is visible from a great distance. Among the dozens of House Sparrows that feed on grain spilled around the elevator, there are almost always a few Eurasian Tree Sparrows.

MONTICELLO
Robert Allerton Park

Robert Allerton Park, owned and operated by the University of Illinois, is an attractive estate of 1,685 acres along the Sangamon River in east-central Illinois. Though much of the Park is covered with mature upland forest, predominantly of white oak and hickory, there are other habitats, such as a woods of bur oak, sycamore, hackberry, silver maple, and many other kinds of trees, shrubby forest edges, open fields, and landscaped gardens.

Nesting species, especially varied because of the diversity of habitats, include the Green Heron, Wood Duck, Red-shouldered Hawk, American Woodcock, Barred Owl, Whip-poor-will, Red-bellied Woodpecker, Carolina Wren, Blue-gray Gnatcatcher, White-eyed Vireo, Bell's Vireo, Prothonotary Warbler, Yellow-breasted Chat, Baltimore Oriole, Indigo Bunting, and Henslow's Sparrow. The Park is notable for its abundance of transient passerines in April and early May.

Just south of the Monticello business district, a paved road branches west from State 105 to Allerton Park, 3.5 miles distant. All its habitats are accessible by roads with many pulloffs, and by trails.

MORRIS
Goose Lake Prairie Nature Preserve

Owned by the State Department of Conservation, the **Goose Lake Prairie Nature Preserve,** just east of Morris and 50 miles southwest of Chicago, harbors 1,513 acres of native prairie, one of the largest of such tracts remaining in Illinois. Besides prairie, other habitats are marshes with willow-bordered potholes, extensive thickets of hawthorn and prairie crabapple, and at least two groves of trees. Birdlife is consequently rich in variety.

Among birds nesting in the dry or wet grasslands are the Northern Harrier, Common Bobwhite, Upland Sandpiper, (Prairie) Horned Lark, Sedge Wren, Bobolink, Eastern Meadowlark, Red-winged Blackbird, Dickcissel, Savannah Sparrow, Grasshopper Sparrow, Henslow's Sparrow, Vesper Sparrow, and Field Sparrow. Pied-billed Grebes, Blue-winged Teal, King Rails, Marsh Wrens, and Swamp Sparrows reside in the marshes and marshy borders of potholes. Yellow-billed Cuckoos, Traill's Flycatchers, Bell's Vireos, Yellow Warblers, and Common Yellowthroats occupy the woody thickets. The time to visit the Preserve is June and early July, when all these species are nesting and the passerines are in full song.

The Preserve is reached by exiting west from I 55 toward Morris, continuing for 3 miles on Lorenzo Road, then turning north on Jugtown Road for 0.75 mile to the entrance and headquarters. Obtain directions here for trails to the best vantage points.

MT. CARMEL
Beall Woods Nature Preserve

In southeastern Illinois, along the Wabash River, which separates the state from Indiana, 290 acres of virgin upland and bottomland forests are owned and managed by the State Department of Conservation as the **Beall Woods Nature Preserve.** This is reached from Mt. Carmel by driving southwest on State 1 for 7 miles to Keensburg, then turning off east at the edge of town for 0.5 mile to the Red Barn Nature Center at the entrance. Here obtain a map of trails leading through Beall Woods and a list of birds one is likely to see.

Beall Woods is a splendid surviving example of the deciduous forest that once covered much of southern Illinois. Thus a walk through the woods is a walk back through time beneath untouched giants: some of the red oaks, green ashes, and sweet gums are the largest in the state. For the bird finder who has never explored a virgin forest, the walk is a unique experience. Birdlife, he discovers, lacks both variety and density. Moreover, big birds look smaller. The Pileated Wookpecker, a year-round resident, seems hardly larger than a creeper on a tree with a girth of 12 feet. The birds of the thickly foilaged upper canopy are so distant that they are difficult to discern.

The summer-resident birds are there, at least to be heard: the vireos, Yellow-throated, Red-eyed, and Warbling; the warblers, Northern Parula, Cerulean, and Yellow-throated; and the tanagers, Scarlet and Summer. Even the summer residents of the lower canopy, such as the Great Crested Flycatcher, Acadian Flycatcher, Eastern Pewee, and Wood Thrush, seem remote.

MURPHYSBORO
Cave Creek Valley

For a fine variety of breeding parulids, including the Swainson's Warbler, explore **Cave Creek Valley** in extreme southwestern Illinois. Drive south 9 miles from Murphysboro on State 127 into Shawnee National Forest; turn west (right) and go 2 miles to the village of Pomona; turn north (right) and proceed 1.0 mile on a dirt road paralleling a railroad as far as Cave Creek Bridge and park. Walk east on an old logging road that cuts through deciduous-wooded bottomlands, pine plantations, and dense growths of cane.

Listen and look for Swainson's Warbler in the canebrakes. Some of the other warblers in suitable habitats are the Prothonotary, Worm-eating, Blue-winged, Northern Parula, Yellow, Cerulean, Yellow-throated, Pine, Prairie, Louisiana Waterthrush, Kentucky, Common Yellowthroat, Yellow-breasted Chat, Hooded, and American Redstart.

NEWTON
Ralph E. Yeatter Prairie Chicken Sanctuary

In this section of the state—southeastern Illinois—Greater Prairie Chickens live in old pastures and fields, especially those producing redtop grass. The booming grounds are ordinarily in pastures and meadows, where the grasses are short, and are used year after year. Display activities begin here the first of February and last until June, but the period of greatest intensity is from about the last week of March to mid-April.

About 90 per cent of all Greater Prairie Chickens in Illinois are in sanctuaries in this part of the state, purchased from private, corporate, and public sources specifically for the breeding and restoration of this endangered species. One particularly productive setting is the **Ralph E. Yeatter Prairie Chicken Sanctuary** (77 acres), owned by the State Department of Conservation and managed by the Illinois Natural History Survey. To reach it from Newton, drive west on State 33 for 1.5 miles, then south on a blacktop road for 4 miles, and left 1.0 mile to the Sanctuary in the northeast corner of the next square mile on the right. No entry is permitted, but there are booming grounds that one may view from a car just outside.

The bird finder must park at the viewing site *before* daylight and stay in the car throughout the whole performance, which begins at the first light of day and lasts for about two hours. Remain quiet. At the slightest disturbance, the birds will disperse and not reappear that day. Furthermore, any disturbance such as getting out of the car, or leaving the site before the birds have ceased performing, may discourage their return later.

If one wishes to camp out the night before so as to be close by for a short pre-dawn drive to the site, free camping is permitted at the Chauncey McCormick Prairie Chicken Sanctuary, reached by continuing south on the east side of the Yeatter Sanctuary, then east 1.0 mile to the McCormick Sanctuary entrance on the right. If the bird finder is seriously interested in close-up observations and photography from blinds, he may make inquiry and reservations for their use by writing, well in advance of his visit, to the Office of the Illinois Natural History Survey, Effingham, Illinois 62401.

The prairie chicken sanctuaries have additional ornithological features. Throughout winter and into April they are one of the most reliable areas in the state for flocks of Smith's Longspurs. Later in the season the sanctuaries are breeding areas for a few Upland Sandpipers and many (Prairie) Horned Larks and Henslow's Sparrows.

OLIVE BRANCH

Horseshoe Lake Conservation Area and Nature Preserve

One of the greatest concentrations of transient and wintering Canada Geese anywhere in North America is in the **Horseshoe Lake Conservation Area** (7,901 acres), far down in Illinois near its southwesternmost extremity. From October through March, populations approaching an estimated hundred thousand have been observed. At the same time, Bald Eagles are often present in impressive numbers.

The Conservation Area, owned and administered by the State Department of Conservation, includes 1,200-acre Horseshoe Lake, where the geese congregate. To reach the Conservation Area, proceed southeast from Olive Branch on State 3. After 0.75 mile the highway parallels the eastern boundary of the Area. From here one may see immense numbers of geese on the Lake or in flight overhead as they move to and from their feeding grounds in the grain fields outside the boundary. At about 2 miles, a large sign on the right marks the entrance road. Follow the road to the edge of the Lake, where the birds may be easily viewed from the car.

The Conservation Area embraces 1,400 acres of timber, 494 acres of which are set aside in two tracts as the **Horseshoe Lake Nature Preserve.** One tract, on Horseshoe Island in the middle of the Lake, has virgin beech, sugar maple, and swamp chestnut oak, with bald cypresses and swamp tupelos standing in sloughs or on the Island's borders. A notable ornithological feature is the presence of Swainson's Warbler during summer. Fish Crows are present all year long. Horseshoe Island is accessible only by boat. Inquire at the ranger's office near the entrance about the use of one.

OTTAWA
Starved Rock State Park | Matthiessen State Park

In the north-central part of the state, **Starved Rock State Park** (2,366 acres) stretches for 5 miles along the southern bank of the Illinois River. Its scenic beauty and well-developed recreational facilities make it the most popular of Illinois' many state parks. Despite the fact that it is visited annually by thousands of people, there are many undisturbed areas in or near the Park that are productive ornithologically.

The Park comprises a long, narrow strip of wooded bluffland with sheer sandstone cliffs overlooking the River. One cliff, Starved Rock, rises 125 feet above the water. Cutting back from the River are more than twenty gorges, each with its own stream and often a waterfall. Though the vegetation is sparse on the blufftops, it is quite thick and luxuriant on the slopes and in the gorges. White pine, ironwood, maple, hickory, and oaks—black, white, red, and bur—are the predominant trees. The shrubby areas are composed of sassafras, sumac, cedar, shadbush, and, occasionally, redbud.

Bird finding is excellent in late April and early May, when many transient landbirds, especially warblers, pass through the area. The Connecticut Warbler is often present at this time. Species breeding regularly in the Park include the Wood Duck, Sharp-shinned Hawk, American Woodcock, Yellow-billed Cuckoo, Whip-poor-will, Red-bellied Woodpecker, Carolina Wren, Marsh Wren (in a cattail-edged lagoon at the eastern end of the Park), Wood Thrush, Blue-gray Gnatcatcher, Louisiana Waterthrush (in the gorges), Kentucky Warbler, Baltimore Oriole, and Rose-breasted Grosbeak.

On the Illinois River just opposite Starved Rock, where the water is backed up by locks, waterfowl gather in both spring and fall. Of the ducks, one of the commonest is the Lesser Scaup; others include Mallards, American Black Ducks, Common Goldeneyes, and Common Mergansers. In late summer, Great Blue Herons, Great Egrets, Little Blue Herons, and Black-crowned Night Herons are present. On the lowlands at the River's edge, shorebirds congregate in late April, May, and September. Common species are the Killdeer, Spotted Sandpiper, Solitary Sandpiper, Lesser Yellowlegs, and Least Sandpiper. A wood-bordered pond in the western end of the Park often attracts Common Loons in fall—until it freezes over—and again in spring.

To reach Starved Rock State Park from Ottawa, drive west on State 71 for 10 miles to the entrance on the right.

Two miles farther west, on the left side of State 71, is **Matthiessen State Park** (1,531 acres) on the Vermilion River. On this once-private estate the plant life and animal life have had many years of care and protection. There is only one gorge, but otherwise the terrain and vegetation are much the same as in Starved Rock Park.

SAVANNA

Savanna Army Ordnance Depot

For 14 miles along the Mississippi River, south of Lock and Dam Number 12 and within the confines of the **Savanna Army Ordnance Depot,** Bald Eagles appear during winter to catch fish in the ice-free water. From Savanna, drive north on State 84 for 9 miles, then turn left (west) onto the Depot's access road (marked by a sign) and proceed 1.5 miles to the Depot gate. Obtain permission to enter, and ask for directions to the River Road. This parallels the main channel of the Mississippi and affords the best vantage point for observing Bald Eagles.

Mississippi Palisades State Park, just north of Savanna on State 84, has observation points. Although generally less productive, the bird finder may wish to try them on his way to the Ordnance Depot.

SPRINGFIELD

Washington Park | Lake Springfield | Lincoln Memorial Garden and Nature Center | Carpenter Park | Lincoln's New Salem State Park

Washington Park, in the southwestern part of this capital city, is recommended for transient passerine birds. Comprising 150 acres of rolling terrain, largely forested with oaks and hickories, the area abounds in flycatchers, kinglets, thrushes, vireos, warblers, and fringillids during their migrations in April, early May, September, and early October. To reach Washington Park, take US Business 66 south from City Square, turn west on State 54 (South Grand Avenue), then north on MacArthur Boulevard for three blocks to the entrance on the left. Follow the winding road to the interior of the Park where there are two ponds. Park near either one and walk to the other. The woods between the two is the most productive area in the Park.

Artificial **Lake Springfield,** south of the city, is a must for bird finding. Its 4,270 acres and marginal lands are owned by the city and reserved for water supply and recreational purposes. Created in 1935 by the construction of Spaulding Dam across Sugar Creek (which flows into the Sangamon River), Lake Springfield was at once declared a wildlife refuge with no hunting allowed, and it has since added greatly to the ornithological importance of this landlocked prairie country by attracting vast numbers of waterbirds, waterfowl, and shorebirds. Extensive plantings of trees (sweet gums in wet ravines, cypresses in wet places along the shore, Scotch pines in eroded clay pastures), shrubs (dogwoods, shadbushes, redbuds), and many flowering plants also increased by the thousands the numbers of transient and breeding birds.

Mid-November and mid-March are the periods when concentrations of birds on the Lake are greatest. When the winter is mild and the water does not freeze over, many remain from fall to spring. Snow Geese and fewer Canada Geese are present in late October and November. Hordes of ducks gather in November and March. Mallards and American Black Ducks are the most abundant, but well represented are Gadwalls, Common Pintails, Green-winged and Blue-winged Teal, American Wigeons, Northern Shovelers, Ring-necked Ducks, Lesser Scaups, Common Goldeneyes, Buffleheads, Ruddy Ducks, Common Mergansers, and Red-breasted Mergansers. Along with waterfowl come numbers of Common Loons, Horned Grebes, Pied-billed Grebes, Herring Gulls, Ring-billed Gulls, and a few Bonaparte's Gulls. In April come the Terns—Common, occasionally Caspian, and Black.

Much of Lake Springfield's 57-mile shoreline is readily accessible by car. Of the various places for seeing birds, the following are recommended.

Take US Business 66 south from City Square, turn left on Stevenson Drive, crossing I 55, to East Lake Shore Drive, which almost immediately crosses Spaulding Dam. Since no parking is permitted on the Dam, park on either end and walk out onto it. The Dam is a good vantage point for seeing waterbirds and waterfowl on the open water of Lake Springfield. In May and from late April to early October, the algae-covered spillway and the muddy edges of several artificially created pools nearby attract shorebirds, particularly Semipalmated Plovers, a few Piping Plovers, Solitary Sandpipers, Greater and Lesser Yellowlegs, and Pectoral Sandpipers.

Continue on East Lake Drive, which eventually goes south along the east side of the Lake and passes signs marking **Lincoln Memorial Garden and Nature Center** (77 acres) on the Lake side of the road. Leave the car in the parking lot, enter the Garden, stop at the Nature Center Building for information, and walk along one of the trails, all of which lead down to the shore. The Garden, planted with native trees, shrubs, and flowers, holds in late spring and early summer a nesting population of Yellow Warblers, Yellow-breasted Chats, Northern Cardinals, Indigo Buntings, Rufous-sided Towhees, and Field Sparrows. From the shore, which overlooks the broadest part of Lake Springfield, excellent views of geese, ducks, and waterbirds may be obtained in fall and spring. Here the bird finder may hide himself in a thick cover and sometimes make remarkably close observations.

Across Lake Springfield from the Garden, another vantage point for seeing waterfowl and waterbirds is a tiny peninsula extending into the Lake. This may be reached by driving back on East Lake Drive and turning left onto the road to Vachel Lindsay Bridge. After crossing the Bridge, bear left and proceed southwestward along the west side of the Lake to an intersection and sign designating the Island Bay Yacht Club; then bear left again and follow this road to the peninsula. In late afternoon, when the sun is in the west, this peninsula is better for observations than Lincoln Memorial Garden.

North of Springfield on the Sangamon River is a woodland of different composition, a forest preserve of 438 acres called **Carpenter Park.** Here there are oak woods on the hilltops; sycamore, pawpaw, and redbud on the river bottoms. Sandstone outcroppings and steep ravines make the area exceptionally picturesque. It is one of the few places in central Illinois where Pileated Woodpeckers nest.

Other breeding birds are Wood Ducks, Barred Owls, Whip-poor-wills, Tufted Titmice (very common), Wood Thrushes, Warbling Vireos, Red-eyed Vireos, Prothonotary Warblers, Yellow-throated Warblers, Cerulean Warblers, Louisiana Waterthrushes, Kentucky Warblers, Scarlet Tanagers, Rose-breasted Grosbeaks, and Indigo Buntings. In April and May the area is alive with transient warblers and other small birds.

To reach Carpenter Park, drive north from Springfield on US Business 66 for 4 miles. After crossing the long bridge over the Sangamon, take the first left turn. No cars are permitted beyond the short entrance road to a picnic area, but trails lead to all areas. For another view of the River, one may turn left into Riverside Park just before crossing the bridge over the Sangamon, and follow the road along the River for possible glimpses of herons and Wood Ducks.

Lincoln's New Salem State Park (522 acres), 20 miles northwest of Springfield, is reached by taking State 97 northwestward. The Park contains an authentic restoration of the village of New Salem, where Abraham Lincoln spent six years of his life, and occupies a hill overlooking the Sangamon River Valley.

Just south of the village is a fine oak-hickory woods with many winding trails, where in late spring and early summer one may observe a good variety of nesting birds. At night Whip-poor-wills are noisy, and a few Great Horned Owls may be heard. Among diurnal birds appearing regularly are Red-bellied Woodpeckers, Acadian Flycatchers, Tufted Titmice, Wood Thrushes, Warbling Vireos, Ovenbirds, Kentucky Warblers, American Redstarts, Scarlet Tanagers, Northern Cardinals, Rose-breasted Grosbeaks, and Rufous-sided Towhees.

WAUKEGAN
Illinois Beach State Park

Illinois Beach State Park (2,746 acres), stretching 3.5 miles along the Lake Michigan shore between Waukegan and Zion, has a sandy beach backed by low dunes and ridges running parallel to each other. Many of these rises are covered with juniper mixed sparingly with stands of white birch, aspen, and white pine; others are covered with ash and black oak. Among the dunes and ridges are marshes of varying size, often bordered by dense shrubs and drained by sluggish streams.

The southern section of the Park is a Nature Preserve containing some of the finest of its natural features. Eastern Pewees, Red-eyed Vireos, Yellow Warblers, American Redstarts, American Goldfinches, and Rufous-sided Towhees are among the common breeding birds seen in the wooded and shrub-bordered areas; American Bitterns, King Rails, Soras, Common Gallinules, and Red-winged Blackbirds, in the larger marshes.

The Park is exceptionally good for bird finding during migrations in spring and fall—seasons when it is not so greatly disturbed by recreational activities. In April and early May, again in September and early October, transient small landbirds of many kinds are prevalent in woods and thickets. During May and in late summer to mid-October, shorebirds such as Black-bellied Plovers, Ruddy Turnstones, Dunlins, and Sanderlings put in an appearance on the beach. From late August to the first of December, many hawks pass south over the shore, just as they do elsewhere along the west shore of Lake Michigan (*see also under* **Port Washington, Wisconsin**). When winter sets in, flocks of Snow Buntings and northern races of the Horned Lark enliven the open scene on both the beach and dunes.

To reach the Park entrance, drive north from Waukegan on Sheridan Road (State 42) for 4.5 miles and turn right on Beach Road. This leads eastward into the Park toward the Lake Michigan shore, passing north of the Nature Preserve, which is accessible from the Nature Preserve parking lot by a self-guiding nature trail.

Indiana

SANDHILL CRANES

When pussy willows, jack-in-the-pulpits, and spring beauties are blooming in Indiana, hundreds of Sandhill Cranes move northward through the state, lingering for a while at the Jasper-Pulaski State Fish and Wildlife Area. As at the Muleshoe National Wildlife Refuge in Texas, the Malheur National Wildlife Refuge in Oregon, and the Platte River in western Nebraska, here in northwestern Indiana these great birds provide a special treat for the bird finder. Wherever Sandhill Cranes gather, he may see them circle to heights of a mile or more, hear them give their far-reaching, bugle-like calls, and watch them perform their fantastic dances. Few other North American birds offer so stirring an exhibition. Fortunately, this Indiana gathering is remarkably accessible, being only 30 miles south of the Indiana Toll Road crossing the state, and 50 miles southeast of Chicago.

Compared with the majority of the states, Indiana shows few changes in terrain. The northern third varies from level to slightly rolling, and is sprinkled with many lakes. Formerly there were many and extensive cattail marshes and wooded swamps, but few remain today. The middle third of the state is markedly level—so level that one may drive for hours without seeing an elevation greater than a haystack. By contrast, the southern third is strikingly scenic. Running southward and opening out into the Ohio River Valley are valleys deep and sometimes very narrow, and bounded by high bluffs with bold escarpments. Streams course through swiftly, occasionally falling in beautiful cataracts, occasionally swelling into mirrored pools. The slope of Indiana is generally southwest, so that most of the drainage is into the Ohio and Mississippi Rivers. Only in the northernmost reaches of the state is the drainage into the Great Lakes.

Indiana was originally heavily timbered. Oak, beech, maple, walnut, and hickory covered the highland areas; sycamore, sweet gum, and tulip tree were common in the river valleys. A few hemlocks and yews often fringed the valley streams and clung to the shaded slopes. Contrary to a prevailing notion, Indiana was not a prairie state, though in the northwest section there was a small stretch of prairie or grassland. Because Indiana is important agriculturally, its forests are confined to river bluffs, privately owned woodlots, state parks, and public preserves. Birds characteristic of farmlands are more numerous than are birds of deciduous forests. Some of the regular breeding species in the forests and farmlands (fields, wet meadows, brushy lands, orchards, and dooryards) throughout the state are the following:

DECIDUOUS FORESTS

Red-shouldered Hawk
Yellow-billed Cuckoo
Black-billed Cuckoo
Common Screech Owl
Barred Owl
Whip-poor-will
Yellow-shafted Flicker
Red-bellied Woodpecker
Hairy Woodpecker
Downy Woodpecker
Great Crested Flycatcher
Acadian Flycatcher
Eastern Pewee
Blue Jay

Black-capped Chickadee (*northern Indiana*)
Tufted Titmouse
White-breasted Nuthatch
Wood Thrush
Blue-gray Gnatcatcher
Yellow-throated Vireo
Red-eyed Vireo
Warbling Vireo
Ovenbird
American Redstart
Baltimore Oriole
Scarlet Tanager
Rose-breasted Grosbeak (*mainly northern Indiana*)

FARMLANDS

Common Bobwhite
Mourning Dove
Red-headed Woodpecker
Eastern Kingbird
Eastern Phoebe
(Prairie) Horned Lark
Tree Swallow
Barn Swallow
House Wren
Carolina Wren
Northern Mockingbird
Gray Catbird
Brown Thrasher
Eastern Bluebird
Loggerhead Shrike
Common Yellowthroat
Bobolink (*except southern Indiana*)
Eastern Meadowlark
Common Grackle
Northern Cardinal
Indigo Bunting
Dickcissel
American Goldfinch
Rufous-sided Towhee
Savannah Sparrow (*mainly northern Indiana*)
Grasshopper Sparrow
Henslow's Sparrow
Vesper Sparrow (*except southern Indiana*)
Chipping Sparrow
Field Sparrow
Swamp Sparrow (*northern Indiana*)
Song Sparrow

Southern Indiana lies within the breeding range of a number of birds inhabiting southern environments. Some of these breeding regularly are listed below:

Turkey Vulture
Black Vulture
Chuck-will's-widow
Carolina Chickadee
Bewick's Wren
White-eyed Vireo
Prothonotary Warbler
Worm-eating Warbler

Yellow-throated Warbler
Louisiana Waterthrush
Kentucky Warbler
Yellow-breasted Chat
Hooded Warbler
Summer Tanager
Bachman's Sparrow

Where northwestern Indiana borders on Lake Michigan there are the wondrous Indiana Dunes, for many years a source of fascination to both naturalists and laymen. Here are sandy hillocks of varied sizes and shapes, some covered with vegetation and some barren, some stationary and some traveling steadily, burying everything in their path. Here are ponds, marshes, and meadows. Here are masses of wildflowers blooming profusely in May and June. Birdlife is not abundant, but this lack is offset by an extraordinary variety of transient and breeding species installed in an equally extraordinary environment. Let the bird finder visit the Indiana Dunes in May and see for himself.

Waterbirds and waterfowl find suitable breeding grounds in Indiana. Natural lakes, though numerous in the extreme north, are either lacking in food requirements or too greatly disturbed by resort activities. The vast northern swamps, made famous through Gene Stratton-Porter's *A Girl of the Limberlost*, have been almost completely destroyed. For nesting waterbirds and waterfowl, the bird finder must be content with only a few marshy regions such as those in and about Pokagon State Park (*see under* **Angola**).

The Ohio River on the southern border and the Wabash and White River Valleys cutting northward and eastward through the state are natural lanes of migration. Waves of small landbirds can be expected at various points along the bluffs; waterbirds, waterfowl, and shorebirds on the mud flats, floodplain pools, and backwaters. The impounded waters in different parts of the state, some forming reservoirs such as the Geist Reservoir outside Indianapolis, and other marshes, such as the 1,400-acre marsh in the Jasper-Pulaski Fish and Wildlife Area (*see under* **Medaryville**), attract impressive numbers of water-loving birds. These watery places, readily accessible, are usually excellent for bird finding in both spring and fall, and especially good throughout winter, unless they freeze over.

Peak flights of transient birds take place in Indiana within the following dates:

Waterfowl: 1 March–10 April; 15 October–1 December
Shorebirds: 1 May–1 June; 1 August–5 October
Landbirds: 15 April–15 May; 1 September–25 October

During winter, (Northern) Horned Larks, Lapland Longspurs, and Snow Buntings visit with fair regularity the beaches and treeless dunes in the Indiana Dunes region. Wherever there are conifers and fruit-bearing shrubs in the state, Evening Grosbeaks, Purple Finches, and Pine Siskins are prevalent. Throughout winter in all parts of the state, Slate-Colored Juncos, American Tree Sparrows, and White-throated Sparrows are common visitants.

Authorities

Edna Banta, Charles K. Barnes, William B. Barnes, Earl Brooks, Irving W. Burr, Mildred F. Campbell, James B. Cope, Merritt C. Farrar, Dorothy M. Hobson, William R. Overlease, Jon E. Rickert, Iva Spangler, Lawrence H. Walkinshaw, Henry C. West.

Reference

An Annotated Check List of Indiana Birds. By Russell E. Mumford and Charles E. Keller. In *Indiana Audubon Quarterly,* vol. 53 (1975), no. 2. Contact the Indiana Audubon Society through its Mary Gray Sanctuary, Connersville, Ind. 47331.

ANGOLA
Pokagon State Park

Five miles north of this community in extreme northeastern Indiana, **Pokagon State Park** comprises 1,175 acres on the east side of Lake James, and is readily accessible from the Indiana Toll Road. From the Angola Exit go south on I 69 and take either the first or second exit west to the Park. Much of the Park consists of second-growth deciduous woodlands and old farmlands reverting to herbaceous perennials, shrubs, and trees. Among breeding birds are the American Woodcock, Red-bellied Woodpecker, Red-headed Woodpecker, Eastern Kingbird, Wood Thrush, Veery, Warbling Vireo, Yellow Warbler, Baltimore Oriole, Scarlet Tanager, Northern Cardinal, Rose-breasted Grosbeak, Indigo Bunting, American Goldfinch, Rufous-sided Towhee, and Field Sparrow.

Entirely within the Park is Lake Lonidaw, with its surrounding cattail and tamarack marshes—the preferred habitat of American and Least Bitterns, King and Virginia Rails, Soras, Marsh Wrens, Red-winged Blackbirds, and Swamp Sparrows.

BLOOMINGTON
Lake Lemon | Yellowwood Lake

Lake Lemon, an artificial impoundment of 1,500 acres northeast of this city in south-central Indiana, attracts many transient waterbirds and waterfowl in March/April and October/November; some species remain throughout winter. From Bloomington, drive east and north on State 45 for 8 miles to Unionville, turn left on Shuffle Creek Road for 2 miles to the Lake, where the Road forms a T. Park here and scan the open water. Continue by either turning left to the 'lower' part of the Lake or right to the 'upper.' Either way, some of the birds most likely to be in view are the Horned Grebe, Pied-billed Grebe, Canada Goose, American Black Duck, Common Pintail, Green-winged and Blue-winged Teal, American Wigeon, Northern Shoveler, Redhead, Ring-necked Duck, Canvasback, Lesser Scaup, Common Goldeneye, Bufflehead, Osprey, American Coot, Herring Gull, Ring-billed Gull, Bonaparte's Gull, Common Tern, and Black Tern.

Yellowwood Lake, another artificial impoundment, of 147 acres east of Bloomington, is reached by driving east on State 46 for 10 miles to the village of Belmont; continue east on State 46 for 0.6 mile, then turn left on an unmarked gravel road for 1.4 miles to a narrow parking area where the road makes a sharp turn right. Leave the car here and walk around the Lake, passing below the dam, for 1.3 miles. The site affords many vantage points for viewing transient waterbirds and waterfowl, which are roughly the same species as on Lake Lemon, though fewer in the number of individuals. At the same time the road passes through second-growth woods and their shrubby borders, which are excellent for transient warblers in May and September and for many summer-resident birds such as Brown Thrashers, Yellow Warblers, Cerulean Warblers, Prairie Warblers, Common Yellowthroats, Yellow-breasted Chats, Indigo Buntings, and Song Sparrows.

CHESTERTON
Indiana Dunes State Park

The Indiana Dunes border Lake Michigan for 25 miles between Gary and Michigan City and extend from one to three miles inland. About 6,000 acres of this unusual area are given over to the Indiana Dunes National Lakeshore, and of this tract 2,182 acres comprise **Indiana Dunes State Park,** less than 50 miles from Chicago. To reach the Park, take the Chesterton Exit from either the Indiana Toll Road or I 94 and proceed north on State 49 to its terminus in the Park. All parts of the Park are accessible by foot trails leading away from the entrance area. Although the Park serves as a recreational site for an immense metropolitan district, it is so vast that there is no difficulty in obtaining solitude for bird finding.

Three parts of the Park are particularly attractive to birds:

1. The beach between Lake Michigan and the dunes. In May, August, and September, this is good for transient shorebirds such as the Semipalmated Plover, Ruddy Turnstone, Pectoral Sandpiper, Least Sandpiper, Dunlin, Semipalmated Sandpiper, and Sanderling. In June and July a few pairs of Piping Plovers may nest where there is a minimum of disturbance. In late fall, until Lake Michigan freezes over, and in early spring, after the Lake opens up, ducks offshore usually include Common Goldeneyes, Buffleheads, Common Mergansers, Red-breasted Mergansers, and an occasional Oldsquaw and White-winged Scoter.

2. The wooded dunes inland from the barren dunes. White pine, jack pine, red cedar, basswood, red oak, and black oak are the predominant tree growth. In some places the woods are open, with a dense undergrowth of blueberry and other shrubs. Among summer-resident birds are the Red-shouldered Hawk, Common Bobwhite, Red-headed Woodpecker, Great Crested Flycatcher, Veery, Cedar Waxwing, Black-and-white Warbler, Chestnut-sided Warbler, Pine Warbler, Prairie Warbler, Baltimore Oriole, and Northern Cardinal.

3. The wet lowlands—grassy meadows, cattail marshes, cranberry bogs, and ponds—bordered by shrubs and by such trees as tamarack, birch, and maple. Lying between the wooded dunes and extending inland from them, this section is by far the most rewarding because of its combination of environments. From the list of species

known to nest here, the following suggest the possibilities: Pied-billed Grebe, Least Bittern, Green Heron, Blue-winged Teal, Wood Duck, King Rail, Virginia Rail, Traill's Flycatcher, Marsh Wren, Sedge Wren, Common Yellowthroat, Bobolink, Eastern Meadowlark, and Swamp Sparrow. In migration seasons, thrushes, kinglets, vireos, warblers, and fringillids gather in the bordering shrubs and trees; waterbirds and waterfowl congregate on the ponds; and shorebirds—e.g., Solitary Sandpipers and dowitchers—frequent the muddy shallows.

CONNERSVILLE
Mary Gray Bird Sanctuary

Southwest of Connersville in east-central Indiana is the **Mary Gray Bird Sanctuary,** a tract of 684 acres owned by the Indiana Audubon Society and open to the public every day of the year. To reach the Sanctuary, drive south from Connersville on State 121 for 3.5 miles, then right on State 350S to the entrance 3 miles distant. A gravel road winds into the Sanctuary.

About 100 acres of the Sanctuary consist of woodland, with stands of both virgin and second-growth timber, primarily beech-maple climax along with many large tulip trees and a few mature butternuts. The rest of the tract, except for the grounds surrounding headquarters, consists of cultivated fields and pastures dotted with shrubs and brambles.

The Sanctuary has no watery areas large enough to attract waterfowl and shorebirds, but it is very attractive to landbirds. Among the species probably nesting are the Broad-winged Hawk, Common Bobwhite, Whip-poor-will, Pileated Woodpecker, Red-bellied Woodpecker, Great Crested Flycatcher, Acadian Flycatcher, Carolina Chickadee, Carolina Wren, White-eyed Vireo, Cerulean Warbler, Yellow-throated Warbler, Kentucky Warbler, Summer Tanager, and Field Sparrow. Transient flycatchers, thrushes, vireos, warblers, and fringillids are often abundant in early May and in September.

RED-BELLIED WOODPECKER

INDIANAPOLIS

Woollen Gardens Nature Preserve | Geist Reservoir | Holcomb Gardens | Maywood | Sewage Disposal Plant | Eagle Creek Park

This capital city, in the midst of flat agricultural country, offers a variety of birds in habitats ranging from woodland to open water. In the northeastern part of the city are the municipally owned **Woollen Gardens Nature Preserve** and the Indianapolis Water Company's **Geist Reservoir**—both highly recommended for bird finding.

Woollen Gardens is reached from ring-road I 465 by exiting west on 56th Street, then turning north on Brendon Forest Drive to its terminus. From here a foot trail leads into the Gardens. This 44-acre tract along the banks of Fall Creek is forested with giant tulip trees, beeches, maples, and oaks, and carpeted in spring with many wildflowers. A few Wood Ducks nest in the section north of Fall Creek. In spring and fall, waves of migrating small landbirds move through the trees on both sides of Fall Creek.

The Geist Reservoir, formed by damming Fall Creek farther northeast, is reached from the ring-road I 465 by exiting on Shadeland Road (State 100) for 0.5 mile, then turning east on Fall Creek Road, which follows the west bank of Fall Creek to Geist Dam—a good vantage point for viewing a part of the Reservoir. Continuing along Fall Creek Road, one may drive around the Reservoir for additional views. The Reservoir is approximately 7.5 miles long and 1.5 miles across at its widest point. Posted as a wildlife refuge, it is a haven from mid-October to mid-March for Canada Geese, Mallards, American Black Ducks, Gadwalls, Common Pintails, Ring-necked Ducks, Canvasbacks, Lesser Scaups, Common Goldeneyes, Buffleheads, Ruddy Ducks, Common Mergansers, and Red-breasted Mergansers. Pied-billed Grebes and American Coots are often present. A few Sandhill Cranes show up in March and early April. Great Blue Herons, Green Herons, and Great Egrets appear frequently in August and September.

Like the Reservoir itself, the more than 5,000 acres surrounding the Reservoir—and also owned by the Water Company—are posted as a wildlife refuge. Plantations of conifers, deciduous trees, and shrubs induce the nesting of many common species and serve as a wintering habitat for such species as the Purple Finch, Pine Siskin, Slate-colored Junco, American Tree Sparrow, White-throated Sparrow, and other fringillids. Christmas bird counts in the area usually record about seventy species. At least seventeen species of warblers can be noted at the height of the warbler migration from 5 to 15 May.

The Fairview Campus of Butler University (246 acres) has rolling hills forested with beech, maple, oak, and hickory, and is especially inviting to transient small landbirds, particularly warblers. Stand at the top of a hill and watch the birds passing through the treetops below. Within the campus is **Holcomb Gardens,** a formal garden bordering a canal and towpath. This is one of the best places in the area for migrating landbirds. Cuckoos, vireos, and warblers often nest in the Gardens during summer. Drive north from downtown on US 31 and turn west on 52nd Street to the Gardens on the left.

Amos W. Butler, foremost Hoosier ornithologist, considered an area called **Maywood,** on the bank of the White River, an excellent spot for birds, and it is outstanding for spring and fall transients. One may walk along a path on the River bluffs through big woods where Pileated Woodpeckers nest, or along the River looking for geese, ducks, and wading birds. Prothonotary Warblers nest in the area.

Southwest Way Park, a large county park, now lies directly north of Maywood, and access to Maywood is through the Park. This is reached by going southwest from the city on State 67 and turning left (south) on Mann Road, or by exiting from I 465 south on Mann Road.

Although the lagoons at the **Sewage Disposal Plant** are no longer used, bird finding here is still very good and bird finders are welcome. From 30 July to 1 November, shorebirds feed here by the hundreds. Among them are Ruddy Turnstones, Spotted Sandpipers, and Sanderlings. To reach the Plant, drive southwest from downtown on State 67, turn left (south) on South Belmont Street and immediately left on Raymond Street to the entrance, where one must register before entering.

Eagle Creek Park, on the east side of Eagle Creek Reservoir northwest of the city, is a fine place to look for migrating waterbirds and waterfowl. Drive northwest from the city on I 65, exit at Traders Point in the northern tip of the Park, and drive south along the Reservoir to a dam at the southern end. There are parking places overlooking the water all along the way. For one who wishes to look for landbirds, there is a nature center where he may obtain a map of the trails, together with information on the best areas to explore.

MADISON
Clifty Falls State Park

On the bluffs of the Ohio River in southeastern Indiana is 1,200-acre **Clifty Falls State Park,** noted for its rugged charm. At the north end of the Park, Clifty Falls tumbles 70 feet into a deep-cut canyon, which receives sunlight only at midday. Splendid forests of beech and oak cover the bluffs; tall sycamores, tulip trees, and sugar maples stand on the shaded canyon floors.

Clifty Creek Canyon and its tributary canyons are the wintering places of large numbers of Turkey Vultures and Black Vultures. A few pairs of both species remain through summer, nesting on the rock ledges near Tunnel Falls. In the perpendicular walls of fossiliferous limestone near Clifty Falls, Rough-winged Swallows nest in the crystal-lined centers of fossil coral colonies. Protected by overhanging ledge shelves, Eastern Phoebes nest on the surfaces of the same walls.

Among other birds in the Park during nesting season in June and July are the following: *along the cascades above Clifty Falls:* Louisiana Waterthrushes; *in the woods of the canyon floor below Tunnel Falls:* Acadian Flycatchers, Wood Thrushes, and Summer Tanagers; *in the margins of secondary deciduous forests along Park Drive on the bluffs above the canyons:* Red-bellied Woodpeckers, Eastern Pewees, Carolina Chickadees, Carolina Wrens, Blue-gray Gnatcatchers, White-eyed Vireos, Cerulean Warblers, Yellow-breasted Chats, Indigo Buntings, and Rufous-sided Towhees.

The Park entrance is 1.5 miles west of Madison on State 56 and 62. Roads and foot paths within the Park make all good bird-finding areas readily accessible. A resident naturalist is on duty during summer.

MEDARYVILLE
Jasper-Pulaski State Fish and Wildlife Area

The **Jasper-Pulaski State Fish and Wildlife Area,** a preserve of 7,585 acres, where the Sandhill Cranes gather in northwestern Indiana, can be reached by going north from Medaryville on US 421 for 3 miles, or south from the Indiana Toll Road for 30 miles, and then west on State 143 for 1.5 miles to the entrance. At headquarters, obtain a map showing the roads leading to the parking areas for watching the Sandhills and inquire about blinds for close-up observation and photography.

The chief attraction for the Sandhills is the 1,400-acre marsh. Here the great birds begin appearing around 20 March, increase by the hundreds during the remainder of the month, until, by the first week of April, they attain maximum number. Between roughly 10 and 20 April, they all vanish, often within 24 hours.

The best times for observation are early morning and late afternoon. In the morning they fly in flocks from their roosting sites in the marsh for feeding in the adjacent grainfields, or to an extensive open area with pools in the preserve—in view from the parking areas—for grain previously planted. In the evening they return. While watching them (a telescope will be useful), one should stay in the car. Shy birds, they take off immediately at the slightest disturbance. On a sunny day they may depart for no obvious reason, circling skyward as they ride the thermals, and soon disappear out of sight. At any place and at any time, a few birds in a group may suddenly 'dance' in the unique fashion of all cranes, jumping up and down and pivoting with much wing-flopping.

The fall gathering of Sandhills, while impressive, is a third smaller in number of individuals. The population begins building up after mid-October and reaches its greatest size in the first two weeks of November. Unlike the spring transients, many individuals linger for a longer period, a few until as late as the first week of December.

While the Sandhills are at the preserve there are other ornithological attractions. In spring, American Woodcock flight-sing after sundown and again before sunrise. A resident flock of Canada Geese is joined by many transients of their kind. Snow Geese show up and stay a while in early April and appear again in October. Transient waterfowl include also an abundance of Mallards and American Black Ducks, together with Common Pintails, Green-winged and Blue-winged Teal, American Wigeons, Ring-necked Ducks, Lesser Scaups, Common Goldeneyes, and Ruddy Ducks. Frequenting the marsh in migration, though usually later in spring and earlier in fall than the Sandhill Cranes, are Pied-billed Grebes, Soras, Common Snipes, and American Coots. A few stay throughout summer for nesting.

MT. VERNON
Hovey Lake State Fish and Wildlife Area

In one of the 'pockets' of extreme southwestern Indiana are 4,400 acres belonging to the **Hovey Lake State Fish and Wildlife Area.**

Part of the acreage is a refuge and closed to the public; the remainder is a hunting area and open to the public the year round.

Hovey Lake is spring-fed but annually receives inundations from the Ohio River. Bald cypress, pecan, pumpkin ash, swamp privet, and other trees and shrubs more commonly found in regions farther south surround the Lake, giving it a semitropical appearance. Wood Ducks, Pileated Woodpeckers, and Prothonotary Warblers nest regularly; Double-crested Cormorants may breed. Great Egrets and other wading birds are numerous in late summer, and many thousands of geese and ducks over-winter.

The preserve may be reached by driving south from Mt. Vernon on State 69 for 10 miles. For bird finding on Hovey Lake, state-owned boats may be rented.

PAOLI
Pioneer Mothers Memorial Forest

Maintained by the United States Forest Service as part of Hoosier National Forest, **Pioneer Mothers Memorial Forest** (254 acres) in south-central Indiana contains about 116 acres of entirely virgin hardwoods, predominantly oak, beech, and black walnut—the largest such tract in Indiana. For woodland birds typical of the deciduous forests that once covered southern Indiana, the bird finder can do no better than take a trip here.

Locally known as Cox's Woods, the Forest is conveniently reached from Paoli by either of two routes: (1) Drive south from the courthouse for 2.1 miles on State 37 to the Pioneer Mothers Rest Park. (2) Drive east for 1.3 miles on US 150, turn right at the Forest's blacktop driveway—open only from April to mid-October—and proceed to the parking area. From either stopping point, foot trails lead through the Forest.

As in all mature virgin forest stands, with uniformity of habitat, bird species are limited in variety. Year-round residents include the Cooper's Hawk, Red-tailed Hawk, Great Horned Owl, Barred Owl, Whip-poor-will, Pileated Woodpecker, Red-bellied Woodpecker, Hairy Woodpecker, Downy Woodpecker, Carolina Chickadee, Tufted Titmouse, and White-breasted Nuthatch. Among summer residents are the Yellow-shafted Flicker, Great Crested Flycatcher, Acadian Flycatcher, Eastern Pewee, Wood Thrush, Red-eyed Vireo, Black-and-white Warbler, Ovenbird, and Summer Tanager.

ROCKVILLE
Turkey Run State Park

Turkey Run State Park, comprising 2,181 acres in west-central Indiana, along both sides of Sugar Creek for at least two miles, provides good landbird finding in one of the most picturesque spots in the state. The terrain is cut by deep canyons, with walls nearly 100 feet high. Almost all of the area is heavily forested—hemlock and yew on the canyon rims; oak, maple, hickory, beech, and walnut elsewhere on high ground; and sycamore, boxelder, and basswood on the lowlands.

Since there are only a few open areas in the Park, summer-resident birds are largely woodland species. Whip-poor-wills and Wood Thrushes are especially abundant. Among other species are the Acadian Flycatcher, Carolina Wren, Blue-gray Gnatcatcher, Warbling Vireo, Prothonotary Warbler, Worm-eating Warbler, Cerulean Warbler, Yellow-throated Warbler, Prairie Warbler, Louisiana Waterthrush, Kentucky Warbler, Yellow-breasted Chat, American Redstart, Scarlet Tanager, and Summer Tanager.

Bird finding in winter offers an unusual feature—Turkey Vulture roosts. For as long as local people can remember, Turkey Vultures, numbering well over 100, gather by tradition from the surrounding countryside to pass the night in the same sites on the canyon walls. In winter there is also an opportunity to see, in one day, as many as seven woodpeckers: Pileated, Red-bellied, Red-headed, Hairy, and Downy, with one or more Yellow-shafted Flickers that failed to move south, and several Yellow-bellied Sapsuckers that have come from the north.

The Park can be reached from Rockville by driving north on US 41 for 8 miles, then right on State 47 for 1.5 miles to the entrance. Footpaths wind through the Park, giving access to all the good bird-finding areas. During summer a resident naturalist conducts field trips.

SEYMOUR
Muscatatuck National Wildlife Refuge

Another link in the chain of National Wildlife Refuges established for the benefit of migrating waterfowl—and, incidentally, rewarding to bird finders—is **Muscatatuck National Wildlife Refuge** (7,724 acres) in south-central Indiana. Mutton Creek and Storm Creek provide water habitat for waterbirds and waterfowl within the Refuge. The Muscatatuck River forms the southern boundary. Roads lead from Refuge headquarters south across the Wood Duck Trail and to public fishing ponds. The Wood Duck Trail (about an hour's walk) winds between and around ponds and past a small marsh where a display flock of Canada and Snow Geese and Mallards attracts many transient waterfowl during migrations, and guarantees some bird finding at any time of year. Elsewhere on the Refuge are deciduous woods, both fallow and cultivated fields, brushy thickets, and hedgerows that invite a variety of landbirds.

Although the Refuge is too recently established to have developed a comprehensive checklist of birds in all seasons, a recent Christmas bird count, on one mid-December day, showed the presence of as many as seventy species totaling 8,000 individual birds. The most abundant were waterfowl—Canada Geese, American Black Ducks, Gadwalls, Green-winged Teal, American Wigeons, and Wood Ducks—and small landbirds such as Carolina Chickadees, Eastern Bluebirds, Eastern Meadowlarks, Northern Cardinals, American Goldfinches, Slate-colored Juncos, American Tree Sparrows, White-crowned Sparrows, and Song Sparrows.

To reach the Refuge, drive east from Seymour on US 50, then turn right onto the entrance road for 0.5 mile to headquarters and the visitor center.

SOUTH BEND
Four-mile Bridge | Izaak Walton League Sanctuary | Notre Dame University Campus | Baugo Creek | Rum Village Park

Lying in the St. Joseph River Valley in northwestern Indiana, South Bend has a number of sites where many transient waterbirds and waterfowl feed and rest. Migrating landbirds find shelter in the trees and shrubs of landscaped campuses and parks and in protected segments of river bottomland. Many stop off to nest. The following five trips should yield a wide variety of birds.

1. From downtown South Bend, proceed north on US 31 (Michigan Street) to the St. Joseph River. Just before the bridge, turn left on Riverside Drive and follow the River to Pinhook Park and north to **Four-Mile Bridge.** The view of the River along the Drive is unobstructed for most of the way. From fall to spring, as long as the water remains open, look for a variety of waterbirds and waterfowl: Mallards, American Black Ducks, Common Pintails, American Wigeons, Redheads, Lesser Scaups, Common Goldeneyes, Buffleheads, occasional Oldsquaws, Common Mergansers, Red-breasted Mergansers, Herring Gulls, and Ring-billed Gulls; during fall and spring migrations, but less likely in winter, Pied-billed Grebes, Horned Grebes, Canada Geese, and American Coots. Look for wading birds—Great Blue Herons, Green Herons, and others—in late summer and early fall.

2. Cross the Four-Mile Bridge and drive a half-mile east to the entrance to the **Izaak Walton League Sanctuary.** Embracing 80 acres, the Sanctuary is surrounded by several hundred acres of diversified habitat attracting a large number of birds of many kinds. North of the Sanctuary is the Healthwin Hospital, with attractively landscaped grounds. To the south are cultivated fields and deciduous woodlands owned by St. Mary's College and adjoining the campus. A sparsely built-up area lies to the east, and the St. Joseph River on the west. The Sanctuary itself, one of the most popular bird-finding places in the area, contains about 0.25 acre of cattail marsh, besides free-flowing Judy Creek, brush-covered fields, wooded uplands, and some 15 acres of original river bottomland. Barred Owls nest here, as well as many species of passerine birds common in northern Indiana. The woods and shrubs along Judy Creek attract transient warblers in great numbers during May and September.

3. From the Izaak Walton Sanctuary, drive east on Darden Road and turn right on US 31. After passing through Roseland, turn left on Angela Boulevard, drive east to Notre Dame Avenue, and then north to Notre Dame University. For a walk in April/May the **Notre**

Dame University Campus is ideal, with its lawns, shrubs, trees, and two lakes. The same species as in the Izaak Walton Sanctuary are usually found here.

4. From downtown South Bend, proceed east on US 33 (Lincolnway East) through Mishawaka to the Elkhart County line. At this spot, where **Baugo Creek** flows into the backwater of Hen Island Dam in the St. Joseph River, one finds waterbirds and waterfowl similar in variety to those along Riverside Drive.

5. From downtown South Bend, go south on US 31 to Ewing Avenue, then right for about 1.0 mile to **Rum Village Park,** with a playground area at its entrance. The Park comprises several hundred acres of diversified habitat, with mature deciduous forests, brushy fields, rolling prairie, and, in the southwest corner along Bowman Creek, a marsh area where American Bitterns, Least Bitterns, and Soras nest.

NORTHERN MOCKINGBIRD

SPENCER
McCormick's Creek State Park

An area of great natural charm, with a fine variety of landbirds, is **McCormick's Creek State Park** (1,753 acres) in west-central Indiana. Traversing the Park is a mile-long, cliff-walled valley through which McCormick's Creek flows to the White River on the west side. The slopes are covered with a forest of beech, oaks, hickory, and other hardwoods—for the most part secondary growth, though a few virgin stands of beech survive. There are a few open areas for recreational use, and abandoned fields grown to weeds and shrubs.

Pileated Woodpeckers, Prothonotary Warblers, and Worm-eating Warblers regularly nest here. Others nesting are the Yellow-billed Cuckoo, Red-bellied Woodpecker, Acadian Flycatcher, Eastern Pewee, Rough-winged Swallow, Bewick's Wren, Carolina Wren, Northern Mockingbird, White-eyed Vireo, Yellow-throated Warbler, Louisiana Waterthrush, Kentucky Warbler, Yellow-breasted Chat, and Summer Tanager. In early May the valley is an ideal spot for seeing transient warblers.

State 46 passes the entrance to the Park 2 miles east of Spencer.

Kentucky

BURT L. MONROE

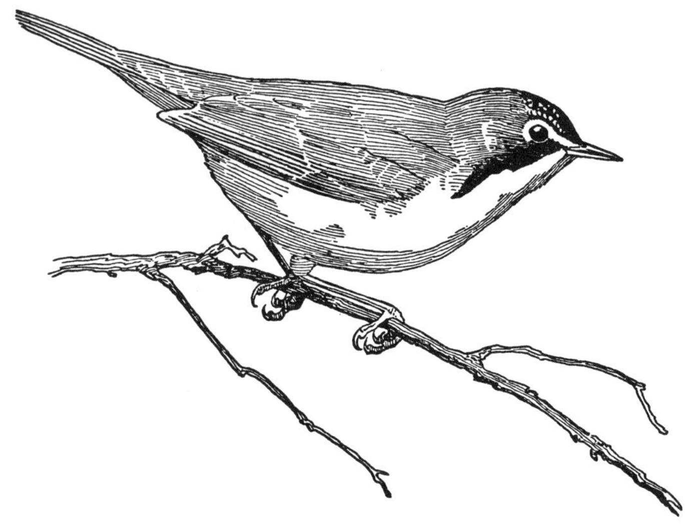

KENTUCKY WARBLER

Since the mid-eighteenth century, when a small band of explorers and surveyors paused on a ridge of the Cumberland Mountains close to the Virginia-Tennessee line and looked through an old water gap to see a luxuriant forest teeming with birds, there have been important written records of the abundant populations of Kentucky birds. Ornithologists have found the state a 'happy hunting ground' for bird study and have sought out fall and winter concentrations of waterfowl at Kentucky Lake, the late-summer and early-fall influx of shorebirds at the famous Falls of the Ohio, spring waves of warblers in the incomparable Mammoth Cave National Park and in Daniel Boone National Forest, and numerous summer residents throughout the state.

No introduction to the subject of bird finding in Kentucky would be complete without mention of John James Audubon, who lived here during the early nineteenth century and whose experiences at Louisville and Henderson are lasting contributions to ornithology. His descriptions of the birds found at the Falls of the Ohio are among the most fascinating in print. His accounts of the enormous flights of Passenger Pigeons and of their roosts near the mouth of the Green River stagger the imagination.

In the spring of 1810 another famous ornithologist, Alexander Wilson, visited Kentucky and described the only bird bearing the name of the state—the Kentucky Warbler. Wilson also encountered the Passenger Pigeon flocks and described the abundance of Greater Prairie Chickens in the 'Barrens' of Warren and Simpson Counties. And when Stephen Collins Foster wrote 'While the birds make music all the day' in 'My Old Kentucky Home,' he was not merely putting words together to make a rhyme—he was describing the incessant songs of Carolina Wrens, Northern Mockingbirds, Gray Catbirds, Brown Thrashers, Wood Thrushes, and others around the old homestead on Federal Hill near Bardstown.

Kentucky is a relatively narrow state from north to south, embracing only 175 miles, or about 2.64 degrees of latitude, with a range of 3,893 feet in altitude; but from east to west, it stretches for more than 400 miles.

In the extreme western part of Kentucky the elevation is 257 feet above sea level, lowest point within the state, and large areas are subject to overflow by the Ohio and Mississippi Rivers. Eastward the land rises steadily to 957 feet at Lexington, near the middle of the state and the approximate center of the extensive rolling area known as the 'Bluegrass.' To the south and east rise the abrupt hills and short ridges of the 'Knobs,' which comprise some of the roughest terrain in the state. Just beyond to the east begins the Cumberland Plateau, with irregular and heavily wooded ridges extending in a northeast-southwest direction. Near the Virginia border, Big Black Mountain, highest point in the state, rises to 4,150 feet (*see under* **Lynch**).

In all, 14,000 miles of streams traverse Kentucky—from the mountain freshets of the eastern region to the lazy, sluggish waters of the western lowlands—and almost every one of the state's 120 counties boasts fresh streams. The Ohio and Mississippi Rivers flow west and south, respectively, to form the main drainage channel. A lengthy segment of the Ohio serves as the northern boundary, and the Mississippi touches the western extremity. The Kentucky, the Tennessee, and the Cumberland—three navigable rivers within the state boundaries—have their source in the eastern mountains.

There are no extensive natural lakes or marshlands in Kentucky, but only man-made impoundments. Kentucky Lake, largest of these, in the southwestern part of the state, was formed by a dam across the Tennessee River at Gilbertsville and has 2,200 miles of shoreline, most of which extends into Tennessee. The Lake has exceptionally good habitats for waterbirds, waterfowl, shorebirds, marsh-loving birds, and eagles. Throughout Kentucky are many farm ponds—referred to locally as 'sky ponds'—and some tiny marshes, many frequented by Green Herons and Red-winged Blackbirds, a few by Pied-billed Grebes, Least Bitterns, King Rails, Wood Ducks, and Hooded Mergansers.

The most scenic part of Kentucky—the eastern mountains—is also the most primitive. Here turbulent rivers cut deep gorges through the forests, and high sandstone escarpments flank the valleys. Most of the ridges and valleys are forested with hardwoods such as oak, maple, ash, dogwood, and beech. Shortleaf, pitch, and scrub pines are plentiful. A few stately white pines dot the region, and hemlocks grow along the streams. Mountain laurel and white rhododendron grow in profusion; in June their lovely flowers blanket the hillsides and the edges of the ravines.

The Knobs, which form a semicircle around the Bluegrass, are for the most part well forested. Pine and mixed growth cover the bases of many of the Knobs; hardwoods such as oak, maple, hickory, and beech dominate the rest of the slope. To the south and west are extensive but scattered ranges of oak forests, with sugar maple and beech in many of the deep, moist ravines. Growing in low places and along the creeks and streams are sweet gum, aspen, willow, and sycamore. The forested areas are broken by farmlands, and the lowland areas surrounding the bases of the Knobs are largely under cultivation. The most abundant breeding bird in the eastern Knobs is the Red-eyed Vireo; the Yellow-throated Vireo is quite common in the more forested areas. In the west, northeast of Louisville, where the Knobs meet the Ohio River, lie the world's largest orchard-grass seed farms; here Henslow's Sparrows nest in numbers.

Since few untouched wild spots are left in the Bluegrass, the birdlife consists mainly of the more familiar farm and dooryard species. In the few wild spots, usually along the rivers, the variety is greater.

Across southern Kentucky, extending from the eastern mountains to the Tennessee River, is Pennyroyal country—in the vernacular of native Kentuckians, the 'Pennyrile'—named after the odorous weed. The Pennyrile contains some of the best bird-finding territory in Kentucky. Much of it is covered with an open-type oak-hickory forest, typically of oak, ash, beech, walnut, hickory, sycamore, black cherry, hackberry, sugar maple, cottonwood, tulip tree, willow, and cedar, with oak predominating. Cedar thickets are plentiful, and there is an undergrowth of redbud, black haw, blackberry, and trumpet vine. Climbing vines are everywhere—poison ivy, wild grape, wisteria, and Virginia creeper. The Pennyrile also has cutover land and abandoned farms; its brushy hillsides, wet meadows, and old fields, splotched with briars, broom sedge, and cane, provide a wide variety of habitats for birds. Species that breed regularly in the deciduous forest and farmlands (fields, wet meadows, brushy lands, orchards, and dooryards) are the following:

DECIDUOUS WOODS

Turkey Vulture
Black Vulture
Red-shouldered Hawk
Broad-winged Hawk
Yellow-billed Cuckoo
Common Screech Owl
Barred Owl
Chuck-will's-widow
Whip-poor-will
Yellow-shafted Flicker
Pileated Woodpecker
Red-bellied Woodpecker
Hairy Woodpecker
Downy Woodpecker
Great Crested Flycatcher
Acadian Flycatcher
Eastern Pewee
Blue Jay
Carolina Chickadee
Tufted Titmouse
White-breasted Nuthatch
Wood Thrush
Blue-gray Gnatcatcher
Yellow-throated Vireo
Red-eyed Vireo
Warbling Vireo
Worm-eating Warbler
Cerulean Warbler

DECIDUOUS WOODS (*Cont.*)

Yellow-throated Warbler
Louisiana Waterthrush
Kentucky Warbler
Hooded Warbler
American Redstart
Summer Tanager

FARMLANDS

Common Bobwhite
Mourning Dove
Red-headed Woodpecker
Eastern Kingbird
Eastern Phoebe
Barn Swallow
Bewick's Wren
Carolina Wren
Northern Mockingbird
Gray Catbird
Brown Thrasher
Eastern Bluebird
Loggerhead Shrike
White-eyed Vireo
Yellow Warbler
Prairie Warbler
Common Yellowthroat
Yellow-breasted Chat
Eastern Meadowlark
Orchard Oriole
Common Grackle
Northern Cardinal
Indigo Bunting
Dickcissel
American Goldfinch
Rufous-sided Towhee
Grasshopper Sparrow
Bachman's Sparrow
Chipping Sparrow
Field Sparrow

Much of extreme western Kentucky is a natural reservation for wildlife and is rich in some of the state's most interesting species of birds. From the cottonfields around the edges of the swamps at the 'Scatters' of Reelfoot Lake in the southwestern corner, to the gently rolling uplands and wide floodplains, punctuated here and there with stream bluffs, cypress swamps, oxbow lagoons, and deeply eroded gullies above Wickliffe, the plant life is similar to that of the Deep South.

Part of the land, once the river bed of the Ohio, is a tangled wilderness of vines and small shrubs extending as far as the eye can see under a canopy of oak, aspen, hickory, and other broadleaf trees. Here the cypress growth is especially fine, and giant trees protrude from the waters. The region is marked by sloughs and a score or more of little lakes, all connected with the creeks and extending intermittently northward, almost to the Ohio. Some contain small areas of deep, open water, but most are shallow areas choked with submerged and surface vegetation. Birds residing here regularly are the Double-crested Cormorant, Great Blue Heron, Cattle Egret, Great Egret, Black-crowned Night Heron, Yellow-crowned Night Heron, and Little Tern.

Vast numbers of birds appear during migrations, since the state lies on the eastern edge of the Mississippi flyway, directly in the path of migrating waterfowl. Waterfowl migrations can best be observed in central and western Kentucky, for most birds swing to the west of the high eastern mountain regions, across the southern part of Ohio, and meet the Ohio River near the eastern boundary of the state; they then follow the Mississippi closely. The shorebirds also follow the Mississippi and Ohio River routes, but may traverse Kentucky at other points, stopping to rest and feed at ponds, lakes, mud flats, and fields flooded by spring and fall rains. The Falls of the Ohio at Louisville is an excellent observation point for fall-migrating shorebirds. During spring flights a good point is the region of 'transient lakes' formed by underground springs after rainy seasons, near Bowling Green. Waves of landbirds cross the state, often being concentrated at the national and state parks and forests, which are fine places for observing transient vireos and warblers. From Ashland to Wickliffe the Ohio River bottomland, between the edge of the River and the bluffs and rolling hills inland, is splendid territory for migrating sparrows.

Here is a Kentucky migration timetable:

NORTHERN KENTUCKY (approximate northern half of state):

Waterfowl: 25 February–10 April; 15 October–25 November
Shorebirds: 30 March–20 May; 10 August–10 October
Landbirds: 5 April–15 May; 1 September–25 October

SOUTHERN KENTUCKY (approximate southern half of state, including extreme west):

Waterfowl: 25 February–5 April; 20 October–1 December
Shorebirds: 20 March–20 May; 10 August–10 October
Landbirds: 1 April–10 May; 5 September–1 November

Kentucky's winter bird population is large and varied. During a fifty-year period, Christmas counts taken in many parts of the state have recorded 135 species, as many as 100 from one small area. Large numbers of geese and ducks spend the winter on the western lakes, notably Canada Geese, which gather at the Land Between The Lakes Recreation Area (*see under* **Cadiz**). Smaller concentrations of Mallards, American Black Ducks, Canvasbacks, Lesser Scaups, Common Goldeneyes, Buffleheads, Oldsquaws, and Common Mergansers winter at select spots along the Ohio River, chiefly in the harbor at Louisville. Gull concentrations, composed largely of Herring and Ring-billed Gulls, with an occasional sprinkling of Bonaparte's Gulls, may also be found at these points and at Kentucky Lake on the Tennessee River.

In midwinter, flights of thousands of Common Crows take place in the central and southern parts of the state; motorists may watch them in late afternoon fly across the highways in seemingly endless lines. Their roosts—shifted almost every year—are usually in the dense cedar thickets and expanses of deciduous woodlands characteristic of these regions. Northern Harriers and Short-eared Owls course the sedge and stubble fields, and, in 'invasion years,' Snowy Owls may appear.

Authorities

Roger W. Barbour, George H. Breiding, John L. DeLime, Victor K. George, L. Y. Lancaster, Harvey B. Lovell, Robert M. Mengel, Burt L. Monroe, Jr., John S. Morse, Edward M. Ray, Robert C. Soaper, C. A. Van Arsdall, Earl Wallace, Gordon Wilson.

References

The Birds of Kentucky. By Robert M. Mengel. Ornithological monographs No. 3 (1965), American Ornithologists' Union. Contact through National Museum of Natural History, Smithsonian Institution, Washington, D.C. 20560.

Kentucky Birds: A Finding Guide. By Roger W. Barbour and others. Lexington: University of Kentucky Press, 1973.

BOWLING GREEN
Chaney Lake | McElroy Lake

One of the best places in Kentucky to observe waterbirds, waterfowl, shorebirds, and marsh-loving birds in spring is south of Bowling Green, in the south-central part of the state, where 'transient lakes' appear almost every year. After prolonged and heavy rainfalls, depressions in the midst of woodlands or cultivated fields may become water-filled, fed by wet-weather springs. In some years they cover hundreds of acres, and the water may remain for weeks. As it recedes, mud flats emerge and shorebirds flock to them, pausing in their spring flight northward. Bird finders should not overlook the importance of these transient lakes.

Chaney Lake, where the water sometimes overflows as much as 275 acres, attracts waterfowl and many other birds. A single day's list here may include sixty or more species, including twenty associated with water. To reach Chaney Lake, go south from Bowling Green on US 31W for 7 miles to the Chaney Farm on the right. The Lake is in sight from the main highway and may even flood it in unusually wet seasons. When the water is receding the marsh area is 0.5 mile from the road and may be reached only on foot.

The major body of water among the 'transient lakes' is **McElroy Lake,** which has been studied since 1912 by the most competent state ornithologists; more than two hundred species have been recorded, of which eighty are generally found near water. The woods near the edge of the Lake, the meadows near the barnyard, the orchards, and the cultivated fields furnish excellent habitat for landbirds, but the species frequenting them vary little from those of other farm areas in the country. Although the water may rise in January—and cover as many as 400 acres—it ordinarily does not appear until February or March. The best time to observe birds here is in late March and April. Among species to record are the Red-necked Grebe, Horned Grebe, Oldsquaw, White-winged Scoter, Surf Scoter, Lesser Golden Plover, Dunlin, and Short-billed Dowitcher.

To reach McElroy Lake, drive south on US 31W for 10 miles, go left on State 240 for about a half-mile, and then left again on State 884 for 3 miles. The Lake is on the right.

CADIZ
Land Between The Lakes Recreation Area

A site famed for its wintering population of Bald and Golden Eagles, as well as for its waterfowl concentrations, is the **Land Between The**

Lakes Recreation Area. About 170,000 acres in Kentucky and Tennessee, managed and operated by the Tennessee Valley Authority, constitute the Land Between The Lakes, of which about one-third lies in Kentucky and was formerly known as the Kentucky Woodlands National Wildlife Refuge.

To reach the Area from Cadiz, drive west on State 80 across Lake Barkley to headquarters in Golden Pond, where one may obtain maps, bird lists, and current information on eagle locations. Three miles west of Golden Pond, State 80 intersects the Trace (State 453), which bisects the Land Between The Lakes from north to south; turn right (north) up the Trace and follow signs on intersecting paved roads to the right for the desired spots.

Most of the Land Between The Lakes is in secondary forest, with some extensive plantings of various pines. The forest supports a native population of Wild Turkeys, which may be spotted by lucky bird finders along any of the roads or trails. Other common forest species abound, including a sizable population of Pileated Woodpeckers.

The main attraction of the Area is, of course, the eagle and waterfowl concentrations, primarily on Lake Barkley at the eastern border. Wintering eagles—generally forty to fifty Bald Eagles with a few Golden Eagles usually present—are most easily observed at Hematite Lake, accessible from the Trace north of State 80 following signs on paved roads running east. Waterfowl and shorebirds may be seen at any access to Lake Barkley, usually the best spots are at Silo Overlook and Taylor Bay, both on well-marked roads. Not only are there thousands of Canada Geese and waterfowl on Lake Barkley in winter, but there is also a breeding population of Canada Geese, as well as nesting Wood Ducks and Hooded Mergansers.

CAVE CITY
Mammoth Cave National Park

If it is landbirds, especially the colorful, fascinating wood warblers, that the bird finder seeks, Kentucky can offer no better territory than **Mammoth Cave National Park.** The Park, cut by the Green River, occupies one of the most rugged sections of south-central

Kentucky. From the Cave City Exit of I 65, drive west on State 70 for 8 miles to the Park boundary, and continue across Mammoth Cave Ridge to Park headquarters at Mammoth Cave. Stop here for a map of roads and trails.

The 80-square-mile tract, especially the northern portion, is almost entirely wooded with original and second-growth timber. The predominant trees are oak, aspen, beech, hickory, walnut, ash, dogwood, and a variety of conifers, including the abundant cedar.

Sixteen warbler species nest within the Park. Some may be located by their songs and call notes—particularly the Louisiana Waterthrush, Kentucky Warbler, Common Yellowthroat, Yellow-breasted Chat, and Hooded Warbler. In the deep woods the Black-and-white Warbler, Worm-eating Warbler, Northern Parula Warbler, Cerulean Warbler, Yellow-throated Warbler, Ovenbird, and American Redstart are common. The Prairie Warbler inhabits the briar-covered fields and slopes; the Prothonotary nests along streams in old willow stubs and other cavities. Spring and fall migrations bring other members of the warbler family. In all, thirty-five species have been listed. The only winter warbler is the Myrtle, which in-

YELLOW-THROATED WARBLER

habits the cedar trees and eats the bountiful supply of poison ivy berries.

Several spots within the Park are particularly rewarding to the bird finder. At the northeast end lies a wild tract known as the Big Woods, which is especially attractive to woodpeckers and tanagers. Territory resembling the eastern mountain region of Kentucky may be explored at Buffalo Creek, reached by driving north from Park headquarters to just outside the Park boundary and turning west. Leave the car in front of the store at Ollie. Here, at night, a serenade of Common Screech Owls, Great Horned Owls, Barred Owls, Chuck-will's-widows, and Whip-poor-wills creates an eerie atmosphere.

If the bird finder desires more primitive country, he should drive to the western end of the Park and visit First Creek, an area of towering sandstone cliffs. Ugly Creek, a tract of similarly rugged terrain on the northern edge, is a deep gorge surrounded by heavily wooded hills. The shy warblers such as the Blue-winged and Northern Parula nest here. The best bird territory in the Park is Mammoth Cave Ridge, within a mile of Mammoth Cave Hotel, where more species can be seen than in any other section of similar size in the Park. Here is one spot where one can almost always see or hear a Pileated Woodpecker.

The best season to visit Mammoth Cave National Park is late spring or early summer, but many species of birds are present at all seasons. The reward of spring and autumn visits will be the migrations of vireos, warblers, and other landbirds; summer will bring the thrill of finding the nests of many of them. During the late afternoons in summer the tinkling, incomparable song of the Bachman's Sparrow comes from the old fields of briars and broom sedge. In winter the permanent residents are joined by eighteen other species, including sizable flocks of American Tree Sparrows at the southern edge of their normal winter range.

CORBIN
Cumberland Falls State Park

Cumberland Falls State Park lies in the southern part of Daniel Boone National Forest, a 900,000-acre wilderness area bordering Cumberland Plateau north and south for more than 100 miles. The entire tract is of ornithological interest, but for non-resident bird finders the 1,098-acre Park is perhaps the best and most accessible place to look.

The part of the Plateau in which the Park is situated lies at an elevation of about 1,100 feet on thick sandstone strata that have been gouged into precipitous, cliff-lined gorges by the Cumberland River and its tributary streams. The uplands are forested with pine and oak, which give way in the ravines to heavy stands of hemlock, maple, and tulip tree. Growths of rhododendron and mountain laurel are particularly impressive. Cumberland Falls, a spectacular cataract 68 feet high, boasts one of the two 'moonbows' in the world.

Faunally this is not a true mountain region: its elevation is too low for the presence of northern species. It is interesting to the bird finder, however, as one of the few vestiges of original Cumberland wilderness; and he may find a number of species which, through preference for either rugged terrain or extensive forest, are rare or lacking in the more settled country to the west. Among the larger birds the Red-tailed Hawk, Ruffed Grouse, Pileated Woodpecker, and particularly the Broad-winged Hawk are common.

Breeding warblers are varied and abundant. Characteristic are the Black-and-white, Worm-eating (secretive), Northern Parula and Black-throated Green (these two favoring shaded ravines and steep slopes in the vicinity of hemlocks), Pine (pine-oak uplands), Ovenbird, Hooded, and Kentucky. This is perhaps the best place in Kentucky to see the Red-cockaded Woodpecker, although discovering it without being familiar with the call notes is largely a matter of luck. Watch for it in the pines along State 90 from the Park to a point 5 or 6 miles east.

On the highway 3 miles east of the center of the Park is an observation point that affords a splendid view of the rugged terrain and is an excellent place to watch for soaring birds.

To reach the Park, go southwest from I 75 south of Corbin on US

25W for 8 miles, and head right (west) on State 90 for 7.6 miles to headquarters.

EDDYVILLE
Kentucky Dam Village State Park

Kentucky Dam Village State Park (1,000 acres), in the western part of the state, is the best spot along Kentucky Lake from which to start on short forays into various types of waterfowl habitat. Boats are available for cruises. To reach the Park, take US 62 west from Eddyville for 17 miles to the entrance just beyond Gilbertsville Dam. Here, at the Dam, the lower section of the Tennessee River is close to the Mississippi, and in this vicinity the Mississippi waterfowl flyway crosses the state. Moderate-to-heavy concentrations of waterfowl appear in the region, and twenty-four species have been noted. The maximum numbers of species and individuals come in during spring and fall migrations; lesser numbers winter here, and even fewer are present during summer.

The habitat most frequently used by geese and ducks is a series of moderate-sized embayments and islands. The Wood Duck is the only waterfowl that nests in numbers near the Lake, but there are occasional broods of Mallards and Hooded Mergansers. Large flights of Blue-winged Teal come in during August and leave before October. In October other species such as the Mallard, American Black Duck, and Lesser Scaup gradually increase in number, and Canada and Snow Geese arrive. Snow Geese do not remain very long, but most of the Canada Geese flocks stay for the winter.

Waterfowl are by no means the only birds at the Park. Tremendous flocks of Herring Gulls and Ring-billed Gulls, as well as a scattering of rarer gull species, arrive in fall and stay until spring. Bonaparte's Gulls, Forster's Terns, Common Terns, Little Terns, and Black Terns use these waters as feeding grounds during spring and fall flights. In summer and again in fall, the water level of the impoundment is lowered by hydroelectric plants, exposing extensive mud flats at the heads of embayments and around low islands along the former banks of the River and its tributaries. Many shorebirds feed and rest in these inviting spots. The flocks are made up largely of 'peeps,' but there is always a chance that the bird finder will see species uncommon or rare for the state.

HENDERSON
Audubon Memorial State Park

Along the Ohio River near Henderson lies the country where Audubon made his world-famous observations and drawings early in the nineteenth century. Here, where he collected and studied birds and cut the timber to build his 'infernal' mill, is the **Audubon Memorial State Park**—520 rolling acres of virgin and cutover timber, shrubs, and vines. Beech, sycamore, dogwood, coralberry, tulip tree, honey locust, sweet gum, dwarf sumac, and wild grape are among the fifty-one varieties of trees, thirteen shrubs, and seven vines that have been catalogued. Two artificial lakes—Recreation Lake (16 acres) and Wildlife Lake (5.5 acres)—beautify the area and attract several species of waterbird and waterfowl to augment the list of birds.

To reach the Park, take US 41 north from Henderson for 2 miles, turn right into a shady lane, and drive for another 2 miles. High on a wooded hill is the massive Audubon Museum, of French Provincial architecture, constructed for the most part of native stone. Covered terraces, balustrades of gray stone ornamented with red brick, and a formal garden, with a huge concrete retaining wall at the back and a front wall of gray stone surround the building. Below, at the far end of the garden, are two ridges of timberland that roll down to flatlands on the west side of US 41. To the north, other timbered ridges form the green bluffs of the Ohio.

Five spacious galleries house the material, which includes forty-six original paintings by Audubon and his sons and from twenty to thirty water colors by Audubon's granddaughter. There are books from his library, the chair in which he sat to paint, his portrait of Daniel Boone, his unfinished portrait of Mrs. Audubon, his silverware, watch, turkey seal, and a quantity of hair jewelry, of which Audubon himself was an accomplished manufacturer. There is a set of Wilson's *Ornithology*, with Audubon's notations, and many works by Audubon's contemporaries. The most valuable of the exhibits is a set of eight of his original bird paintings on tin and wood. A good part of the Museum is devoted to displaying the hand-colored prints from the double elephant folio, among which is the famous print of the Northern Mockingbird's nest being attacked by a rattlesnake.

Henderson has been honoring John James Audubon for a long time. In 1915 the site of his mill on the Ohio banks was converted into a park and named Audubon Mill Park. The stones of the gateway were taken from the old mill itself. A bronze tablet marks the location of Audubon's general store in Henderson's business district.

HICKMAN
Mississippi River

Along the **Mississippi River** in the extreme southwestern corner of Kentucky are extensive bottomlands, mostly under cultivation. A levee parallels the River from Hickman south to below the Tennessee line. Between the levee and the River are extensive stands of cottonwood and willow, breeding sites for Mississippi Kites, Red-headed Woodpeckers, Fish Crows, and Baltimore Orioles. Indeed, this is the only location within the state where one can reasonably expect to find Mississippi Kites and Fish Crows, and the abundance of orioles and woodpeckers is exceeded nowhere else. Also along the levee, where cattle usually graze, Cattle Egrets feed, particularly in spring and fall. During spring rains and occasional floods the agricultural lands between the levee and State 94 south of Hickman develop many temporary ponds or lakes along with their few permanent ones, quite attractive to transient waterfowl and shorebirds. During May this is one of the best places in the state to see White-rumped Sandpipers and Dunlins.

To reach the levee, drive south from Hickman on State 94 for 14 miles to Phillippy, Tennessee, just south of the state line, then right (west) on the only paved road for 2 miles to the levee. At this point a gravel road goes right (north), paralleling the levee back into Kentucky. Take the levee road, or, as is possible in some seasons, drive on the levee itself. All roads eventually lead back to State 94. The best bird finding is usually the first 3 miles north of the Tennessee border.

LOUISVILLE
Falls of the Ohio | Cherokee Park | Seneca Park | Bernheim Forest Park

Although Louisville, largest city in Kentucky, is the hub of a heavily populated area, much of it built up with homes and factories, it affords some excellent localities for bird finding.

The **Falls of the Ohio,** only a few blocks from downtown Louisville, is unquestionably one of the most exciting places in this section of the country for migrating waterbirds, waterfowl, and shorebirds. Audubon's tales, such as those of Snowy Owls fishing at the potholes in the ice and of nesting Swallow-tailed Kites and Ospreys, first drew the attention of ornithologists to this little-known spot. Since the days of Audubon, when the Falls consisted of a series of cascades as the River dropped 26 feet to the mile, the area has continued to attract bird finders. Records of several species new for the state have been obtained here.

Although construction of a dam and a canal for river traffic has shifted somewhat the many sand and gravel banks protruding from the water above and below the Falls, the birds continue to concentrate in the vicinity. The dam, an L-shaped structure 8,626 feet long, holds the Ohio River in a channel for approximately three-quarters of a mile downstream. Water running over the dam fans out across a large area of rocks pitted with potholes. Since the water is laden with sewage, it supplies ample food for shorebirds. Bare rock and sand dunes stand out beyond the water-covered section. At the lower end of the dam there is a sand island large enough to support a thick grove of willows and cottonwoods.

From the top of the dam the bird finder can look down on another world. He will see large expanses of rock, worn and riddled with potholes and pools. There will be sand bars and islands and—as background—the tall buildings of Louisville across the River. One need not leave the dam to obtain a good list of birds, for there are numerous spots along the high banks for a fine view of the whole area; but the adventurer—one with the surefootedness of a mountain goat, since the algae-covered rocks can be very slippery—may descend to these rocks and enter a bird haven of rare possibilities. One can rent a boat at the dam or hire a fisherman to ferry him to the main territory. At low water it is possible to walk to the open expanses at the upper end near the dam.

On all trips to the Falls regardless of season, the bird finder should be alert for new and exciting species. Late August brings Semipalmated Plovers, Piping Plovers (rare), Lesser Golden Plovers, Black-bellied Plovers, Willets (rare), Stilt Sandpipers. From late July to late September, fair numbers of Killdeers, Spotted Sandpipers, Solitary Sandpipers, Greater and Lesser Yellowlegs, Pectoral Sandpipers, and myriads of Least Sandpipers, Semipalmated Sandpipers, and Western Sandpipers invade the area. Scan each group of these regular shorebirds carefully, for occasionally the Ruddy Turnstone, Baird's Sandpiper, White-rumped Sandpiper, and Dunlin appear. In early September there is a real, if slight, chance of seeing a Buff-breasted Sandpiper.

Caspian Terns are present from late July until late September, generally in groups of three or four, but sometimes in flocks of as many as fifteen. The Black Tern, in late summer the most abundant of the tern family, is often accompanied by the Forster's Tern and Common Tern and, more rarely, the Little Tern. Late fall and winter bring hundreds of Herring Gulls and Ring-billed Gulls to course the River and perch on the dikes. Bonaparte's Gulls come about 10 April, with an indiviudal or two appearing in December or January.

Among landbirds attracted to the Falls in late summer are great flocks of Bank Swallows, Rough-winged Swallows, Barn Swallows, Cliff Swallows, and Purple Martins. Water Pipits visit the sand dunes in September. During April and September, Ospreys appear, and sometimes a Peregrine Falcon to bother the shorebirds. The willows and cottonwoods attract warblers during migration. Mourning Doves arrive for water in the late evening; the rattle of the Belted Kingfisher can be heard at times.

To reach the Falls, drive north from Louisville on Second Street, crossing the River on the George Rogers Clark Memorial Bridge. Immediately after reaching the Indiana end of the Bridge, make a U-turn right and follow the street paralleling the Bridge back toward the River. Drive one block, turn right on Market Street until it merges with another road paralleling the River, and continue west to a parking area at a dead end. The distance from the Bridge to the Falls is 1.1 miles. One may also reach the Falls by crossing the

River on the Kennedy Bridge (I 65), taking the first exit on the Indiana side, turning left for two blocks to the end of Clark Bridge, turning left again, and proceeding as above.

Louisville's parks, all with suitable habitat for birds of the deciduous woodlands and open grasslands, ring the city in a loose semicircle. **Cherokee Park,** on the east, is considered one of the most beautiful parks in the United States. This rolling tract of 409 acres, thickly planted with trees and traversed by Beargrass Creek, is an ideal spot for migrating warblers. **Seneca Park** and the airport at Bowman Field, adjacent to Cherokee Park on the east, contain many large open spaces where (Northern) Horned Larks may be seen in winter. To reach the Parks, take I 64 east from downtown, exit at Grinstead Road, go south one block, turn left on Lexington Road and proceed for one block to the entrance to Cherokee Park on the right.

About 25 miles south of Louisville lies **Bernheim Forest Park,** a forest and wildlife sanctuary operated by the Isaac W. Bernheim Foundation. Consisting of some 10,000 acres in the Knobs region of sharp hills, the Forest is superb for bird finding. In addition to its more than 8,000 acres of hardwood forest interspersed with native Virginia pine, there is an extensive cleared area known as the Big Meadow, as well as several lakes. The largest, Lake Nevin, is attractive to transient waterfowl and Ospreys; the smaller lakes now have an established breeding population of Canada Geese. A list of two hundred-plus species of birds recorded in the Park, along with general information on the Park, are available at the nature center and museum. For the botanically inclined there is an arboretum, and most of its extensive plantings are identified and labeled.

To reach the Park, drive south from Louisville on I 65 to the Bernheim Forest Exit and go east on State 245 for 2 miles to the entrance on the right.

LYNCH
Big Black Mountain

Big Black Mountain, northernmost and highest part of the Log Mountain Range, with an average elevation of 3,800 feet for some 15 miles of its length, is recommended as a cool, beautiful area of particular interest to bird finders desiring experience with Appalachian birds. The best time for a visit is from late May to early July. To look for birds, drive southeast from the mining town of Lynch on State 160, which passes over the Mountain en route to Big Stone Gap, Virginia. After climbing for 4.5 miles along a winding road to the summit, turn south onto a poorly paved road and follow it as far as possible. The summit, which is merely a gentle swell in the crest of the Mountain, reaches the highest elevation in Kentucky—4,150 feet—a short distance south of the pass.

That part of Big Black Mountain lying above 3,000 feet is the only area in Kentucky with breeding-bird species of northern affinities; but the Mountain also has typically southern species: in all, a remarkable admixture of northern and southern forms. The Solitary Vireo is present together with the White-eyed Vireo; Black-throated Blue Warblers, Black-throated Green Warblers, Blackburnian Warblers, Chestnut-sided Warblers, and Canada Warblers breed side by side with Swainson's Warblers, Kentucky Warblers, Yellow-breasted Chats, and Hooded Warblers. Rose-breasted Grosbeaks and Slate-colored Juncos are active in the same thickets inhabited by Northern Cardinals. The Veery, whose enchanting song is the most typical June sound on the Mountain, rounds out the list of common northern species.

On Big Black Mountain there is no spruce-fir forest, such as that found in the Great Smokies. Although hemlocks and several species of pines grow in some places, the heavy forest is mainly deciduous—a mixture of beech, maple, oak, birch, and other hardwoods. The ravines are choked with thickets of shrubs and vines, in some cases making bird finding a task requiring patience and careful observation.

Caution: The timber rattlesnake is found even on the summit, but is a less formidable hazard than the abundant blackberry bushes.

PINE RIDGE
Natural Bridge State Park

The principal scenic feature of **Natural Bridge State Park,** in the northern part of Daniel Boone National Forest, is the great natural-sandstone arch—one of many in its vicinity—which crowns its highest ridge. The general characteristics of this area are similar to those of Cumberland Falls State Park (*see under* **Corbin**). The characteristic bird species and plant life are also similar, except that the Red-cockaded Woodpecker is unlikely to be seen here.

In some ways this region is more spectacular than Cumberland Falls. The northern part of the Cumberland Plateau is more heavily cut by streams than the southern, with many towering cliffs and hollowed-out precipices, locally called 'rock houses.' When two rock houses on opposite sides of a narrow ridge meet, a natural arch, a 'lighthouse,' results.

Natural Bridge State Park is readily accessible from Mountain Parkway, which begins at Winchester. To reach the Park, leave the Parkway at the Pine Ridge–Slade Exit and go south on State 11 for 2 miles to the entrance. For the bird finder who wishes to see more of the area, there is a 20-mile drive well worth taking. From Slade, go north on State 715, which loops east and south to rejoin the Parkway near Nada. The road is fairly good, but it is narrow, following the gorgeous valley of the Red River, perhaps the most precipitous and spectacularly beautiful country in eastern Kentucky. The birds are the same as those in Natural Bridge State Park. There is also a chance of seeing Ruffed Grouse. Sky Bridge, considered by many to be more impressive than Natural Bridge, lies on this road.

WICKLIFFE
Swan Lake

In the heart of the Ohio River lowlands slightly northwest of Wickliffe is a chain of lakes famous in western Kentucky as a veritable paradise for waterbirds, waterfowl, and marsh-loving birds. Some of the lakes consist of deep, open water, but others are shallower, and clogged with considerable growth of pickerel weed, mud plantain, and water chinquapin. Sloughs of sluggish, tepid water crisscross the region, and mixed forests of bald cypress, cottonwood, water locust, maple, pecan, aspen, and other trees rise from the saturated swamplands. Black willows and buttonbushes rim some of the ponds, and jungles of tangled vines and shrubs are prevalent. The focal point for an ornithological trip into the territory is **Swan Lake,** largest of the chain, covering 1,000 acres. The bird finder should take US 51 northwest from Wickliffe for 2.3 miles to the junction with a gravel road, where the causeway to the Cairo Bridge makes a sharp left turn. Turn right on the gravel road marked by a sign pointing to Swan Lake, and follow it for a little over a mile. Boats are available for hire.

The bird species are generally typical of those of the more southerly states. The Double-crested Cormorant is present throughout the year but is much more frequent in early spring and fall. Nesting conditions around the lakes are ideal for Wood Ducks. Ospreys may nest in the area, as one or two pairs are sometimes in view around the Lake in summer. The Red-winged Blackbird is an abundant summer resident among the shallow lakes, where it walks on the aquatic vegetation and on partly submerged or floating logs in search of insects. The Prothonotary Warbler feeding among the buttonbushes or examining willow stumps lends even more color to the scene.

Other lakes in the area are First Lake, Hunter's Pond, Long Pond, Minor Lake, Ox Lake, and Prairie Lake. All are attractive to hordes of migrating waterfowl. Many of the lakes are shallow enough for hunters to stand in the water, concealed by the foliage near the shores. Since the lakes are popular for hunting, the bird finder should limit his visits to spring, summer, and early fall.

Maine

COMMON EIDERS

Extensive sea coast, scattered mountains, numerous lakes, and vast tracts of coniferous forest—such are the natural areas of Maine that are most attractive to the visiting bird finder.

Measured in a straight line, the coast is only 230 miles long; but measured to include the shores of its hundreds of islands, deeply cut bays, harbors, and inlets, it is no less than 3,470 miles in extent. Topography is varied and picturesque. From the southern extremity at Kittery Point to Cape Elizabeth near Portland, and northeastward in the vicincity of the Kennebec River mouth, the coast is low-lying, with broad salt marshes sheltered by sand dunes and sandy beaches. Elsewhere the islands, headlands, and indentations are rugged. The shores are walled sometimes with forbidding cliffs, sometimes with loose slabs of rock piled one upon the other, and sometimes with mountains. On Mount Desert Island, Cadillac Mountain rises from the sea to an elevation of 1,532 feet, the loftiest mountain on the Atlantic seaboaard north of Rio de Janeiro in Brazil.

The coastal islands offer the bird finder a great opportunity. Between mid-May and the first of July, he can view Atlantic Puffins and Razorbills nesting at Matinicus Rock (*see under* **Rockland**), the only place on United States territory where both species breed, or at Canadian-owned Machias Seal Island, 'down east' in the Bay of Fundy (*see under* **Machias**). Other species he can see nesting, often in impressive numbers, are the following:

Leach's Storm Petrel
Double-crested Cormorant
Common Eider
Greater Black-backed Gull
Herring Gull
Common Tern
Arctic Tern
Roseate Tern (*local*)
Black Guillemot

The islands are attractive ornithologically for other reasons. Where there are cliffs and rocky promontories, Ospreys and Northern Ravens sometimes nest; where there are treeless, grassy areas, Spotted Sandpipers and Savannah Sparrows are abundant; where there are lighthouses and dwellings, Cliff Swallows congregate in large colonies. At any time, even in midsummer, transient shorebirds feed and rest among the rocks regularly bathed by the tides.

A trip to the sea-bird islands is an adventure as well as a quest, for the islands are isolated by strong tides, by erratic sea action, and frequently by curtains of fog; furthermore, they either are entirely uninhabited or have lighthouses only, and they are without sheltered harbors. Transportation must be directed by an experienced navigator.

The mountains of Maine comprise a northeastward extension of the Appalachians from the state's western boundary to the north-central interior. The majority are mere foothills, though scattered among them are over a hundred peaks that range from 3,000 to 5,000 feet in height. The highest, Katahdin, stands seemingly alone in the north-central part of the state. Lakes, well over two thousand, are fairly evenly distributed; fresh-water bogs and marshes are numerous and large. The chief river valleys bear directly southward to the ocean. The most prominent, from west to east, are the Saco, Androscoggin, Kennebec, Penobscot, St. Croix, and St. John.

Extreme southwestern Maine is primarily sand plain interrupted by low hills, small lakes, and cool sphagnum bogs. Pitch pine, several species of oak, and occasionally white cedar are characteristic forest growth. This area extends northward to an imaginary line drawn from Acton near the state's western border through Sebago Lake to Cape Elizabeth, but isolated areas exist northward and northeastward for 50 miles. Among nesting species characteristic of this area of Maine are the Great Crested Flycatcher, Veery, Yellow-throated Vireo, Pine Warbler, Rufous-sided Towhee, and Field Sparrow.

Roughly 88 per cent of Maine, comprising an area of the state larger than West Virginia, is forested—and grows more forested as farmland after farmland is abandoned and reverts to timber. Much of northern and western Maine—well over half the state—is a continuous wilderness of trees and lakes, neither inhabited by man nor incorporated within any form of local government. Here, then, Maine offers the bird finder another great opportunity.

In the forests of northern and western Maine as well as in those elsewhere on the higher mountains and along the coast northeast of Penobscot Bay, spruce, fir, and hemlock predominate where the atmosphere is cool and moist, but are mixed with white pine, sugar maple, beech, and birch in warmer areas. These conditions favor an unusually large variety of birds. From late May to early July the bird finder may observe fifteen or more species of breeding warblers, as well as the following breeding species:

Northern Goshawk
Spruce Grouse
Black-backed Three-toed Woodpecker
Yellow-bellied Flycatcher
Olive-sided Flycatcher
Gray Jay
Boreal Chickadee
Red-breasted Nuthatch
Brown Creeper
Winter Wren
Hermit Thrush
Swainson's Thrush
(Bicknell's) Gray-cheeked Thrush
Golden-crowned Kinglet
Solitary Vireo
Pine Siskin
Purple Finch
Slate-colored Junco
White-throated Sparrow

Elsewhere in Maine, in woods with beech, sugar maple, oaks, and other deciduous trees predominating, and on farmlands (fields, wet meadows, brushy lands, orchards, and dooryards), the following breeding birds are characteristic:

DECIDUOUS WOODS

Red-shouldered Hawk
Broad-winged Hawk
Ruffed Grouse
Barred Owl
Whip-poor-will
Yellow-shafted Flicker
Hairy Woodpecker
Downy Woodpecker
Least Flycatcher
Eastern Pewee
Blue Jay
Black-capped Chickadee
Tufted Titmouse (*southern Maine*)
White-breasted Nuthatch
Red-eyed Vireo
American Redstart
Baltimore Oriole
Scarlet Tanager
Rose-breasted Grosbeak

FARMLANDS

Eastern Kingbird
Eastern Phoebe
Tree Swallow
Barn Swallow
Gray Catbird
Brown Thrasher (*southern Maine*)
Northern Mockingbird (*southern Maine*)
Eastern Bluebird
Yellow Warbler
Chestnut-sided Warbler
Common Yellowthroat
Bobolink
Eastern Meadowlark
Northern Cardinal (*southern Maine*)
Indigo Bunting
House Finch (*southern Maine*)
American Goldfinch
Savannah Sparrow
Vesper Sparrow
Chipping Sparrow
Song Sparrow

Breeding American Woodcock are probably more common in eastern Maine than in any other part of the United States. Young stands of second-growth deciduous trees mixed with conifers, near moist ground, are their favorite nesting habitat.

The coast and river valleys are important flyways in migration. In

AMERICAN WOODCOCK

spring, birds first appear from the coastal region of New Hampshire and thence proceed northeastward. If their breeding grounds are in northern Maine, Quebec, or the Arctic, they pass up through the river valleys, especially the Kennebec and Penobscot Valleys; but if their destination lies in the Maritime Provinces or the regions of the Gulf of St. Lawrence and Labrador, they continue along the coast to the Bay of Fundy. Swans, geese, ducks, shorebirds, and the majority of landbirds follow these routes and return over them in fall. The following dates indicate the main migratory flights:

Waterfowl: 1 April–25 April; 1 October–10 November
Shorebirds: 5 May–5 June; 25 July–20 September
Landbirds: 25 April–1 June; 15 August–1 October

Inland, winter birdlife is distinguished by the more or less regular appearance of Northern Shrikes (a few), Evening Grosbeaks, Common Redpolls, American Tree Sparrows, and Snow Buntings, and by sporadic incursions of Snowy Owls, Pine Grosbeaks, Red Crossbills, and White-winged Crossbills. Along the coast, countless loons, ducks, and gulls gather in harbors and river mouths, where they find both food and shelter. Sea birds, such as Razorbills, Thin-billed and Thick-billed Murres, Dovekies, and Black Guillemots, come close to shore in blustery weather. Purple Sandpipers frequent the rocky shores of headlands and islands, where at low tide they feed among the rockweed.

Authorities

Richard B. Anderson, James Bond, Carl W. Buchheister, Edward F. Dana, Alfred O. Gross, Charles E. Huntington, Howard L. Mendall, Daisy Dill Norton, Christopher M. Packard, Wendell Taber.

Reference

Enjoying Maine Birds. An Aid to Finding, Studying, and Attracting Birds in Maine. Edited by Olin Sewall Pettingill, Jr.; revised by Richard B. Anderson and Irving Richardson. Maine Audubon Society (headquarters at Gilsland Farm, Falmouth, Me. 04105), 1972.

BANGOR
Bangor Bog

Probably one of the best and most readily accessible places in central Maine for observing a good variety of small landbirds is **Bangor Bog.** From Bangor, drive north on State 15, turn right on Stillwater Avenue, and proceed for about 4 miles until the road starts to pass over a quarter-mile of conspicuous swampland. This is a small section of the Bog; the main part lies to the left and extends 10 miles northwestward.

Much of the Bog has a sphagnum mat with tamarack, black spruce, and deciduous shrubs, but there are many areas of open water and many 'islands' supporting alders, gray birches, and aspens. A dense growth of alders surrounds the Bog, while on the higher margins there are woods of mixed growth.

In this Bog, Ora W. Knight, one of Maine's pioneer ornithologists, made many of his notable life-history studies, including that of the Palm Warbler. Look for this bird, which still breeds in the open part of the Bog.

To observe the birdlife of the Bog and vicinity under the most favorable conditions, visit it in early morning between late May and early July. From the roadside there is an excellent opportunity of seeing, or at least hearing, American Bitterns, Ruffed Grouse, Traill's Flycatchers, Hermit Thrushes, Veeries, Chestnut-sided Warblers, Palm Warblers, Northern Waterthrushes, Wilson's Warblers, Canada Warblers, and Swamp Sparrows. Look for Common Goldeneyes on the open water and try to flush an American Woodcock from among the alders.

BAR HARBOR

Mount Desert Island | Acadia National Park | Little Duck Island | Ship Harbor | Seawall Bog | Bass Harbor Marsh

If there is one area typifying the rugged grandeur of 'the rock-bound coast of Maine' and at the same time promising highly productive bird finding, it is **Mount Desert Island.** Here granite-domed Cadillac Mountain looks down on lakes and ponds that mirror the forests around them, and, far beyond, to deeply cut harbors, bold headlands battered by crashing surf, and the open, cold-blue sea scattered with spruce-topped islands to the horizon. A very large island, Mount Desert comfortably accommodates a long-famous summer resort while providing ample area besides for the largest of six units among the 65 square miles belonging to **Acadia National Park**—the other units are on the mainland and outer islands.

On reaching Mount Desert Island, visit Park headquarters in Bar Harbor on the east side of the Island for a map showing the roads and the excellent system of trails, for a list of native birds, and for information on transportation to three of the outlying islands—Little Duck, Ironbound, and Ship—that have sea-bird colonies. **Little Duck Island,** 5 miles southeast of Mount Desert, has probably the broadest representation of species: Leach's Storm Petrels, Double-crested Cormorants, Common Eiders, Herring and Greater Black-backed Gulls, and Black Guillemots. The ideal time for a visit is late June or early July. Choose a day when the sea is calm, as landing on the Island is otherwise hazardous. Little Duck Island has been a sanctuary of the National Audubon Society since 1934.

One of the best places on Mount Desert Island for small landbirds in June and early July is the vicinity of **Ship Harbor,** a tiny inlet just southeast of McKinley in the southern part of the Island. Inland from its rocky shores, the forest, chiefly of conifers, has a remarkable population of breeding warblers. As many as eighteen species have been recorded in the summer.

The Yellow Warbler is common in brushy areas around McKinley. To find other warblers, go half a mile south on State 102A, until it meets the road to Bass Harbor and swings east. From this junction to the head of Ship Harbor, three-quarters of a mile distant, investigate the woods on both sides of the road. Along the south side in a few stands of spruce not extensively lumbered, Northern Parula Warblers, Magnolia Warblers, Black-throated Green Warblers, and Bay-breasted Warblers are common, and careful searching may yield a few Blackburnian Warblers and possibly a pair or two of Cape May Warblers. The first nest of the Cape May Warbler recorded in the United States was found here in 1936.

Elsewhere near the road the woods are of a mixed deciduous-coniferous growth varying in height and density. In them are a fair number of Black-and-white Warblers, Nashville Warblers, Ovenbirds, Canada Warblers, and American Redstarts. Where there is low, marshy growth with alders—particularly, far south of the road near the ocean—there are Tennessee Warblers and Wilson's Warblers. Sharing some of these warbler habitats are Yellow-bellied Flycatchers, Traill's Flycatchers, Golden-crowned Kinglets, Hermit Thrushes, Swainson's Thrushes, and White-throated Sparrows.

A half-mile east of the Ship Harbor parking lot and opposite the entrance gate to 'Wonderland' is **Seawall Bog,** sometimes called 'Big Heath.' This large area, less than 100 yards from the road, has an extensive sphagnum mat with deciduous shrubs such as sheep laurel and leatherleaf, and thick clumps of black spruce. Palm Warblers and Common Yellowthroats are summer residents here, and it is the only area on Mount Desert Island where Lincoln's Sparrows are known to breed. In the spruces on and around the Bog, look for Yellow-bellied Flycatchers, Boreal Chickadees, and Ruby-crowned Kinglets.

Bass Harbor Marsh, a salt marsh not far away, has Sharp-tailed Sparrows nesting along with Savannah Sparrows. To reach the Marsh, drive north from McKinley on State 102A for one mile and turn right on State 102. After about a quarter-mile the Marsh is clearly visible on both sides of the road.

Reference

Native Birds of Mount Desert Island. By James Bond. 2d rev. ed. Academy of Natural Sciences of Philadelphia (19th Street and the Parkway, Philadelphia, Pa. 19103), 1971.

BATH
Popham Beach | Sugar Loaf Islands | Heron Islands | Reid State Park

Drive south from Bath on State 209 to **Popham Beach,** near the tip of a small peninsula extending eastward between the Kennebec River and the ocean. On the way the road passes a grassy salt marsh through which several creeks meander and in which Sharp-tailed Sparrows nest in June and early July. They are easily seen only in early morning when giving their flight songs.

At Popham Beach, notice two dome-shaped islets in the mouth of the River: the northern one, **Upper Sugarloaf,** the southern, **Lower Sugarloaf.** Several hundred pairs of Common Terns and fewer pairs of Roseate and Arctic Terns nest there. Engage a boat—a rowboat will do if the oaring is efficient and strong—and investigate the islands. Make the trip when the tide is not running lest the boat be unmanageable in the strong currents.

The **Heron Islands,** a group of closely adjoining ledges not far out in the ocean south of Popham Beach, are nesting sites for many Herring Gulls and smaller numbers of Double-crested Cormorants and Greater Black-backed Gulls. For reaching the Islands a powerboat is recommended unless the sea is unusually calm.

In May and September, many transient shorebirds feed on the wide, sandy beaches of the eastern and southern shores of the peninsula. Some species to be expected are the Semipalmated Plover, Black-bellied Plover, Ruddy Turnstone, Whimbrel, Greater and Lesser Yellowlegs, Least Sandpiper, Dunlin, Semipalmated Sandpiper, and Sanderling.

Reid State Park (792 acres) offers good opportunities for bird finding, along with some of Maine's finest coastal headland scenery. There is a variety of habitats—salt marshes, sand beaches, rocky coast, and woods, both coniferous and mixed deciduous-coniferous—and all are fairly close together. Although the Park is often crowded in summer, few visitors venture into the best bird-finding areas.

In summer, look for Pileated Woodpeckers, Olive-sided Flycatchers, Red-breasted Nuthatches, Brown Creepers, Magnolia Warblers, and Blackburnian Warblers in the woods; Sharp-tailed Sparrows in the salt marshes; and Northern Gannets, Common Eiders, summering scoters, and Common and Arctic Terns offshore from Griffith's Head. In late summer the flats in the salt marshes and along the Little River near Todd's Point are especially good for shorebirds, including such possibilities as a Willet or Hudsonian Godwit. (Ipswich) Savannah Sparrows winter in the dune grasses.

To reach the Park from Bath, take US 1 across the Kennebec River to Woolwich, turn right on State 127 for 13 miles, then right again on a well-marked road to the entrance. Ask here for a map of the area.

BIDDEFORD
Biddeford Pool | Fletcher's Neck

For transient shorebirds, drive south from Biddeford on US 1 and take State 9 east for about 8 miles to the junction with State 208. Proceed on State 208 for about a mile to **Biddeford Pool,** a tidal bay enclosed by barrier dunes and ledges with just one outlet. When the tide recedes through this single channel, it exposes mud flats unusually rich in organic matter attractive to transient shorebirds and making the Pool one of the best places in Maine for seeing a wide variety of species. From mid-July through September, the usual 'peeps' to be expected on the coast in migration appear in impressive numbers, along with such strong possibilities as Piping and Black-bellied Plovers, Whimbrels, Willets, Short-billed Dowitchers, and Hudsonian Godwits. Try to be at the Pool when an incoming tide forces the birds to the edge near the road. Other birds that may show up are terns—the Common, Arctic, and Roseate—and Bonaparte's Gulls.

Close by the Pool is **Fletcher's Neck,** a peninsula extending into the ocean. This is reached by continuing on State 208 about 0.2 mile, turning right at the first fork in the road, and then right again on the shore road, which makes a loop along the rocky coast. From the Neck, look offshore any time of year for shearwaters, jaegers, and occasional Northern Gannets. During winter, look for large rafts of eiders and scoters and a few Harlequin Ducks. In late September, Purple Sandpipers begin to appear on the ledges and are usually present from then on through winter.

BRUNSWICK
Merrymeeting Bay

For congregations of spring-migrating waterfowl, drive north from Brunswick on State 24 across the Androscoggin River to Topsham, follow State 24 for about a mile, under a railroad overpass, and turn right on Foreside Road. Continue for about 4 miles, then turn right again on the road to Pleasant Point, a small peninsula projecting into **Merrymeeting Bay.** This body of water, formed by the confluence of the Kennebec and Androscoggin Rivers, is a stopping point for enormous numbers of Canada Geese from late March, when the ice goes out, until late April. Smaller numbers of other waterfowl may be present, among them Whistling Swans, Snow Geese, American Black Ducks, Blue-winged Teal, Greater Scaups, Common Goldeneyes, and Red-breasted Mergansers.

CALAIS
Moosehorn National Wildlife Refuge

To witness the American Woodcock's unique flight song, a striking display of aerial maneuvers coupled with vocal and mechanical sounds, visit the **Moosehorn National Wildlife Refuge** (22,665 acres), 'way down east' in Maine. During the height of the singing season—between 20 April and 15 May—the birds perform for about 45 minutes after sundown and again before dawn.

The Refuge, established in 1937 primarily for woodcock and presently conducting management programs for woodcock and waterfowl, has two units about 20 miles apart. The Baring Unit (16,065 acres), site of the Refuge headquarters and visitor center, borders on the St. Croix River south of Calais. The smaller Edmunds Unit borders on the tidal waters of Cobscook Bay, farther south of Calais along US 1 between Dennysville and Whiting.

Although the Refuge embraces a few small lakes, fresh-water marshes, bogs, and some abandoned farmland, most of the area is forested with second-growth stands of beech, maple, birch, aspen, and pine on the drier uplands, and spruce, fir, and tamarack in lowland situations and bogs. Over 200 species of birds have been identified in these and other habitats on the Refuge and 130 have been found nesting. These include the American Black Duck, Green-winged and Blue-winged Teal, Wood Duck, Ring-necked Duck, Common Goldeneye, Hooded and Common Mergansers, Northern Goshawk, Ruffed Grouse, Saw-whet Owl, Gray Jay, Northern Raven, Boreal Chickadee, Philadelphia Vireo, Rusty Blackbird, Lincoln's Sparrow, and no less than twenty-three warblers.

From late May to October, information on birds is available at the Baring Unit visitor center, 3 miles from Calais at the junction of US 1 and the Charlotte Road. Anyone wishing to observe flight-performing woodcock should take US 1 north from Calais, pass the Refuge boundary marker, cross over the marshy Magurrewock Stream, and take the first left just beyond. From this point, which is near the entrance to the visitor center, stop where there are clearings along the road for the next 8 miles, and listen. Provided the time is right—soon after sunset or about an hour before sunup—one can be practically certain to hear the flight songs. At the same time, as well as throughout the day, one may also hear Common Snipes winnowing.

For headquarters of the Edmunds unit, where woodland and coast merge, take US 1 south from Calais to just 5 miles beyond Dennysville. Double-crested Cormorants and many gulls are usually present here, and in winter, American Black Ducks, Common Goldeneyes, Buffleheads, and Common Mergansers are numerous as well.

GRAND MANAN, NEW BRUNSWICK

Grand Manan Island | Kent Island | Yellow Murr Ledge

Grand Manan Island at the mouth of the Bay of Fundy is the largest—15 miles long and 7 miles wide—of a closely neighboring group comprising the Grand Manan Archipelago. Although politically a part of New Brunswick, the Archipelago is geographically closer to Maine, its nearest part being less than 6 miles from Quoddy Head.

The Archipelago offers a rich variety of 'down east' birds in a setting of great charm. Write to Coastal Transport Limited, P.O. Box

26, Saint John, New Brunswick, for information about ferry service to Grand Manan Island.

The western frontage of Grand Manan Island rises abruptly for 50 to 400 feet above the high-tide mark in spectacular reddish cliffs cut by several fjord-like indentations such as Dark Harbour. The eastern frontage is low with many coves and harbors. Here most of the Archipelago's 2,500 human inhabitants live; fishing is their main source of livelihood. North Head, where the ferry docks, is the chief port of entry. Improved gravel roads lead from North Head to four other villages and to the outer reaches of the Island. Thus cars are a great convenience, and visitors frequently bring them on the ferry.

The interior of the Island is generally high and heavily wooded. Toward the eastern frontage there are mound-like hills and poorly drained lowlands with sphagnum bogs and small marshy ponds; toward the western frontage the Island is plateau-like, with occasional deep gulleys. A few original stands of spruce and fir remain in pocket areas, but for the most part these have been replaced by a secondary growth of beech, aspen, birch, maple, and other deciduous varieties. Near the edge of the western cliffs and at Northern Head and Southern Head there is a thick scrubby growth almost wholly spruce.

Several persons have given considerable attention to the birdlife of this Archipelago, with the result that over 275 species have been recorded and at least 75 have been found to nest. Grand Manan Island itself offers a wide assortment of land species, many of which are characteristic of the northern coniferous forests. By car, visit the following areas in June or early July:

1. *Castalia Marsh.* This is easily reached from the Castalia-Woodward's Cove road, which passes the Marsh. A small colony of Sharp-tailed Sparrows is here, and sometimes, near the brushy edges, a few pairs of Lincoln's Sparrows.

2. *Dark Harbour.* Drive along the road that begins near the Castalia Marsh, goes west through the interior of the Island, and eventually passes along the side of a ravine leading to a beautiful fjord. Almost anywhere along the road or near the stream at the foot of the ravine are Yellow-bellied Flycatchers, Red-breasted

Nuthatches, Winter Wrens, Golden-crowned Kinglets, Red-eyed Vireos, Northern Parula Warblers, Magnolia Warblers, Black-throated Blue Warblers, Black-throated Green Warblers, Bay-breasted Warblers, American Redstarts, and (irregularly) White-winged Crossbills.

3. *Deep Cove.* In the scrubby, thick growth west of the road running from Deep Cove to Southern Head, Boreal Chickadees are as common as they are anywhere in the Archipelago, and Blackpoll Warblers are sometimes numerous. White-throated Sparrows and Slate-colored Juncos reside where there are brushy pastures.

4. *Southern Head.* In addition to the grandeur of the high cliffs that drop 200 feet into deep water, and the strange rock formation called the 'Southern Cross,' is a sizable number of Black Guillemots, some of which undoubtedly nest in the crevices of the cliffs. While there are nesting colonies of the species that can be viewed to better advantage elsewhere in the Archipelago, here the observer has an unusual opportunity to watch, from a high elevation, these birds diving, 'flying' under water, and performing various courtship antics.

Although many smaller islands of the Archipelago have a fair number of species of nesting birds, **Kent Island,** lying off the southeastern coast of Grand Manan Island, has the greatest variety. At Seal Cove on Grand Manan, find a fisherman willing to take passengers to Kent Island in his powerboat. Since there is a sheltered harbor, landing is always safe, even in the roughest weather. Owned by Bowdoin College, Kent Island is the site of the Bowdoin Scientific Station, a permanent facility for ornithological and other biological investigations during the summer months. Bird finders are welcome. If one contemplates a visit, write for further information and arrangements to the Director, Bowdoin Scientific Station, Department of Biology, Bowdoin College (Brunswick, Maine 04011), in winter; or Kent Island, Grand Manan, New Brunswick, in summer.

Kent Island is quite low-lying, with shores of ledges and loose rock; one part is thickly forested with spruce, the rest is open and pasture-like. The Island has one of the largest colonies of Herring Gulls on the Atlantic coast. Seemingly everywhere in the soft soil are scattered the burrows of Leach's Storm Petrels. Other seabirds are Common Eiders and Black Guillemots. Early in spring, one or two pairs of Northern Ravens nest on the higher ledges or in the trees; in summer they patrol the shores. In the spruces, Myrtle Warblers, Blackpoll Warblers, and Golden-crowned Kinglets are common. Brown Creepers are sometimes present and have been known to nest.

Many Razorbills nest on the **Yellow Murr Ledge,** 6 miles south of Kent Island. Except for a few pairs nesting on Machias Seal Island (*see under* **Machias**), and Matinicus Rock (*see under* **Rockland**), this is the only known breeding population near any coast of the United States. The birds nest among crevices formed by loose rocks resting on the Ledge. Possibly the fisherman taking passengers to Kent Island will consent to make a trip to this isolated islet. Only in mildest weather can the trip be undertaken. Navigation is extremely dangerous, for there are many adjacent ledges that are just slightly submerged at certain tide levels. Moreover, Yellow Murr Ledge has no sheltered lee or cove; when landing, a boat has to be drawn up directly on the somewhat steep incline of the shore.

LEWISTON
Thorncrag

On the northern outskirts of this city is **Thorncrag,** a sanctuary owned by the Stanton Bird Club. Visitors are welcome. Thorncrag consists of 209 acres of varied habitat: a meadow, a large brook, and a small marsh; high ground with bare ledges; extensively wooded areas, some supporting mixed deciduous trees and other supporting coniferous trees, principally white pines and hemlocks. There are many trails. The sanctuary is particularly good for both breeding and transient landbirds.

The sanctuary is easily reached from the center of the city by driving north on US 202, turning right on Montello Street, and staying on it until it passes over Montello Heights. A conspicuous sign marks the entrance to the sanctuary.

MACHIAS
Machias Seal Island | Old Man Island

The best opportunity for observing Atlantic Puffins south of the Gulf of St. Lawrence is on **Machias Seal Island,** 10 miles offshore in the Bay of Fundy.

Much of the Island's 15 acres of surface is grass-covered soil. There are no trees. On the highest part stand a lighthouse and a fog-signal station with accessory buildings, owned and operated by the Canadian Government. For an extended stay, obtain permission from the Department of the Marine, St. John, New Brunswick.

From mid-May to mid-August a thriving colony of Atlantic Puffins, together with Razorbills, occupies the ledges and loose rock near the southwestern shore. Nesting over the Island in places not occupied by Government buildings are a few Common Terns, many

ATLANTIC PUFFIN

Arctic Terns, and, deep in burrows, countless Leach's Storm Petrels.

To reach Machias Seal Island, drive east from Machias on US Route 1 for 4 miles and then right on State Route 191 to the small fishing village of Cutler, where one may hire a fisherman with a powerboat for a trip to the Island—though landing is possible only in a calm sea.

When going to or from Machias Seal Island, ask the fisherman to stop at **Old Man Island,** locally called 'The Old Man,' a smaller island of 10 acres near the mainland and halfway between the entrance to Cutler Harbor and Machias Bay. As on Machias Seal Island, one can land only on a calm day. Although the Island's sides are rocky, precipitous, and devoid of vegetation, its higher surfaces, some 40 feet above the high-tide mark, have thick turf, a rich growth of small plants and bushes, and scattered stands of stunted spruce and fir. The trees, most of them dead, their trunks and branches naked and bleached, give the Island an awesome aspect. A deep, 5-foot-wide chasm cleaves the Island, and a boulder wedged in the upper part provides a natural bridge from one side to the other. Nesting in trees and on the barren high rocks are Double-crested Cormorants; also nesting are Common Eiders, Greater Black-backed Gulls, and Herring Gulls. One can have a good view of these birds from the boat even when landing is impossible.

MILLINOCKET
Baxter State Park

Less than 20 miles northwest of this paper-making town in north-central Maine looms Katahdin, the state's highest mountain, amid a vast wilderness of lesser peaks, woods, lakes, and streams. **Baxter State Park,** the gift of one man, Percival P. Baxter, to the people of the State of Maine, includes Katahdin and much of the area immediately surrounding it—216 square miles in all. Automobile roads lead to the Park boundaries; within it are many miles of improved foot trails. The Park's natural grandeur and solitude have been reserved for those willing to leave their cars and walk. Anyone wishing to find a great variety of coniferous-forest birds while hiking, mountain climbing, and camping has here the best opportunity afforded in Maine, if not in the eastern United States.

Baxter Peak, Katahdin's highest point, is 5,267 feet above sea level; six subordinate peaks are all over 4,600 feet. A peculiar feature of the mountain is a broad plateau, 3 miles in length, called 'the Tableland.' The surface is scattered with loose granite blocks and slabs, between which there is some soil that supports coarse grasses and wiry sedges. Biologically it shows subalpine, if not true alpine, affinities. The mountain has several 'peninsulas,' or arms, that reach outward and partly enclose enormous glacial basins, the most spectacular of which is the Great Basin. Its floor, at an elevation of 2,910 feet, is thickly forested with spruce intermixed with some white birch. A mirror-like body of water, Chimney Pond, covers approximately 8 acres and greatly enhances the scenic beauty of the Basin. From the floor to the rim of the Basin rise sheer granite cliffs with patches of spruce that become increasingly scrubby and finally disappear before reaching the top. The 'lowlands' surrounding the mountain were once subjected to lumbering operations and fires in wide areas, with the result that they have developed a secondary growth of aspen, gray birch, and maples. Nevertheless, there are still large, dense stands of conifers south and west of the mountain, particularly within the confines of the Park.

In order to observe the birdlife at its best on or near Katahdin, plan to be here between 1 June and 15 July. If possible, schedule the visit in early June before the Park is crowded with visitors, and be prepared for hordes of mosquitoes and blackflies. Write to Baxter State Park (116 Aroostook Avenue, Millinocket, Maine 04462) for a guide to the Park, together with rules and regulations. Field equipment should include an up-to-date edition of the Appalachian Mountain Club's *Maine Mountain Guide*, obtainable at the organization's clubhouse, 5 Joy Street, Boston, Massachusetts 02101. Its map of Katahdin and instructions are indispensable when entering the area. Anyone staying overnight in the Park must bring his own food and camping equipment.

From Millinocket, drive northwest over the Park Road (Sourdnahunk Road) to Millinocket Lake, a distance of 8 miles. From here the road continues northwestward as a causeway between Millinocket and Ambajejus Lakes, enters a once severely burned-over area, and soon passes, on the left, a road to a private camp. Part of the burned area is low-lying and boggy, while the remainder is somewhat hilly; among the tall black stumps stand a few live pines and spruces and a new deciduous growth. An early-morning walk along the lake shores, the causeway, and the road through the burned area is well worthwhile. Some species to look for are the American Black Duck, Ring-necked Duck, Common Merganser, Pileated Woodpecker, Black-backed Three-toed Woodpecker (occasional), Traill's Flycatcher, Olive-sided Flycatcher, Northern Raven, Nashville Warbler, Palm Warbler, Wilson's Warbler, and Rusty Blackbird.

Proceed along the Park Road, which turns right about 7 miles from the causeway and soon passes between Upper and Lower Togue Ponds to Park headquarters and gatehouse at Togue Pond, 18 miles from Millinocket. After registering here, take the Roaring Brook Road to the right and go the 8 miles to its end at the Roaring Brook Campground. (To stay here, one must make a reservation well in advance of the trip.) Leave the car and walk up the Basin-Chimney Ponds Trail about 2 miles to Basin Ponds. The woods along this trail, a mixed growth of spruce, maple, birch, and beech, are excellent for Northern Goshawks, Yellow-bellied Sapsuckers, Solitary Vireos, Philadelphia Vireos, and Black-throated Blue Warblers. The Mourning Warbler resides in more open areas, especially where there are raspberry patches; the Black-backed Three-toed Woodpecker has been observed here.

From Basin Ponds the trail climbs steeply, then leisurely, to Chimney Pond in the Great Basin. The Pond is 1.5 miles from Basin Ponds and about 2.5 hours' walking time from Roaring Brook. There are lean-tos and a bunkhouse. Because the Great Basin region is particularly rewarding, an overnight stop is advised so that one will have time for searching and will be present for the impressive evening and early-morning chorus of bird songs. Among species in the area are the Yellow-bellied Flycatcher, Boreal Chickadee, Red-breasted Nuthatch, Swainson's Thrush, (Bicknell's) Gray-cheeked Thrush, both Ruby-crowned and Golden-crowned Kinglets, Bay-breasted Warbler, Blackpoll Warbler, Pine Siskin, and occasionally the Red Crossbill and White-winged Crossbill.

If the bird finder wishes to climb Katahdin from Chimney Pond,

and turn right on US 1) to Oak Hill; turn off left (east) on State 207 to Prouts Neck; at 4.2 miles, turn right on Ferry Road, which immediately crosses a golf course and dead-ends in a parking area overlooking a tidal basin from the northeast side of Prouts Neck. From the parking area, walk toward the ocean on the shore of Prouts Neck where it converges with Pine Point across the basin, forming the Prouts Neck–Pine Point Narrows, through which there is a rip tide.

When the tide is out, there are extensive sand flats along the basin to the Narrows. From mid-April to June and from late August to late September these flats are ideal feeding grounds for many shorebirds such as Semipalmated Povers, Lesser Golden Plovers, (late summer only), Black-bellied Plovers, Ruddy Turnstones, Greater Yellowlegs, Lesser Yellowlegs, Least Sandpipers, Dunlins, Short-billed Dowitchers, Semipalmated Sandpipers, Sanderlings, and an occasional Whimbrel, Red Knot, and Hudsonian Godwit.

Waterbirds and waterfowl of many kinds are attracted to the basin and Narrows from October through March; when the weather is severe, they appear in great numbers. One can expect Common Loons, Greater Scaups, Common Goldeneyes, Buffleheads, Oldsquaws, Common Eiders, White-winged Scoters, Surf Scoters, Black Scoters, Red-breasted Mergansers, Greater Black-backed Gulls, Herring Gulls, and Ring-billed Gulls. Look also for Red-throated Loons, Glaucous Gulls, Iceland Gulls, and Bonaparte's Gulls.

Backtrack on State 207 to Oak Hill and continue southwest on US 1 to Dunstan; turn left on State 9 (Pine Point Road) to the Scarborough Marsh Nature Center, a mile farther along on the left. The Center, operated by the Maine Audubon Society and the State Department of Inland Fisheries and Wildlife, has a year-round observation deck that overlooks a broad section of the Scarborough Salt Marshes.

Through the Scarborough Marshes the Nonesuch and other rivers meander and eventually reach the tidal basin and pass through the Prouts Neck–Pine Point Narrows to the sea. For finding birds there is no better section of the Marshes than the vicinity of the Nature Center. In June and July, Willets and Sharp-tailed Sparrows are among the breeding species; in fall and early spring, Mallards and American Black Ducks are common, and Common Pintails, Blue-winged Teal, American Wigeons, and Northern Shovelers sometimes put in an appearance. During spring migration, in late March and April, large numbers of Canada Geese and a few Snow Geese stop by. In summer, Green Herons and Black-crowned Night Herons are abundant; Great Blue Herons, Snowy Egrets, and Glossy Ibises show up regularly, and Little Blue Herons and Great Egrets on occasion.

From the Nature Center one may continue east on State 9 to the village of Pine Point and turn left to Pine Point for a view of the tidal basin and Prouts Neck–Pine Point Narrows from the southwest side.

ROCKLAND

Matinicus Island | No Man's Land
Matinicus Rock | Monhegan Island

Three times a week a mail boat leaves Rockland on the west shore of Penobscot Bay, passes seaward between Owls Head and Vinalhaven, and in about two hours reaches the little fishing hamlet of Matinicus on **Matinicus Island,** approximately 15 miles from the nearest point on the mainland, and largest of eight islands comprising the Matinicus Archipelago. The boat returns the same day. Write the Rockland Chamber of Commerce for the summer schedules. Take this trip in June or July, preferably in late June, and stop off at Matinicus for a few days to explore the Archipelago. Several homes offer overnight lodging and meals.

Matinicus Island, nearly 2 miles long and a mile wide, is low-lying and partly forested with thick-growing spruces. Here are excellent opportunities for finding typical coniferous-forest birds such as Swainson's Thrushes, Golden-crowned Kinglets, Northern Parula Warblers, Myrtle Warblers, Black-throated Green Warblers, Bay-breasted Warblers, and White-throated Sparrows. Yellow-bellied Flycatchers and Olive-sided Flycatchers have been seen nesting. Of special interest are the sea-bird colonies on neighboring islands, which the bird finder may reach by employing a fisherman to take him in a powerboat. If there is a severe ground swell and wave action, landing may be impossible, since their shorelines are rocky and precipitous, without protected harbors. Still, the birds may be readily observed if the boat circles the islands close to shore.

Maryland and the District of Columbia

CHANDLER S. ROBBINS

ORCHARD ORIOLE

Chesapeake Bay is without question the outstanding area in Maryland from an ornithologist's point of view. With its adjoining estuaries and tidal marshes, the Bay is a focal point for more than a million migrating waterfowl annually. Hundreds of thousands of ducks and geese and many thousands of swans are attracted by the extensive beds of wild celery, sago pondweed, redhead grass, widgeon grass, and eel grass, and the lush stands of wild rice and three-square. Most sections of the Bay also teem with animal food in the form of fish and of mollusks, crustaceans, and other invertebrates. Numerous birds besides waterfowl—among them loons, grebes, cormorants, herons, rails, coots, shorebirds, gulls, and terns—take advantage of the abundant food and concentrate here in large numbers. The arrivals and departures of the great flocks of migratory waterfowl furnish some of the most spectacular ornithological sights in America.

Stretching from the Allegheny Plateau to the Atlantic Ocean, Maryland boasts the largest variety of breeding birds of any state its size. Owing to its peculiar shape and position in relation to the ranges of birds with southern and northern affinities, rare and remarkable combinations of northern and southern breeding species are included in the Maryland avifauna. The westernmost county, for example, although not rising more than 3,360 feet above sea level, has such breeding birds as the Saw-whet Owl, Hermit Thrush, Nashville Warbler, Northern Waterthrush, Mourning Warbler, and Purple Finch. In southeastern Maryland, on the other hand, breeding birds include southern species such as the Louisiana Heron, Wilson's Plover, Gull-billed Tern, Chuck-will's-widow, Brown-headed Nuthatch, Swainson's Warbler, and Boat-tailed Grackle. Few states can exceed Maryland's record of twenty-eight species of regularly nesting warblers.

There are four principal physiographical provinces in Maryland: the Allegheny Plateau, the Ridge and Valley Province, the Piedmont Plateau, and the Coastal Plain. The Allegheny Plateau west of Frostburg, in the westernmost part of the state, is the highest, averaging about 2,500 feet above sea level. It is an undulating plateau crossed diagonally, northeast to southwest, by several ridges that rise some 500 feet above it. The highest point in the state is here, on Backbone Mountain (3,360 feet). Traveling eastward on US 40 between Frostburg and Frederick, one passes through the Ridge and Valley Province, where numerous parallel ridges, oriented north-northeast, range up to 2,000 feet. Most of its valleys are narrow and little used for farming. To the east is the fertile Hagerstown Valley, an extension of Virginia's Shenandoah Valley, which is much wider and intensively farmed. The Piedmont extending eastward from the base of Catoctin Mountain near Thurmont to the Fall Line, which passes through Washington, D.C., Baltimore, and Elkton, is gently rolling, with elevations of from 300 to 800 feet, and consists mainly of agricultural lands with scattered woodlots. The Coastal Plain comprises all state territory below the Fall Line. It is divided by Chesapeake Bay into the Eastern Shore (part of the Delmarva Peninsula)—flat and low, with elevations under 100 feet, and the Western

On the Allegheny Plateau, communities of plants and animals are much more northern in character than elsewhere in the state. Many bird species appearing in the oak-pine region are also present here; others are for the most part restricted to this western province. A few mountain (northern) species, however, appear also on the higher ridges and in the cooler ravines of the Ridge and Valley Province. The majority of the forest trees are those characteristic of the hemlock-northern hardwood region, with nearly pure stands of hemlock in some ravines.

Scattered through the Plateau are boreal bogs inhabited by several species of birds not found elsewhere. Vegetation in these bogs consists of sedge meadows and bog heaths interspersed with patches of taller shrubs, such as alder and rhododendron, and trees, usually hemlock, red spruce, yellow birch, and red maple. Even the agricultural areas are inhabited by a few nesting birds that are rare in or absent from the oak-pine region: the Cliff Swallow, Bewick's Wren, Bobolink, and Savannah Sparrow. Henslow's Sparrows are beginning to colonize abandoned fields. Other breeding birds restricted largely to the Allegheny Plateau are:

WOODLAND AND EDGE

Ruffed Grouse
Wild Turkey (*local*)
Black-billed Cuckoo
Yellow-bellied Sapsucker
Least Flycatcher
Black-capped Chickadee
Hermit Thrush (*local*)
Solitary Vireo
Golden-winged Warbler
Magnolia Warbler
Black-throated Blue Warbler
Black-throated Green Warbler
Blackburnian Warbler
Chestnut-sided Warbler
Mourning Warbler (*local*)
Canada Warbler
Rose-breasted Grosbeak
Slate-colored Junco (*local*)

BOREAL BOG

Saw-whet Owl
Traill's Flycatcher (*local*)
Golden-crowned Kinglet (*local*)
Nashville Warbler (*local*)
Northern Waterthrush
Purple Finch
Swamp Sparrow

At the opposite end of the state, the loblolly pine region occupies the southern quarter of the Western Shore and the southern half of the Eastern Shore. Loblolly pine is the predominant tree, growing either in pure, even-aged stands or mixed with many of the same hardwoods that are characteristic of the oak-pine region. In swamps along many of the streams, however, are other southern species, including bald cypress, water oak, and red bay. Most breeding birds of the oak-pine region figure also in the loblolly pine region, where a few more southern species are associated with them. These include the Little Blue Heron, Cattle and Great Egrets, Chuck-will's-widow, Brown-headed Nuthatch, Swainson's Warbler (local), and Yellow-throated Warbler.

Associations of plants and breeding birds in tidewater habitats are quite different from those of other regions, and each of numerous types of tidal marshes along the bays and estuaries has a distinct avifauna. Fresh and brackish marsh types include American three-square, Olney three-square, river bulrush, cattail, wild rice, phragmites (quillreeds), big cord grass, and switch grass. Salt-marsh types are salt-marsh cord grass, salt-meadow cord grass, salt grass, needle-rush, salt-marsh bulrush, blackrush, and glasswort. Characteristic breeding birds of tidewater habitats are:

TIDAL MARSHES

American Bittern
Least Bittern
American Black Duck
Gadwall (*local*)
Blue-winged Teal
Northern Harrier
King Rail
Clapper Rail (*salt marshes*)
Virginia Rail
Black Rail
Common Gallinule
Willet (*salt marshes*)
Marsh Wren
Sedge Wren (*local*)
Eastern Meadowlark
Red-winged Blackbird
Savannah Sparrow (*local*)
Henslow's Sparrow (*local*)
Sharp-tailed Sparrow (*salt marshes*)
Seaside Sparrow (*salt marshes*)
Swamp Sparrow (*local*)

gheny Plateau. The Black Vulture and Red-bellied Woodpecker can be seen throughout the Coastal Plain and the Piedmont, and the Brown-headed Nuthatch, Sedge Wren (locally), and Boat-tailed Grackle (locally) from Blackwater Refuge southward. Other species that regularly winter as far north as the Eastern Shore include the Eastern Phoebe, Gray Catbird, Brown Thrasher, Water Pipit, Pine Warbler, Palm Warbler, Common Yellowthroat, and Chipping Sparrow. Northern visitants, such as the Rough-legged Hawk, Purple Sandpiper (Ocean City), Evening Grosbeak, and Snow Bunting, appear regularly in small numbers. In the Ridge and Valley and Piedmont areas the White-crowned Sparrow winters regularly. Hawks are particularly conspicuous on the Eastern Shore, where the bird finder may observe thirty individuals of six or more species on a day's trip. Eastern Meadowlarks, Red-winged Blackbirds, and Common Grackles winter abundantly on the Eastern Shore, more sparingly or locally elsewhere. Watch for the stray Yellow-headed or Brewer's Blackbird among them.

Authorities

Irston R. Barnes, Danny Bystrak, Carl W. Carlson, William Julian, C. Haven Kolb, Jr., Frances Pope, Robert E. Stewart, Clark G. Webster, Hal Wierenga.

ANNAPOLIS
Sandy Point State Park

Seven miles northeast of Annapolis on US 50, at the west end of Chesapeake Bay Bridge, lies **Sandy Point State Park** (813 acres). Strategically located on a point that juts out into Chesapeake Bay, the Park is the most productive bird-finding spot in the Annapolis area and perhaps the very best on the Western Shore. The great variety of upland, marsh, and shore habitats attracts a large number of transients, as well as both diving and surface-feeding ducks, throughout the winter. Shorebird concentrations are spectacular at times from May through October, depending upon habitat management and precipitation. Landbird finding can be superb in the early morning when the winds are out of the west. Hawk flights often take place in April, and from September through mid-November. Snow

Buntings are most likely to appear in or near the south parking lot in November, and occasionally a Lapland Longspur is among them.

The Park also encompasses a 204-acre preserve, the Corcoran Environmental Study Area, which features a 2.7-mile nature trail. This heavily wooded habitat, both deciduous and coniferous, provides an excellent showcase for woodland birds in any season. Permission necessary to enter the Corcoran tract is available to bird finders upon request at Park headquarters, along with directions to other bird-finding areas and nature trails in the Park.

BALTIMORE

Druid Lake | Montebello Lake | Robert E. Lee Park | Loch Raven Reservoir | Gunpowder State Park | Patapsco State Park

The best bird-finding areas in the vicinity of Maryland's largest city are various parks and reservoirs. Although the reservoirs cannot compare for waterfowl concentration with Chesapeake Bay, by checking them frequently the bird finder can build up a fine local list and even expect a few rarities—especially during stormy weather. **Druid Lake** and **Montebello Lake** are generally the city's most productive reservoirs. When water levels are low during migration periods, they can also be a mecca for transient shorebirds. For Druid Lake in Druid Hill Park, go north from the center of the city on State 139 (Charles Street) and turn left on 29th Street, which runs into Druid Park Lake Drive. For Montebello Lake in Herring Run Park, go north on Calvert Street and then right on 33rd Street for about 1.5 miles.

Just north of the Baltimore city line and east of Falls Road (State 25) is **Robert E. Lee Park** (450 acres) with Lake Roland. This Park has long been a favorite bird-finding spot for Baltimoreans because of its easy access and large variety of habitats. It is especially fine during migration, and is one of the most dependable places in Maryland to find the Yellow-crowned Night Heron.

Loch Raven Reservoir, north of the city, offers a bird finder many opportunities, as do various sections of **Gunpowder State Park** (10,914 acres), above and below the Reservoir. Most of the typical breeding species of the Piedmont are found in this area, and

ing the Marsh, stop and search the pine islands in the 'sea' of marsh grass; during migration these are often profitable landbird 'traps.'

Hooper Island, at the end of State 335, 20 miles southwest of Church Creek, is a concentration point in fall for exceptionally large numbers of migrating Sharp-shinned Hawks, American Kestrels, and Yellow-shafted Flickers. The greatest flights take place during late September and October. In winter one can look out over Chesapeake Bay from the road and see enormous congregations of swans, geese, and ducks resting safely beyond the hunters' firepower.

Elliott Island, east of Hooper Island, may be reached from Cambridge by driving east on US 50 to Vienna and turning south for 18 miles on an unnumbered dead-end road. With the aid of a detailed road map, one can take paved back roads from Blackwater Refuge and pick up the Elliott road at Henrys Crossroads—a route far more productive than US 50. On this road, in winter, stop at Bestpitch Bridge over the Transquaking River and look for Rough-legged Hawks by day and Short-eared Owls at dusk. Between Henrys Crossroads and the salt marshes, the road goes through mixed, wet forest in which Summer Tanagers are almost common. In spring migration there is a wide variety of warblers. Henslow's Sparrows nest in the weedy, brushy abandoned fields along the road, especially in the last mile before Savannah Lake; but they are sparsely distributed and require searching.

The extensive brackish marshes that line the Elliott road are alive with herons, waterfowl, shorebirds, and rails before and after the waterfowl hunting season. Black Rails, the Elliott Island specialty, are rarely seen, but may be heard on calm nights, especially between 10 P.M. and 2 A.M. in May and June. They live in the short salt-meadow cord grass (*Spartina patens*) marshes, particularly in the vicinity of the Pokata Creek Bridge, about 5 miles north of Elliott.

CHESTERTOWN

Remington Farms | Eastern Neck National Wildlife Refuge

Chestertown, on the upper Eastern Shore, is in the heart of Canada Goose country. There are two superb waterfowl areas southwest of

BLACK RAIL

town, and both are excellent for other birds as well. **Remington Farms** is reached from Chestertown by driving 9 miles west on State 20, turning left at the Remington Farms sign onto Ricauds Branch–Langford Road, and proceeding to the main entrance and waterfowl sanctuary. Among thousands of Canada Geese that feed and rest here from early October through April, one can generally find Snow Geese (both white and blue phases) and occasionally a stray bird of one of the other goose species. There are always hundreds of surface-feeding ducks, primarily Mallards, but one has a good chance of spotting numerous Common Pintails and a half-dozen or more other species, except during freeze-up. A self-guided wildlife habitat tour is open from March to mid-October past fields, woods, and ponds, where a variety of landbirds including the Bald Eagle and Prothonotary Warbler may be seen. Check at the office for tips on recent observations.

From Remington Farms, return to State 20, turn left to Rock Hall, then go south on State 445 for 16 miles to **Eastern Neck National**

Valley offers excellent bird finding. From I 70 south of Frederick, exit south on New Design Road. Open fields along this road, 6 miles south of I 70, may yield nesting Ring-necked Pheasants, Upland Sandpipers, and Dickcissels. Orchard Orioles, Baltimore Orioles, and Scarlet Tanagers frequent the forest patches. Many abandoned farms offer excellent places to search for sparrows in fall. The west slope of the Valley, which becomes the foothills of Catoctin Mountain, appears to be a secondary, but very rewarding, migration route where local bird finders seek wintering Short-eared Owls and transient shorebirds and finches.

In the Monocacy area the most productive bird finding is at **Lily Pons,** probably the largest goldfish hatchery in the country. Innumerable pools are carved in the floodplain for goldfish culture, and usually there are many drained ponds whose muddy bottoms attract a wide variety of shorebirds. Such rarities as Baird's Sandpipers and Wilson's and Northern Phalaropes have been recorded. In shallow ponds, Great Blue Herons, Green Herons, and Great Egrets gather to feed on the helpless fish. Landbird and waterbird finding can be exciting because of frequent rarities.

Drive southeast from Frederick on I 270 (I 70S on old maps) for 8 miles, take State 80 west for a short distance, bear left on Park Mills Road, and watch for a sign on the right that marks the entrance. Stop at the office in the small white house and ask for permission to look for birds.

LA PLATA

Rock Point | Cobb Island | Allens Fresh | Zekiah Swamp | Cedarville and Doncaster State Forests

South of La Plata, in southern Maryland, are **Rock Point** and adjacent **Cobb Island** on the tip of a peninsula between the Potomac and Wicomico Rivers, both first-rate places for waterfowl and landbird concentrations. Hundreds of ducks and American Coots spend the winters as well as the migration seasons here. Most common ducks are Gadwalls, American Wigeons, Canvasbacks, and Ruddies, but many other species are present in smaller numbers, as are Whistling Swans and Canada Geese. The Common Loon and Horned Grebe appear during migration.

A considerable variety of habitat marks the area, including marshes, wooded swamps, upland deciduous forest, stands of loblolly pine, and several types of fields and edge—with a correspondingly large number of species. Southern birds, such as the Chuck-will's-widow and Yellow-throated Warbler, are common in summer. In winter, many half-hardy species that are rare farther north may be expected, including the Water Pipit, Rusty Blackbird, and Savannah Sparrow. Black Vultures are regular at all seasons and may be as numerous as Turkey Vultures.

To reach Rock Point, drive south from La Plata on US 301 for about 13 miles, and then southeast on State 257 to its end. For Cobb Island, turn south from State 257 on State 254 for a mile before Rock Point.

On the way back to La Plata, make a one-mile detour east on State 234 to **Allens Fresh,** where the great **Zekiah Swamp** drains into the head of the Wicomico River. The marshes, though not extensive, attract Common Snipes throughout winter, and other shorebirds during migration. The Swamp, also crossed by State 6 farther north, is difficult to traverse, but even around the edges it will reward the summer bird finder with glimpses of Prothonotary and Kentucky Warblers and perhaps a Pileated Woodpecker.

In the extensive woodlands of this area, nesting warblers are present in great profusion. Look for all the Coastal Plain species except the Swainson's and Cerulean. Two very productive woodlands to explore are the **Cedarville State Forest,** reached by going north from La Plata on US 301 to Waldorf, turning right on State 5, and left on State 382; and **Doncaster State Forest,** reached by going west from La Plata on State 6.

OAKLAND

Swallow Falls State Forest | Herrington Manor | Cranesville Bog | Deep Creek Lake | Roth Rock Fire Tower | Negro Mountain | Wolf Swamp

Oakland, in extreme western Maryland, makes an excellent base for exploring Maryland's 'north country'—the Allegheny Plateau. Restricted remnants of former spruce-hemlock bogs still harbor large numbers of such northern breeding species as Northern Water-

219 to Red House, then east on US 50 for 2.5 miles, go south on a paved road to the top of Backbone Mountain for 1.0 mile, and then right on a steep, stony road that climbs for 1.0 mile to the **Roth Rock Fire Tower.** If season and weather are suitable, a substantial hawk flight may be observed from the Tower. In spring migration, from 1 to 20 May, the ridge is often alive with woodpeckers, thrushes, and warblers, easily seen in the leafless branches. At any season a Northern Raven may pass by. Nesting around the foot of the Tower are Slate-colored Juncos. A half-mile hike down the road from the Fire Tower in the early morning may yield such species as the Pileated Woodpecker, Black-capped Chickadee, Golden-winged Warbler, Black-throated Blue Warbler, Black-throated Green Warbler, Chestnut-sided Warbler, Canada Warbler, and Rose-breasted Grosbeak.

Most of these same birds reside along the summit of **Negro Mountain.** Drive north from Oakland on US 219 for about 35 miles, then east on US 40 for about 2 miles, and park in a picnic area near the spot marked as the highest point on the route east of the Rockies. Climb southward for 0.5 mile on a gravel road to the Negro Mountain fire tower. Lumber trails, extending southward along the ridge for 0.5 mile, pass through mature and lumbered hardwood forest, where northern species such as Black-throated Blue and Canada Warblers nest in close proximity with their southern relatives, the Yellow-breasted Chat and Hooded Warbler.

Most of the species attracted to Cranesville Bog (*see above*) appear also in **Wolf Swamp.** Drive north from Oakland on US 219, then east on US 48 to the Avilton–New Germany interchange (3 miles beyond the Casselman River Bridge at Grantsville), and exit south, keeping to the right, for 1.0 mile. At this point a fair-weather road leads west for 0.4 mile to private property at the edge of the Swamp. For the most productive part of the Swamp, which includes the best stand of red spruce left in the state, one must go south along the Swamp's eastern edge for about a half-mile. The Northern Waterthrush and Canada Warbler are exceptionally abundant here, the Golden-crowned Kinglet and Blackburnian Warbler are common, and a few Hermit Thrushes, Nashville Warblers, Purple Finches, and Slate-colored Juncos are present. During spring and fall migrations, a great variety and abundance of birds appear, with particularly large numbers of such species as the Yellow-bellied, Traill's, Least, and Olive-sided Flycatchers, Cliff Swallows, Tennessee and Wilson's Warblers, and Rose-breasted Grosbeaks.

OCEAN CITY
Assateague Island National Seashore

The environs of Ocean City, facing the Atlantic on a barrier island on the Eastern Shore, are excellent for migrating waterbirds, especially those associated with salt water. In general, more species come into the coastal bays, such as Chincoteague, Sinepuxent, Isle of Wight, and Assawoman, which are protected by the barrier islands, but many unusual birds frequent the oceanside. Although all the following areas can be reached by boat from Ocean City, many sections are accessible by car, in particular those along State 528 north from Ocean City to the Delaware line.

Three good areas for shorebirds are the north tip of Assateague Island, the vicinity of the ruined west end of the 'old bridge' in West Ocean City, and, at low tide, the mud-flat islands in the bay opposite 4th Street. At various points in the bay (e.g., 26th Street) similar islands are in view. For shorebirds such as Ruddy Turnstones, Red Knots, and Purple Sandpipers, the jetty and nearby beach at the south end of Ocean City are usually rewarding. During migration it is not unusual to see, over the ocean, flock after flock of migrating Common Loons, Red-throated Loons, Horned Grebes, Double-crested Cormorants, Brant, Canada Geese, Greater and Lesser Scaups, White-winged, Surf, and Black Scoters, Red-breasted Mergansers, and Herring Gulls. Other species appearing regularly, and sometimes in large numbers, during these periods include the (Greater) Snow Goose, Common Goldeneye, Bufflehead, Oldsquaw, and Greater Black-backed, Ring-billed, and Bonaparte's Gulls. In winter, one can try for Common and King Eiders, Black-headed Gulls, and Little Gulls at the inlet.

Late summer—August and September—is a prime time for observing gulls and terns and the oceanic shearwaters and petrels. Large numbers of Herring, Ring-billed, and Laughing Gulls, and Forster's, Common, and Little Terns appear on the bays and along

other groups, jaegers (mostly Pomarines), Great Skuas, and a few alcids are generally observed, as well as hundreds of Northern Gannets and Black-legged Kittiwakes.

Landbirds wintering on the barrier beaches include (Northern) Horned Larks, Savannah Sparrows, (Ipswich) Savannah Sparrows, Snow Buntings, and, more rarely, Lapland Longspurs. The salt marshes are comparatively barren in winter, although one may flush a few Sharp-tailed Sparrows and occasionally other marsh birds. Certain landbirds that are either absent or at least uncommon at this season in other sections of the state appear regularly in the upland habitats on the relatively warm mainland west of Ocean City. These include the Eastern Phoebe, Gray Catbird, Palm Warbler, and Vesper and Chipping Sparrows.

Assateague Island National Seashore, reached by going west from Ocean City on US 50 for 2 miles, then left on State 611 to Assateague Island, comprises most of the barrier beach south of Ocean City Inlet. It is not only one of the best sites for observing herons, shorebirds, gulls, terns, and transient landbirds, but also is outstanding in late September and early October for Peregrine Falcons. During the migration peak one may see as many as a dozen of these magnificent birds in a single day—an experience that can be equaled in very few places in North America. Merlins are common transients here, principally in the brush and patches of woods back of the sand dunes.

PERRYVILLE

Susquehanna National Wildlife Refuge | Susquehanna State Park | Elk Neck State Park

The Susquehanna Flats, in the extreme upper section of Chesapeake Bay from the mouth of the Susquehanna River to Spesutie Island, have extensive beds of wild celery, the chief food of the transient waterfowl that concentrate here in spectacular flights—Whistling Swans, Canada Geese, and ducks. Canvasbacks, in particular, stop here in great numbers. The American Black Duck, Common Pintail, Redhead, and Common Goldeneye often gather by the hundreds. Others present include the American Wigeon, Lesser

Scaup, and Ruddy Duck. A scattering of Common Loons and Horned Grebes, and occasional flocks of American Coots are also present. The **Susquehanna National Wildlife Refuge** (18,413 acres, all water) occupies the central part of the Flats.

Although fall is the best time for large concentrations of waterfowl, many species are common also in spring, and a few large flocks remain for the winter. The drive along the shore on the grounds of the Perry Point Veterans Administration Hospital, just south of Perryville, is one of the best vantage points for observing these birds.

Other rewarding areas in this region are the floodplain forests of the Susquehanna River and Elk Neck, a peninsula extending into the northeastern corner of Chesapeake Bay. On the floodplain reside most of the typical breeding bottomland-forest birds, including the Cerulean Warbler; and the Prothonotary and Yellow-throated Warblers are rare but regular breeding birds. Blue-winged Warblers are fairly common in damp second-growth thickets along the upper slopes of the Susquehanna Valley.

To reach the most extensive floodplains along the Susquehanna River, drive north from Perryville on US 222, which runs close to the River between Port Deposit and Conowingo. Cross the River on US 1 at Conowingo and go left on State 161 to **Susquehanna State Park** (1,751 acres) and other good areas.

For **Elk Neck State Park** (1,575 acres), drive east from Perryville on State 7 to the town of North East and turn south on State 272, which bisects the Park. This is a good place for the characteristic upland birds of the region. Many forest birds, including Worm-eating and Hooded Warblers, are common here.

POCOMOKE CITY

Pocomoke Swamp | Shad Landing State Park | Milburn Landing State Park

The flora and fauna of the wooded swamp along the Pocomoke River and its tributaries in southern Maryland are in many respects typical of the swamp lands in the states farther south. In fact, **Pocomoke Swamp** might be considered the northernmost of the true southern swamps. Southern trees such as cypress, loblolly pine, water oak, and red bay are numerous here, and several southern shrubs and vines are also represented.

Breeding birds include species residing near the northern limit of their breeding ranges, as well as others near their southern limit. Among characteristic breeding birds of the Swamp are the Great Blue Heron, Great Egret, Wood Duck, Red-shouldered Hawk, Barred Owl, Pileated Woodpecker, Red-bellied Woodpecker, Acadian Flycatcher, White-breasted Nuthatch, Yellow-throated Vireo, Black-and-white Warbler, Prothonotary Warbler, Swainson's Warbler, Northern Parula Warbler, Yellow-throated Warbler, Louisiana Waterthrush, Kentucky Warbler, Hooded Warbler, American Redstart, and Scarlet Tanager.

Most of these birds are distributed throughout the Swamp. Swainson's Warbler, however, inhabits only outlying sections adjacent to upland habitats. Certain birds, including the Pileated Woodpecker, Blue-gray Gnatcatcher, and Prothonotary Warbler, are much more common in the Pocomoke Swamp than elsewhere in Maryland, and Swainson's Warbler is restricted to this area. The Worm-eating Warbler is common locally in the Swamp wherever mountain laurel grows. Many other forest birds, among them the Whip-poor-will, Pine Warbler, Ovenbird, and Summer Tanager, reside regularly in the better-drained woods of the upland.

The Swamp is most accessible from Pocomoke City, Snow Hill, and US 50. From Pocomoke City drive northeast on US 113 to Pocomoke State Forest. In the Forest, many one-lane dirt 'fire roads' run west to the Pocomoke River and provide fine bird finding away from cars and people. Visit **Shad Landing State Park** (545 acres), about 10 miles north of Pocomoke City and within the Forest. The open areas and brushy edges in this Park are especially good for winter landbirds, and the whole Forest can be 'alive' in spring migration. For close views of Prothonotary Warblers, rent a canoe in the Park.

Continue north on US 113 through Snow Hill to just beyond Berlin and turn left on US 50, which crosses the Swamp. For another section of the Swamp, drive west from Pocomoke City on US 13 and, just after crossing the River, turn right on State 364 for **Milburn Landing State Park** (370 acres), where a much more open forest permits good views of treetop warblers such as the Northern

Parula, Yellow-throated, and Pine. For directions to the northern part of this Swamp, *see under* **Selbyville** in the Delaware chapter.

ST. MARYS CITY
Point Lookout State Park | St. George Island

St. Marys City, in extreme southern Maryland, served as Maryland's first capital. Thirteen miles south of the city, State 5 ends at **Point Lookout State Park** (515 acres), where the Potomac River estuary joins Chesapeake Bay. In fall, winter, and spring, this is one of the best spots in Maryland to see a large variety of waterfowl. The most common in winter are the Whistling Swan, Canada Goose, Mallard, Canvasback, Greater and Lesser Scaups, Common Goldeneye, Bufflehead, Oldsquaw, White-winged Scoter, and Ruddy Duck. On cool mornings during fall migration, when winds are from the northwest or west, many thousands of transient flycatchers, thrushes, vireos, warblers, and sparrows take refuge in the limited cover on the Point and in nearby woods.

During winter a small flock of Great Cormorants is often in view on the pilings in the Potomac River off **St. George Island.** To reach the Island, go north from St. Marys City on State 5 to Callaway, then left on State 249 to its end.

STEVENSVILLE
Kent Island

Just east of the Chesapeake Bay Bridge lies historic **Kent Island,** largest island in Chesapeake Bay and site of the first settlement in Maryland. Kent Island is most attractive to the bird finder from September to early April. Two or three thousand Whistling Swans, Canada Geese, and American Coots, one or two hundred Horned Grebes, and over a dozen species of ducks in large numbers may be found in its many coves and tidal creeks from November through March. Most numerous are the American Black Duck, American Wigeon, Redhead, Canvasback, Common Goldeneye, Bufflehead, and Oldsquaw.

BUFFLEHEAD

One may profitably spend a whole day investigating the many passable roads that give access to the long points of land extending like fingers into Eastern Bay. The pier at Romancoke, south from Stevensville on State 8 to its end, is one of the best vantage points for waterfowl. Many waterbirds can also be observed east of the Island from US 50, where it crosses Kent Narrows and the tidal marsh. This is also a fairly good shorebird area in late spring and early fall.

Among species besides waterfowl that appear in the Kent Island area, especially during summer, are scattered pairs of Blue Grosbeaks in hedgerows, wood margins, and other brushy habitats; Henslow's Sparrows in abandoned fields of broom sedge; Pine Warblers and occasional pairs of Yellow-throated Warblers in the stands of loblolly pine. On calm summer evenings the distinctive songs of both the Chuck-will's-widow and Whip-poor-will may be heard along State 8 in the southern half of Kent Island. Warbling Vireos and Baltimore Orioles nest in towns and cottage developments on the Island or in adjacent sections of the mainland. Both Sharp-tailed and Seaside Sparrows breed no farther north in Chesapeake Bay than in the extensive tidal marshes at Kent Narrows, while Ospreys, an occasional Bald Eagle, and Fish Crows frequent almost all of the tidewater habitats.

In migrations, when a strong east wind pushes the night migrants west to the Bay shore, a remarkable concentration results. Lesser Golden Plovers and Black-bellied Plovers seem particularly apt to appear in this way. Kent Point, at the southern extremity of the Island, is noted for concentrations of kestrels, jays, and transient warblers in September and October, when winds are from a northerly quadrant.

WASHINGTON, DISTRICT OF COLUMBIA
Chesapeake and Ohio Canal | Rock Creek Park

Washington, the nation's capital, no longer provides quality bird-finding opportunities during nesting season, but there are several parks worth visiting in the city and suburbs. The **Chesapeake and Ohio Canal,** extending for 184 miles from Georgetown in the District to Cumberland, Maryland, has a hiking and bicycling trail along its entire length, and can be reached from any public road—except the Interstates—that goes to the Maryland banks of the Potomac. The most popular bird-finding areas along this Canal are those in Maryland, 3 or 4 miles on either side of Seneca Creek—reached from State 190 northwest of the District—where there is a good variety of habitats and a minimum of traffic noise.

Prothonotary and Cerulean Warblers are among the locally breeding specialties. The dense cover along the Canal attracts large numbers and a fine variety of wintering landbirds, and grebes, ducks, and gulls on the River add to the bird-finding opportunities.

Paralleled in the city by US 29, **Rock Creek Park** stretches from the Potomac in the District north into Maryland and attracts transient songbirds, especially in April, May, September, and October.

Massachusetts

LITTLE TERN

Few areas have been searched for birds with greater regularity and enthusiasm or over a longer period of time than the Bay State—and no area has yielded more impressive results. Approximately 290 bird species occur regularly, and easily 70 per cent of all the species in conterminous United States have been seen here at one time or another. For a region that is relatively small as well as thickly settled (the ratio is 700 people to a square mile), this record seems remarkable. But it is not so remarkable when one considers the circumstances. First, there are abundant habitats for birds. Despite its population density, nearly two-thirds of the land area of the state is farmland and forest. In addition, it has the coastal area, with salt marshes, protected bays and harbors, and wide beaches. Second, hundreds of acres of the most suitable bird habitats are public and private preserves and sanctuaries for the protection and encouragement of birdlife. Third, the number of persons who watch birds and report their findings is unusually great. The chance that a rare bird might go unnoticed in Massachusetts is very slight!

Eastern Massachusetts is primarily coastal plain, low-lying and somewhat rolling, with the usual elevation much less than 500 feet. The shoreline between New Hampshire and Rhode Island is some 300 miles in total length, much of it accounted for by the long, elbowed arm of Cape Cod—locally referred to as 'the Cape'—and the many deep indentations such as Buzzards Bay and the harbors of Newburyport, Boston, and Plymouth.

The Cape and the islands that lie off its southeastern coast—notably Nantucket, Martha's Vineyard, and the Elizabeth group—have sandy soil with scattered accumulations of boulders and other glacial debris. Their interiors are generally bleak, with grassy heaths, pitch pine woods, and scrub oak barrens, while their outer extremities, with sand dunes and beaches, have an appearance of wild desolation. It was among the scrub oaks of Martha's Vineyard that the Heath Hen made its last stand for over sixty years before becoming extinct in 1932 or soon thereafter.

The birdlife of the Cape and its adjoining islands has many peculiarites. Pine and Prairie Warblers are more frequent than elsewhere in Massachusetts, nesting in the pitch pines and oaks, respectively, and Rufous-sided Towhees are very numerous; but most warblers and other landbirds are poorly represented. Easily a hundred species that breed commonly in the interior of the state are rare or absent on the Cape. Even during migration they ignore the area by passing directly north or south far westward, though exceptions do occur when strong westerly winds blow them off their course. On the other hand, waterbirds and shorebirds abound. On undisturbed sand spits and small islands, gulls and terns nest in varying numbers; chief among them are the following:

Greater Black-backed Gull
Herring Gull
Laughing Gull
Common Tern
Arctic Tern
Roseate Tern
Little Tern

North of the Cape the coast varies. Between Boston Harbor and Cape Ann the shoreline is irregular, with protected harbors, pebbly beaches and coves, and bold, rocky headlands; there are many outlying ledges and rocky islands. Between Cape Ann and Newburyport Harbor the wide, straight, sandy beaches are backed by dunes and interrupted by river mouths. Inland are tidal creeks and flats and extensive salt marshes. Probably the characteristic breeding bird is the Sharp-tailed Sparrow, which nests exclusively in the salt marshes. Because of its secretive habits, this bird is invariably a challenge to the bird finder.

In the interior of the state there are two uplands. One, in the central part of the state, is a southward extension of the White Mountains of New Hampshire and has a maximum elevation of only 2,006 feet. The other, near the western border, is a southward continuation of the Green Mountains of Vermont. Familiarly called the Berkshires, it is in reality composed of several ranges, the westernmost being the Taconic Range. The highest peak of the Berkshires, Mt. Greylock, is also the highest in the state, at 3,491 feet. Partly isolated among the ranges of the Berkshires are scenic valleys where dairy farming is predominant. The largest is the Housatonic Valley, directly east of the Taconic Range.

Between these two uplands the broad Connecticut River Valley cuts from the north to the south boundary. Though it has several thriving cities and many wooded ridges—some, such as Mt. Tom, attaining modest heights—it is primarily agricultural country, with a mild climate and extensive fields whose reddish soil produces valuable commodities, the most unexpected being onions and tobacco.

Scattered here and there, especially on the higher elevations in the western part of the state, are 'conifer islands,' where the following birds breed regularly:

Yellow-bellied Sapsucker
Olive-sided Flycatcher
Red-breasted Nuthatch
Brown Creeper
Winter Wren
Hermit Thrush
Swainson's Thrush
Golden-crowned Kinglet
Solitary Vireo
Nashville Warbler
Magnolia Warbler
Black-throated Blue Warbler
Myrtle Warbler
Black-throated Green Warbler

Blackburnian Warbler
Northern Waterthrush
Canada Warbler
Purple Finch
Slate-colored Junco
White-throated Sparrow

The forests that originally covered the interior of the state were—with the exception of those on the higher elevations—typically deciduous, consisting of maple, beech, and oak, mixed, sometimes heavily, with ash, hickory, birch, and white pine. The woods of today are almost entirely secondary, although the same in composition. Characteristic breeding birds of these woods and the adjoining farmlands (fields, wet meadows, brushy lands, orchards, and dooryards) are listed below:

DECIDUOUS WOODS

Broad-winged Hawk
Ruffed Grouse
Black-billed Cuckoo
Barred Owl
Whip-poor-will
Yellow-shafted Flicker
Hairy Woodpecker
Downy Woodpecker
Great Crested Flycatcher
Least Flycatcher
Eastern Pewee
Blue Jay
Black-capped Chickadee
Tufted Titmouse
White-breasted Nuthatch
Wood Thrush
Yellow-throated Vireo
Red-eyed Vireo
Warbling Vireo
Ovenbird
American Redstart
Baltimore Oriole
Scarlet Tanager
Rose-breasted Grosbeak

FARMLANDS

Common Bobwhite
Eastern Kingbird
Eastern Phoebe
Tree Swallow
Barn Swallow
Northern Mockingbird
Gray Catbird
Brown Thrasher
Eastern Bluebird
Yellow Warbler
Chestnut-sided Warbler
Common Yellowthroat
Bobolink
Eastern Meadowlark
Common Grackle
Northern Cardinal

FARMLANDS (*Cont.*)

Indigo Bunting
House Finch
American Goldfinch
Rufous-sided Towhee
Savannah Sparrow
Vesper Sparrow
Chipping Sparrow
Field Sparrow
Song Sparrow

In addition, the Yellow-billed Cuckoo, Golden-winged and Blue-winged Warblers, and Grasshopper Sparrow breed in the Connecticut Valley and similar lowlands.

There are three primary migration routes in Massachusetts. The first, along the coast, is used for the most part by waterbirds, waterfowl, and shorebirds, as well as by many landbirds. The second and third pass north and south through the Connecticut and Housatonic Valleys, respectively. Large numbers of landbirds follow those same routes, and small numbers of ducks and shorebirds frequently take advantage of them. Times of the year recommended for finding migrating birds:

Waterfowl: 25 March–20 April; 5 October– 15 November
Shorebirds: 1 May–1 June; 1 August–25 September
Landbirds: 20 April–25 May; 20 August–10 October

Chief ornithological attractions in winter are the concentrations of loons, grebes, and ducks in northeastern harbors; of eiders, scoters, and Canada Geese on the outlying shoals of the Cape, Martha's Vineyard, and Nantucket; and of various alcids and other oceanic birds, which feed and rest off the rocky shores north of Boston Harbor. Landbirds usually appearing in winter include (Northern) Horned Larks, Evening Grosbeaks, Pine Siskins, American Tree Sparrows, Fox Sparrows, and Snow Buntings. Others turning up irregularly are Pine Grosbeaks, Common Redpolls, and Red and White-winged Crossbills.

Authorities

C. E. Addy, Francis H. Allen, Oliver L. Austin, Sr., Aaron Moore Bagg, G. W. Cottrell, Jr., Ruth P. Emery, Mrs. Gilbert F. Fernandez, Robert Grayce, Ludlow Griscom, Joseph A. Hagar, Bartlett Hendricks, Edith M. Johnson, Gordon W. Johnson, C. Russell Mason, Edwin A. Mason, Gordon M. Meade, Alvah W. Sanborn.

Reference

The Birds of Cape Cod, Massachusetts. By Norman P. Hill. New York: William Morrow and Company, 1965.

BOSTON

Because of the intricate highway system in and around the Boston metropolitan area, the places for bird finding are described under the nearest towns—namely, Concord, Gloucester, Ipswich, Lincoln, Lynnfield Center, Newburyport, North Scituate, Plymouth, and Sharon. All are within easy reach of State 128, a belt road west of Boston, or I 95, which crosses the city.

CONCORD
Great Meadows National Wildlife Refuge

Great Meadows National Wildlife Refuge (4,000 acres) lies on the floodplain of the Concord River, surrounded by some of the country's famous historical sites. Embracing more than 2,600 acres of marsh and river bottomlands, the Refuge is one of the best inland places in the state for finding waterbirds and waterfowl.

In summer, Great Blue Herons, Least Bitterns, Wood Ducks, and Northern Harriers nest in or frequent the marsh where the water level is controlled by a series of dikes; and Ruffed Grouse, Yellow-shafted Flickers, Downy Woodpeckers, Gray Catbirds, Wood Thrushes, Red-eyed Vireos, Black-and-white Warblers, Yellow Warblers, Ovenbirds, Common Yellowthroats, American Goldfinches, Rufous-sided Towhees, Swamp Sparrows, and Song Sparrows reside in the 48 acres of woodland or along its edges. In spring and fall, Canada Geese, Mallards, American Black Ducks, and Blue-winged Teal are among the waterfowl that stop to feed and rest in the marsh.

The northern section, particularly recommended for bird finding, is reached by driving east from Concord on State 62 for about a mile and turning left on Monsen Road. One may walk along the dikes into the marsh. On one of the dikes a high platform provides an excellent observation point. For information about the Refuge, drive west on State 62 from the intersection of State 62 and 2A for about a half-mile and turn left on Sudbury Road to Refuge headquarters on the right.

FALL RIVER

Acoaxet | Gooseberry Neck | Demarest Lloyd Memorial State Park

Unusual birds sneak across the Rhode Island border into the southeastern corner of the state, south of Fall River, and the local bird finders are ready for them.

The most reliable specialty is the only colony of Ospreys nesting in Massachusetts. Other specialties, not quite so reliable, are Carolina Wrens, White-eyed Vireos, and Hooded Warblers singing during the nesting season. The combination of ocean, marsh, woods, scrub growth, fresh- and salt-water ponds, and beach in this coastal region practically guarantees good bird finding all year and excellent bird finding during migration season.

Drive south from Fall River on State 81 through Rhode Island; shortly after passing Adamsville and a sign indicating entrance to Westport (Massachusetts), bear left at a fork on River Road. Continue until there is an open view of the River. The Osprey nests, on islands, are easy to see from this point and from other points south along River Road to **Acoaxet.** Carolina Wrens, White-eyed Vireos, and Hooded Warblers may be singing here. If not, return to the fork and take the right branch, Old Harbor Road, south, stopping now and again to listen.

Continue south on Old Harbor Road to Acoaxet at the head of Westport Harbor. Members of both the Rhode Island and Massachusetts Audubon Societies often schedule field trips to this area and find it very productive. Two other rewarding places in the vicinity are **Gooseberry Neck** and **Demarest Lloyd Memorial State Park.**

To reach Gooseberry Neck, drive east from Fall River on I 195 and turn south on State 88. Shortly after passing over water, State 88 continues as John Reed Road. An extensive marsh lies to the left, and Horseneck Beach to the right. Continue on this road to the ocean and turn right to Gooseberry Neck, a narrow peninsula extending to an inlet across from Acoaxet. Drive over the causeway to Gooseberry Neck, park the car, and explore the area. Gooseberry Neck is a fine stopping place for migrating birds, and over the years some unusual 'strays' have turned up. In fall, one can almost count on Short-eared Owls and Yellow-breasted Chats; in winter, American Kestrels, Purple Sandpipers, an occasional Snowy Owl, and, offshore, a variety of waterbirds and waterfowl.

Drive east from Gooseberry Neck along East Horseneck Beach and then north on Horseneck Road in Westport past Horseneck Road in Dartmouth on the right to the next right, which is Allen's Neck Road. Turn here and continue on the road for 2-scant miles to the sign marking the entrance to Demarest Lloyd State Park. If the gate is locked, leave the car and enter on foot. The Park (221 acres) has habitats—woodland, swamp, pond, shore of a tidal river, and ocean beach—attractive to many different birds. The Green Heron, Red-tailed Hawk, Common Screech Owl, Blue-gray Gnatcatcher, Common Yellowthroat, and American Redstart are among nesting species. During migration seasons the Park rivals Gooseberry Neck for the number and variety of birds that it attracts. Throughout winter it is greatly favored by sparrows—Savannah, American Tree, Field, White-throated, and Song.

GLOUCESTER

Halibut Point | Andrews Point

This famous fishing city shares with Rockport the whole of Cape Ann, a granite peninsula bounding Massachusetts Bay on the north. If there is any place in the state that can be called 'best' for wintering waterbirds, waterfowl, and Purple Sandpipers, it is Cape Ann's two closely adjoining promontories, **Halibut Point** and **Andrews Point.**

From Gloucester, State 127 makes a loop around Cape Ann and returns to the city. Take this route by way of Lanesville and Folly

Cove. Soon after passing Folly Cove—about 6 miles from Gloucester—turn left on Gott Avenue, and drive to a parking lot. Leave the car and proceed along a footpath for 0.5 mile to the jagged headland of Halibut Point. To reach Andrews Point, return to State 127, continue east for about 0.5 mile, and turn left. Andrews Point is about 300 yards beyond. Return to Gloucester through Rockport, and from here, if time permits, take 127A along roads that parallel the coast. Though this trip is roundabout, there will be opportunities to investigate such bird-observation points as Brace Cove, the lighthouse at Eastern Point, and the fishing wharves at Gloucester Inner Harbor.

Chances for seeing waterbirds and waterfowl are good any time from mid-October to mid-April, and even better after violent 'northeasters.' Some of the species appearing with greatest regularity off Halibut Point and Andrews Point are Common Loons, Horned Grebes, Common Goldeneyes, Buffleheads, Oldsquaws, the three species of scoters, Red-breasted Mergansers, and Greater Black-backed Gulls. Sometimes there are Red-throated Loons, Red-necked Grebes, Northern Gannets, Great Cormorants, Harlequin Ducks, Common Eiders, and Black-legged Kittiwakes. At low tide, Purple Sandpipers frequently feed in the exposed seaweed on the ledges.

The greatest concentrations of waterbirds occur from mid-November through February. Alcids—Razorbills, murres, Dovekies, and Black Guillemots—and 'white-winged' gulls—Glaucous, Iceland, and (Kumlien's) Iceland—are present in midwinter, from December to March. Gulls are recorded not only at the Points but also between Rockport and Gloucester at the several places previously mentioned.

HOLYOKE
Mt. Tom State Reservation

The Peregrine Falcons that once nested on Mt. Tom's western cliffs are gone now, but the hawk migration in fall is still often spectacular, and warblers still nest in the woods around Bray's Lake.

Drive north from Holyoke on US 5 for about 4 miles to Smith's

PEREGRINE FALCON

Ferry and turn left through the entrance to the **Mt. Tom State Reservation** (1,679 acres). At Bray's Lake, on the left, breeding birds include Pileated Woodpeckers, Magnolia Warblers, Black-throated Blue Warblers, and Louisiana Waterthrushes. Continue on this road to the summit ridge and the intersection with Christopher Clark Road. In late spring, turn left on Clark Road to Bray Tower and look for the Slate-colored Juncos that are reported to nest on the summit (1,202 feet). In early fall, turn right to Goat Peak Tower, an observation point for Northern Goshawks, Sharp-shinned, Cooper's, Red-tailed, Red-shouldered, Broad-winged, and Rough-legged Hawks, Bald Eagles, and Ospreys.

IPSWICH
Crane's Beach | Ipswich River Wildlife Sanctuary

Crane's Beach is a 2-mile stretch of beautiful sand beach, backed by a ridge of dunes and salt marsh, at the mouth of the Ipswich River and south across an inlet from Plum Island (*see under* **Newburyport**). Since it is a part of the Richard T. Crane Memorial Reservation, owned and maintained by the Trustees of Public Reservations, and very close to a densely populated area, bird finding is not recommended until the end of summer. Migrating landbirds seek shelter in the thickets on the dunes in April/May and September/October. Migrating shorebirds begin to come toward the end of July. Terns rest on the beach. The species are about the same as those on Plum Island.

For Crane's Beach, drive east from Ipswich on an unnumbered but well-marked road. Park the car and walk north. The best bird finding is on the northern tip.

The **Ipswich River Wildlife Sanctuary** (2,300 acres), owned and operated by the Massachusetts Audubon Society, is in the village of Topsfield, west of Ipswich.

The Sanctuary includes mixed woods, fields, marshes along the Ipswich River, and uplands. These natural areas plus extensive plantings of a fine variety of trees and shrubs attract many warblers and small landbirds in spring migration. No less than ten of the warblers—Black-and-white Warbler, Golden-winged Warbler, Yellow Warbler, Black-throated Green Warbler, Blackburnian Warbler, Chestnut-sided Warbler, Ovenbird, Common Yellowthroat, Canada Warbler, and American Redstart—remain to nest. There are at least four species of hawks that nest—Cooper's, Red-tailed, Red-shouldered, Broad-winged—as well as the American Kestrel.

Some other breeding birds may be the following: Green-winged Teal, Blue-winged Teal, Wood Duck, Ruffed Grouse (permanent resident), American Bittern, American Woodcock, Yellow-billed and Black-billed Cuckoos, Whip-poor-Will, White-breasted Nuthatch (permanent resident), House Wren, Sedge Wren, Hermit Thrush, Bobolink, and Rufous-sided Towhee.

The easiest directions to the Sanctuary are from US 1 in Topsfield. Exit from I 95 or US 1 on State 97 and drive east about a half-mile from US 1 to the first through street, Perkins Row. Turn left on Perkins Row and watch for the Sanctuary entrance on the left. Like all the Massachusetts Audubon Society facilities, it has a fine system of trails.

LENOX
Pleasant Valley Wildlife Sanctuary

The **Pleasant Valley Wildlife Sanctuary** (660 acres), owned and operated by the Massachusetts Audubon Society, is 2.5 miles from this beautiful little town deep in the Berkshires. In an area of primarily rolling woodland interrupted by several small streams and meadows, the trees are predominantly sugar maple, white ash, and birches—white, gray, and yellow—mixed with white pine and hemlock. A beaver colony has converted a mile-long alder swamp into a series of dams and ponds. The summer birdlife is especially attractive, even though the altitude is not great enough to encourage many montane or northern types; nearly seventy species nest more or less regularly.

In addition to such eye-catchers as Baltimore Orioles, Scarlet Tanagers, Rose-breasted Grosbeaks, and Indigo Buntings, there are Traill's Flycatchers, Wood and Hermit Thrushes, Veeries, and fourteen warblers, including the Golden-winged, Nashville, Black-throated Blue, Black-throated Green, Blackburnian, and Canada. Since the establishment of the beaver colony, Green Herons, American Bitterns, American Black and Wood Ducks, and Northern Waterthrushes have found conditions suitable for nesting. All areas where these birds are found are readily accessible by 10 miles of trails. Between 120 and 130 species are observed annually at the Sanctuary. The fall migration, which is outstanding, accounts for the unusually large number.

To reach the Sanctuary, drive north from the edge of Lenox on US 7A for about a mile, turn left opposite a Holiday Inn, and follow signs to the Sanctuary.

LINCOLN
Drumlin Farm Sanctuary

Drumlin Farm Sanctuary (220 acres), headquarters of the Massachusetts Audubon Society, consists of woods, fields, ponds, and lawns of a former farm-estate. Drive east from State 128 on US 20 (Exit 49) for 0.9 mile, then take a sharp left on State 117, go back (west) across State 128 and watch for the entrance on the right, just beyond the Codman Road fork. There are nature trails, exhibits, and a New England farm demonstration area open to the public. And there are the Society's personnel at headquarters, ready to answer questions and relay information on birds in the state. Some species nesting commonly in the Sanctuary are the Tree Swallow, Barn Swallow, Gray Catbird, Eastern Meadowlark, Scarlet Tanager, Northern Cardinal, Rose-breasted Grosbeak, and Indigo Bunting.

LYNNFIELD CENTER
Lynnfield Marsh

Near this little community is **Lynnfield Marsh,** an excellent example of the fresh-water marshes formerly common in the state. Once a lake basin, it has now become thickly matted with sedges, sweet gales, and willows. There is little open water other than meandering creeks, except when the Marsh is flooded in spring. The section near Lynnfield Center holds a typical association of marsh-birds, which are at their noisy best from mid-May to mid-June.

At the intersection of US 1 and State 128, drive east from Lynnfield on State 128 to Exit 31 and turn right to the village of Lynnfield Center. At the railroad crossing, turn left to the railroad station and park. Walk south along the tracks, which begin to cross the Marsh on an embankment about 0.4 mile beyond. The embankment is a good point for observing the Marsh close at hand without getting wet feet. Birds to be seen or heard are the American Bittern, Virginia Rail, Sora, Marsh Wren, Red-winged Blackbird, and Swamp Sparrow.

The other end of this section of the Marsh can be reached by turning off State Route 128 onto Charles Avenue. At the end of Charles Avenue, walk across the golf course to the edge of the Marsh. Pied-billed Grebes, Least Bitterns, King Rails, and Common Gallinules nest here regularly. In addition American Black Ducks, Wood Ducks, and a pair or two of Blue-winged Teal are frequently seen.

MARTHA'S VINEYARD
Island of Martha's Vineyard

Five miles southeast of Woods Hole, off the southern coast of Cape Cod, lies the island of **Martha's Vineyard,** 20 miles long from east to west and 10 miles in greatest width. The eastern half is low and flat, but, toward the western end, rolling hills terminate in the brilliantly colored clay cliffs of Gay Head on the southwestern tip. With the exception of a few stands of pitch pine, vegetation consists chiefly of scrub oak in the east and grasses on the open heaths in the west. Wide sand beaches surround the island.

Along the southern side is a series of ponds of varying size, separated from the ocean by narrow sand strips. Though these ponds usually contain fresh water, storms with high winds and excessively high tides cause the sea to break through the sand strips at times, making the water brackish. There is also Poncha Pond on Chappaquiddick, an island just off the eastern shore of Martha's Vineyard.

The ponds are the foremost attraction to the bird finder. Beginning in November and continuing throughout winter, they are important gathering places for many waterfowl, especially when the fresh-water ponds on the mainland freeze over. From east to west, the larger ponds and the waterfowl they attract are: *Poncha Pond on Chappaquiddick,* Green-winged and Blue-winged Teal, American Wigeons, Redheads, Canvasbacks, Greater Scaups, Common Goldeneyes, and, rarely, a Eurasian Wigeon; *Edgartown Great Pond,* Canada Geese, White-winged and Surf Scoters, and, rarely, a Whistling Swan; *Tisbury Great Pond,* Greater Scaups, Common Goldeneyes, Buffleheads, and other diving ducks; *Squibnocket Pond at Gay Head,* Canada Geese, Redheads, Canvasbacks, Ruddy Ducks, and a variety of surface-feeding ducks. Of the several ponds, probably Squibnocket attracts the greatest number and variety of waterfowl; but when any of the ponds becomes extremely brackish and the duck food dies, few ducks appear for a long time thereafter.

In winter—a pleasant season on Martha's Vineyard—Harlequin Ducks can sometimes be seen from the cliffs at Squibnocket. From the cliffs at Gay Head, also during winter, there are good opportunities for observing alcids—Razorbills, Thick-billed Murres, Dovekies, Black Guillemots, Atlantic Puffins—as well as Northern Gannets, Black-legged Kittiwakes, and quite often the Common Loon, Red-throated Loon, Red-necked Grebe, Common Eider, King Eider, White-winged Scoter, Surf Scoter, Black Scoter, and Red-breasted Merganser. Off Chappaquiddick, at the opposite end of the island, huge flocks of scoters often appear in the late fall and winter.

When driving across the interior in winter, the bird finder should be constantly alert for Red-tailed and Rough-legged Hawks, which frequent the scrub oak barrens near Chilmark. There is always the possibility of seeing Short-eared Owls here in winter, but the chances are greater in spring, when more of them have returned from the south. On Menemsha Neck, east of Gay Head, is a large nesting colony of Black-crowned Night Herons. As one drives between Oak Bluffs and Edgartown, the sight of gulls dropping clams and other shellfish on the road to break them open is often spectacular. This activity, common throughout the year, increases in winter.

Car-carrying ferries run from Woods Hole to Martha's Vineyard. The bird finder who does not bring his own car can hire one in Vineyard Haven, where the ferries dock. Good roads connect the three principal towns—Vineyard Haven, Oak Bluffs, and Edgartown—and extend to Gay Head. A mini-ferry goes to Chappaquiddick from Edgartown. Roads of varying quality lead to the many ponds, inlets, and beaches. Katama Road leads south from Edgartown to South Beach, and one may walk west along South Beach to the shores of Edgartown Great Pond. South Beach itself is a good place for shorebirds beginning in late July.

Warning: Much of the property around the ponds is privately owned and posted. Be sure to obtain permission before entering any private lands.

NANTUCKET

Island of Nantucket | Muskeget Island

Thirty miles southeast of Woods Hole, off the southern coast of Cape Cod, lies the island of **Nantucket,** about 15 miles long from east to west and about 6.5 miles in greatest width. Inland from its western and southern sand beaches and bluffs are extensive heaths or moors. Elsewhere on this generally low-lying island are small patches of pine and oak scrub, small fresh-water ponds, and tidal flats flanking the harbors. Nearly all parts of the island are accessible by car, and ferries carry cars from Woods Hole to Nantucket. The bird finder without a car can hire one in the town of Nantucket at the ferry dock.

In summer, looking for birds on Nantucket is a strange experience, since many species that one might expect do not materialize or are very rare—as, e.g., the Prairie Warbler and Rufous-sided Towhee. A few other species, such as the Short-eared Owl, appear regularly. The Savannah Sparrow is abundant on the moors. There are at least two colonies of Black-crowned Night Herons. And, after mid-July, migrating shorebirds appear on the tidal flats along the south shore of Nantucket Harbor, at the end of Washington Street east of town; on the south shore of Sesachacha Pond, near Polpis Road; on the shores of Polpis Harbor; along the south beach, and at many similar localities. For observing migrating and wintering oceanic birds and waterfowl there is probably no better place than the waters off Nantucket Harbor and Maddaket Harbor.

There is an enormous gull colony on **Muskeget Island** off the western end of Nantucket and difficult to reach because of the surrounding shoals and strong, unpredictable tidal currents. In Nantucket, hire a boatman familiar with the eccentricities of the waters around Muskeget, to make the trip in a suitable craft.

Muskeget belongs to Nantucket and is considered a sanctuary. More than a mile in greatest length and low-lying with a surface of approximately 500 acres, it has sand dunes rising as high as 15 feet, two or three brackish ponds, and from 15 to 20 acres of tidal salt marsh. The vegetation is chiefly beach and marsh grasses.

Muskeget has a large colony of Herring Gulls and a few pairs of Greater Black-backed Gulls. The peak of the egg-laying period is variable, ranging from early to late June; chicks are abundant by July.

NEW BEDFORD
Penikese Island | Ram Island | Bird Island

A number of islands in Buzzards Bay off New Bedford have nesting colonies of Herring Gull and/or both Common and Roseate Terns, the number of breeding pairs varying from year to year. The peak of egg-laying for gulls is around 10 May, for terns about 1 June. Anybody visiting the islands between 15 June and 15 July will see various stages of nesting activities. Described below are three islands that are fairly accessible.

Penikese Island (90 acres), one of the Elizabeth Islands, can be reached from New Bedford by ferry to Cuttyhunk Island and by charter boat from Cuttyhunk. Penikese is a state-owned sanctuary nearly three-quarters of a mile in greatest length. The shoreline is an almost unbroken bluff above a beach strewn with large boulders. The interior rises to a height of 60 to 70 feet, with vegetation consisting of grasses and sorrel, one small stand of introduced Scotch pine, and scattered clumps of elderberry and other shrubs. Besides large colonies of Herring Gulls and both Common and Roseate Terns, Penikese has a small colony of Leach's Storm Petrels. The Manx Shearwater has been found nesting here.

Drive north from New Bedford on US 6 to Mattapoisett and charter a boat for the 1.5-mile trip to **Ram Island,** which has large colonies of Common and Roseate Terns. The Island is circular, with a diameter of about a half-mile. Although it is state-owned and posted, persons with a genuine interest in birds are usually granted permission to land.

Drive north from New Bedford on US 6 to Marion and charter a powerboat for a 4-mile trip to circular **Bird Island** (8 acres). In calm weather the Island can be reached also by canoe or skiff from Butlers Point. Despite the Island's small surface, it usually has a large population of Common and Roseate Terns.

NEWBURYPORT
Plum Island | Parker River National Wildlife Refuge | Newburyport Harbor | Joppa Flats

Newburyport is in Essex County, one of the most famous ornithological areas in eastern United States—a reputation justified by the impressive variety of species recorded here and the scores of ornithologists who frequent its woods, ponds, rivers, marshes, and ocean beaches. If there is one spot that can be recommended above all others in the state, it is **Plum Island,** east and southeast of the city.

Plum Island, a long, narrow strip of land, parallels the coast from the entrance to Newburyport Harbor south to Ipswich Bay, 9 miles distant. With the exception of a summer resort on the northernmost part, Plum Island is a division of the **Parker River National Wildlife Refuge** embracing 3,891 acres. A broad ocean beach runs the length of its eastern side, and on the west there is a vast expanse of salt marsh, with ponds and creeks whose muddy bottoms become 'clam flats' at low tide. Sand dunes, often high and steep and covered with vegetation, form a ridge between the beach and marshes.

From Newburyport, drive east on the Plum Island Turnpike. The road follows the southern side of **Newburyport Harbor,** affording an excellent view of the open water and **Joppa Flats,** and crosses the Plum Island River to Plum Island. The first road on the right goes south in back of the sand dunes to the Island's southern tip. The second road on the left leads to the northern part and to Refuge headquarters. Bird finding is good here in winter. Watch for (Ipswich) Savannah Sparrows on the dunes and Purple Sandpipers on the ledges at low tide. Because the area is strategically located on the Atlantic flyway, it is most remarkable as a stopping point and wintering ground.

Be sure and stop at Joppa Flats on the way to Plum Island. This is definitely the best place for shorebirds from mid-July to October, with peak in September. Some species appearing are Semipalmated Plovers, Black-bellied Plovers, Ruddy Turnstones, Whimbrels, White-rumped Sandpipers, Least Sandpipers, Short-billed Dowitchers, Semipalmated Sandpipers, and Greater and Lesser Yellow-

(IPSWICH) SAVANNAH SPARROW

legs. Usually in view at the same time are Greater Black-backed Gulls, Herring Gulls, Ring-billed Gulls, and Bonaparte's Gulls.

In October and November, scan the Harbor and Flats for Canada Geese, American Black Ducks, Greater Scaups, Common Goldeneyes, Buffleheads, Oldsquaws, White-winged Scoters, and Red-breasted Mergansers. In late November, look for Snowy Owls, (Northern) Horned Larks, Lapland Longspurs, and Snow Buntings.

Once on the Island and headed south, watch all the pools beside the road for shorebirds from mid-July through September. In August and September, Great Blue Herons, Great Egrets, Snowy Egrets, and Black-crowned Night Herons stalk prey along the edges of the ponds and creeks.

Stop often and walk along the beach where, from mid-July through September, there may be Dunlins and Sanderlings; from mid-October to mid-November, Water Pipits and (Ipswich) Savannah Sparrows; and in November, (Northern) Horned Larks, Lapland Longspurs, and Snow Buntings. In the marshes at low tide, Whimbrels and Hudsonian Godwits feed on the clam flats. In early fall, swallows concentrate in great flocks on the Island. The main land-bird migration takes place in April/May and September/October.

A winter trip is bound to be rewarding, not only for ducks on the open water in Newburyport Harbor, but also because of the possibilities of a Snowy Owl on the Flats, Purple Sandpipers on the rocky ledges of Bar Head and Emersons' Rocks on the beach, and Common Loons, Red-throated Loons, Red-necked Grebes, Black-legged Kittiwakes, Thick-billed Murres, and Dovekies on the outlying ocean together with Glaucous, Iceland, and (Kumlien's) Iceland Gulls. And there is always the possibility of seeing at least one rare bird, perhaps a grebe, duck, or gull that has wandered far beyond its range. No area on the Atlantic Coast has more unusual records than Plum Island.

NORTH ADAMS
Mt. Greylock State Reservation

On the slopes of Mt. Greylock (3,491 feet), the state's highest peak in **Mt. Greylock State Reservation** (8,600 acres), is a rich assortment of summer-resident birds. Most noteworthy are the (Bicknell's) Gray-cheeked Thrushes and Blackpoll Warblers. Other species include the Olive-sided Flycatcher, Red-breasted Nuthatch, Brown Creeper, Winter Wren, Hermit Thrush, Swainson's Thrush, Golden-crowned Kinglet, Magnolia Warbler, Myrtle Warbler, Blackburnian Warbler, Mourning Warbler, Slate-colored Junco, and White-throated Sparrow. In addition, Saw-whet Owls, Yellow-bellied Flycatchers, Pine Siskins, White-winged Crossbills, and Red Crossbills are sometimes reported and possibly nest.

The lower slopes, up to 2,500 feet, are forested with beech, maple, yellow and white birches, and some hemlock. Above 2,500 feet, spruce and fir become dominant. Whether walking or riding to the summit, stop often to investigate the birdlife inhabiting the various tree associations. The best time of year to ascend the mountain is between 15 June and 5 July, when the singing reaches its maximum intensity. Listen for the (Bicknell's) Gray-cheeked Thrushes on the summit, particularly in early evening, and, at this time, watch for their flight songs, a display of aerial spirals and tumbles accompanied by musical, reedy notes that may be prolonged for fully 30 seconds.

One can drive to the summit of Mt. Greylock, or walk the Appalachian Trail. The road to the summit turns south from State 2 about a mile west of North Adams; the Trail crosses State 2 just beyond.

NORTHAMPTON
Arcadia Wildlife Sanctuary

Three miles from the center of Northampton is the **Arcadia Wildlife Sanctuary** (300 acres), owned and operated by the Massachusetts Audubon Society. Many trails wind through the Sanctuary, which consists of deciduous woodland with small islands of coniferous growth, open fields, and marshes, cut by the Mill River and bordering an old oxbow of the Connecticut River, New England's foremost inland migration flyway. The Sanctuary's strategic location, together with its varied habitats, attract an unusually large variety of birds during the course of the year. Over two hundred species have been recorded and at least seventy, including Wood Ducks, American Kestrels, Common Screech Owls, and Northern Waterthrushes, nest within its boundaries. One bird watcher described the fall migration of waterfowl as 'tremendous.'

To reach the Sanctuary, drive south on State 10, turn left on Lovefield Street, then left again on Clapp Street, and take a left fork on Old Coach Road to the entrance and parking lot.

NORTH SCITUATE
Minot | Musquashcut Pond | Third Cliff

One of the best areas between Boston and Cape Cod for transient shorebirds is along the coast from Minot to Third Cliff. From North Scituate, drive east to the summer resort of **Minot,** a good observation point in winter for Purple Sandpipers on the rocks, and Common Eiders and scoters offshore.

Drive south from Minot and stop to look over **Musquashcut Pond** on the left for shorebirds and ducks in the proper season.

Continue south through Scituate to **Third Cliff** and park the car. The beach here is rough shingle with some sandy area, including a sandy back beach with pools at high tide. Many transient shorebirds stop by, including thousands of Red Knots in August. The beach extends as a hook of low dunes covered with beach grass, along the mouth of the North River, and salt marshes line the River. In late May, Little Terns begin forming a small colony in the dunes, the surest place to find them nesting within easy reach of Boston.

PLYMOUTH

North Point | Manomet Point

South of Plymouth, in a picturesque setting at the tip of **North Point** (sometimes called Plymouth Point), one may observe colonies of terns—Common, Arctic, Roseate, and Little. The area they occupy—a sandy peninsula extending 3 miles into Cape Cod Bay—is owned by the town of Plymouth and patrolled by a warden during the nesting season. Free access is granted people seriously interested in birds.

Drive south from Plymouth on State 3A, turn left at the Plymouth Beach bathhouse just on the edge of town, leave the car in the northernmost parking area, and walk along a beach-buggy road to the Point. The nests are spread along the terminal half of the last mile of dunes, blanketed with beach grass in varying degrees of thickness. The Point is best for terns between 1 June and 1 August. While walking down the beach, watch for Piping Plovers that may be nesting, Ruddy Turnstones, (Prairie) Horned Larks, and Sharp-tailed Sparrows. Black Skimmers occasionally nest near the tip. After the last of July the Point is one of the best places in southeastern Massachusetts for migrating shorebirds.

Continue south on State 3A and turn left to **Manomet Point,** about 8 miles from Plymouth. At the Point a terminal moraine affords excellent views of wintering waterbirds and ducks, including Red-necked Grebes, Common Eiders, and alcids. The eiders and alcids are best seen at high tide.

PRINCETON
Wachusett Meadows Wildlife Sanctuary

Wachusett Meadows Wildlife Sanctuary (750 acres), owned and operated by the Massachusetts Audubon Society, is an area of upland forest and meadows just south of Mt. Wachusett (2,006 feet). There are many trails, including one that crosses a maple swamp on a boardwalk. Bird finding here is best in summer, when nesting birds include the Northern Goshawk, Cliff Swallow, Canada Warbler, and Rose-breasted Grosbeak. The Sanctuary may be reached by driving west from Princeton on State 62 for 0.7 mile, then going right on Goodnow Road to the entrance.

SHARON
Moose Hill Sanctuary

Moose Hill, the Massachusetts Audubon Society's oldest sanctuary, lies 18 miles south of Boston. The Sanctuary (250 acres) has a wildflower and fern garden and a pond. There are eight miles of trails, and the view from the Fire Tower Trail is one of the best in southeastern Massachusetts.

Well over a hundred bird species are recorded here annually, and at least sixty of these nest regularly. In the woods, which are largely of oak with a sprinkling of white and pitch pines, nesting birds include Cooper's and Broad-winged Hawks, Ruffed Grouse, American Woodcock, Wood and Hermit Thrushes, Veery, Solitary Vireo, Black-and-white Warbler, Black-throated Green Warbler, Chestnut-sided Warbler, Pine Warbler, Prairie Warbler, Ovenbird, Canada Warbler, and American Redstart. A few Swamp Sparrows nest in the region bordering the pond.

To reach the Sanctuary from Boston, drive south on I 95 or US 1 and then east on State 27. Turn right on Moose Hill Parkway, and left on Moose Street to the entrance.

WELLFLEET

Wellfleet Bay Wildlife Sanctuary | Marconi Station | Cape Cod National Seashore | Monomoy National Wildlife Refuge

Cape Cod, from Chatham at the bend of the 'elbow' to Provincetown at the tip, has many good areas for bird finding. Before stopping at any of the places listed below, visit or write the Wellfleet Bay Wildlife Sanctuary (South Wellfleet, Massachusetts 02663), for current information on conducted tours.

The **Wellfleet Bay Wildlife Sanctuary** (650 acres), formerly the Austin Ornithological Research Station, is now owned and operated by the Massachusetts Audubon Society. The Sanctuary includes large salt marshes, tidal inlets, and a beach on Massachusetts Bay. Pine Warblers nest by the parking lot, Clapper Rails in the salt marsh, and Piping Plovers on the beach. Drive south from Wellfleet on US 6; just beyond South Wellfleet, watch for the Sanctuary entrance on the right.

The entrance to the **Marconi Station** is also just off US 6 in South Wellfleet. From the parking lot, follow a well-marked trail for 0.5 mile through a woods of scrubby pine and oak to a white cedar swamp made accessible by elevated boardwalks. Species nesting in the swamp include the Whip-poor-will, Eastern Pewee, Brown Creeper, Hermit Thrush, Black-and-white Warbler, Northern Parula Warbler, Pine Warbler, Prairie Warbler, and Common Yellowthroat. On the fringes, (Prairie) Horned Larks, Vesper Sparrows, and occasionally Northern Mockingbirds nest.

The **Cape Cod National Seashore** includes about 40 miles of ocean beach from Chatham Light on the elbow of the Cape to Long Point Light in Provincetown Harbor, and 14 miles of bay beach. Massachusetts ornithologists agree that Nauset Beach is an excellent place for migrating shorebirds, beginning in late July. Drive south from Wellfleet to Eastham, turn left on Nauset Road to the old Coast Guard station, park the car, and walk south along the Beach for about a mile. The broad outer Beach is backed by a ridge of sand dunes that protects extensive salt marshes. Shorebirds congregate both on the Beach and in the marshes. Look for such species as Black-bellied Plovers, Willets, Red Knots, Greater and Lesser Yellowlegs, Short-billed Dowitchers, and Semipalmated Sandpipers.

Monomoy National Wildlife Refuge (2,698 acres) includes Monomoy Island, the place to see impressive numbers of transient waterbirds, waterfowl, and shorebirds. Extending south for almost 10 miles into the Atlantic from near the elbow of the Cape, Monomoy is virtually one long barrier sand beach, and behind it are sand dunes, shrubby dune hollows, ponds, and fresh- and salt-water marshes. Until 1945, when the Island was cut from the mainland by the sea, the hundreds of ornithologists who visited it were able to drive the entire length. Now one must go by boat. Information on transportation to Monomoy is available from the Refuge office in Chatham or from the Wellfleet Bay Wildlife Sanctuary in Wellfleet. To reach Chatham, drive south from Wellfleet on US 6, exit at Orleans, and continue on State 28. From the west, go south from US 6 at the exit to East Harwich.

Any time of year is exciting for bird finders at Monomoy. In summer, Canada Geese, American Black Ducks, occasionally Green-winged and Blue-winged Teal, Piping Plovers, (Prairie) Horned Larks, Savannah Sparrows, and Sharp-tailed Sparrows are among nesting species, and there are also breeding colonies of Common, Arctic, Roseate, and Little Terns. From late July until October, great numbers of shorebirds—notably, Semipalmated and Black-bellied Plovers, Ruddy Turnstones, Whimbrels, Greater and Lesser Yellowlegs, Red Knots, Pectoral, Least, and Semipalmated Sandpipers, Dunlins, and Sanderlings—stop off in their southward migration. In winter, Monomoy is justly famous for the thousands of Common Eiders and White-winged Scoters off the beach, together with Common Loons, Cómmon Goldeneyes, Buffleheads, and Red-breasted Mergansers. And in fall or winter there is always a possibility of spotting offshore such oceanic birds as shearwaters, Northern Gannets, jaegers, Black-legged Kittiwakes, Razorbills, Thick-billed Murres, and Dovekies.

Michigan

KIRTLAND'S WARBLER

Anyone who aspires to find all the different kinds of birds in the United States must sooner or later seek the Kirtland's Warbler. Consequently he will go to Michigan, the only place in the world where it is known to nest—unless he prefers to take his chances of seeing it on its wintering grounds in the Bahamas or on its migratory journeys.

While having the only breeding population of the Kirtland's Warbler is Michigan's chief claim to ornithological distinction, this is but one of the state's several attractions. Its intimate relation to the Great Lakes, its hundreds of lakes, marshes, and bogs, and its coniferous and deciduous forests present a wide and often unexpected variety of species.

Michigan is divided by the Straits of Mackinac into the Upper

Peninsula and the Lower Peninsula (sometimes called Upper and Lower Michigan), whose combined Great Lakes shoreline, including various harbors, inlets, and outlying islands, extends for 3,177 miles. Although the two Peninsulas have many lakes, bogs, and marshes and are well drained by many rivers, they differ sufficiently in topography and biota to warrant separate consideration.

On the Upper Peninsula the shoreline has a few beaches and dunes, but it is generally rocky and abrupt. Along Lake Superior, in various places from the Keweenaw Peninsula westward, it is dramatically precipitous, and is a favorite nesting site for Northern Ravens. The interior of the eastern, central, and southwestern part of the Upper Peninsula is relatively low-lying. There are large bogs and wet lowlands with mats of tamarack, black spruce, cedar, small shrubs, and sedges; there are also upland areas, some sandy, with white, red, and jack pines, and some with rich soils capable of producing deciduous trees, chiefly sugar maple and beech, and hemlock. Only in the northwestern part, where the terrain is high and rugged, are there large forests of the northern coniferous type, the dominant trees being white spruce and balsam-fir. Here the Porcupine Mountains, with an elevation of 2,023 feet, reach the highest point in the state.

On the mitten-shaped Lower Peninsula the shoreline has gradually sloping sand beaches backed by high dunes. Where the beaches are broad, a few Piping Plovers nest. The interior is moderately low-lying, except for scattered hills and a few uplands. Terrain, soils, and vegetation of the northern half of the Lower Peninsula constitute a continuation of conditions found in adjacent parts of the Upper Peninsula. In the southern half, however, bogs become less extensive and fewer in number; white and red oaks, with scattered shagbark and pignut hickories, take the place of pine in the sandy regions; and red maple, ash, and butternut dominate the deciduous forests, particularly those in the lowland areas.

No other state has suffered greater destruction of its forest resources than Michigan. Fires, lumbering operations, and farming have laid waste the forests that originally covered most of the state. Only in isolated areas are there remnants that can be called virgin; and it is doubtful whether these areas are sufficiently extensive to hold the population of breeding birds that was once typical. Probably the coniferous forests in the cool bogs and lowlands of the Upper Peninsula and the northern half of the Lower Peninsula have undergone the least modification. In any case it is in these coniferous forests, and their edges intermingled with deciduous growth, that the birds listed below breed regularly. Species marked with an asterisk breed regularly in the Upper Peninsula only.

* Spruce Grouse
Yellow-bellied Sapsucker
* Black-backed Three-toed Woodpecker
Yellow-bellied Flycatcher
Olive-sided Flycatcher
* Gray Jay
* Boreal Chickadee
Red-breasted Nuthatch
Brown Creeper
Winter Wren
Hermit Thrush
Swainson's Thrush
Golden-crowned Kinglet
Ruby-crowned Kinglet
Solitary Vireo
Nashville Warbler
Northern Parula Warbler
Magnolia Warbler
Black-throated Blue Warbler
Myrtle Warbler
Black-throated Green Warbler
Blackburnian Warbler
* Palm Warbler
Northern Waterthrush
Mourning Warbler
Canada Warbler
Purple Finch
* Pine Siskin
Slate-colored Junco
White-throated Sparrow
* Lincoln's Sparrow

For the most part, secondary growth timber and farmlands have replaced the original pine and deciduous forests. In the secondary growth, aspen is frequently prominent, and there is often a mixture of such forms as oak, maple, beech, white birch, and occasionally pines. Juneberry, blueberry, sumac, and other shrubs now cover the many abandoned farms in the northern part of the Lower Peninsula. Listed below are the birds breeding regularly in secondary deciduous woods and farmlands (fields, wet meadows, brushy lands, orchards, and dooryards). Species marked with one asterisk breed regularly in the Lower Peninsula only; species marked with two asterisks breed regularly in the southern half of the Lower Peninsula only.

DECIDUOUS WOODS

* Red-shouldered Hawk
Broad-winged Hawk
Ruffed Grouse
* Yellow-billed Cuckoo
Black-billed Cuckoo
** Common Screech Owl
Barred Owl
Whip-poor-will
Yellow-shafted Flicker
** Red-bellied Woodpecker
Hairy Woodpecker
Downy Woodpecker
Great Crested Flycatcher
** Acadian Flycatcher
Least Flycatcher
Eastern Pewee
Blue Jay
Black-capped Chickadee
** Tufted Titmouse
White-breasted Nuthatch
Wood Thrush
** Blue-gray Gnatcatcher
* Yellow-throated Vireo
Red-eyed Vireo
* Warbling Vireo
** Cerulean Warbler
Ovenbird
American Redstart
Baltimore Oriole
Scarlet Tanager
Rose-breasted Grosbeak

FARMLANDS

** Common Bobwhite
* Mourning Dove
* Red-headed Woodpecker
Eastern Kingbird
Eastern Phoebe
(Prairie) Horned Lark
Tree Swallow
Barn Swallow
House Wren
** Carolina Wren
* Northern Mockingbird
Gray Catbird
Brown Thrasher
Eastern Bluebird
Yellow Warbler
Common Yellowthroat
Bobolink
Eastern Meadowlark
Brewer's Blackbird (*mainly Upper Peninsula*)
Common Grackle
* Northern Cardinal
Indigo Bunting
American Goldfinch
Rufous-sided Towhee
Savannah Sparrow
* Grasshopper Sparrow
* Henslow's Sparrow
Vesper Sparrow
Chipping Sparrow
* Field Sparrow
Song Sparrow

A noteworthy feature of the Lower Peninsula is the large number of marshes in which cattails, bulrushes, and sedges grow luxuriantly. A few exceptionally large marshes cover hundreds of acres, as on the delta of the St. Clair River in Lake St. Clair and on the western shore of Lake Erie. Species that nest regularly in most of these marshes are the following:

Pied-billed Grebe
American Bittern
Least Bittern
King Rail
Virginia Rail
Sora
Common Gallinule
American Coot
Black Tern
Marsh Wren
Red-winged Blackbird
Swamp Sparrow

During spring and fall migrations, concentrations are sometimes impressive in a number of localities. Small landbirds such as thrushes, kinglets, vireos, warblers, and fringillids tend to 'bunch up' on the thickly wooded points of land that extend into the surrounding Great Lakes. Waugoshance Point on Lake Michigan (reached from Mackinaw City) and Point Pelee in Ontario (reached from Detroit) are notable examples. In spring, hawks moving north through Michigan sometimes rest in large numbers on the south side of the Straits of Mackinac before crossing to the Upper Peninsula. And in fall, many hawks from Canada migrate south along the St. Clair and Detroit Rivers on both Michigan and Ontario sides. Shorebirds, swans, geese, and ducks are best seen in numbers on the eastern side of the state: the shorebirds, for instance, at Tawas Point, and the swans, geese, and ducks on the Shiawassee Flats, the St. Clair Flats, and the western shore of Lake Erie.

In Michigan, peak flights may be expected within the following dates:

UPPER PENINSULA

Waterfowl: 1 April–25 April; 1 October–10 November
Shorebirds: 5 May–5 June; 25 July–20 September
Landbirds: 25 April–1 June; 15 August–1 October

LOWER PENINSULA

Waterfowl: 25 March–20 April; 5 October–15 November
Shorebirds: 1 May–1 June; 1 August–25 September
Landbirds: 20 April–25 May; 20 August–10 October

Michigan is visited more or less regularly during winter by (Northern) Horned Larks, Evening Grosbeaks, Common Redpolls, Red Crossbills, White-winged Crossbills, American Tree Sparrows, Lapland Longspurs, and Snow Buntings. Occasionally, Snowy Owls, Pine Grosbeaks, and Bohemian Waxwings appear in considerable numbers. Because the St. Clair and Detroit Rivers do not freeze over entirely, they attract a fascinating variety of waterbirds and waterfowl, some of which are quite unusual in this section of the country. No winter goes by without local bird finders reporting a rare duck or gull.

Authorities

David E. Baker, H. Lewis Batts, Jr., Elizabeth B. Beard, Edward M. Brigham, Jr., Betty Darling Cottrille, W. Powell Cottrille, F. N. Hamerstrom, Jr., Harry W. Hann, G. Stuart Keith, Harold Mayfield, Clarence J. Messner, Hazel W. Messner, Theodora Nelson, Walter P. Nickell, William Sheldon, Arthur E. Staebler, George M. Stirrett, George Miksch Sutton, Josselyn Van Tyne, Lawrence H. Walkinshaw, George J. Wallace, Harold F. Wing.

References

Enjoying Birds in Michigan. A Guide and Resource Book for Finding, Attracting and Studying Birds in Michigan. Edited by William L. Thomson. 3d ed. Michigan Audubon Society, Kalamazoo Nature Center (7000 North Westnedge, Kalamazoo, Mich. 49007), 1970.

Ornithology at the University of Michigan Biological Station and the Birds of the Region [Northern tip of Lower Peninsula]. By Olin Sewall Pettingill, Jr. Kalamazoo Nature Center, 1974.

ANN ARBOR
Nichols Arboretum | Huron River Overflow

Nichols Arboretum (135 acres), owned by the University of Michigan, is excellent for migrating warblers from 1 May to 25 May and from 1 September to 1 October. The area is hilly, and its native vegetation is mixed with introduced plantings. Warblers concentrate in many places, particularly among the willows and other trees bordering the north side along the Huron River. From downtown Ann Arbor, in southeastern Michigan, drive east on I 94B and turn left on Geddes Avenue. After passing Forest Hill Cemetery, take the second left, Glen Drive, which goes through a gate into the Arboretum. From Glen Drive, several roads and trails permit easy access to all sections.

To reach the **Huron River Overflow,** a fine place for migrating waterbirds and waterfowl, continue east on Geddes Avenue for 1.5 miles to a bridge over the Huron River. Do not cross the bridge; instead, turn right and follow along the south side of the Overflow for the best views. From 15 March to 1 April and 1 October to 15 November there should be Mallards, American Black Ducks, American Wigeons, Common Pintails, Blue-winged Teal, Wood Ducks, Redheads, Canvasbacks, Common Goldeneyes, Hooded Mergansers, and Common Mergansers. In late spring and early fall, if the water level is low, the Overflow attracts several kinds of herons and shorebirds.

BATTLE CREEK
W. K. Kellogg Bird Sanctuary | Bernard W. Baker Sanctuary

A well-known focus for transient waterfowl in abundance, especially Canada Geese, is Wintergreen Lake in the **W. K. Kellogg Bird Sanctuary** (500 acres), about 12 miles northwest of Battle Creek. The Sanctuary was established in 1927 by W. K. Kellogg and later given to Michigan State University. Wintergreen Lake, covering only 30 acres, is bordered in part by marshes. Here in late March and early April and from mid-October through November the geese and ducks representing as many as sixteen species stop by. A small number of geese reside either at the Sanctuary or in adjacent areas. In the woods, fields, and marshes and in tree and shrub plantings surrounding the visitor center, lakeside museum, and aviaries, are found other species such as Mourning Doves, Red-winged Blackbirds, Purple Martins, Tree Swallows, and Eastern Bluebirds. To reach the Sanctuary, drive west from Battle Creek on State 89 for about 10 miles, north on County 184 (40th Street) for about 2.5 miles, then left on C Avenue to the entrance.

The **Bernard W. Baker Sanctuary** (571 acres), owned by the Michigan Audubon Society, contains an exceptionally fine marsh, noted particularly for nesting and transient Sandhill Cranes. The marsh has cattail and sedge associations, both extensive enough to hold considerable numbers of American Bitterns, Mallards, American Black Ducks, Virginia Rails, Soras, Common Snipes, Marsh and Sedge Wrens, Red-winged Blackbirds, Henslow's Sparrows on the grassy edges, and Swamp Sparrows. Of special note are the Sandhill Cranes, King Rails, and Common Gallinules occupying the cattails in the southern part, and Northern Harriers and Short-eared Owls sometimes nesting among the sedges in the northern, drier section.

Bordering the marsh are bushes, chiefly dogwood, mixed with tamarack, basswood, ash, and maple, where birds such as Red-tailed Hawks, Ruffed Grouse, American Woodcock, Veeries, and Blue-winged Warblers spend the summer. Grasshopper Sparrows nest in the fields outside the Sanctuary. During migration additional numbers of Sandhill Cranes appear. Yellow Rails regularly stop among the sedges. Locating Yellow Rails is a real challenge, since they are usually both silent and reluctant to fly up when approached.

Visit the Sanctuary in the evening any time in April, to hear the calls of Sandhill Cranes and witness the flight performances of American Woodcock and Common Snipes; from the second week of May to mid-June, to observe marsh birdlife when most vociferous; the last two weeks in March and mid-October, to see transient cranes; and the last two weeks in April, to search for Yellow Rails.

To reach the Sanctuary, drive north from Battle Creek on State 66 for about 7 miles, then right on State 78 for 6 miles to Bellevue. Turn right at the stop light in Bellevue on Junction Road and continue for about 4 miles to the entrance, marked by a low cinder-block building, on the right.

CHEBOYGAN
Cheboygan Marsh | Reese's Bog

The **Cheboygan Marsh,** stretching west from a lakeside park within the city limits, consists of 60 acres of cattails and sedges, interrupted here and there by small patches of open water and flanked on its northern side by alders, willows, and bushes. To reach the Marsh, drive west on US 23 from its intersection with State 27 for one block, then turn right and proceed north a short distance to the park. Leave the car and walk along the side of the Marsh. If it is June—the best time for a visit—listen and look for Least and American Bitterns, Virginia Rails, Soras, Common Gallinules, American Coots, Black Terns, Marsh Wrens, and Swamp Sparrows.

Reese's Bog, a place for 'northern' birds in the Southern Peninsula, is a 450-acre lowland forest of black spruce, balsam-firs, tamaracks, and white cedars on the north shore of Burt Lake. Summer residents include the Pileated Woodpecker, Yellow-bellied Sapsucker, Yellow-bellied Flycatcher, Olive-sided Flycatcher, Red-breasted Nuthatch, Brown Creeper, Winter Wren, Veery, both Golden-crowned and Ruby-crowned Kinglets, Solitary Vireo, Black-and-white Warbler, Nashville Warbler, Northern Parula Warbler, Magnolia Warbler, Cape May Warbler, Myrtle Warbler, Black-

BLACK TERN

throated Green Warbler, Blackburnian Warbler, Canada Warbler, Purple Finch, and White-throated Sparrow. In June they are all in full song.

To reach the Bog, drive west from Cheboygan on County 64 to a crossroad 3 miles beyond the Riggsville Exit from I 75. The road on the right leads to the University of Michigan Biological Station; the road to the left goes to the Bog, after crossing Hogback Road, and through it to the Burt Lake shore. Park the car by the Lake and walk back along the road through the Bog, or follow roads to the left and right that lead to cottages on the lakeshore. While in the vicinity, stop at the Biological Station and ask for the latest information on bird finding in the area.

CLARE
Prairie Chicken Management Area

In the **Prairie Chicken Management Area** (1,280 acres), owned by the State Department of Natural Resources, one may see some of the few Greater Prairie Chickens remaining in Michigan. The time to visit the Management Area is between mid-March and May, preferably in the last half of April, when the birds gather before sunup on their booming grounds to perform nuptial 'dances.'

To reach the Area, drive north from Clare on US 27, in central Michigan, west on State 61, and north on State 66 for 3 miles beyond Marion; then east on a gravel road bordering the south side of the Area to a parking space. Leave the car and walk northward to a point about 150 yards from the booming ground ahead, which is posted against entry. From this point, with a telescope or good binoculars, one can watch the dances at sunup.

DETROIT
Belle Isle | St. Clair Flats | Point Pelee National Park | Holiday Beach Provincial Park

Despite the fact that this great city sprawls over a vast territory, there are exceptional opportunities for observing birds in its vicinity.

During winter months a good place to see swans and ducks is off the east and west shores of **Belle Isle,** a city park on an island in the Detroit River. From I 375 in center city, drive north on Jefferson Avenue, turn right on East Grand Boulevard, and cross a bridge to the island. From the road encircling Belle Isle there are fine views of the River and the large concentrations of wintering ducks: Redheads, Canvasbacks, Lesser Scaups, Common Goldeneyes, all three mergansers, now and then some Oldsquaws, and sometimes a few Buffleheads. When the River is partly iced over and the birds concentrate in the remaining areas of open water, bird finding is best; when the River stays open, the ducks scatter and are harder to see. Watch the gulls. In spring and fall, Bonaparte's Gulls are sometimes quite common.

The **St. Clair Flats** consist of several islands in the northwest part of Lake St. Clair, where the St. Clair River enters from the north. Geologically, the islands comprise a delta of the St. Clair, with a maximum width of about 9 miles. They are characteristically open and marshy and have rich growths of bulrushes, cattails, and sedges through which numerous channels meander. On the three largest islands—American-owned Dickinson and Harsen's and Canadian-owned Walpole—the terrain is high enough above water level to permit woods, hayfields, and summer resorts. For the bird finder, Harsen's is the most convenient, since it is easily reached by car and has 18 miles of roads that pass some of the best places in the Flats for observing birds.

To reach Harsen's Island from Detroit, drive north on I 94, turn east on State 29, follow it around Anchor Bay to 2 miles beyond the village of Pearl River, and take the car ferry to the Island. Here one has a choice of several roads, any one of which leads to good views of the surrounding water and marshes and to places for exploration on foot.

Bird finding here is invariably rewarding in all seasons. During winter, when the Flats are frozen over, it is always possible to find small spots of open water, as one drives around, where Redheads and Canvasbacks concentrate in large numbers, and other ducks are present in smaller numbers. From early March to early April, a prime attraction is the great number of Whistling Swans that linger two or three weeks during their northward migration. Often several hundred gather in rafts on the open water, feed in the shallow channels, or rest on the shore.

During this same period, many species of ducks are common, though the spring concentrations are usually not so great as those in fall from late September to mid-November. Species that one will surely see are the Mallard, American Black Duck, American Wigeon, Common Pintail, Blue-winged Teal, Redhead, Ring-necked Duck, Canvasback, Lesser Scaup, Common Goldeneye, Hooded Merganser, and Common Merganser. From mid-April to the first of July, the Common Snipe frequently gives its flight display. Among marsh dwellers during the breeding season are the Pied-billed Grebe, American Bittern, King Rail, American Coot, Black Tern, Marsh Wren, and Swamp Sparrow. Great Blue Herons, Green Herons, and Black-crowned Night Herons are present any time of year except winter.

About 30 miles southeast of Detroit, in Canada, a slender peninsula extends 9 miles south into Lake Erie—by name Point Pelee, familiar to ornithologists for its spectacular aggregations of migrating birds in spring. The peninsula, low-lying and with a ridge of higher ground running like a spine almost its entire length, ends in a sand spit. Beaches edge the western side of the ridge, which has some vegetation, mostly of hackberry and red cedar; marshes stretch to the east. **Point Pelee National Park** (6 square miles) occupies the southern part, where migrating birds are best observed.

A Point Pelee timetable: *March* for Whistling Swans, many diving ducks, Horned Larks, and Eastern Bluebirds; *late March and early April* for Red-winged Blackbirds; *April* for Horned Grebes, Double-crested Cormorants, surface-feeding ducks, American Coots, Yellow-shafted Flickers, Eastern Phoebes, Brown Creepers, Golden-crowned Kinglets, Slate-colored Juncos, and Field, Fox, and Song Sparrows; *early May* for Common Terns, Black Terns, Ruby-crowned Kinglets, White-crowned Sparrows, and White-throated Sparrows; *second and third weeks in May* for shorebirds, flycatchers, swallows, wrens, thrushes, warblers, Bobolinks, Orchard and Baltimore Orioles, and Scarlet Tanagers.

Mid-May is truly the peak period for migration, and the tip of the Point, in early morning, is the place for enjoying it. Provided the weather conditions were right the night before, the area will swarm with small familiar landbirds, sometimes in strange places. For instance, one may see Red-headed Woodpeckers, Eastern Kingbirds, American Redstarts, and Baltimore Orioles resting on the beaches or in the grass nearby. Out on the sand spit there should be Ruddy Turnstones, Pectoral Sandpipers, and other shorebirds, and, just offshore, Double-crested Cormorants and rafts of Red-breasted Mergansers. Rare and stray species are more apt to show up in May than in any other month.

Now and then there is a reverse migration during which one sees such birds as Red-headed Woodpeckers, swallows, Cedar Waxwings, and American Goldfinches flying south over the Lake. The commonly accepted explanation is that the birds, driven by southerly winds, have overshot their destination and are returning to it.

Anyone planning to visit Point Pelee in mid-May should watch the weather maps carefully and plan to be at the tip of the Point in early morning during, or just after, the arrival of a cold front from the west. At such times small landbirds crossing Lake Erie with the southerly winds suddenly meet the northerly winds of the cold front and drop down on the first land they come to—the tip of Point Pelee—where they 'bunch up' exhausted. Waterbirds and waterfowl seem to move north regardless of fronts.

In addition to its exciting migration season, Point Pelee has fine areas worth exploring for birds in any season. The Woodland Nature Trail, starting at the interpretive center a mile north of the tip of the peninsula, winds 1.25 miles through an abandoned apple orchard, brushy tangles, and a mature hardwood forest with shaded pools. Breeding birds that one may see or hear in suitable habitats along the way include the Wood Duck, Great Crested Flycatcher, Tufted Titmouse, Bewick's and Carolina Wrens, Gray Catbird, Wood Thrush, Blue-gray Gnatcatcher, Prothonotary Warbler, Yellow-breasted Chat, Orchard Oriole, Indigo Bunting, Rufous-sided Towhee, and Field Sparrow.

The Boardwalk Nature Trail, beginning 1.5 miles south of the Park entrance, stretches for more than a half-mile over an unspoiled marsh with extensive cattail, bulrush, and sedge habitats that attract summer-resident American and Least Bitterns, Virginia Rails, Soras, Common Gallinules, American Coots, Black Terns, Marsh and Sedge Wrens, and Swamp Sparrows. The boardwalk terminates in a

12-foot-high observation tower overlooking ponds and channels where migrating ducks and other waterfowl gather from mid-March to early May.

To reach Point Pelee, cross the Detroit River to Windsor, Ontario, by tunnel or bridge, drive east on Provincial 3 to Leamington and thence south to the Park.

One of the most convenient places to watch migrating hawks in fall as well as migrating waterfowl in the spring is **Holiday Beach Provincial Park,** southeast of Amherstburg, Ontario. After crossing the Detroit River, drive south and east on Provincial 18, then south on Provincial 18A for 2 miles to the entrance. The best views of hawks are from the parking lot nearest the shore and close to a marshy area. The hawks move across Provincial 18 and the Park between 10 September and 1 October. Most of them are Broad-winged Hawks; others are Sharp-shinned Hawks, Northern Harriers, American Kestrels, and Ospreys. Sometimes there are considerable numbers of Blue Jays.

Reference

The Spring Birds of Point Pelee National Park. By George M. Stirrett. Rev. ed. Ottawa: Indian and Northern Affairs, National and Historic Parks Branch, 1973. Available by mail from Information Canada, Ottawa, Ont. K1A 0S9.

ERIE
Erie State Game Area

The western part of Lake Erie is shallow, and its shores are low, usually consisting of a narrow sand beach behind which, on even lower ground, a marsh extends inland for a half-mile or more. A good example of such territory lies within the **Erie State Game Area** in the southeastern corner of the state. To reach the Area, drive east from the Erie Exit from I 95 for about 0.75 mile to a large powerhouse on the lakeshore. A public lane, suitable for cars in good weather, leads through the utility's property to the beach. For the best bird finding, walk south along the shore toward the southern tip of North Cape, 4 miles distant, with the open lake on the left and a marsh on the right. The first half-mile is game refuge, the remainder a public hunting ground.

In the strip of trees and shrubs lining the shore, especially where the cover is heaviest, landbirds are prevalent at any time of year but in greatest numbers during migration seasons. Hermit Thrushes, Myrtle Warblers, Rufous-sided Towhees, White-throated Sparrows, and Swamp Sparrows are abundant in fall, sometimes lingering until December. After the ice breaks up on Lake Erie—usually around the first of March—waterfowl and gulls appear. Outstanding attractions are the flight of Whistling Swans, numbering in the thousands and often remaining until 15 April, and the Greater Black-backed Gulls, which are never very common in the Great Lakes region. On days in May, August, and September when strong westerly winds lower the water level, shorebirds gather on the exposed mud flats and sand bars. In late summer, Great Egrets become common along the shore and marsh. Large concentrations of ducks are offshore in late fall. At sunset, when they pass from the marsh and fields where they have been feeding, the flights are spectacular.

EWEN
Connecticut Warbler Area

The Connecticut Warbler, known to most ornithologists only as a spring and fall transient, successfully concealed its Michigan nesting area for a long time. Finally an astute bird finder tracked it to a moist woods north of the village of Ewen in the western part of the Northern Peninsula where, in early June, he heard six singing males.

To reach this area, drive east from Ewen on State 28 for 1.0 mile; turn left and drive north for about 4 miles to just beyond a cemetery; left again for 1.0 mile, passing three farms on the right; then north to the first farm on the right. Park here and walk along the road for about a mile to an aspen woods dotted here and there with spruce and balsam-fir. The singing is most intense in early morning.

Around the farm and along the road, watch for Northern Ravens, Red-breasted Nuthatches, Slate-colored Juncos, Golden-winged Warblers, and Clay-colored Sparrows.

FLINT
Seven Ponds Nature Center

The **Seven Ponds Nature Center** (245 acres), owned and operated by the Michigan Audubon Society, embraces a very pleasant segment of Michigan's countryside. Gently rolling hills, clear ponds lined with cattails, cedar swamps, mature deciduous woods, and open fields—a variety of habitats that means a variety of birds. The Center is almost equidistant from Flint and Pontiac. To reach it from Flint, drive east on State 21 to Lapeer, south on State 24 for 9 miles, left on Dryden Road, and then right on Calkins Road (marked by a sign) to the Center. Drive directly to the interpretive building, where the latest information on birds as well as maps of the trails are available.

GRAYLING
Areas for the Kirtland's Warbler | Dead Stream Swamp

The Kirtland's Warbler, Michigan's own bird and an endangered species, nests within an area 60 by 100 miles in the northern Lower Peninsula mainly east of Grayling. Its habitat requirements for breeding—usually young stands of jack pines 6 to 18 feet high, with intervening open spaces—limit the distribution of the species. As the maturing trees become too tall, the birds forsake them for lower stands.

Kirtland's Warblers are best located from mid-May to mid-July by listening for their loud, clear, almost explosive songs, often delivered from the top of a pine or a high dead stub. Remarkably tame for parulids, they may sometimes be seen closely. By August they cease singing altogether and are almost impossible to find. Their nests are on the ground, well concealed by grasses and other low plants, and generally under one or more young jack pines, occasionally red pines.

In an effort to perpetuate suitable nesting habitat, the Michigan Department of Natural Resources maintains three 'management areas' near Lovells, Luzerne, and West Branch, and the United States Forest Service maintains another near Mio. Here certain plots of maturing jack pines are cut or burned and then re-seeded so as to assure the prevalence of young growth. In an effort of another sort, to reduce loss of the Kirtland's eggs and young, Brown-headed Cowbirds, which heavily parasitize the warbler nests, are trapped and removed from breeding areas. Despite these measures, the total adult population of the Kirtland's Warbler, estimated in 1961 to consist of 1,000 individuals, decreased inexplicably by an ominous 60 per cent. Consequently all known nesting areas have been closed and posted against public entry.

Two guided tours to see the Kirtland's Warbler are available from mid-May to mid-July: one from the District Office of the Michigan Department of Natural Resources in Grayling (telephone: (517) 348–6371), the other from the District Ranger Station, Huron National Forest, in Mio (telephone: (517) 826–3717). To reach the District Office by car, exit west from I 75 on Business I 75 to Grayling; at the traffic light in town, turn east and proceed 0.5 mile to the Office in the State Fish Hatchery Building. To reach the Ranger Station, drive east from Grayling on State 72 to Mio; in town, turn north on State 33 to the Station immediately on the right. While in Mio, note the statue of the Kirtland's Warbler in the central park.

To see Kirtland's Warblers in a natural, unmanaged habitat, drive north from Grayling on I 75, east from the Lewiston–Frederic Exit on County 612 for 7.5 miles, then south on Stephan Bridge Road, which, after 4 miles, traverses an area of jack pines burned in 1955. From the Road, listen for several males singing in a wide stretch of jack pines that have since seeded themselves. Do not harass the birds by playing tape recordings of their songs. Photography is prohibited. While on the Road, listen also for Clay-colored and Field Sparrows among low shrubs, and Prairie Warblers in scrub oaks. Other birds in the region are American Kestrels, Common Nighthawks, Eastern Bluebirds, and Slate-colored Juncos. Since the Road passes through a military reservation, it may be closed during maneuvers.

Dead Stream Swamp, an impoundment of some 2,000 acres, is a fine place to observe nesting Ospreys and a variety of marsh birds. Though the dead stubs of countless trees killed by the flooding confer a rather grim aspect on the Swamp, they provide perches and nesting sites for Ospreys and a great many Tree Swallows. And the lush growth of cattails, bulrushes, and sedges gives ample cover for such breeding birds as Pied-billed Grebes, Least Bitterns, Mallards,

American Black Ducks, Ring-necked Ducks, Common Goldeneyes, King and Virginia Rails, Soras, Common Gallinules, American Coots, Black Terns, Marsh Wrens, Red-winged Blackbirds, and Swamp Sparrows.

To reach the Swamp, drive south from Grayling on I 75, bear right on new US 27, and exit west on State 55. Take the next right on County 303, turn right again at the next paved intersection, then left almost immediately at a sign indicating Mickelson's Landing, a boat livery on the edge of Dead Stream Swamp. Exploring the Swamp is unsatisfactory without a boat. If the bird finder does not bring his own, he can hire one here.

ISLE ROYALE
Isle Royale National Park

The bird finder looking for northern birds beyond the reach of automobiles and tourists will enjoy the isolation and solitude of Isle Royale in northwestern Lake Superior. Thirteen miles or more from Ontario, the nearest mainland, which it parallels for 45 miles, Isle Royale is a roadless wilderness. Though no wider at any point than 8 miles, it nonetheless covers about 210 square miles. Together with its two hundred outlying islands and islets, it was established in 1940 as **Isle Royale National Park.**

Isle Royale consists of low ridges and valleys—many with lakes and bogs—running in a northeast-southwest direction. At the northern and southern extremities especially, the ridges dip intermittently below the level of Lake Superior, leaving a chain of small islands and peninsulas that form snug coves and fjord-like harbors. The main island and even the smallest islands are alike forested with both coniferous and deciduous stands. Balsam-fir and white spruce predominate in many upland areas, sugar maple in others; and in all areas yellow birch is prominent. In particularly dry localities there are small stands of oak and pine. Black spruce, white cedar, and tamarack thrive in bogs and other poorly drained spots. A large central and southwest area, burned many years ago, supports principally white birch and aspen.

Isle Royale birdlife, though unquestionably typical of northern forests and waterways, has not been thoroughly investigated. A peculiarity is the presence of the Sharp-tailed Grouse, usually in the burned portions, and the absence of both Spruce and Ruffed Grouse despite their existence on the nearby mainland. The Hawk Owl and Saw-whet Owl have been reported breeding. The Gray Jay is a common camp symbiont, and the Northern Raven is a conspicuous resident. Common Loons, Common Goldeneyes, and Common Mergansers are prevalent summer residents in sheltered coves and inland lakes, as are Red-breasted Mergansers along the Lake Superior shores.

The primary reward for summer bird finding is the variety of small landbirds, notably warblers. Besides species known to breed regularly in Michigan's Upper Peninsula there are probably Tennessee, Cape May, Bay-breasted, Blackpoll, and Wilson's Warblers in summer residency. An excellent time for bird finding here is late summer, when the local population of warblers and other small birds is augmented by vast numbers of passerines migrating from Canada and temporarily 'trapped' before moving south over Lake Superior.

Isle Royale is known less for its birdlife than for its moose and wolves, which are presumed to have reached the island from the nearby mainland in 1906 and 1948 respectively. Moose are seen readily but wolves only with luck.

Except for a few scientists conducting special field studies, no one resides at Isle Royale in winter. During summer, overnight accommodations are available at Rock Harbor and Washington Harbor. From both places many miles of foot trails lead to good spots for birds. Isle Royale may be reached by boat from Copper Harbor and Houghton, Michigan, and Grand Portage, Minnesota, and by floatplane from Houghton only. For information on accommodations and transportation, write the Superintendent, Isle Royale National Park, Houghton, Michigan 49931.

JACKSON

Portage Marsh | Schumacher Marsh

Portage Marsh and **Schumacher Marsh,** northwest of Waterloo in south-central Michigan, are segments of the extensive Waterloo

State Recreation Area. Many marsh birds nest in these wetlands, and in fall Sandhill Cranes concentrate here.

The Marshes, over 500 acres each with some open water, are covered for the most part with cattails, scattered tamaracks, and shrubs. Nesting marsh birds, including the King Rail and Common Gallinule, are best observed from mid-May to mid-June. A few pairs of Sandhill Cranes nest regularly in remote parts during early spring. Because Sandhills are very wary, people are urged not to enter the Marshes to search for them lest the birds desert their nests. Rather, visit the Marshes in September and October, when two dozen or more Sandhills congregate and can easily be seen from a road around Portage Marsh.

To reach the Marshes, drive east from Jackson for 17 miles on I 94; exit and drive north to about 3 miles beyond Chelsea, then turn left on the Waterloo-Munith road. About 3 miles beyond Waterloo, stop at a bridge over Portage Creek. Portage Marsh is in view on the left; Schumacher, on the right.

KALAMAZOO
Kalamazoo Nature Center

The **Kalamazoo Nature Center** (397 acres) for research, education, conservation, and pure enjoyment of the out-of-doors, lies about 5 miles north of central Kalamazoo on the Kalamazoo River in south-central Michigan. The varied habitats—a mature forest of beech and maple with some oak, American hornbeam, and tulip trees; a bog-marsh cut by a stream; thickets of hawthorn and sumac; farmlands, both cultivated and abandoned; and the River—hold many different species of birds. Summer residents include Traill's Flycatchers, Gray Catbirds, Cedar Waxwings, Common Yellowthroats, Bobolinks, Northern Cardinals, Indigo Buntings, and American Goldfinches. In winter, Mallards, American Black Ducks, Black-capped Chickadees, Tufted Titmice, and American Tree Sparrows are common. A visit to the Center is worthwhile at any season.

To reach the Center from downtown Kalamazoo, drive north on North Park Street and, shortly after it joins Westnedge Street, watch for the well-marked entrance drive on the right. Check in at the interpretive center for current information on birds and a map of trails leading to various parts of the sanctuary.

LANSING
Rose Lake Wildlife Experiment Station

For a variety of birds typical of the secondary woods and farmlands of the Lower Peninsula, visit the **Rose Lake Wildlife Experiment Station** (2,567 acres), a state-owned tract in south-central Michigan, 12 miles northeast of Lansing on State 78. A Department of Wildlife Resources sign indicates the side road on the left leading to the Station. The tract contains many fields and woodlots, together with several marshes, small lakes, and streams, nearly all accessible by roads and lanes. The trees are principally oak and hickory except for ash, red maple, and some tamarack in the moist lowlands.

About eighty species of birds breed regularly. (Prairie) Horned Larks nest on almost all cultivated farm plots; Golden-winged Warblers, Blue-winged Warblers, and Louisiana Waterthrushes find suitable habitats in the vicinity of Vermilion Creek. Occasionally a pair of Sandhill Cranes nests in the marsh bordering Rose Lake, where the Least Bittern, Virginia Rail, Sora, Marsh Wren, Swamp Sparrow, and other marsh species are common.

MACKINAW CITY
Straits of Mackinac | Wilderness State Park

The spring hawk movements north across the **Straits of Mackinac** separating the Upper and Lower Peninsulas of Michigan can be impressive between 12 and 30 April. In some years, when the weather is clear with favoring winds, the birds filter through unnoticed; but when the weather is foggy or rainy and windless they perch on fence posts, utility poles, and trees on the south side of the Straits, until the weather changes. As the skies clear they depart, spiraling high in the air and then peeling off in a beeline across the 5-mile stretch of water to the Upper Peninsula. The majority are Broad-winged Hawks, with Red-tailed Hawks next in number. The best position for watching their movements is from 1.0 to 2 miles west of Mackinaw City along Wilderness Park Drive, which begins at the last exit from I 75 before Mackinac Bridge.

Wilderness State Park, fronting on Lake Michigan 11 miles west of Mackinaw City, has extensive, often mixed coniferous-deciduous, forests that hold over seventy-five species of breeding birds, many of northern affinities. Included in the Park's total area of 8,035 acres is Waugoshance Point, a peninsula jutting westward 3 miles into Lake Michigan and notable for its concentrations of transient birds in late summer and early fall.

From Mackinaw City, leave I 75 at the last exit before Mackinac Bridge and proceed west on Wilderness Park Drive to the Park entrance; continue westward through the Park on a winding paved road that eventually, after passing picnicking and camping areas, leaves the Park through a private settlement with a store and comes to an unpaved road, again in the Park, and marked by a sign indicating Waugoshance Point. Follow this, the Waugoshance Point Road, through 2.8 miles of forest to its end in a parking area at the base of the Point. Explore the Point on foot.

To find summer-resident birds during June and early July, make the following stops on the way to Waugoshance Point:

1. Between picnicking and camping areas, turn left from the main Park road to the outdoor center, park the car, and walk along the Swamp Line Trail that begins on the west side of Goose Pond and shortly goes through, or around, spruce-tamarack bogs, cedar woods, and shrubby places. Among species one may expect to hear or see are the Traill's Flycatcher, Olive-sided Flycatcher, Nashville Warbler, Mourning Warbler, Canada Warbler, White-throated Sparrow, and Swamp Sparrow.

2. After driving 2.3 miles on Waugoshance Point Road, walk north on a road to the Burnt Cabin Site and investigate the adjacent coniferous-deciduous woods for the Pileated Woodpecker, Yellow-bellied Sapsucker, Red-breasted Nuthatch, Brown Creeper, Winter Wren, Swainson's Thrush, Golden-crowned Kinglet, Solitary Vireo, Black-throated Blue Warbler, Black-throated Green Warbler, and Blackburnian Warbler. Walk west through the woods bordering the Lake Michigan shore, where, in the pines and balsam-firs edging the woods, there are Northern Parula Warblers, Magnolia War-

OLIVE-SIDED FLYCATCHER

blers, Myrtle Warblers, and Pine Warblers. Look for Parula nests in the *Usnea* lichen suspended from the balsam-firs.

Waugoshance Point, from the parking area to its tip, is generally low-lying and open. Its northern shore and outlying bars—partly sand and partly loose rock and coarse gravel—occasionally attract during nesting season a pair or two of Piping Plovers, and during migration season—particularly in May, August, and September, when the Great Lakes water level is low—small numbers of Semipalmated and Black-bellied Plovers, Ruddy Turnstones, Red Knots, Least and Semipalmated Sandpipers, and Sanderlings. Immediately inland from the north shore and low dunes are three beach pools where a few transient Greater and Lesser Yellowlegs, Pectoral Sandpipers, and sometimes a Solitary Sandpiper feed and rest.

In the interior, about midway between the parking area and tip, is a large bulrush marsh edged with cattails and sedges—breeding habitat for American Bitterns, Virginia Rails, Common Snipes,

Sedge Wrens, Red-winged Blackbirds, and Swamp Sparrows. By far the outstanding features of the Point for bird finders are the large 'islands' of black spruce, red pine, and cedar where thousands of transient passerines gather at night and pass the daylight hours. At the height of migration in September, it is possible to identify at least sixty species of thrushes, vireos, warblers, and fringillids.

NEWBERRY

Sleeper Lake | Seney National Wildlife Refuge | Adams Trail | Kingston Plains

The following information for bird finding in Michigan's north country applies to the period between 1 June and 10 July, when most birds are in full song.

The vicinity of **Sleeper Lake,** in the east-central part of the Northern Peninsula, holds numerous breeding birds representative of the north country. From Newberry, drive north on State 123 for 4.5 miles, then left on County 37 for 11.2 miles—just 1.2 miles beyond Wolf Tavern on the right—and then right, on an obscure road, for about 1.5 miles to an open area, once the site of Bailey's Saw Mill. Park here and walk east along an old lumber road that traverses a bog, crosses a creek, and wanders over and around wooded sand ridges for about 2 miles to Sleeper Lake itself—tiny and unimportant for birds. En route one may hear singing, simultaneously, the Hermit Thrush, Swainson's Thrush, and Veery.

At the outset the road traverses an open bog covered with sedges and shrubs such as leatherleaf and Laborador tea and then passes through a dense stand of black spruce, tamarack, and cedar. Along the brushy edges of the open bog, listen for the melodious songs of Lincoln's Sparrows and look for Olive-sided Flycatchers calling from the tops of isolated trees and dead stubs.

Once in the stands of conifers, be alert for Spruce Grouse, Yellow-bellied Flycatchers, Gray Jays, Boreal Chickadees, Winter Wrens, Golden-crowned and Ruby-crowned Kinglets, Solitary Vireos, Northern Parula Warblers, Blackburnian Warblers, and Magnolia Warblers. When crossing the creek, listen for Traill's Flycatchers and Canada Warblers in the thickets along its banks.

The sand ridges over and around which the road wanders before reaching Sleeper Lake are partly open and partly forested with pines and aspens, and here Nashville Warblers, Myrtle Warblers, Pine Warblers, Purple Finches, Pine Siskins, and Slate-colored Juncos are common. Occasionally there are Palm Warblers and Evening Grosbeaks. Below the ridges, stretching far southward, are open bogs—such as at the start of the walk—interrupted by sparse stands of conifers. Sedge Wrens are numerous here, as are Swamp Sparrows. High above the bogs, Common Snipes frequently perform their aerial maneuvers. From the tops of the higher ridges, soon after sunrise, scan the bogs for Sandhill Cranes. These wary birds, usually seen only in the distance, arrive in the bogs during the evening and depart well before midmorning.

The **Seney National Wildlife Refuge** (95,455 acres) embraces forests and a remarkable scramble of sedge marshes, sandy knolls topped by tall red pines and jack pines, and grass-covered dikes impounding big pools of water. The Refuge is noteworthy for its large numbers of nesting Canada Geese. From Newberry, drive south on State 123, west on State 28 to the village of Seney, and south on State 77 for about 5 miles to the entrance on the right. Stop at the visitor center for information on a self-guided tour in one's own car on the dike roads.

Canada Geese, first introduced in 1936, are established not only within the Refuge but also in suitable locations in the surrounding countryside. In late June and early July, entire families appear around the visitor center and cruise about on the pools, the goslings escorted by both parents. Later in the season, small flocks of geese—probably family groups—fly from one pool to another.

Sandhill Cranes breed regularly, but because they nest in inaccessible sections of the Refuge, visitors observe very few of them from the dikes early in the season. The best time to see Sandhills from the dikes is in September and early October, when flocks sometimes numbering from fifteen to twenty individuals feed and rest in the marshes, along the shallow margins of the pools, and on the sandy knolls. Among ducks present, Mallards and American Black Ducks are the most common. Blue-winged Teal, Wood Ducks, Ring-necked Ducks, and Hooded and Common Mergansers are visible in small numbers.

HOODED MERGANSERS

Other summer residents often in view from the dikes are Common Loons, Pied-billed Grebes, Great Blue Herons, American Bitterns, and Northern Harriers. A few pairs of Bald Eagles nest—the aeries of at least two pairs are in view from the dikes. In early summer the bird finder should search for the elusive Le Conte's Sparrow in the sedge marshes, where it may nest. On the same occasion he should listen for the ticking notes of the Yellow Rail, rare yet possible.

The **Adams Trail,** a wide, unimproved road to Munising, swings west from State 77 about 15 miles north of the village of Seney. While driving on State 77, note the well-marked Schoolcraft–Alger County Line and the Rustic Cabins beyond on the right. Between the County Line and the Cabins lies an extensive spruce-tamarack bog on the left of the highway—a fine habitat for such north-country birds as the Yellow-bellied Flycatcher, Ruby-crowned Kinglet, Black-and-white Warbler, Nashville Warbler, and Canada Warbler. On the opposite side of the highway, the mixed coniferous-deciduous woods is worth investigating for the Ruffed Grouse, Pileated Woodpecker, Yellow-bellied Sapsucker, Red-breasted Nuthatch, Brown Creeper, Winter Wren, Solitary Vireo, and Black-throated Green Warbler. Drop in at the Rustic Cabins, a favorite lodging for bird finders, for the latest information on birds in the area.

In its first 10 miles the Adams Trail passes through a forest of mature beech, maple, and hemlock and two open bogs with scattered tamarack and spruce. Leave the car and follow one or another lumbering road leading off the Trail. Along almost any such road, listen for the Great Crested Flycatcher, Eastern Pewee, Swainson's Thrush, Scarlet Tanager, and Rose-breasted Grosbeak. Listen especially for Black-throated Blue Warblers and Mourning Warblers wherever a new growth of saplings and low shrubs conceal recent lumbering operations. Investigate the two open bogs for Olive-sided Flycatchers, Purple Finches, and Lincoln's Sparrows. Watch for Black-backed Three-toed Woodpeckers on dead stubs. Possibly a wandering family of Boreal Chickadees will show up among the spruces.

After 10 miles the Adams Trail crosses the rolling **Kingston Plains,** site of a magnificent pine forest long since lumbered and burned and now open country, grass-and-bracken-covered, with dark gray stumps looming starkly as far as the eye can see. Only here and there is the scene relieved by a grove of young pines or a lake in one of the depressions.

Among birds inhabiting the Plains are Upland Sandpipers, Common Nighthawks, both Eastern and Western Meadowlarks, and Brewer's Blackbirds. American Kestrels and Eastern Bluebirds nest in the higher stumps. Sandhill Cranes pass over the area in evening and early morning, going to and from the larger lakes where they pass the night. All the lakes are worth looking over for Pied-billed Grebes, American Black Ducks, Blue-winged Teal, Ring-necked Ducks, Common Goldeneyes, and Hooded Mergansers.

ROGERS CITY
Calcite Plant Breakwall

On the **Calcite Plant Breakwall,** projecting into Lake Huron from Rogers City on the northeastern edge of the Michigan 'mitten,' is a large gull colony, probably one of few such colonies readily accessible anywhere by car. Herring Gulls nest on the outer part of this man-made peninsula; Ring-billed Gulls, nearer the mainland. Both species have young nearly ready to fly by late June.

The Breakwall lies within the confines of a calcite plant owned and operated by the United States Steel Corporation and closed to the public. Permission, necessary to visit the Breakwall, may be obtained by writing in advance to the Superintendent, United States Steel Corporation, Rogers City, Michigan 49779. To reach the main gate for instructions about entering, drive north on State 68 from its intersection with US 23 and continue north through town on Erie Street until it dead-ends at Lake Street; turn right on Lake Street and keep bearing left on the road following the Lake Huron shore. The calcite plant at the end of this road soon looms up ahead.

SAGINAW
Shiawassee National Wildlife Refuge | Shiawassee River State Game Area

A joint project of the United States Fish and Wildlife Service and the Michigan Department of Natural Resources consists of the **Shiawassee National Wildlife Refuge** (8,850 acres) and the **Shiawassee River State Game Area** (8,000 acres). The two holdings together embrace the Shiawassee Flats where five rivers converge to form the Saginaw River, which eventually empties into Saginaw Bay. Hunting is permitted on parts of the state area.

The Shiawassee Flats, with river marshes, water impoundments, and agricultural lands, attract great numbers of migrating waterfowl for feeding and resting. Whistling Swans appear in spring, often in immense flocks. In both spring and fall, Canada Geese and fewer numbers of Snow Geese feed during the day in fields planted with supplementary food for them, and spend the night on the open water. Mallards, American Blacks Ducks, Blue-winged Teal, and Green-winged Teal are usually the commonest ducks in both seasons.

The bottomland woods and brushlands on the Refuge provide good bird finding in late spring and summer for common species such as the Red-tailed and Broad-winged Hawks, Great Crested Flycatcher, Rough-winged Swallow, Gray Catbird, Brown Thrasher, Yellow Warbler, Common Yellowthroat, American Redstart, Eastern Meadowlark, Rose-breasted Grosbeak, Rufous-sided Towhee, and Song Sparrow.

For National Refuge headquarters, drive south from Saginaw on

State 13 for 6 miles and go right on Curtis Road for 0.5 mile. For state headquarters, drive west from Saginaw on State 46, go left on State 52 to St. Charles, and then left on a well-marked route.

TAWAS CITY
Tawas Point | Oscoda Point | Au Sable Woods

Tawas Point, extending south at the northern extremity of Saginaw Bay, and **Oscoda Point,** on Lake Huron at the mouth of the Au Sable River, are perfect 'traps' for migrating shorebirds in spring and fall, and small landbirds as well.

Tawas Point, at the end of a pine-oak peninsula, has sand beaches on its lakeside and exposed mud flats, marshes with sedges and rushes, and beach pools on its bayside. Common shorebirds include the Black-bellied Plover, Ruddy Turnstone, Whimbrel, White-rumped Sandpiper, Baird's Sandpiper, Short-billed Dowitcher, and Sanderling. In spring, gulls and terns appear—a few Glaucous Gulls, Bonaparte's Gulls, and Caspian Terns, as well as many Herring and Ring-billed Gulls—together with flocks of scaups and mergansers. Among birds regularly nesting on the peninsula are Least Flycatchers, Bank Swallows, Common Yellowthroats, and American Redstarts.

The sandy shores of Oscoda Point have nesting Piping Plovers, Spotted Sandpipers, Herring Gulls, and Common Terns. Pine Warblers nest in the white pines and jack pines along the shore—the same trees that provide shelter for migrating warblers in spring and fall.

Along the Au Sable River west of the villages of Oscoda and Au Sable is **Au Sable Woods,** an area of jack pine woods and brushy lands with nesting Red-tailed Hawks, Olive-sided Flycatchers, Wood Thrushes, Hermit Thrushes, Cedar Waxwings, Prairie Warblers, Ovenbirds, Scarlet Tanagers, Rufous-sided Towhees, and Clay-colored Sparrows.

To reach these areas, drive east from Tawas City on US 23 and turn right on a well-marked road to the Point; continue on US 23, with a stop at Au Sable Point, to Oscoda on the north side of the Au Sable River, and walk along the shore; drive west from the village of Au Sable on an unnumbered road along the River to the Woods.

Mississippi

CHUCK-WILL'S-WIDOW

Over the years Mississippi has been one of the states least explored ornithologically, and now bears the dubious distinction of being the only state that has never had a publication—book, pamphlet, or scientific treatise—covering all the birdlife within its boundaries. This is happenstance. Today Mississippi is enjoying considerable ornithological activity and can anticipate a statewide publication before long.

Good opportunities for bird finding in Mississippi range from its scenic northeastern sector, west to the lakes and swampy forests in the immense Yazoo-Mississippi Delta, and south to the Gulf coast with its sandy beaches and offshore islands.

Except for the Tennessee River Hills in its northeastern corner, Mississippi lies entirely in the Coastal Plain. Southern Mississippi—

roughly, south of Jackson—is low and flat to slightly rolling country. Its sandy, well-drained uplands were once covered with open forests of longleaf pine, mixed occasionally with oak; in the river bottomlands and swamps were various hardwoods—evergreen magnolia, red bay, water tupelo, water ash—and scattered stands of cypress and slash pine. Few forests remain, having been replaced by farmlands or by smaller second-growth trees. In De Soto National Forest north of Gulfport a few stretches of uncut timber may be considered typical of original conditions.

Where Mississippi borders on the Gulf, the shoreline is consistently sandy, artificially maintained, and interrupted only by bays and river mouths that are often fringed with salt and brackish marshes. Inland from the shore, characteristic vegetation consists of live oak with a dense undergrowth of wax-myrtle and other shrubs. Birds breeding on the shore are scarce, because of two factors. Most of the original sloping beaches have been leveled and landscaped close to the water's edge, and a concrete sea wall with a highway back of it has been constructed, spoiling the nesting habitats; and homes and recreational facilities have appeared in almost unbroken succession along the waterfront, causing too much disturbance. In the end, the chief attractions of the Mississippi coast for bird finders are the many waterbirds, waterfowl, and shorebirds that visit its beaches, bays, and outlying waters, and not infrequently remain for long periods.

Offshore in the Gulf and separated from the mainland by Mississippi Sound are four long, narrow barrier islands more or less paralleling the mainland, by name (from east to west) Petit Bois (the eastern part of which belongs to Alabama), Horn, Ship, and Cat. The first three are incorporated in the Gulf Islands National Seashore. In Mississippi Sound, close to the mainland, is Deer Island. All these islands have fine beaches and extensive salt marshes; all have open groves of slash pine; Ship and Cat have, in addition, numerous groves of live oak.

Breeding-bird populations on the islands are large, but the species represented are few as compared with those in mainland areas of similar size and composition. A few species, such as Royal Terns, Sandwich Terns, and Gray Kingbirds, nest on the islands but rarely on the mainland; on the other hand, Marsh Wrens and many landbird species on the mainland do not nest on the islands. All the islands provide equally good opportunities for bird finding because their birdlife is quite similar, but Deer Island is the most strongly recommended since it is so readily accessible. (For further information, *see under* **Biloxi.**)

Along its western side, northern Mississippi has the state's most distinctive physiographical feature—the Yazoo-Mississippi Delta, an immense alluvial floodplain of the Yazoo and Mississippi Rivers, stretching from Vicksburg northward 190 miles to the Tennessee line. The Mississippi forms the Delta's western boundary, and the eastern boundary is marked by a range of rockless hills—the Bluff, or Loess, Hills. About midway between Vicksburg and the Tennessee line the distance between the two boundaries—the width of the Delta—is as great as 60 miles but gradually diminishes northward and southward as the Bluff Hills draw close to the Mississippi, causing a convergence of the boundaries. Although the greater portion of the Delta is under cultivation, it has many meandering streams, river cutoffs, bayous, sloughs, and lakes, all bordered by stands of cypress and of various hardwoods such as oak and tupelo. Next to the Mississippi coast and the offshore islands, these waterways and their wooded margins comprise the most productive bird-finding area in the state.

Some of the species nesting regularly, in suitable habitats, are the Green Heron, Least Bittern, Wood Duck, Hooded Merganser, Mississippi Kite (uncommon), King Rail, Purple Gallinule (uncommon), Common Gallinule, and Dickcissel. Occasionally the Swainson's Warbler can be found breeding. In brushy places along the highways the Painted Bunting is a fairly common summer resident. Of the many fine spots in the Delta for investigating birds, Moon Lake (reached from Clarksville), Legion Lake (reached from Rosedale), and Lake Washington (reached from Greenville) are the most highly recommended.

In the northeastern corner of Mississippi the Tennessee River Hills constitute a region of rather abrupt hills with narrow, deep ravines. One of them, Woodall Mountain near Iuka, has an elevation of 800 feet, highest in the state. Hardwood forests once covered the entire area,

The remainder of northern Mississippi—the territory extending from the Tennessee River Hills to the Bluff Hills of the Delta and the longleaf pine country south of Jackson—is moderately undulating. Formerly the area had two treeless prairies, one a crescent skirting the western border of the Tennessee River Hills, the other passing across the state along the northern edge of the longleaf pine country. Elsewhere it was forested, chiefly with shortleaf and loblolly pine interspersed with oak and hickory. Hardwoods, however, predominated on the Bluff Hills; cypress, water ash, sycamore, and tupelo were characteristic growth on the river bottomlands. As in southern Mississippi, agricultural activities changed these original conditions. Wherever there were soils suitable for cultivation, farming was initiated. Now most of the farms have been abandoned and the lands have reverted to second-growth forest, generally of loblolly pine. Yet there are a few forested tracts with conditions similar enough to the original to attract birdlife that was once characteristic. One such tract is in Tishomingo State Park (reached from Iuka); another is in the Noxubee National Wildlife Refuge (reached from Starkville).

In both northern and southern Mississippi, the following species breed more or less regularly in the pine and hardwood forests and in the farmlands (fields, brushy areas, hedgerows, orchards, and dooryards):

PINE AND HARDWOOD FORESTS

Turkey Vulture
Black Vulture
Red-shouldered Hawk
Broad-winged Hawk
Yellow-billed Cuckoo
Common Screech Owl
Barred Owl
Chuck-will's-widow
Yellow-shafted Flicker
Pileated Woodpecker
Red-bellied Woodpecker
Hairy Woodpecker
Downy Woodpecker
Red-cockaded Woodpecker (*pine forests*)
Great Crested Flycatcher
Acadian Flycatcher
Eastern Pewee (*except near coast*)
Blue Jay

PINE AND HARDWOOD FORESTS (*Cont.*)

Fish Crow
Carolina Chickadee
Tufted Titmouse
White-breasted Nuthatch (*mainly in north*)
Brown-headed Nuthatch (*pine forests*)
Wood Thrush
Blue-gray Gnatcatcher
Yellow-throated Vireo
Red-eyed Vireo
Black-and-white Warbler (*mainly in north*)
Prothonotary Warbler
Swainson's Warbler
Northern Parula Warbler
Yellow-throated Warbler
Pine Warbler (*pine forests*)
Louisiana Waterthrush (*except near coast*)
Kentucky Warbler
Hooded Warbler
American Redstart
Baltimore Oriole (*mainly in north*)
Summer Tanager (*except near coast*)

FARMLANDS

Common Bobwhite
Mourning Dove
Red-headed Woodpecker
Eastern Kingbird
Eastern Phoebe (*mainly in northeast*)
Barn Swallow
Bewick's Wren (*mainly in north*)
Carolina Wren
Northern Mockingbird
Gray Catbird (*mainly in north*)
Brown Thrasher
Eastern Bluebird
Loggerhead Shrike
White-eyed Vireo
Prairie Warbler
Common Yellowthroat
Yellow-breasted Chat
Eastern Meadowlark
Orchard Oriole
Common Grackle
Northern Cardinal
Blue Grosbeak
Indigo Bunting
Painted Bunting (*mainly in Delta*)
Dickcissel (*except near coast*)
American Goldfinch (*mainly in northeast*)
Rufous-sided Towhee
Grasshopper Sparrow (*mainly north*)
Bachman's Sparrow (*uncommon*)
Chipping Sparrow (*except Delta*)
Field Sparrow (*mainly north*)

The Delta, lying in the path of the Mississippi flyway, is an ideal spot for finding transient birds. Here, north of the flyway's terminus in Louisiana, have converged the migratory routes of many species breeding in Canada and northern United States—in western Quebec, Ontario, and the eastern Prairie Provinces; in Ohio, Indiana, Michigan, Wisconsin, Minnesota, and the Dakotas. Thus waterbirds, waterfowl, landbirds, and shorebirds are at times abundant.

The migratory movements of landbirds on the Mississippi coast deserve special mention. In fall, from late July until after the first of November, with peaks in September/October, southbound thrushes, vireos, warblers, and other passerines appear regularly prior to taking off across the Gulf. At the same time a few species such as Scissor-tailed Flycatchers and Vermilion Flycatchers, bound for Florida from western United States, appear also, sometimes in appreciable numbers. In spring, northbound passerines show up from early March through early May, but with considerable irregularity, owing to weather conditions. When the days are pleasant—clear and windless, with mild temperatures—northbound transients arriving from across the Gulf are scarce. At the height of the migration season, several days or even weeks may pass with few, if any, birds in evidence. Presumably they continue inland undetected, without stopping on the coast, to settle down in a widely dispersed fashion. An entirely different situation exists immediately after bad weather—heavy rains, strong winds, and a drop in temperature. Then transients traversing the Gulf appear in astonishing numbers, having been forced to seek land at the first opportunity. Deer Island (reached from Biloxi) is one of the best places to witness such a phenomenon.

The following timetable applies to the principal flights passing over Mississippi:

Shorebirds: 15 March–15 May; 15 August–20 October
Landbirds: 15 March–1 May; 15 September–10 November

In addition to serving as a feeding and resting area for a great many migrating birds, the Delta is also a much-used wintering ground. Although the variety of species is more notable than their populations, a few species such as the Ruddy Duck are present in impressive numbers. Christmas bird counts, taken in the vicinity of Moon Lake (*see under* **Clarksdale**) and Lake Washington (*see under* **Greenville**) usually total from seventy to eighty-five species. The Gulf coast is undoubtedly the state's most rewarding place for winter bird finding. Loons, grebes, cormorants, ducks, shorebirds, gulls, terns, and skimmers can always be seen from vantage points along US 90 (*see under* **Gulfport**), which follows the beach front for many miles.

Authorities

Otis W. Allen, Thomas D. Burleigh, Ben B. Coffey, Jr., Jerome A. Jackson, Anthony V. Ragusin, Henry M. Stevenson, William H. Turcotte, M. G. Vaiden, Burton S. Webster.

BILOXI
Deer Island | Ship Island | Petit Bois Island | Horn Island

In Mississippi Sound, less than 0.5 mile offshore from this coastal city, is **Deer Island,** about 9 miles long and 0.5 mile wide—its length roughly paralleling the mainland. Unless the sea is rough, one can reach it by hiring a boat at the Biloxi wharves, directly opposite. Though the Island is privately owned, visits may be made at any time without obtaining permission.

Deer Island has broad stretches of sandy beach where Wilson's Plovers, Common Nighthawks, and Little Terns nest; extensive salt marshes providing habitat for Clapper Rails, Willets, Boat-tailed Grackles, and Seaside Sparrows; open woods of slash pine containing a resident population of Brown-headed Nuthatches and Pine Warblers; and groves of live oaks. The breeding season for Pine Warblers begins as early as February: for Brown-headed Nuthatches, in March and early April; for birds of beaches and marshes, usually in May. Among other species nesting on Deer Island are the Mourning Dove, Red-headed Woodpecker, Eastern Kingbird, Great Crested Flycatcher, Fish Crow, Carolina Wren, Northern Mockingbird, Brown Thrasher, Blue-gray Gnatcatcher, Loggerhead Shrike, and Orchard Oriole.

For observing transient and wintering birdlife, Deer Island is excellent. Gulls, terns, and shorebirds gather along the beach, waterbirds and waterfowl in the surrounding waters of the Sound. Immediately following a 'norther' in April or early May, the groves of huge live oaks and even the open pine woods and shrubby growth swarm with unbelievable numbers of migrating flycatchers, vireos, warblers, and other small landbirds. At such times it is possible to see, in one day, twenty species of warblers, as well as the other passerine transients appearing regularly on the Gulf Coast at this time of year.

Ship Island, Petit Bois Island and **Horn Island,** between Mississippi Sound and the Gulf of Mexico, are 12 to 16 miles offshore. Included in the Gulf Islands National Seashore, these strips of land have habitats with birdlife similar to Deer Island's. Additional ornithological attractions are nesting Ospreys and Snowy Plovers. Access to Ship Island from April through October is by excursion boats operated by National Park concessionaires from the Biloxi Small Craft Harbor, 15 blocks east of the Biloxi lighthouse. Access at other times of year, and to Petit Bois and Horn Islands all year, is by chartered or private boat.

CLARKSDALE
Moon Lake

Highly recommended for year-long bird finding in northwestern Mississippi is **Moon Lake,** a crescent-shaped meander cutoff of the Mississippi River about 8 miles long, west of US 61, 17 miles north of Clarksdale and 40 miles south of the Tennessee line.

During winter one may see on Moon Lake a variety of waterbirds and waterfowl that includes the following: Horned Grebe, Mallard, Ring-necked Duck, Canvasback, Lesser Scaup, Ruddy Duck (usually the most common duck), and Hooded Merganser. One can see these birds easily from the road along the east shore of the Lake, reached by going north on US 61 and west on US 49 for 2 miles. Just after crossing railroad tracks, turn left on a paved road that soon parallels the eastern lakeshore for about 7 miles and then intersects State 1. The bird finder may return to US 61 by turning right on State 1, which follows the Mississippi levee north, and turning right on US 49.

On the batture (elevated river bed) west of the levee are stands of cypress, cottonwood, and willow, thickets, and sloughs that are invariably productive. In winter, look for a combination of permanent residents and northern visitants such as Red-bellied Woodpeckers, Carolina Chickadees, Winter Wrens, Northern Mockingbirds, Hermit Thrushes, Loggerhead Shrikes, Myrtle Warblers, Red-winged Blackbirds, American Goldfinches, Savannah Sparrows, Vesper Sparrows, Slate-colored Juncos, Field Sparrows, White-throated Sparrows, Fox Sparrows, Swamp Sparrows, and Song Sparrows.

GREENVILLE
Lake Washington | Yazoo National Wildlife Refuge

Paralleling the west side of State 1 between Erwin and Hampton is **Lake Washington,** 24 miles south of Greenville. One of the most beautiful lakes in the Yazoo-Mississippi Delta, it is a narrow body of water running north and south; the extreme northwestern and southwestern ends are shallow, marshy, and bordered by willows, cypresses, and cottonwoods. During late spring and summer, among birds accounted for on or around the Lake are Great Blue Herons, Green Herons, Little Blue Herons, Great Egrets, Purple Gallinules, Common Gallinules, Little Terns, Fish Crows, Carolina Wrens, Northern Mockingbirds, White-eyed Vireos, Warbling Vireos, Red-winged Blackbirds, Orchard Orioles, and Baltimore Orioles. In May and September, when the low water level exposes muddy areas, shorebirds are present in large numbers.

Along State 1 between Greenville and Hampton, Indigo and Painted Buntings reside in brushy places during spring and summer.

A few miles east of Lake Washington lies the **Yazoo National Wildlife Refuge** (12,470 acres), managed primarily as a winter haven for waterfowl. While retaining woodlands and swamps with their stands of oak, pecan, sweet gum, cottonwood, and cypress, and preserving several small lakes in their natural condition, the Refuge has cleared and modified much of the land for impounding water and raising crops—corn, grasses, soybeans, and various small grains—attractive to geese and ducks.

Only a few geese, mostly Canadas, appear regularly in fall to pass the winter, but ducks, chiefly Mallards, Gadwalls, Common Pintails, Green-winged Teal, American Wigeons, Northern Shovelers, and Ring-necked Ducks, begin appearing by mid-November in large numbers and stay until early March. Already present are Wood Ducks, common year-round residents and, except for a captive decoy flock of Canada Geese, the only waterfowl rearing broods on the Refuge. In spring and summer, Cattle Egrets, Yellow-crowned Night Herons, Prothonotary Warblers, Eastern Meadowlarks, and Grasshopper Sparrows are common, and most of the species on Lake Washington or in its immediate vicinity appear also on the Refuge.

Refuge headquarters may be reached from Lake Washington by turning northeast at Hampton on State 436, or directly from Greenville by going east on US 82 for 8 miles, south on US 61 for 22 miles, and west on State 436. Either way, there are signs on State 436 indicating the direction of Refuge headquarters. Nearly all parts of the Refuge are accessible by roads, although in wet winter months they may be impassable for conventional automobiles.

GULFPORT
Harrison Experimental Forest

From Gulfport east to Biloxi and west to Bay St. Louis, US 90 closely follows the shoreline of Mississippi Sound. In the distance are the open waters of the Gulf of Mexico and the intervening islands. Rarely is the highway more than 50 yards from the water's edge, and it thus permits an unobstructed view of the many kinds of waterbirds, waterfowl, and shorebirds that frequent this section of Gulf Coast.

Among birds to look for are the following: Common Loon and Horned Grebe (winter); Double-crested Cormorant (early October–late March); Lesser Scaup, Common Goldeneye, Ruddy Duck, and Red-breasted Merganser (usually early November–late April); Semipalmated Plover, Black-bellied Plover, Ruddy Turnstone, Least Sandpiper, Semipalmated Sandpiper, and Sanderling (throughout year, except for a period during June/July); Short-billed Dowitcher and/or Long-billed Dowitcher (early July–early October); Piping Plover, Killdeer, and Dunlin (usually early November–early April); Ring-billed and Herring Gulls (early October–late April); Laughing Gull (throughout year, but scarce in midwinter and late spring); Bonaparte's Gull (late November–mid-May); Black Tern (mid-July–late September); Little, Royal, and Caspian Terns (throughout year); Forster's Tern (throughout year, but scarce from mid-December to early March); Black Skimmer (throughout year, abundant in winter).

On larger ponds and bayous immediately inland from the highway, American Coots pass the winter in large numbers, appearing in October and remaining until April. Many Blue-winged Teal are present in migration from mid-September to mid-October and from mid-March to mid-April.

In Gulfport, particularly good vantage points for observing waterbirds are the piers at the northern end of the Harbor and Ship Canal, on the Sound between 26th and 30th Avenues.

North of Gulfport, in the **Harrison Experimental Forest,** a part of De Soto National Forest, are typical open pine woods—one of the few places near the Mississippi Coast where one may find the Brown-headed Nuthatch, Pine Warbler, and Bachman's Sparrow. Other birds sharing this environment during breeding season are the Common Nighthawk, Eastern Pewee, and Summer Tanager. Also in the Experimental Forest are typical bottomland woods where slash pine, evergreen magnolia, red bay, water tupelo, and water ash are the predominant trees. One area, known as 'the Hammock,' is especially worth visiting for such breeding birds as the Barred Owl, Pileated Woodpecker, Red-bellied Woodpecker, Acadian Flycatcher, Carolina Wren, Wood Thrush, Prothonotary Warbler, Northern Parula Warbler, Kentucky Warbler, and Hooded Warbler.

To reach the Experimental Forest from Gulfport, go north on US 49 for about 20 miles to Saucier, then turn right on State 67 and proceed 4 miles to the entrance. Drive to ranger headquarters, not far from the entrance, and obtain directions to a typical section of open pine woods and to the Hammock.

IUKA
Tishomingo State Park

Tishomingo State Park is a 1,541-acre tract of highland forest in northeastern Mississippi near the Alabama line. Situated in the so-called Tennessee River Hills and crossed by Bear Creek as it flows northward to the Tennessee River through a narrow valley with sandstone escarpments, the Park is one of the most ruggedly scenic areas in the state. To reach it from Iuka, drive south on State 25 past Woodall Mountain on the right, a heavily wooded ridge that is the highest point in the state. Go through the town of Tishomingo to a marked turnoff to Tishomingo State Park, 14 miles from Iuka. Turn left on this paved road and proceed 2 miles to the entrance.

Except for a sloping field, a meadow, small clearings with cabins, campgrounds, and picnic areas, and man-made Lake Haynes, the Park is covered with woods. Loblolly and scrub pines, oaks, and

BLUE-GRAY GNATCATCHER

hickories constitute the predominant tree growth. A few cypresses stand along Bear Creek, and tupelos grow thickly on the western slope of the valley. Shrubs are abundant, the oak-leaved hydrangea and the white and pink azaleas being conspicuous when flowering in the spring. Elevations in the Park range from 460 to 600 feet above sea level.

Among birds breeding regularly in the Park are the Broad-winged Hawk, Yellow-billed Cuckoo, both Chuck-will's-widow and Whip-poor-will, Pileated Woodpecker, Great Crested Flycatcher, Eastern Phoebe (nesting under bridges), Acadian Flycatcher, Cliff Swallow (nesting under bridges), Carolina Chickadee, Tufted Titmouse, Wood Thrush, Blue-gray Gnatcatcher, White-eyed Vireo, Yellow-throated Vireo, Red-eyed Vireo (probably the most abundant breeding species), Black-and-white Warbler, Yellow-throated Warbler, Pine Warbler, Prairie Warbler, Kentucky Warbler, Hooded Warbler, Summer Tanager, Indigo Bunting, Bachman's Sparrow, and Field Sparrow. For many woodland species, the area along Bear Creek in the vicinity of the swinging bridge is particularly productive.

JACKSON
Ross Barnett Reservoir

Bird finding in the area of this capital city centers on the **Ross Barnett Reservoir,** covering about 30,000 acres northeast of its outskirts.

Leaving the city northward on I 55, take Tougaloo Exit (37) east on County Line Road to Old Canton Road, turn left to Charity Church Road, then bear right to the Reservoir Dam and cross it to the waterside parks by turning north on Lakeshore Drive. On the Dam are pulloffs for viewing loons, grebes, diving ducks—many of them Ring-necked Ducks and Lesser Scaups—gulls, and terns from October to April. Look for the Bald Eagles that winter occasionally along the Reservoir.

Backtrack across the Dam, bear right to the Main Harbor Marina, circle it for more views of waterbirds and waterfowl, then leave the Marina for Natchez Trace Parkway via Old Canton Road–Rice Boulevard. Go northeast on the Parkway, which parallels the Reservoir on its northern side, and traverses upland areas that offer good bird finding. The area called River Bend, 19 miles from the Rice Boulevard entrance and well marked, is particularly recommended.

Continue east on the Parkway for 3 miles to a pine grove on the northwest side, where there is a clan of Red-cockaded Woodpeckers. The best time to see the birds is just before sunset, when they are gathering to roost.

Backtrack on the Parkway, turn south on State 43, and take the first blacktop road to the left into the state waterfowl refuge. Consult with the manager about viewing ducks from refuge roads, since they are often impassable when the refuge is flooded.

Return to State 43 and continue south a short distance, turning off left on Pipeline Road, an unmarked, all-weather gravel road. This passes through bottomland and upland hardwoods as well as marshes and grassy places, all supporting a rich assortment of birds in any season. Early in the year, beginning in January, American Woodcock perform their evening flight songs above adjacent open areas at the end of Pipeline Road.

ROSEDALE
Legion Lake | Lake Bolivar | Lake Concordia

The many lakes in the vicinity of this northerly Mississippi River town attract large numbers of birds using the Mississippi flyway; in addition, the woods and brushy thickets bordering the River and the lakes bring together a fine variety of breeding birds.

South of Rosedale near the Mississippi is a series of small bodies of water that become at flood time one large lake of approximately 1,000 acres. **Legion Lake,** the name given this group, is reached by driving 1.5 miles south from downtown on State 1 and turning right onto a gravel road that leads 500 yards to the crown of the levee. From here there is a good view of the Lake, the surrounding woods, and the Mississippi. One can return to Rosedale by driving northward atop the levee, a picturesque trip all the way.

The woods adjacent to Legion Lake consist principally of cypress and pecan, with willow and buttonbush near the water's edge.

Among birds commonly nesting here in May and early June are White-eyed Vireos, Prothonotary Warblers, Northern Parula Warblers, and American Redstarts. In the cattails and other aquatic vegetation growing in the shallow water, Pied-billed Grebes and Red-winged Blackbirds breed. Countless numbers of shorebirds gather on the mud flats in May and September, the most common species being the Spotted Sandpiper, Solitary Sandpiper, Lesser Yellowlegs, Pectoral Sandpiper, Least Sandpiper, Semipalmated Sandpiper, and Western Sandpiper.

Seventeen miles south of Rosedale, near the town of Scott, State 1 passes **Lake Bolivar,** normally covering 800 acres but in flood time 1,200 acres. Cypresses, cottonwoods, pecans, willows, and hackberries, with some persimmons and black walnuts, surround it. From late October to March, nearly all species of ducks using the Mississippi flyway may be found on the Lake. These include the Common Pintail, Green-winged Teal, Blue-winged Teal, Northern Shoveler, Wood Duck, Ring-necked Duck, Canvasback, and Lesser Scaup. In March there are frequently large concentrations of Snow Geese.

A smaller lake, comparable to Lake Bolivar in its attractiveness to waterfowl, is **Lake Concordia,** covering 700 acres. To reach it, drive north from Rosedale on State 1 for 6 miles; at Waxhaw Plantation turn left onto a gravel road and proceed westward to the levee, from which the Lake is in view.

STARKVILLE
Noxubee National Wildlife Refuge

A fine area for bird finding in east-central Mississippi is the **Noxubee National Wildlife Refuge** (45,762 acres), a flat to slightly rolling upland area interrupted by bottomlands along the Noxubee River and adjacent creeks. Bluff Lake, a water impoundment, covers about 1,000 acres of the Refuge, and Loakfoma Lake, another impoundment, covers about 600 acres; abandoned fields reverting to brush comprise about 7,000 acres. The remainder of the Refuge is largely forested: on the uplands, with tracts of loblolly pines sometimes mixed with stands of such hardwoods as oaks and hickories; on the bottomlands, with cypresses, water oaks, and other trees requiring moist conditions.

RED-COCKADED WOODPECKER

Bluff Lake and Loakfoma Lake are the centers of ornithological interest in the Refuge. In their fringe of dead timber and nearby swamp forest, Wood Ducks and Prothonotary Warblers nest commonly. Hundreds of Great and Snowy Egrets, together with several other species of wading birds, including Wood Storks, are present in late August and September, feeding in the shallows, resting in the trees, and flying in every direction. By mid-October, ducks begin to come in from the north and are present until late March. The wintering population between mid-December and late February approximates several thousand, commonest species being the Mallard, Gadwall, Common Pintail, Green-winged Teal, American Wigeon, Wood Duck, and Ruddy Duck. American Coots are abundant transients in the fall and spring but are much less numerous as winter residents.

Elsewhere on the Refuge birdlife varies according to the cover. Some breeding species are the following: *on abandoned farmlands,* the Common Bobwhite, Prairie Warbler, Orchard Oriole, Blue

Grosbeak, Indigo Bunting, Dickcissel, Rufous-sided Towhee, and Field Sparrow; *in forests,* the Red-shouldered Hawk (mainly bottomland forests), Wild Turkey, Barred Owl (mainly bottomland forests), Chuck-will's-widow, Red-bellied Woodpecker, Red-cockaded Woodpecker (open pine woods; several clans), Great Crested Flycatcher, Acadian Flycatcher, Brown-headed Nuthatch (open pine woods), Wood Thrush, Yellow-throated Vireo, Northern Parula Warbler (bottomland forests), Yellow-throated Warbler, Pine Warbler (pine woods), Kentucky Warbler (bottomland forests), Hooded Warbler, American Redstart, Summer Tanager, and Bachman's Sparrow (open pine woods; uncommon).

To reach the Noxubee Refuge, starting from the junction of US 82 and State 12 in Starkville, proceed southwest on State 12 for 1.2 miles to Spring Street. Turn left here where a sign points to the Refuge. One mile from State 12, at a Y intersection with a store and gas station between the forks, take the right fork, which leads southward and becomes Oktoc Road. At 12.4 miles from State 12 past the small community of Oktoc, Oktoc Road comes to a T intersection. Turn right here, following a sign pointing to Bluff Lake. After 0.9 mile the road curves to the left past the Refuge entrance sign. Bluff Lake soon appears on the right. Refuge headquarters, on the opposite (west) side of the Lake, is reached by continuing south around the Lake and keeping to the right.

New Hampshire

BAY-BREASTED WARBLER

New Hampshire is called mountainous, and justifiably so. Yet the 9,304 square miles that compose this small triangular state offer, in addition to mountains, 18 miles of coast and several sea islets, large stretches of forest, agricultural lands, and thirteen hundred lakes and ponds.

The White Mountains, which occupy 1,200 square miles in the northern part of the state, surpass in height and magnificence all other Appalachian peaks north of North Carolina and Tennessee. The dominating peaks belong to the Presidential Range, of which Mount Washington attains the greatest altitude—6,288 feet. Grouped about them are some eighty lesser peaks belonging to subordinate ranges and separated by deep valleys, or 'notches.'

Above 5,000 feet the mountains are treeless and strewn with

boulders. Conditions are typically subalpine and therefore rigorous. Sedges and a few species of alpine flowering plants grow in the scanty soil between the boulders, but birdlife is scarce. Only Slate-colored Juncos and White-throated Sparrows nest regularly under these bleak conditions.

From 5,000 feet to a level of approximately 4,000 feet, the mountains are belted by a timber line consisting of densely growing, stunted spruce and fir. Here, not only Slate-colored Juncos and White-throated Sparrows nest regularly, but also Boreal Chickadees, (Bicknell's) Gray-cheeked Thrushes, Golden-crowned Kinglets, Myrtle Warblers, and Blackpoll Warblers. Occasionally Spruce Grouse, Red Crossbills, and White-winged Crossbills nest.

From 4,000 down to 3,000 or even 2,000 feet, the forests are of tall spruce and fir, with scattered white and yellow birches. Because clouds linger persistently, these forests are cool and moist, with mossy carpets. Yellow-bellied Flycatchers, Northern Parula Warblers, Bay-breasted Warblers, and Pine Siskins breed here frequently; Northern Goshawks, Gray Jays, Red Crossbills, and White-winged Crossbills nest here more often than elsewhere in the mountains.

Other birds nesting regularly in the coniferous forests, usually with greater frequency in the areas between 4,000 and 2,000 feet than in areas higher up, are the following:

Yellow-bellied Sapsucker
Olive-sided Flycatcher
Red-breasted Nuthatch
Brown Creeper
Winter Wren
Hermit Thrush
Swainson's Thrush
Solitary Vireo
Magnolia Warbler
Black-throated Blue Warbler
Black-throated Green Warbler
Blackburnian Warbler
Northern Waterthrush
Mourning Warbler
Canada Warbler
Purple Finch

At about 2,000 feet the forests are usually of fir, spruce, and considerable hemlock mixed with beech, maple, and birch. Conditions are warmer and drier than at higher elevations, except along the deeply shaded banks of tumbling brooks and streams. Birds species nesting here are a mixture of those nesting regularly in the coniferous woods above and in the deciduous woods at lower elevations.

Below 2,000 feet the forests often contain white pine, but more generally second-growth beech, birch, maple, basswood, and red oak. Birds in these forests are almost identical with those found in deciduous forests elsewhere in the state.

The Appalachian Mountain Club, with the co-operation of a dozen or more local clubs, maintains fine trails reaching nearly all important summits in the White Mountains. It maintains also a number of huts and shelters open to anyone. The bird finder who wants to 'rough it' in a mountain wilderness, and observe for himself the striking differences in birdlife at various altitudes, should consult the *A.M.C. White Mountain Guide* (available from the Appalachian Mountain Club, 50 Joy Street, Boston, Massachusetts 02101) for descriptions of trails, locations of huts and shelters, and maps, which may be purchased separately.

North of the White Mountains is New Hampshire's so-called 'North Country,' embraced entirely by Coos County. Forested primarily with spruce and fir, it is comparable in environmental conditions and birdlife to the White Mountain forests between the levels of 3,000 and 5,000 feet. Settlements are few.

From the White Mountains east, south, and west to the state's borders, the country is generally hilly, with a few isolated peaks and small ranges. Immediately south of the Mountains is the 'Lakes District' dominated by Winnipesaukee, the state's largest lake. Farther south is the broad, low-lying valley of the Merrimack River; and to the west, along the state's western border, are the rich bottomlands of the Connecticut River. Originally the country was completely forested, especially with coniferous trees such as white pine, but its character has changed. In the river valleys there are many farms and settlements of varying size; along the lakeshores there are thriving summer resorts. Though much of the country, being unsuited to agriculture, is still forest-covered, the coniferous trees have been decimated and replaced in large areas by a mixed deciduous growth. Breeding birds characteristic of farmlands (fields, wet meadows, brushy lands, orchards, and dooryards) and deciduous woods are listed below. Species marked with an asterisk breed regularly in southern New Hampshire only.

FARMLANDS

Eastern Kingbird
Eastern Phoebe
Tree Swallow
Barn Swallow
* Northern Mockingbird
Gray Catbird
Brown Thrasher
Eastern Bluebird
Yellow Warbler
Chestnut-sided Warbler
Common Yellowthroat
Bobolink
Eastern Meadowlark
Common Grackle
* Northern Cardinal
Indigo Bunting
* House Finch
American Goldfinch
Savannah Sparrow
Vesper Sparrow
Chipping Sparrow
Field Sparrow
Song Sparrow

DECIDUOUS WOODS

Broad-winged Hawk
Ruffed Grouse
Black-billed Cuckoo
Barred Owl
Whip-poor-will
Yellow-shafter Flicker
Pileated Woodpecker
Hairy Woodpecker
Downy Woodpecker
Great Crested Flycatcher
Least Flycatcher
Eastern Pewee
Blue Jay
Black-capped Chickadee
* Tufted Titmouse
White-breasted Nuthatch
Wood Thrush
Red-eyed Vireo
Ovenbird
American Redstart
Baltimore Oriole
Scarlet Tanager
Rose-breasted Grosbeak

Where the state borders on the Atlantic Ocean there are salt marshes, sandy beaches, harbors, and rocky promontories. Nine miles offshore are the Isles of Shoals, a group of six small islands two of which are politically a part of New Hampshire. Treeless, with many ledges and patches of grass-covered turf, all held colonies of gulls and terns until encroached upon by various human agencies.

The nesting season in New Hampshire varies with species and location. American Woodcock normally have nests by mid-April,

Ruffed Grouse by May first. Most landbirds breeding below the 2,000-foot elevation start nesting by the last weeks of May; those above 2,000 feet begin two weeks later. Gulls and terns have young by the last week of June.

In May, August, and early September, transient shorebirds appear in great numbers on the coastal beaches; in April, October, and November, transient waterfowl can be seen to advantage in the Portsmouth area and in the many tiny coves of the Isles of Shoals, where they feed and take refuge in severe weather. Most landbirds migrate along the river and mountain valleys, whose trend is north and south. This is especially true in the Connecticut River Valley, where, in May and September, kinglets, vireos, warblers, and sparrows move through in impressive waves.

During winter, bird visitants in the interior lowlands include the Northern Shrike, Evening Grosbeak, Common Redpoll, American Tree Sparrow, and occasionally Pine Grosbeak and Red and White-winged Crossbills.

Authorities

Francis H. Allen, Annie H. Duncan, Kimball C. Elkins, Edith M. Halberg, Henry N. Halberg, Vera H. Herbert, C. F. Jackson, Henry M. Parker, Tudor Richards, Wendell Taber, Douglas E. Wade.

CAMPTON
Waterville Valley (White Mountains)

From Campton Exit of I 93, take State 49 north through intermontane country to **Waterville Valley.** Good trails lead from the area of the comfortable inn up four 4,000-foot mountains. In the Valley and on the low, forested slopes, watch in summer for the Yellow-bellied Sapsucker, Olive-sided Flycatcher, Brown Creeper, Solitary Vireo, Philadelphia Vireo, Northern Parula Warbler, Northern Waterthrush, Mourning Warbler, Rusty Blackbird, and Evening Grosbeak. At all altitudes except the very highest, look for Yellow-bellied Flycatchers, Boreal Chickadees, Red-breasted Nuthatches, Winter Wrens, Swainson's Thrushes, Golden-crowned Kinglets, Nashville Warblers, Bay-breasted Warblers, and Pine Siskins. On the heights, watch for Blackpoll Warblers, (Bicknell's) Gray-cheeked Thrushes, and, possibly, Spruce Grouse.

COLEBROOK

Dixville Notch | Lake Umbagog | Pontook Reservoir

State 26, east from Colebrook in northern New Hampshire, rises to 1,870 feet as it passes through ruggedly scenic **Dixville Notch,** probably the lowest elevation in New Hampshire for hearing the (Bicknell's) Gray-cheeked Thrush—even without emerging from the car. Yellow-bellied Flycatchers and Blackpoll Warblers nest in the woods along the highway through the Notch.

Farther east on State 26, **Lake Umbagog,** headwaters of the Androscoggin River, extends along the New Hampshire–Maine line. The Lake is well known from the ornithological studies of William Brewster, who described it in 1907 as 'too narrow, tortuous, shallow, and muddy to be in itself especially admirable, yet like many a precious stone, has beauty and charm due to the effectiveness of its perfect setting amid the majestic mountains and virgin forests that surround it closely on every hand.'

For seeing birds on Lake Umbagog today, there are two convenient areas not far from Errol.

1. Drive northward from Errol on State 16 for about 2 miles, to just above the dam across the Androscoggin. From here, with a canoe, one can explore the wetlands beyond, and paddle up a backwater that parallels the Androscoggin to 'Floating Island,' a bog of at least 100 acres described by Brewster as 'an immense raft of vegetation, literally floating on water everywhere six or eight feet in depth, and supporting the roots as well as the stems of not only thickets of large, vigorous-looking alders, but also of hundreds of short, stunted, scraggy but living larches [tamaracks] hung with blackish Usnea.' Among many summer-residents birds in these wetlands are Pied-billed Grebes, American Bitterns, Ring-necked Ducks, Soras, Common Snipes, Traill's Flycatchers, Tennessee Warblers, Common Yellowthroats, Wilson's Warblers, Rusty Blackbirds, and both Lincoln's and Swamp Sparrows.

2. East from Errol, State 26 runs along the southern part of Lake Umbagog, where in the spring or fall linger a variety of transient waterbirds and waterfowl such as Red-throated Loons, Red-necked and Horned Grebes, Canada Geese, Lesser Scaups, Buffleheads, Oldsquaws, scoters, and mergansers.

South from Errol, about 16 miles, State 16 follows the west side of **Pontook Reservoir,** a dammed-up portion of the Androscoggin, excellent for transient waterbirds and waterfowl, all easily seen from the car. Common Goldeneyes and Common Mergansers are permanent residents. Watch for Red-tailed and Broad-winged Hawks and an occasional Bald Eagle and Osprey.

CONCORD

Great Turkey Pond | St. Paul's School Woodland | Turee Marsh | Turtle Town Pond | Bear Brook State Park

The several hundred acres around **Great Turkey Pond,** just west of this capital city, comprise a portion of **St. Paul's School Woodland,** consisting of white pine with scattered oaks and other deciduous trees. During a walk in the Woodland around the Pond on a day in late June, one ornithologist observed fifty-four species, the majority characteristic of most woodlands and woodland edges in the southern part of the state. On his list, for example, were the Ruffed Grouse, Yellow-billed and Black-billed Cuckoos, Least Flycatcher, Eastern Pewee, Black-capped Chickadee, House Wren, Gray Catbird, Brown Thrasher, Wood Thrush, Veery, Red-eyed and Warbling Vireos, Black-and-white Warbler, Northern Parula Warbler, Yellow Warbler, Black-throated Blue Warbler, Chestnut-sided Warbler, Pine Warbler, Ovenbird, American Redstart, Baltimore Oriole, Scarlet Tanager, Rose-breasted Grosbeak, Indigo Bunting, American Goldfinch, Rufous-sided Towee, and Song Sparrow.

Immediately south of the Pond is **Turee Marsh,** a nesting habitat for American Bitterns and Swamp Sparrows. The Pond itself attracts summer-resident American Black Ducks and Wood Ducks and transient waterfowl such as Canada Geese, Green-winged Teal, Ring-necked Ducks, Greater Scaups, Common Goldeneyes, Buffleheads, and, less commonly, Ruddy Ducks, and White-winged and Black Scoters.

Audubon House, headquarters of the Audubon Society of New Hampshire, situated as it is at the eastern edge of the Woodland, makes a good starting point for a walk through the Woodland and around Great Turkey Pond. To reach Audubon House from I 93, turn off west on I 89 and immediately take the Bow Exit, turn toward Concord on South Street, left on Iron Works Road, left on Clinton Street, left once more under I 89, and, finally, right on Silk Farm Road to Audubon House, marked by a sign. In Audubon House, inquire about participating in the field trips conducted by the Society and visiting its sanctuaries for bird finding.

Turtle Town Pond, in East Concord, is a good area in summer for finding American Black Ducks, Wood Ducks, Ring-necked Ducks, Great Blue Herons, American Bitterns, Virginia Rails, and Marsh Wrens and in migration for Common Loons, Pied-billed Grebes, several species of surface-feeding ducks and diving ducks including both Hooded and Common Mergansers. To reach the Pond from I 93, take Exit 16 to Shawmut Street, then left on Oak Hill Road, which soon passes the Pond on the right.

Bear Brook State Park (7,300 acres), although primarily a recreational area, is the site of a nature center with a summer program sponsored by the Audubon Society of New Hampshire. A full-time naturalist conducts the program, which includes field trips, open to everyone, along some 20 miles of woodland trails. The Park, in Allenstown, can be reached from Concord by driving south on US 3 to Suncook, then left on State 28 to the Park.

DURHAM

College Woods | Great Bay

Adjacent to the University of New Hampshire campus on the south is a stand of old pines and hemlocks called the **College Woods,** an officially designated natural area. Nesting birds include the Eastern Pewee, Brown Creeper, Hermit Thrush, Black-throated Blue Warbler, Black-throated Green Warbler, Blackburnian Warbler, Magnolia Warbler, Canada Warbler, and Scarlet Tanager.

The best place in the state for transient and wintering waterfowl is

Great Bay, New England's largest body of tidewater, between Durham and Portsmouth. Transient waterfowl include the Common Pintail, Green-winged Teal, Blue-winged Teal, and American Wigeon. Many stay for the entire winter, when their numbers are augmented by Canada Geese, American Black Ducks, Greater Scaups, Common Goldeneyes, Buffleheads, Common Mergansers, and Red-breasted Mergansers. And always there are gulls.

From Durham, State 108, State 101, and US 4 circle Great Bay. Best approaches are via side roads from State 108 between Durham and Newmarket. A turnoff from US 4 toward Newington can also be rewarding.

HINSDALE
Lake Wantastiquet | Spofford Lake

In extreme southwestern New Hampshire, two bodies of water in the Connecticut River Valley are particularly attractive to migrating birds. One of them, **Lake Wantastiquet,** between Hinsdale and Brattleboro (Vermont) is actually a stretch of the River that has been impounded by Vernon Dam. Although fairly good for spotting transient waterbirds and waterfowl, it is excellent for shorebirds, especially when low water exposes mud flats and sand bars. Semipalmated and Black-bellied Plovers, Solitary Sandpipers, both yellowlegs, Pectoral, Least, and Semipalmated Sandpipers, and Short-billed Dowitchers are some of the northern shorebirds to be expected. State 119, north from Hinsdale, follows the riverbed closely, and a railroad track between the highway and the water's edge provides fine vantage points. In late summer, swallows and blackbirds gather by the thousands in the cattails of the islands, where they roost for the night.

Spofford Lake, rather ordinary in appearance and ringed by cottages, is notable for its congregating waterbirds and waterfowl from mid-April to early May and from late September to late December. Look for Common Loons, Horned Grebes, American Coots, Herring, Ring-billed, and Bonaparte's Gulls, Black Terns, Canada and Snow Geese, both scaups, Common Goldeneyes, Buffleheads, Oldsquaws, all three scoters, and all three mergansers.

Spofford Lake is reached from Hinsdale by driving north on State

63 through Chesterfield to the intersection with State 9 at the south end of the Lake. Here one may turn east on State 9 for vantage points along the south side of the Lake, or continue north on State 63 for other good sites on the west side of the Lake.

JAFFREY
Mt. Monadnock

About 10 miles north of the Massachusetts line in southwestern New Hampshire, **Mt. Monadnock** rises alone to a barren, austere summit at 3,165 feet. The bird finder wishing to explore the mountain has a choice of walking one of several trails or driving up on a toll road. In early summer on the lower, forested slopes, watch for Red-tailed and Broad-winged Hawks and expect to see or hear, among a variety of birds, the Winter Wren, Hermit Thrush, Swainson's Thrush, Golden-crowned Kinglet, Solitary Vireo, Magnolia Warbler, Myrtle Warbler, and Canada Warbler. Look for Rufous-sided Towhees in the scrubby growth below the open ledges of the summit.

Mt. Monadnock lies within a public reservation of about 4,100 acres, part of which comprises Mt. Monadnock State Park. To reach it and some of the trails, drive west from Jaffrey on State 124 for about 2 miles to Jaffrey Center and turn north on a road leading directly to the Park. To reach the toll road, continue west from Jaffrey Center on State 124 for about 4 miles.

LACONIA
Mt. Belknap

Mt. Belknap (2,378 feet) in central New Hampshire has a good variety of birds in summer. Some species are the Golden-crowned Kinglet, Solitary Vireo, Red-eyed Vireo, Nashville Warbler, Black-and-white Warbler, Magnolia Warbler, Black-throated Blue Warbler, Black-throated Green Warbler, Myrtle Warbler, Chestnut-sided Warbler, Canada Warbler, Ovenbird, Common Yellowthroat, and American Redstart.

From US 3 on the north edge of Laconia, turn east on State 11A to Guilford, then south on an unnumbered road, and, after about 3 miles, watch for a road and trail up the Mountain.

MANCHESTER
Massabesic Lake

Massabesic Lake, a large, irregularly shaped body of water just east of this industrial city in southern New Hampshire, has Common Loons nesting along its western shore. Herring and Greater Black-backed Gulls, often joined by a few Glaucous and Iceland Gulls, concentrate in the eastern part of the Lake during spring and fall and spend the winter on the Merrimack River in the city. In migration seasons, large numbers of Common Mergansers and smaller numbers of Hooded Mergansers, as well as Horned Grebes and Common Goldeneyes, may be expected on the Lake. A good vantage point for observing many of these birds is from State Bypass 28, where it crosses the Lake.

NEWPORT
Mt. Sunapee State Park

Drive east from Newport in southwestern New Hampshire on State 103 to the village of Mt. Sunapee and turn right to **Mt. Sunapee State Park** (1,787 acres). From back of the parking area, take a trail up heavily forested Mt. Sunapee (2,743 feet), past the top of the chair lift, and on to Lake Solitude (2,400 feet elevation), highest lake in the southern part of the state and the most southerly known nesting site of the Rusty Blackbird. Among birds expected in summer along the trail as it ascends through the woods—largely mixed coniferous-deciduous—are the Pileated Woodpecker, Yellow-bellied Sapsucker, Yellow-bellied Flycatcher, Olive-sided Flycatcher, Winter Wren, Swainson's Thrush, Solitary Vireo, numerous warblers including the Magnolia, Black-throated Blue, Blackburnian, and Canada, and the Scarlet Tanager.

NORTH CONWAY
Zealand Trail | Mt. Washington

Drive north and west from North Conway on US 302 to Zealand Campground, just beyond Fabyan House, then south on Zealand Road for 3.6 miles, park, and walk 2.7 miles on **Zealand Trail** to the Zealand Falls Hut of the Appalachian Mountain Club. From this hut, trails lead west over the Twin and Franconia Ranges, east to the Willey Range and Crawford Notch, and south to Zealand Notch.

Both the Philadelphia and Red-eyed Vireos are summer residents along the Trail, the Philadelphia being the more numerous. Watch for Rusty Blackbirds around the small ponds; Spruce Grouse and (Bicknell's) Gray-cheeked Thrushes in higher reaches; Boreal Chickadees, Yellow-bellied Flycatchers, Blackpoll Warblers, Bay-breasted Warblers, and Pine Siskins among dense evergreens; and Mourning Warblers in cutover or wind-thrown areas. Three-toed woodpeckers are possible, as are Gray Jays and crossbills.

For the Tuckerman Ravine Trail up **Mt. Washington** (6,288 feet), drive north from North Conway on US 302 for about 6 miles, then right on State 16 for about 13 miles to Pinkham Notch, a charming valley separating the Carter-Moriah Range on the east and the Presidental Range on the west. Along the highway and on the lower slopes, the forests and associated birdlife are almost identical to those in Franconia Notch (*see under* **North Woodstock**). Park at Pinkham Notch Camp and take the Trail to the summit, 4.1 miles distant.

Hermit Lake, about half-way up at 3,800 feet, is a deep, dark tarn surrounded by thick coniferous woods. The following species are regular in June and early July: Yellow-bellied Flycatcher, Olive-sided Flycatcher, Boreal Chickadee, Red-breasted Nuthatch, Brown Creeper, Swainson's Thrush, Golden-crowned Kinglet. Occasionally Northern Three-toed Woodpeckers and Gray Jays may be observed.

NORTH WOODSTOCK
Lost River Gorge Reservation | Mt. Moosilauke | Beaver Lake Franconia Notch | Cannon Mountain | Lonesome Lake Kancamagus Highway | Church Pond Bog

In early summer there are many areas for productive bird finding within easy reach of this small resort town in the White Mountains.

Drive 5.5 miles on State 112 west of the village through rugged Kinsman Notch to the **Lost River Gorge Reservation,** 900 acres at 2,300 feet elevation, owned by the Society for the Protection of New Hampshire Forests. Here it is possible to observe a variety of species, a few of which appear more commonly at higher elevations.

If possible, visit the Reservation during an afternoon in late June or early July. First, walk through the Nature Garden, with its splendid collection of indigenous wildflowers, and enjoy a trip below the enormous rocks, where the Moosilauke River 'loses' itself; then, linger till sundown for a chorus of bird songs unexcelled in the White Mountains. The best point for listening is the porch of the Reservation's administration building, which overlooks the narrow valley. From the nearby walls of the Notch, as well as from the valley floor, come the songs of Winter Wrens, Purple Finches, Slate-colored Juncos, White-throated Sparrows, and at least eight different warblers, including the Magnolia Warbler, Blackpoll Warbler, Ovenbird, Northern Waterthrush, and Canada Warbler. Most impressive of all is the chorus of five thrushes—the Wood, Hermit, Swainson's, (Bicknell's) Gray-cheeked, and Veery.

A quarter-mile beyond the Reservation, State 112 meets, on the

WHITE-THROATED SPARROW

left, the Beaver Brook Trail to the summit of **Mt. Moosilauke** (4,810 feet), strongly recommended for mountain bird finding. Slate-colored Juncos and White-throated Sparrows nest on the several hundred acres above the timber line; Spruce Grouse, Boreal Chickadees, (Bicknell's) Gray-cheeked Thrushes, and Blackpoll Warblers in the stunted spruces and firs just below the timber line. Species nesting in the tall, coniferous-deciduous timber of the lower slopes include the Northern Goshawk (rare), Red-tailed Hawk, Broad-winged Hawk, Ruffed Grouse, Pileated Woodpecker, Red-breasted Nuthatch, Winter Wren, Wood Thrush, Hermit Thrush, Swainson's Thrush, Veery, Northern Parula Warbler, Magnolia Warbler, Black-throated Blue Warbler, Myrtle Warbler, and Blackburnian Warbler.

Half a mile beyond the Reservation, State 112 passes **Beaver Lake,** a forest-bordered tarn where Rusty Blackbirds are summer residents.

Four miles north of North Woodstock, US 3 enters **Franconia Notch,** part of the Pemigewasset Valley. Within this deep cut extending 6 miles between the Franconia Range on the east and the Kinsman Range on the west is the profile of the Old Man of the Mountains ('Great Stone Face'), New Hampshire's best-known natural wonder; there are also several beauty spots such as Profile and Echo Lakes, the Flume, the Pool, and the Basin. Here the forests are largely deciduous, with scattered stands of white pine and hemlock, replaced by spruce and fir in the Notch proper.

Among birds one may observe regularly in June and early July amid these scenic attractions are the Yellow-bellied Sapsucker, Winter Wren, Hermit Thrush, Solitary Vireo, Nashville Warbler, Northern Parula Warbler, Magnolia Warbler, Black-throated Blue Warbler, Myrtle Warbler, Black-throated Green Warbler, Blackburnian Warbler, Northern Waterthrush, Canada Warbler, and White-throated Sparrow. At Profile Lake, for example, Winter Wrens and Canada Warblers may be heard singing in the densely wooded slopes rising from the Lake to the shoulder of Cannon Mountain, which bears the stern features of the Old Man. At Echo Lake, look for Wood Ducks.

The summit of **Cannon Mountain** (4,077 feet), reached by cable car from the Notch, then by a quarter-mile walk, is a good vantage point for viewing Northern Ravens. In the stunted forest below are

Spruce Grouse, Yellow-bellied Flycatchers, and (Bicknell's) Gray-cheeked Thrushes. One may walk back to the Notch by way of **Lonesome Lake** (2,700 feet), habitat of Rusty Blackbirds, with Ruby-crowned Kinglets, Philadelphia Vireos, and Tennessee Warblers in the immediate vicinity, and on to US 3 about 2 miles south of the start of the tramway. The trail from the top of Cannon Mountain, including a walk around the Lake by way of a tamarack bog, is about 6.5 miles.

The very scenic **Kancamagus Highway,** State 112, runs for 32 miles from Lincoln to Conway, rising to 2,850 feet at the height of land. Drive from North Woodstock east on State 112. Stop at Kancamagus Pass, where there should be Blackpoll Warblers and (Bicknell's) Gray-cheeked Thrushes. Stop also at Passaconaway Campground, about 19 miles from Lincoln, and take a loop trail that passes around and through **Church Pond Bog,** one of the few places in the White Mountains where the Lincoln's Sparrow breeds. Other species breeding in or near the Bog or in the wooded section near the Campground may include the Boreal Chickadee, Ruby-crowned Kinglet, Tennessee Warbler, Bay-breasted Warbler, Mourning Warbler, and Rusty Blackbird.

PITTSBURG
Black Lake | East Inlet | 'Moose Pasture'

From this northernmost town in the state, drive north on US 3 through 'the North Country,' a wilderness region of spruce and fir forests with the so-called 'Connecticut Lakes.' Where the woods are dense, there is the best opportunity north of the White Mountains for finding in any season such birds as Spruce Grouse, Saw-whet Owls, Gray Jays, Northern Ravens, Boreal Chickadees, Pine Grosbeaks (irregular), and White-winged Crossbills, and the opportunity here is better than in the White Mountains for observing both Black-backed Three-toed Woodpeckers and Northern Three-toed Woodpeckers. Watch for them on dead stubs. Lumbered areas in the forest are in varying stages of re-growth with shrubs, aspens, birches, spruces, and firs, and traversed by logging roads and trails that are often impassable in wet weather. Among summer-resident

BLACK-BACKED THREE-TOED WOODPECKER

birds, in suitable habitats, are the American Woodcock, Yellow-bellied and Olive-sided Flycatchers, Swainson's Thrush, Philadelphia Vireo, Tennessee Warbler, Bay-breasted Warbler, Mourning Warbler, Wilson's Warbler, Rusty Blackbird, Pine Siskin, and Swamp Sparrow.

Three places merit special investigation. A canoe, though not necessary, can be useful.

Black Lake, 3 miles north of Pittsburg. Pause here to look for waterbirds and waterfowl, and then drive west on a road that passes Middle Pond and Moose Pond—both good places for Common Loons, Ring-necked Ducks, Common Goldeneyes, Hooded Mergansers, and Common Mergansers.

East Inlet and **'Moose Pasture.'** About 3 miles beyond the entrance to Camp Idlewild on Second Lake, turn right about a half-mile to the Fire Warden's Cabin and check for the condition of the road, which starts south of the Cabin, to the bridge over East Inlet. From here take a canoe or walk a very rough road to Moose Pasture, a bog west of the bridge on the north side of East Inlet. Black-backed Three-toed Woodpeckers nest in this area, together with Common Snipes, Ruby-crowned Kinglets, and Lincoln's Sparrows.

PORTSMOUTH
Isles of Shoals

This historic city in southern New Hampshire is within easy reach of migrating and wintering waterfowl (*see under* **Durham**), migrating shorebirds (*see under* **Seabrook**), and the open ocean.

From June to September, a steamer leaves the Market Street Wharf twice daily for the **Isles of Shoals.** Here is an opportunity for seeing nesting colonies of gulls, providing the trip is made before the breeding season ends in early summer. The boat goes to Star Island, where Herring Gulls nest and where one can arrange for transportation to Duck Island, a large, desolate ledge, the nesting place for thousands of Herring Gulls and lesser numbers of Greater Black-backed Gulls. Gulls nest also on nearby Eastern, Shag, and Mingo Rocks, discouraging the Common and Roseate Terns that formerly nested here.

The Isles constitute one of the best localities on the New Hampshire coast for observing transient and wintering birdlife. From mid-April to late May and from early August to early October, landbirds and shorebirds frequently stop here. From October through March the rocky shores are favorite feeding grounds for Purple Sandpipers, and the lees and tiny coves are havens for such species as Northern Gannets, Great and Double-crested Cormorants, Common Eiders, and Black Guillemots. Northern Phalaropes are abundant in late May.

Members of the Audubon Society of New Hampshire, which has several field trips to Star Island in late fall, winter, and early spring, often find the rare or accidental landbirds—flycatchers, thrushes, warblers, and sparrows that take refuge in the bushes on Star Island—the most rewarding part of the trip. At one time or another they have recorded many remarkable 'strays.'

Since the steamer runs only during the summer, one must charter a boat in off-season from one of the fishermen who visit the vicinity.

This boat trip may prove to be productive for 'offshore' birds such as Razorbills, Thick-billed Murres, Dovekies, Black Guillemots, and Atlantic Puffins.

SEABROOK
US 1A from Seabrook to Rye

US 1A follows the coast closely for 15 miles from Salisbury Beach, on the Massachusetts line, to Portsmouth. Heavily built up though it is, this stretch, especially between Seabrook and Rye, is good for birds the year round. Because of the dense traffic in midsummer, however, bird finding is pleasanter from September to May.

In general, one should scan the ocean at every opportunity for offshore birds, look over all surf-splashed rocks for Purple Sandpipers in winter, watch for Sanderlings on beaches, and other transient shorebirds on the mud flats in fall and spring, the bushes and thickets for migrating landbirds, and the marshes for visiting wading birds, waterbirds, and waterfowl. Beginning at the state line and going north:

1. Inspect the sand dunes immediately on the left. Green Herons nest in the thickets; many small landbirds stop by in migration.

2. *Seabrook-Hampton Harbor.* The extensive mud flats here attract migrating shorebirds from mid-May to early June and from late July to early October. Some species to be expected include the Semipalmated Plover, Piping Plover, Black-bellied Plover, Ruddy Turnstone, Red Knot, White-rumped Sandpiper, Least Sandpiper, Dunlin, Short-billed Dowitcher, Semipalmated Sandpiper, and Sanderling. Gulls, including Bonaparte's, and terns, including Common, Arctic, Roseate, and Little, often show up here.

3. *Marshes west of the Harbor* have visiting Great Blue Herons, Green Herons, Great Egrets, Snowy Egrets, Black-crowned Night Herons, Greater and Lesser Yellowlegs, and Short-eared Owls. The bird finder wishing to work these marshes should take US 1 from the state line to Hampton Falls, and follow a short road

east from Hampton Falls Common to its end at a railroad station. US 1 crosses the marshes north of Hampton Falls, but heavy traffic makes stopping perilous.

4. *Hampton Beach State Park* (50 acres). Transient Black-bellied Plovers appear on the beach, wintering Horned Larks in the mowed fields.

5. *Hampton Beach.* Scan the ocean in colder months for Common Loons, Red-necked Grebes, Great Cormorants, Double-crested Coromorants, Common Goldeneyes, Oldsquaws, Common Eiders, White-winged Scoters, Surf Scoters, Black Scoters, and Red-breasted Mergansers.

6. *Great Boar's Head,* which divides Hampton Beach from North Beach, is a fine outlook for seabirds. Watch the rocks here for wintering Purple Sandpipers.

7. *Meadow Pond,* behind Hampton Beach, has Marsh and Sedge Wrens nesting in the bordering cattails and tall grasses. This is a good place for migrating waterfowl—Blue-winged Teal, Green-winged Teal, Ringed-necked Ducks, and mergansers.

8. *Eel Pond,* on the left of the highway in Rye, has nesting American Black Ducks and Wood Ducks and is excellent for migrating waterfowl.

9. *Rye Harbor and Marsh.* Watch for visiting Snowy Egrets and other wading birds. Shorebirds stop here during migration.

WARNER
Mt. Kearsarge

One of the New Hampshire peaks standing apart far south from the Presidential Range is **Mt. Kearsarge** (2,937 feet) within Winslow State Park (2,918 acres). In the mixed deciduous-coniferous forests on its slopes, summer-resident birds include the Red-tailed Hawk, Olive-sided Flycatcher, Red-breasted Nuthatch, Winter Wren, Hermit and Swainson's Thrushes, Golden-crowned Kinglet, Nashville,

Magnolia, Black-throated Blue, Myrtle, Blackburnian, and Canada Warblers, Slate-colored Juncos, and White-throated Sparrows. (Bicknell's) Gray-cheeked Thrushes and Blackpoll Warblers nest near the summit. The birdlife, while not exceptional in variety and abundance, has the distinction of being easily observed, amid some of the finest scenery in the state, from a toll road that comes to within a half-mile of the summit. To reach the toll road, exit from I 89 to Warner, turn north from State 103, and follow directional signs.

WHITEFIELD
Pondicherry Wildlife Refuge

Within a triangle formed by the towns of Whitefield, Meadows, and Jefferson in northern New Hampshire lies the **Pondicherry Wildlife Refuge** (310 acres), administered by the Audubon Society of New Hampshire and the New Hampshire Fish and Game Department. The Refuge contains two ponds, Big Cherry Pond and Little Cherry Pond, the best places near the White Mountains for birds preferring open water and marshes. The tamarack-spruce bogs around the Ponds and along their connecting stream attract a fine variety of landbirds.

The list of summer residents around the Ponds includes the Pied-billed Grebe, American Bittern, American Black Duck, Green-winged Teal, Wood Duck, Ring-necked Duck, Northern Harrier, and Rusty Blackbird. In the bogs, one may find breeding the Yellow-bellied Flycatcher, Gray Jay, Boreal Chickadee, Ruby-crowned Kinglet, Tennessee Warbler, Wilson's Warbler, and Rusty Blackbird.

Unfortunately—or perhaps fortunately for the wildlife—the Refuge is not readily accessible. The best approach is by driving along the old railroad right-of-way from the Whitefield Airport as far as the car can go, then walking the remaining mile. A short distance beyond where the railroad bed passes Big Cherry Pond, look for a marker on a white birch indicating the trail—often very wet—to Little Cherry Pond, about two-thirds of a mile distant.

New Jersey

BRANT

In no other place in North America is the fall migration more spectacular than in New Jersey. From late July, when shorebirds begin to collect on Long Beach Island, and August, when Tree Swallows assemble in droves at Cape May, to November, when legions of Brant make their appearance in Barnegat Bay, there are avian congregations of amazing proportions. To watch their mass movements is to acquire new and lasting impressions of migration behavior.

From north to south, New Jersey shows distinct variations in physiographical conditions. North New Jersey—roughly that part of the state north of an imaginary line drawn between Trenton and South Amboy—has many rocky ridges, low mountains, and lakes. Within this area, near the northwestern boundary, are the Kittatinny Mountains, which reach elevations up to 1,803 feet, highest in the state. The forests, now largely confined to the mountainous regions, contain oak, beech, maple, hickory, ash, and birch, mixed with a few stands of white and pitch pines, hemlock, and—more rarely—spruce, cedar, and tamarack.

Elements of northern birdlife are represented in the summer-resident populations, chiefly the Brown Creeper, Hermit Thrush, Solitary Vireo, Black-and-white Warbler, Nashville Warbler, Magnolia Warbler, Black-throated Blue Warbler, Black-throated Green Warbler, Blackburnian Warbler, Northern Waterthrush, and Canada Warbler.

In contrast with northern New Jersey, southern New Jersey has extensive sandy plains. In the central interior are the famous pine barrens—a sparsely inhabited country of flatlands and low hills covered with pitch pines, shortleaf pines, and scrub oaks; of meandering streams bordered by gums and other hardwoods; and of swamps with white cedars. Characteristic breeding birds are Pine Warblers, Prairie Warblers, and Rufous-sided Towhees in the pinelands, White-eyed Vireos and Hooded Warblers in the lowland woods along streams. In the extreme south, on the tip of Cape May, which extends southward to the same latitude as Washington, D.C., birdlife is notably southern in composition. The presence in summer of the Chuck-will's-widow is a good example.

Birds breeding more or less regularly throughout New Jersey in deciduous woods and on farmlands (fields, wet meadows, brushy lands, orchards, and dooryards) are the following:

DECIDUOUS WOODS

Turkey Vulture
Red-shouldered Hawk
Broad-winged Hawk
Ruffed Grouse (*local*)
Yellow-billed Cuckoo
Black-billed Cuckoo
Common Screech Owl
Barred Owl
Whip-poor-will
Yellow-shafted Flicker
Red-bellied Woodpecker
Hairy Woodpecker
Downy Woodpecker
Great Crested Flycatcher
Least Flycatcher
Eastern Pewee
Blue Jay
Black-capped Chickadee (*mainly northern New Jersey*)
Carolina Chickadee (*southern New Jersey*)
Tufted Titmouse

White-breasted Nuthatch
Wood Thrush
Yellow-throated Vireo
Red-eyed Vireo
Warbling Vireo
Ovenbird
American Redstart
Baltimore Oriole
Scarlet Tanager
Rose-breasted Grosbeak

FARMLANDS

Common Bobwhite
Mourning Dove
Eastern Kingbird
Eastern Phoebe
Tree Swallow
Barn Swallow
House Wren
Carolina Wren
Northern Mockingbird
Gray Catbird
Brown Thrasher
Eastern Bluebird
White-eyed Vireo (*southern New Jersey*)
Yellow Warbler
Chestnut-sided Warbler (*northern New Jersey*)
Common Yellowthroat
Bobolink
Eastern Meadowlark
Orchard Oriole
Common Grackle
Northern Cardinal
Indigo Bunting
House Finch
American Goldfinch
Rufous-sided Towhee
Grasshopper Sparrow
Vesper Sparrow
Chipping Sparrow
Field Sparrow
Song Sparrow

Except for the 50-mile northern boundary extending between the Hudson and Delaware Rivers, New Jersey is surrounded by water: the Hudson River and Atlantic Ocean on the east, the Delaware River and Delaware Bay on the west. Ornithologically, the coastal area, which borders the eastern shore from Raritan Bay south to Cape May and from the Cape northward and westward along Delaware Bay, is of paramount interest. Here are barrier sand beaches and vast salt marshes, the latter, locally called meadows, covered with a luxuriant growth of cord grasses, bulrushes, salt grasses, and glassworts. A peculiarity of the eastern coastal area from upper Barnegat Bay south to the Cape is the location of beaches on narrow, sometimes very long sand islands, and the division of the marshes into separate sections by numerous sounds, bays, inlets, and channels.

Nesting on the beaches and sand islands are Piping Plovers and widely scattered colonies of Common Terns, sometimes with a few Roseate Terns, Little Terns, and Black Skimmers. Clapper Rails are the characteristic birds of the salt marshes, where both Sharp-tailed and Seaside Sparrows and colonies of Laughing Gulls also reside. Willets breed abundantly in the marshes around Fortescue on Delaware Bay.

In addition to salt marshes, New Jersey has brackish marshes near the tidewater areas, and widely distributed fresh-water marshes in which the following birds are usually summer residents:

Pied-billed Grebe
American Bittern
Least Bittern
Mallard
American Black Duck
Blue-winged Teal
King Rail
Virginia Rail
Sora (*mainly northern New Jersey*)
Common Gallinule
American Coot (*except southern New Jersey*)
Marsh Wren
Swamp Sparrow

Of the fresh-water marshes, probably the most famous is a cattail marsh called Troy Meadows, near Caldwell in northern New Jersey. The boardwalk that crosses it permits close observation of its rich birdlife without the necessity of wading.

New Jersey is sandwiched between two important flyways—the Atlantic on the east, and the Delaware River Valley on the west. It has, in addition, the hawk flyway along the mountain ridges in the north and the extraordinary 'migration block' at Cape May Point in the south.

Waterbirds tend to follow the Atlantic flyway, gathering on the bays, sounds, and inlets. The bird finder may see such birds as Red-throated Loons, Horned Grebes, Northern Gannets, and Double-crested Cormorants by taking some vantage point along the coast, preferably Long Beach Island east of Manahawkin. Brant and Snow Geese also follow the Atlantic flyway, stopping by the hundreds or thousands—the Brant in Barnegat Bay (viewed from Long Beach

Island) and the Snow Geese in Delaware Bay (reached from Bridgeton). Although Greater Scaups and a few other ducks favor the Atlantic, the great majority—Mallards, American Black Ducks, American Wigeons, Common Pintails, Green-winged Teal, Blue-winged Teal, Northern Shovelers, Wood Ducks, Ruddy Ducks—prefer the Delaware Valley. Both flyways are used by shorebirds, with the exception of Ruddy Turnstones, Whimbrels, Red Knots, and Sanderlings, which prefer the Atlantic.

A short distance from New York City are the ridges of the Watchung Mountains (reached from Montclair), along which an impressive parade of hawks takes place every spring and fall. Many species, some of them rare to the average bird finder, pass within a range of a few yards. Eclipsing the hawk flights is the fall spectacle at Cape May. To this southernmost tip of the Cape May Peninsula come the southward flights of landbirds. Perhaps they converge from the Atlantic and Delaware Valley flyways, or perhaps they are forced to this location by strong northwest winds. In any event, here they 'bunch up' in countless numbers as if blocked by the open water of Delaware Bay. The congestion is so great that it has to be seen to be believed.

The following timetable indicates the periods of the heaviest migratory movements in New Jersey

Waterfowl: 1 March–10 April; 15 October–1 December
Shorebirds: 1 May–1 June; 1 August–5 October
Landbirds: 15 April–15 May; 1 September–25 October

During winter, bird finding can be remarkably worthwhile in New Jersey, especially along the eastern barrier beaches and on the Cape May Peninsula. Long Beach Island (*see under* **Manahawkin**), the best spot for watching Brant between December and March, offers (Northern) Horned Larks and the possibility of such rarities as a Northern Shrike, (Ipswich) Savannah Sparrow, and Lapland Longspur. The stone jetty on Five Mile Point (*see under* **Stone Harbor**) presents additional rewards: Purple Sandpipers and the chance of a few alcids. Delaware Valley Ornithological Club members find the Cape May Peninsula a haven for a great variety as well as vast numbers of winter birds. Each year the members make a Christmas census of the area, usually recording over 130 species and obtaining totals as high as 207,115 individuals. Brant, American Black Ducks, Dunlins, Herring and Ring-billed Gulls, Common Crows, European Starlings, Myrtle Warblers, House Sparrows, Red-winged Blackbirds, Common Grackles, and Brown-headed Cowbirds contribute substantially to the total of individuals; the five species of loons and grebes, eight species of wading birds, twenty species of ducks, and ten species of shorebirds combined with many species of permanent-resident and visiting landbirds account for the remarkably long list of species.

Authorities

Robert S. Arbib, Jr., Raymond J. Blichartz, Clarence D. Brown, Gilbert Cant, Dale R. Coman, Robert C. Conn, Alfred E. Eynon, David G. Fables, Quintin Kramer, Edward B. Lang, Jeannette Middlebrook, Charles H. Rogers, Edward I. Stearns, Phillips B. Street.

Reference

The Birds of New Jersey. Their Habits and Habitats. By Charles Leck. New Brunswick: Rutgers University Press, 1975.

ASBURY PARK
Shark River Marshes | Lake Como | Wreck Pond | Manasquan Inlet

For an excellent tour along the coast in search of waterbirds and waterfowl in spring and fall, drive south from Asbury Park on State 71, turn right on State 33, and, just beyond Neptune, turn left at a sign 'Shark River Hills' for 1.0 mile to the **Shark River Marshes.** Clapper Rails nest here in the summer. Turn left on the first road beyond the Marshes and turn left twice more, keeping close to the River. Look for loons, grebes, wading birds, and ducks from vantage points along the way.

Continue east to Ocean Avenue in Belmar, turn right, and proceed south parallel to the Atlantic. On the right, near the gates to Spring Lake, is **Lake Como,** where, from March through early April and from October through early December, one may see such ducks as the Redhead, Ring-necked Duck, Canvasback, and Ruddy Duck.

At this point also, scan the ocean for Northern Gannets near the fishing nets.

Continue south on Ocean Avenue to **Wreck Pond** on the right, a place especially good for gull concentrations in April.

Beyond Wreck Pond, turn right from Ocean Avenue, left on State 71, left on State 34 across **Manasquan Inlet,** and left again to the village of Point Pleasant Beach. The Inlet, in view from several places, always attracts a wide variety of waterbirds and waterfowl. When visiting this area in winter, follow roads leading northeastward from Point Pleasant Beach to the mouth of the Inlet, where there are often Purple Sandpipers on the stone jetties.

ATLANTIC CITY
Brigantine Island | Brigantine National Wildlife Refuge

Transient shorebirds in May and from late July to late September, and wintering Brant from mid-December to mid-March, appear in impressive numbers near this famous resort.

From Atlantic City, drive west on US 30, then north on State 87 opposite South Carolina Avenue, crossing Absecon Inlet by bridge and causeway to **Brigantine Island.** Continue to Lighthouse Circle, then go southeast on 34th Street to Ocean Avenue and follow it north and south to its end on the beach. Walk south along the beach to a point that compares favorably with the tip of the Holgate Peninsula (*see under* **Manahawkin**) in variety of shorebirds, but concentrations are lighter owing to the lack of mud flats.

Brigantine National Wildlife Refuge, northwest of Atlantic City, embraces 19,233 acres consisting largely of salt marshes adjoining Great, Little, and Reeds Bays. About 1,600 acres of these marshes have been converted to fresh-water habitats by dikes. For headquarters on Lily Lake, take US 30 west from Atlantic City, turn right on State 9 to Oceanville, then right again on Great Creek Road for 0.9 mile.

At headquarters, pick up a leaflet explaining a self-guided tour by car. This 8-mile drive—highly rewarding—begins at headquarters and goes one way on dikes overlooking, on the right, salt marshes interrupted by bays and tidewater creeks and, on the left, fresh-water pools and marshy islands. The drive terminates in the western part of the Refuge, an upland area covered with pines and mixed hardwoods. Readily in view throughout warmer months from either side of the dikes are wading birds: Great Blue, Green, Little Blue, and Black-crowned Night Herons, Cattle, Great, and Snowy Egrets, Glossy Ibises, and occasionally Louisiana Herons and Yellow-crowned Night Herons.

In early spring, Canada Geese nest on the dikes, and by the first of May appear with their goslings along the drive. In June and early July, broods of other waterfowl—Mute Swans, Mallards, American Black Ducks, Gadwalls, Green-winged and Blue-winged Teal, Northern Shovelers, and Ruddy Ducks—show up from the dikes.

Before the marsh vegetation has begun to grow in early spring, Clapper Rails are conspicuous and easily watched as they forage. Later in early summer they nest commonly in the rank grasses and sedges of the salt marshes, especially along tidewater creeks, where the plants attain greater height. The nests can sometimes be spotted by the arches of live grasses that the occupant birds have pulled together over them, but the birds themselves are difficult to see. Their loud clatter, however, reveals their presence. Also nesting in the marshes at the same time are a few Forster's Terns, Marsh Wrens, and both Sharp-tailed and Seaside Sparrows.

Conspicuous by May first are many Willets performing their flight displays with much vocalizing above the marshes, where they are establishing territories for eventual nesting. In May, and again in August/September, vast numbers of transient shorebirds—chiefly Dunlins, Short-billed Dowitchers, Least and Semipalmated Sandpipers along with Semipalmated and Black-bellied Plovers, Whimbrels, and Greater and Lesser Yellowlegs—feed and rest on the tidal mud flats.

By October, transient or wintering ducks, including Common Pintails, American Wigeons, Greater Scaups, Common Goldeneyes, Buffleheads, Oldsquaws, scoters, and mergansers, begin to show up. By mid-November, hundreds of Brant and Snow Geese start appearing on the Refuge; the population may total many thousands by mid-December, when all have arrived for the winter. Throughout winter, many will be seen at one time in flight between the bays and pools, sometimes alighting on the water in full view

WHIMBREL

from the dikes. Although a few Brant and Snow Geese linger in spring until May, most non-breeding waterfowl leave before April.

BRIDGETON
Fortescue

From this small city on State 49 in southern New Jersey, drive east on State 49 for 3 miles and south on County 553 to the town of Newport. From Newport, continue directly south on an unnumbered road for about 5 miles to **Fortescue** on Delaware Bay, where there is a vast, undrained salt marsh. Willets nest here from May through July, but one can see them best after the mid-June, when they have young and are consequently more vociferous when approached. The marsh, extending east to the Maurice River and northwest to Cohansey Creek and covering approximately 30,000 acres, was once estimated to contain ten thousand pairs of these big shorebirds. A few Northern Harriers and perhaps a few pairs of Short-eared Owls are among other species nesting in the marsh.

One of New Jersey's great ornithological shows is the enormous concentration of Snow Geese in Delaware Bay in late March. Because they are usually a mile or more offshore—so far away that they appear as a thin white line—one should charter a local powerboat to move closer to them. The flock will eventually rise like a great, shimmering white cloud, a sight never to be forgotten. The Snow Geese arrive in early November and remain until the ice forces them south. They reappear in late February or early March and stay into early April.

CALDWELL
Troy Meadows

A famous bird-finding area in northern New Jersey is **Troy Meadows,** where Troy Brook joins the Whippany River. Here, within an hour's drive from New York City, are 3,000 acres of open cattail marsh bordered by wooded swamp. To reach Troy Meadows, take Roseland Avenue south from the center of Caldwell to Roseland, right on Eagle Rock Avenue for about 2 miles, and left on Ridgedale Avenue to about 0.5 mile beyond Swinefield Bridge across the Passaic River. Park near the power lines and walk northeast on a boardwalk that runs through the marsh under the power lines, crosses the Whippany River and Troy Brook, and comes out on I 80 just west of Pine Brook. Waterproof footwear is recommended, since the pathway to the dry boardwalk is quite soggy in spring.

In March and April, surface-feeding ducks, especially Common Pintails, are common at the height of migration. From mid-May into July, visits at dawn will bring a close acquaintance with the vocalizations of American and Least Bitterns, Virginia Rails, Soras, Common Gallinules, Marsh and Sedge Wrens, and Swamp Sparrows. Infrequently, the King Rail may also be heard. As one goes along the boardwalk during the day, he may flush Least Bitterns close by, and see Traill's Flycatchers regularly on the tops of bushes scattered over the marsh.

The wooded margins of the marsh and the surrounding farms are excellent for landbirds. In mid-May, when many transients pass through, bird finders have listed over a hundred species before mid-morning by covering first the marsh from the boardwalk and then the wooded margins.

CAPE MAY
The Rips | Lake Lily | Higbee Beach

No place on the Atlantic Coast has, through the years, given bird finders greater satisfaction than the southern part of the Cape May Peninsula, familiarly known as Cape May. The focal point is the village of Cape May Point on the very tip of the Peninsula. Off its shores the waters of Delaware Bay and the Atlantic Ocean meet and churn, producing what is locally referred to as **The Rips.**

Cape May Point has a luxuriance of shade trees and garden shrubbery. On the northern edge of the village is tree-bordered **Lake Lily.** To the east and northeast are open fields and pastures with intervening dense hedgerows of wild plum and bayberry bound together by green briars and grapevines. Here and there are low woodlots, principally of pine. North of the village along Delaware Bay are beaches, the least disturbed being **Higbee Beach.** Back from the beaches are sand dunes, those close to the beaches mostly devoid of vegetation, those farther inland covered either with stands of pine, oak, holly, and cedar, or with dense, almost impenetrable thickets of bayberry entangled with green briars, poison ivy, and other vine-like growth.

To reach Cape May Point, drive west from the city of Cape May on Sunset Boulevard for 2 miles, then south on Cape Avenue to Lake Lily. A road leads from Cape Avenue around the Lake. Continue past Lake Lily to the lighthouse where, from the beach, The Rips may be viewed. To reach Higbee Beach, backtrack on Cape Avenue to Sunset Boulevard and return east on it to Bayshore Road, turn north for 1.8 miles, then west on the second paved road marked 'Dead End.' Where the pavement ends, park the car and walk along an unimproved road winding westward through woods and thickets to the mouth of the Cape May Canal on Delaware Bay. Higbee Beach extends north from here.

There are good opportunities for bird finding at Cape May in spring, when many northward-bound small landbirds linger awhile upon reaching the Cape May Peninsula from across Delaware Bay. In late April and early May a walk around Lake Lily and a search of the woods, thickets, and shrubby fields back of Higbee Beach are almost certain to turn up transient kinglets, vireos, warblers, and fringillids.

The most exciting time to visit Cape May is the last half of the year, when there are wide varieties of migrating birds and phenomenal flights of a few species. Almost without exception, the large flights come with a strong northwest wind and a dip in temperature. Thus the best dates for observations cannot be predicted with certainty.

Midway in summer, usually in early August, Tree Swallows begin to arrive in droves, appearing everywhere: swirling over bodies of water, lining up on wires, resting on highways, crowding onto the crowns of trees and shrubs. Once the migration is well under way, the concentration increases daily through September; though some individuals depart, greater numbers arrive. Other species of swallows gather to swell the concentration, but they play a secondary role. By October the peak of abundance has passed.

In mid-August or soon thereafter come the Eastern Kingbirds, seemingly the larger part of the total kingbird population of the northeastern states and Canadian provinces. In flocks ranging from a few to fifty or more, they settle down at Cape May Point from high overhead, perching on trees, wires, and hedgerows. When they have rested, the flocks depart; but many more arrive hour after hour to take their place. The peak is reached during the last few days of August.

From mid-August through late September, Bobolinks arrive in their 'reed bird' plumage. The flocks stay only a short while. For the most part, concentrations do not build up unless there are strong northwest winds, in which case the numbers increase to impressive proportions until the winds subside.

The flights of Yellow-shafted Flickers, taking place from late September to early October, constitute one of the most striking

MERLIN

spectacles of all. These birds arrive during the night, and one can watch them in the morning as they move along the shore of the Bay from Cape May Point *north,* not in flocks, but singly, forming a continuous stream. Several counts, made one morning on the Bay, showed the birds passing at the rate of thirty to fifty per minute. To watch the flights to best advantage, stand on a high dune back from Higbee Beach.

Hawk flights usually start with the first 'northwesters' in September, or even in late August. Over 90 per cent of the individuals are estimated to be birds-of-the-year. Ospreys and American Kestrels compose the vanguard, and other species soon come—some dozen altogether. 'Wide-winged' species are in the majority when the flights are going strong in mid-October. With them are Sharp-shinned and Cooper's Hawks and a few Peregrine Falcons and Merlins. By mid-November the hawk flights have almost ceased. The best vantage points for viewing them are the high dunes back of Higbee Beach.

The flights of American Woodcock usually take place during the first two weeks of November, though they may be earlier or later in some years. Like the flickers, the woodcock arrive at night; unlike the flickers, they pass the day in hedgerows east and northeast of the village or in thickets back from the dunes along Delaware Bay. Upon investigating these places after a nocturnal flight, the bird finder will be startled repeatedly as one bird after another flushes almost at his feet, zigzags upward on whistling wings until, clear of the vegetation, it speeds forward to alight in some secluded spot not far away, but usually beyond view.

A special treat is in store for the bird finder who will rise before daybreak, conceal himself near a hedgerow, and remain quiet. With the first evidences of dawn, he will see woodcock silhouettes moving in from the surrounding fields and settling down in the hedgerows. At such times, he will probably hear a few incomplete flight songs. If the bird finder will similarly locate himself in the early evening, just before total darkness, he will see woodcock leaving the hedgerow to feed or to continue their migration southward.

Coinciding with the woodcock flights, but beginning a little earlier in fall, are the tremendous flights of American Robins, often accompanied by flights of Eastern Bluebirds. Watch the robins especially from the dunes back from Higbee Beach as they proceed northward in loose flocks.

In September and early October, while the big flights are in progress, the woods, thickets, and hedgerows at Cape May Point are usually alive with passerine species—Gray Catbirds, Brown Thrashers, Ruby-crowned Kinglets, Palm Warblers, White-throated Sparrows, Fox Sparrows—which have arrived during the night. But the populations are apparently not any larger than they are in many other spots on the continent.

Waterbirds, waterfowl, and shorebirds (except American Woodcock) appear in no spectacular aggregations at Cape May Point, but the numbers and variety are nevertheless impressive. Shorebirds are present from late July to early September. Semipalmated Plovers, Ruddy Turnstones, Willets, Red Knots, Semipalmated Sandpipers, and Sanderlings on the beaches; Black-bellied Plovers, Whimbrels, Greater Yellowlegs, Lesser Yellowlegs, and Short-billed Dowitchers in marshes back from the Bay. The Rips are favorite feeding grounds for transient gulls and terns, which, in late September and early October, are occasionally harried by one or more Parasitic Jaegers. Northern Gannets show up, usually quite far offshore, in November.

By December, all flights taper off, and birdlife attains a winter stability. This is not to imply, however, that birdlife is lacking in variety. The annual Christmas census, taken by the Delaware Valley Ornithological Club over a recent period of twenty years, recorded as many species as 153 and as few as 114, with an average of 136.8.

CHATHAM
Great Swamp National Wildlife Refuge

Several states have 'great swamps,' but New Jersey's, south of Chatham and only a few miles west of Newark, is distinguished by boardwalks running into its very heart. Altogether 5,327 acres of the Swamp, which is about 6 miles long and 3 miles wide, are incorporated in the **Great Swamp National Wildlife Refuge.** This includes wooded uplands with mature oaks and beeches, mountain laurels, rhododendrons, and other plants of both northern and southern affinities; cattail marshes; and open water with an abundance of such aquatic plants as duckweed and pond lily.

Among birds breeding in suitable habitats during spring and early summer are the American Bittern, Canada Goose, Red-tailed Hawk, Ruffed Grouse, Virginia Rail, Sora, American Woodcock, Red-bellied Woodpecker, Traill's Flycatcher, Marsh Wren, both Wood Thrush and Veery, Blue-gray Gnatcatcher, Blue-winged and Chestnut-sided Warblers, Scarlet Tanager, Rose-breasted Grosbeak, Rufous-sided Towhee, and Swamp Sparrow. Wood Ducks are numerous the year round. Common Pintails and Green-winged and Blue-winged Teal are among the transient ducks stopping in March and April. On early spring evenings one may hear Common Screech, Great Horned, and Barred Owls.

The wildlife observation center, maintained by the Refuge on its west side, is excellent for birds typifying the Swamp. To reach the center, take Fairmount Avenue out of Chatham for about 4 miles to Meyersville Road; turn right and proceed 2 miles to the hamlet of Meyersville and bear right on the New Vernon–Long Hill Road, which, after 1.0 mile, crosses White Bridge Road and, after another mile, passes the entrance to the center on the left. Park the car near the entrance and take the half-mile boardwalk to the large observation building, or take a shorter walk, passing rest rooms on the right to a small observation building. To reach Refuge headquarters, return to the car and backtrack to White Bridge Road, turn right and go 1.25 miles to Pleasant Plains Road, turn right again and go 1.0 mile.

On the east side of the Swamp, the Morris County Park System operates an outdoor education center from which trails with boardwalks over the wet places lead through productive spots for birds. Drive west from Chatham on Fairmount Avenue for 1.5 miles to Southern Boulevard, turn right and continue for 1.0 mile to the center's entrance on the left, indicated by a directional sign.

HIGHLANDS
Sandy Hook State Park

Sandy Hook, a long barrier peninsula projecting northward into New York Harbor, has beaches on both the oceanside and the side facing Sandy Hook Bay, as well as tidal marshes and mud flats on the bayside. Inland are stands of holly mixed with red maple, hackberry, red cedar, and other woody growth. The northern two-thirds of the peninsula is occupied by Fort Hancock and closed to the public; **Sandy Hook State Park,** 468 acres, occupies the southern third. To reach the Park, drive east from Garden State Parkway Exit 117 on State 36 through Highlands, then north a short distance.

The Park's marshes have nesting Clapper Rails and Soras and its beaches offer good shorebirding in May and in August through September. During November, Northern Gannets appear far out from the ocean beach. Throughout winter, rafts of diving ducks, including 'sea ducks,' show up on the open ocean. Red-necked and Horned Grebes, Canada Geese, American Black Ducks, Common Pintails, and American Wigeons gather in the Bay.

LAKEHURST
Pine Lake | Lake Horicon

For a typical section of the New Jersey pine barrens with their characteristic birdlife, proceed southeast from Lakehurst on State 37 for 3 miles, then bear left on a road to the summer community of Pine Lake Park. The **Pine Lake** area offers the best opportunities.

In late spring and summer a walk around **Lake Horicon** at Lakehurst is worthwhile. Among species likely to be seen are the Great Blue Heron, Great Egret, Little Blue Heron, Carolina Chickadee, White-eyed Vireo, Black-and-white Warbler, Pine Warbler, and Prairie Warbler.

MANAHAWKIN
Long Beach Island | Barnegat Lighthouse State Park | Holgate Unit of Brigantine National Wildlife Refuge

One of the best places for year-round bird finding on the New Jersey coast is **Long Beach Island,** a narrow barrier island with a beach on the oceanside and mud flats and salt marshes on the mainland side, extending for 18 miles between Barnegat Inlet on the north and Beach Haven Inlet on the south. Though separated from the mainland by the Intracoastal Waterway, one can reach it from Manahawkin by taking State 72 east for 6 miles over a bridge and causeway to Ship Bottom.

The reward for a trip any time between mid-December and mid-March is one of New Jersey's greatest ornithological spectacles—thousands of Brant. No particular vantage point is necessary, for they stay anywhere in the Waterway in view from the causeway or from the road that runs north and south on the Island from Ship Bottom. Frequently they fly over in their peculiar, uneven formations.

To see other birds, drive north from Ship Bottom. Along the way, watch for Northern Shrikes; one or two may be on the utility wires. Stop now and then to (1) scan the Ocean and Waterway for Common and Red-throated Loons, Red-necked and Horned Grebes, Greater Scaups, Great Cormorants, Buffleheads, and Oldsquaws; (2) look up and down the beach for Sanderlings and oceanward for rare alcids; (3) identify Greater Black-backed Gulls and Ring-billed Gulls among the omnipresent Herring Gulls; and (4) explore the sand dunes, where there is always a possibility of scaring up (Ipswich) Savannah Sparrows.

When the road ends in **Barnegat Lighthouse State Park** (17 acres), leave the car and walk toward the Waterway (Barnegat Bay), first investigating a cedar and holly thicket behind the lighthouse—the only woods on the Island—which is often, especially in migration, a haven for a large assortment of passerines. Return to the car and drive south. Turn west on 21st Street and go out on a sand fill, an ideal point for viewing Barnegat Bay with its Brant and other waterfowl. Near the Bay side and sand fill, look for flocks of Horned Larks (northern races), Snow Buntings, and an occasional Lapland Longspur.

A trip to Long Beach Island in April will be rewarded by the sight of large numbers of transient Northern Gannets and Double-crested Cormorants offshore on the oceanside. To see them in fall migration, one should make the trip in October for cormorants and in November for gannets.

For transient shorebirds, the time for a trip is May and, better still, late July to early October; for breeding birds, the time is June or early July. The best place for all of them is the extreme southern end of the Island, the Holgate Peninsula, reached from Ship Bottom by driving south about 10 miles through the community of Beach Haven Inlet to the **Holgate Unit of Brigantine National Wildlife Refuge,** which embraces the 256-acre Peninsula. Entrance, by foot travel only, is at the end of Bay Avenue near the Coast Guard tower.

Look for transient Piping Plovers and Sanderlings on the ocean beach; Semipalmated Plovers, Black-bellied Plovers, Willets (in fall), Greater Yellowlegs, Lesser Yellowlegs, Red Knots, Pectoral Sandpipers (fall), Least Sandpipers, Dunlins, Short-billed Dowitchers, Semipalmated Sandpipers, and Western Sandpipers (fall) on the mud flats of the Waterway side. A few Hudsonian Godwits appear occasionally in late summer.

The salt marshes on the Waterway side are the nesting habitat for Clapper Rails and the elusive Black Rail, as well as for both Sharp-tailed and Seaside Sparrows. On the oceanside, the higher portions of the beach are nesting sites for a few pairs of Piping Plovers, perhaps a pair or two of Wilson's Plovers, and sometimes small colonies of Little Terns; back from the beach among the dunes are nesting sites for (Prairie) Horned Larks and Savannah Sparrows. Formerly, colonies of Common Terns, with a few Roseate Terns intermingled, and Black Skimmers occupied parts of Holgate Peninsula, but human incursions have driven most, if not all, of them to islands offshore in the Waterway.

MONTCLAIR
Montclair Hawk Lookout Sanctuary

One of the best spots for watching hawk flights in northeastern New Jersey is a clearing on the crest of First Watchung Mountain above an abandoned traprock quarry within the **Montclair Hawk Lookout Sanctuary,** owned by the New Jersey Audubon Society. To approach the Mountain from Montclair, drive north from the center of town on Valley Road, north on Upper Montclair Avenue, left on Bradford Avenue for 2 blocks, right on Edgecliffe Road to the top of the hill past the aforementioned quarry for 0.25 mile. Park at the corner of Crestmont Road and climb east a short distance to the crest, no more than 600 feet above sea level. In view are the other Watchungs, all low ridges that run northeast-southwest and along which the hawks pass.

In fall the hawks either follow the ridges southwest or cross the ridges from the east. In spring they tend to follow the ridges northeast. The fall flights are heaviest when there are cool westerly winds preceded by climatic disturbances. Usually the flights begin in late August with small movements of Sharp-shinned Hawks, Broad-winged Hawks, and American Kestrels. By the third week of September, the Sharp-shins and Broad-wings have reached their peak of abundance, and other species are flying: the Cooper's Hawk, Red-tailed Hawk, Red-shouldered Hawk, Bald Eagle, Northern Harrier, Osprey, and, much less frequently, Merlin and Peregrine Falcon. By mid-October the Red-tailed and Sharp-shinned Hawks dominate the flights, and the other species become fewer, with the exception of the Red-shouldered Hawk, which is frequently common in early November. Thereafter the hawk migration definitely begins to wane. In spring the hawk flights start in late February and continue into May. The sequence of the species is generally reversed, the Red-shouldered and Red-tailed Hawks appearing first.

NEWTON
Stokes State Forest

In the Kittatinny Mountains of northwestern New Jersey, 12,428 acres of forest land comprise **Stokes State Forest.** The elevation above sea level ranges from 420 feet near the southern boundary to 1,640 feet on Sunrise Mountain in the northern part. US 206 leading north and west from Newton for 11 miles crosses the Forest through Culver's Gap (elevation 931 feet) and thus makes the area readily accessible. Although there are many places in this New England-like tract where bird finding is productive, the following are outstanding in late May, in June, and in early July:

1. *Tillman Ravine,* in the southern section, is crossed by Tillman Brook, whose clear water cascades over mossy rocks and whose precipitous banks are covered with rhododendrons and ferns, and canopied by magnificent hemlocks. Magnolia Warblers, Black-throated Blue Warblers, Black-throated Green Warblers, Blackburnian Warblers, Northern Waterthrushes, and Canada Warblers are characteristic breeding birds, and Ruffed Grouse are permanent residents in the area. A well-marked road goes left from US 206 near the Stokes Forest office to the Tillman Ravine parking area, 4 miles distant. A path leads from there to the Ravine.

2. *Lake Ocquittunk,* covering about 9 acres in the northern section of the Forest, has in its area stands of tamarack and pine with extensive shrubby clearings. Though the Lake is the center of a recreational development, there remain habitats for breeding Black-and-white Warblers, Golden-winged Warblers, Nashville Warblers, Yellow Warblers, and Chestnut-sided Warblers. The open water and the campsites, together with the woods, make the area attractive to a wide variety of woodpeckers, flycatchers, swallows, vireos, and fringillids. To reach the Lake, drive north on US 206 from the Forest office for 2.5 miles, then right on Flatbrook Road for 3 miles. Investigate the fields along Flatbrook Road for Henslow's Sparrows.

3. *Sunrise Mountain,* a fine lookout for fall-migrating hawks, is forested with mixed hardwoods and conifers and has the Ruffed Grouse and Pileated Woodpecker among its permanent residents. The road to Sunrise Mountain leaves US 206 just north of Culver's Gap. The Appalachian Trail, which comes through Stokes Forest near Sunrise Mountain, provides access to rewarding spots, as do other trails.

PRINCETON
Princeton Wildlife Refuge

In eleven of twelve consecutive years the Connecticut Warbler has been observed in the **Princeton Wildlife Refuge** during fall migration, and in all these years at least thirty-five species of warblers have been recorded here during spring migration.

The property of the Elizabethtown Water Company, and maintained by the Princeton Open Space Commission, the Refuge embraces 49 acres of marsh, woodland, and shrubby areas, crisscrossed by dirt roads and lanes. To reach the Refuge, leave US 1 toward Princeton on Alexander Street about 0.4 mile south of the Princeton traffic circle; drive 0.7 mile and turn left on West Drive for 0.2 mile to the entrance, clearly marked; continue past the observation tower, overlooking the marshy area, to the pumping station, and then walk the path beginning on the left of the station and paralleling Stony Brook.

CONNECTICUT WARBLER

Along this path during September and the first half of October—or better, from mid-September to the second week of October—look for the Connecticut Warbler in the growth of blackberry, catbriar, wild grape, honeysuckle, and other shrubs and vines that form a dense tangle. Try 'squeaking' to excite the bird and bring it into view. There are similar habitats where the species has been seen along paths elsewhere in the Refuge and in the 500-acre woods of the Princeton Institute for Advanced Studies adjoining the Refuge on the west.

Both the Refuge and the woods of the Institute abound in transient warblers during the first three weeks of May. Among the ninety-two bird species reported nesting in the Refuge are the Black-and-white and Prothonotary Warblers and four vireos: the White-eyed, Yellow-throated, Red-eyed, and Warbling. The Kentucky Warbler nests in the Institute woods.

SALEM
Killcohook National Wildlife Refuge | Mannington Creek

Surrounding Fort Mott on a bend in the Delaware River, not far northwest of Salem, is the **Killcohook National Wildlife Refuge.** The Refuge, covering 1,362 acres, is a wet area largely overgrown with quillreeds, which appeared after filling operations that accompanied the dredging of the Delaware. During the fall hunting season the two remaining ponds usually contain many thousands of ducks, especially Mallards, American Black Ducks, Common Pintails, Green-winged Teal, together with a few American Wigeons, Blue-winged Teal, and Northern Shovelers, but almost none at any other time, since there is little natural food. Tremendous flocks of Red-winged Blackbirds and Common Grackles roost in the quillreeds after breeding season, the congregation being most spectacular in late summer. The huge masonry wall around nearby Fort Mott National Cemetery is an excellent vantage point for viewing ducks and blackbirds, especially toward evening when they are particularly active.

To reach the Cemetery from Salem, drive northwestward on State

49 for 3 miles to Harrisonville, then turn left on a road (no route number) that heads westward to a fork. Take the left fork south, passing the entrance to Fort Mott on the right, to the entrance to the Cemetery, also on the right.

From Salem take State 45 north for 2 miles to three forks, all of which—State 45, State 540, and Sharptown Road—provide excellent vantage points when crossing **Mannington Creek** ahead. Very broad and choked extensively by spatterdock and other aquatic plants, the Creek, during warmer months, attracts wading and marsh-dwelling birds such as Pied-billed Grebes, Great Blue Herons, Green Herons, Great Egrets, Snowy Egrets, Black-crowned Night Herons, and often Little Blue Herons, Yellow-crowned Night Herons, and Glossy Ibises. In spring and fall, Mallards, American Black Ducks, Common Pintails, Green-winged Teal, American Wigeons, Northern Shovelers, and Wood Ducks gather in fair numbers.

STONE HARBOR

Wetlands Institute | Stone Harbor Bird Sanctuary | Nummy Island | Five Mile Point

The ocean side of Cape May Peninsula has long, paralleling barrier islands with sand beaches backed by dunes that slope westward to salt marshes. Cut off from the mainland by the Intracoastal Waterway, these islands are reached by bridges and causeways. Although now greatly developed as resorts, the islands still have a few undisturbed areas for birds.

Drive east from Garden State Parkway (Exit 10) over Stone Harbor Boulevard to Stone Harbor, a resort community on Seven Mile Beach. Just before crossing the Intracoastal Waterway, 3 miles from the Parkway, turn in at the **Wetlands Institute** on the right overlooking an extensive salt marsh. Operated for the study of wetlands ecology, the Institute owns 34.5 acres of marsh, adjacent to the gray-shingled headquarters building, and has the use of, for scientific investigation, the outlying 5,000 acres of marsh owned and under the protection of the State of New Jersey. Ask permission to ascend the observation tower atop headquarters building for a fine view of the salt marsh. Scan the marsh with a telescope for wading birds, Willets, gulls, and terns.

Continue across the Intracoastal Waterway on Stone Harbor Boulevard, then turn right on Ocean Drive (Third Avenue; State 585 on some maps) to the **Stone Harbor Bird Sanctuary** (21 acres), between 111th and 117th Streets near the southern edge of town. Here in a dense grove of cedars, hollies, and other low growth, surrounded by a fence, is the site of a thriving nesting aggregation of wading birds: Green Herons, Little Blue Herons, a few Cattle Egrets, Great Egrets, Snowy Egrets, Louisiana Herons, Black-crowned Night Herons, a few Yellow-crowned Night Herons, and Glossy Ibises. The birds begin arriving at the Sanctuary in mid-March, and nesting is well under way by May first. Even though nesting is over by mid-July, the birds remain well into fall. Entering the Sanctuary is prohibited, but the activities may be easily observed from outside the fence. A good time for viewing is within a half-hour before sundown, when large numbers fly in from their feeding grounds elsewhere—all except the night herons, which leave at that time for their feeding grounds.

Continue south on Ocean Drive (State 585) from Seven Mile Beach. Here the highway crosses **Nummy Island,** with mud flats that are feeding grounds for transient shorebirds, and salt marshes that are nesting habitats for Clapper Rails, Willets, Sharp-tailed Sparrows, and Seaside Sparrows. When looking for birds, park the car well off the pavement of this busy highway.

Again, continue south on Ocean Drive (State 585) through the Wildwood resort communities of Five Mile Beach until the highway bears west, eventually crossing the bridge to the mainland. Leave the car in a parking area near the beach off the highway to the left, and walk south along the beach to **Five Mile Point,** the end of the island; then seaward on a stone jetty which, with its fellow across the water on Sewell's Point, protects the entrance to Cape May Harbor. The end of the jetty, about a mile from shore, is an excellent vantage point for seeing Wilson's Storm Petrels in summer, usually beginning in late July; Northern Gannets in late March, April, and November; and an occasional Parasitic Jaeger in late September and October. If the bird finder is hardy enough to make the trip to the end of the jetty after mid-December, he may possibly see Red-throated Loons, Oldsquaws, White-winged Scoters, Surf Scoters, Black Scoters, Bonaparte's Gulls, and perhaps a rare alcid or two, but his major reward will be a few Purple Sandpipers. Preferring as they do rocky shores to sandy beaches, Purple Sandpipers are attracted with remarkable regularity to these rocks. The birds may be overlooked, unless searched for carefully, since they have a habit of moving around among the rocks and are frequently out of sight.

SUSSEX
High Point State Park

High Point State Park in the northwestern corner of New Jersey embraces 11,135 acres of the Kittatinny Mountains, including High Point (1,803 feet), highest elevation in the state. State 23 crosses the Park 9 miles north of Sussex and 6 miles south of Port Jervis (New York). At High Point summit, marked by a monument, turn east on a road to the lodge for information. Like Sunrise Mountain (*see under* **Newton**), High Point is fine as a lookout for hawks in fall.

One place in the Park noteworthy for bird finding is Cedar Swamp—actually a bog—just 0.5 mile north of the High Point monument on the west side of the road. Although the predominant vegetation consists of hemlock with an understory of high rhododendron, there is also black spruce, characteristic of more northern situations, and southern white cedar. Among the twenty-nine species of breeding birds recorded in Cedar Swamp are the Sharp-shinned Hawk, Wood Thrush, Veery, Black-and-white Warbler, Worm-eating Warbler, Golden-winged Warbler, Scarlet Tanager, Rose-breasted Grosbeak, and Rufous-sided Towhee.

TRENTON
Trenton Marsh

Trenton Marsh within the John A. Roebling Memorial Park (300 acres) provides a lake, a marsh, lowland and upland woods, and

ROSE-BREASTED GROSBEAK

open fields. Birds here regularly in spring and summer—the best time for bird finding—include the Pied-billed Grebe, Great Blue Heron, Green Heron, Black-crowned Night Heron, American Bittern, Least Bittern, Wood Duck, Barred Owl, King Rail, Virginia Rail, Sora, Traill's Flycatcher, Tree Swallow, Purple Martin, Marsh Wren, Blue-gray Gnatcatcher, Blue-winged Warbler, Yellow Warbler, Common Yellowthroat, American Redstart, Rose-breasted Grosbeak, and Indigo Bunting. Large numbers of Common Gallinules are present the year round, as are Carolina Wrens. Common Snipes and Winter Wrens are numerous from October to April. Since 1959 one observer has recorded 192 species, 34 of them warblers.

Lying south of Trenton, the Marsh is reached by driving south on US 206 (South Broad Street) to the 1800 block, turning right on Sewell Avenue to a dead end, and then walking or driving down a minor road on the left. The Marsh is at the bottom of the hill, and part of it is accessible by power-line boardwalk.

TUCKERTON
Tuckerton Meadows

Tuckerton Meadows, between Little Egg Harbor and Great Bay on the southern coast of New Jersey, are productive salt marshes for the bird finder. From Tuckerton, drive west on US 9 to the edge of town, then turn off south on Great Bay Boulevard, which crosses Tuckerton Meadows and ends abruptly on a point overlooking Little Egg Inlet.

The road to the Inlet passes many brackish pools and mud flats that are often alive with shorebirds at low tide in May and from late July to late September. Whimbrels are sometimes numerous in two or three spots near the end of the road; flocks of Red Knots and occasionally American Oystercatchers feed along the edge of the Inlet. Despite the lack of a rocky shoreline, Purple Sandpipers in breeding plumage often appear in May.

In fall, always be alert for a Merlin, particularly after a northwest wind. In winter the road is an excellent vantage point for Brant and diving ducks, either in Little Egg Harbor to the north or in Great Bay to the south. Just before dusk, watch for a Short-eared Owl flying low over the marshes.

New York State

RUFFED GROUSE

Few states offer more diverse environments than New York: Montauk Point probing the realm of the Atlantic; the Adirondacks and Catskills, with their coniferous forests; the forbidding cliffs of the Hudson and the inviting beaches of the Great Lakes; the sluggish tidewaters of Long Island Sound and the torrents of the Niagara River. Stretching between these extremes is the deeply undulating, lush countryside of rural New York, with its hillside pastures and fields, its valley meadows and settlements, its woodlands and fruit orchards—and here and there a gorge, a stream, a lake, or a marsh.

In view of the varied environments, the number of bird species inhabiting New York is unusually large. Though no species is outstanding, certainly the Ruffed Grouse is the one most commonly associated with the state in the minds of ornithologists. For many years this bird of the upstate woodlands was the most comprehensively investigated and most voluminously written about of any wild creature in the world.

The loftiest mountains of New York are the Adirondacks in the northeast. Mt. Marcy, the highest, with an elevation of 5,344 feet, and several other peaks nearly as high have lower slopes clothed with spruce and fir mixed with beech, white and yellow birches, sugar maple, and scattered stands of hemlock and white pine. Some virgin timber remains, although much of it has been cut over or burned. Above 3,000 feet, spruce and fir predominate, becoming dwarfed toward the timber line between 4,500 and 5,000 feet. Above the timber line the mountains are treeless and rocky. Through the Adirondack valleys the forests consist largely of a secondary growth of maple, birch, and aspen, with occasional stands of white pine and hemlock, but they are by no means continuous, being interrupted by lakes, tamarack-sphagnum bogs, beaver meadows, clearings, and settlements.

The Catskills, in the southeast, show a somewhat lower elevation; Slide Mountain, the highest peak, has an altitude of 4,204 feet. Though similar to the Adirondacks in many respects, the Catskills are more rocky and have a more mixed forest cover. In general, they resemble the mountain ridges of Pennsylvania, whereas the Adirondacks are more like the Green Mountains in Vermont and the White Mountains in New Hampshire. On the lower slopes of the Catskills are black oak, chestnut oak, tulip tree, and sweet gum, with some stands of beech, sugar maple, hemlock, and white pine, and often a thick undergrowth of mountain laurel and rhododendron. At higher elevations, spruce and fir predominate, particularly on the northern slopes and on the summits, where they become somewhat stunted. In the valleys, usually narrow, are farms with pastures extending up the slopes to the forests.

The higher elevations of both the Adirondacks and the Catskills attract Olive-sided Flycatchers, (Bicknell's) Gray-cheeked Thrushes, Blackpoll Warblers, and Pine Siskins, species not nesting elsewhere in the state. Furthermore, the Adirondacks are the only mountains in the state to attract nesting Boreal Chickadees and Bay-breasted Warblers.

The Catskills represent the northeastern margin of the Allegheny

Plateau, which extends westward across the state and is continuous with the plateau of the same name in Pennsylvania. Though certain elevations near Pennsylvania exceed 2,000 feet, the average height is about 1,600 feet. The Plateau is broken up by deep valleys whose sides are cut by streams meandering through shaded gorges and glens. The Finger Lakes—Cayuga Lake, for example—lie in several of the valleys running north and south. The original forests of white pine and hemlock with scattered beech, sugar maple, and yellow birch once covering the Plateau, have been for the most part completely stripped, their place being taken by farmland and second-growth woods. In a few places, such as on extremely high hills, in deep, cool gorges, and in glens where forest conditions remain similar to their original state, the avifauna includes a number of breeding species characteristic of northern coniferous forests and swampy thickets. Some of them, listed below, breed also in the Adirondacks and Catskills.

Yellow-bellied Sapsucker
Red-breasted Nuthatch
Brown Creeper
Hermit Thrush
Golden-crowned Kinglet
Solitary Vireo
Nashville Warbler
Northern Parula Warbler
Magnolia Warbler
Black-throated Blue Warbler
Myrtle Warbler
Black-throated Green Warbler
Blackburnian Warbler
Northern Waterthrush
Mourning Warbler
Canada Warbler
Purple Finch
Slate-colored Junco
White-throated Sparrow

Directly west of the Adirondacks is the Tug Hill Plateau, rising to about 1,700 feet. This is a wilderness area, penetrated by few roads.

The extensive lowland areas of upstate New York comprise the St. Lawrence Valley, running along the state boundary north of the Adirondacks; the Lake Champlain–Hudson River Valley, passing south through the eastern side of the state; the Mohawk River Valley, cutting eastward through the central part of the state between the Adirondacks and Catskills to join the Hudson River Valley; the Lake Ontario Plain, extending 30 to 40 miles south of Lake Ontario; and the Lake Erie Plain, reaching eastward for about five miles along Lake Erie. All are densely settled and have many farmlands (fields, wet meadows, brushy lands, orchards, and dooryards) and deciduous woods. Birds regularly nesting in these particular lowlands, as well as in the valleys of the Adirondacks, Catskills, and Allegheny Plateau, are the following:

FARMLANDS

Mourning Dove
Eastern Kingbird
Eastern Phoebe
(Prairie) Horned Lark
Tree Swallow
Barn Swallow
House Wren
Northern Mockingbird (*rare in Adirondacks*)
Gray Catbird
Brown Thrasher
Eastern Bluebird
Yellow Warbler
Chestnut-sided Warbler
Common Yellowthroat
Bobolink
Eastern Meadowlark
Common Grackle
Northern Cardinal (*rare in Adirondacks*)
Indigo Bunting
House Finch (*southern lowlands and valleys*)
American Goldfinch
Rufous-sided Towhee (*rare in Adirondacks*)
Savannah Sparrow
Grasshopper Sparrow
Vesper Sparrow
Chipping Sparrow
Field Sparrow
Song Sparrow

DECIDUOUS WOODS

Red-shouldered Hawk
Broad-winged Hawk
Ruffed Grouse
Yellow-billed Cuckoo (*rare in Adirondacks*)
Black-billed Cuckoo
Common Screech Owl
Barred Owl
Whip-poor-will
Yellow-shafted Flicker
Hairy Woodpecker
Downy Woodpecker
Great Crested Flycatcher
Least Flycatcher
Eastern Pewee
Blue Jay
Black-capped Chickadee
Tufted Titmouse (*rare in Adirondacks*)
White-breasted Nuthatch
Wood Thrush
Yellow-throated Vireo

DECIDUOUS WOODS (*Cont.*)

Red-eyed Vireo
Warbling Vireo
Ovenbird
American Redstart
Baltimore Oriole
Scarlet Tanager
Rose-breasted Grosbeak

On the plains south of Lake Ontario and east of Lake Erie, and more locally in central New York, Golden-winged Warblers and Blue-winged Warblers regularly breed in shrubby second growth and occasionally in swampy thickets. In these same parts of the state several species of southern affinities nest: the Red-bellied Woodpecker, Carolina Wren, Blue-gray Gnatcatcher, Cerulean Warbler, and Hooded Warbler in deciduous woodlands that are often mixed with white pine and hemlock; the Prothonotary Warbler and Louisiana Waterthrush in swampy woodlands near streams.

In central and western New York many marshes and wooded swamps provide some of the outstanding bird-finding areas in the state. The larger marshes, when they support cattails, bulrushes, and sedges, usually contain the following birds:

Pied-billed Grebe
American Bittern
Least Bittern
Virginia Rail
Sora
Common Gallinule
American Coot
Black Tern
Marsh Wren
Red-winged Blackbird
Swamp Sparrow

Swamps with a growth of deciduous trees, such as red maple, slippery elm, and black ash mixed with clumps of cedar and tamarack, are particularly fascinating because of the unexpected combination of species, of both northern and southern affinities—for example, the Brown Creeper, Magnolia Warbler, Black-throated Green Warbler, and Canada Warbler, together with the Prothonotary Warbler and Cerulean Warbler.

The principal migration routes in upstate New York lie along the Great Lakes and the Lake Champlain–Hudson River Valleys. The Lake Erie shore is an exciting place from late March to June, when many hawks move northeastward parallel with it, when shorebirds gather on its beaches, and when small landbirds pass in waves through its bordering woodlands and thickets. Certain spots on the Ontario shore are equally good. Fall migration is less exciting and is better observed on the north shore of Lake Erie in Ontario. The Niagara River and the harbors at Buffalo are favorite feeding and resting areas for transient waterbirds and waterfowl in both spring and fall.

Although the Lake Champlain–Hudson River Valley is used by many birds, there are no known places for impressive concentrations in eastern upstate New York. In western New York the Iroquois and Montezuma National Wildlife Refuges embrace wetlands where vast numbers of transient waterbirds, waterfowl, and shorebirds gather in spring and fall. Peak flights may be expected in upstate New York within the following dates:

Waterfowl: 25 March–20 April; 5 October–15 November
Shorebirds: 1 May–1 June; 1 August–25 September
Landbirds: 20 April–25 May; 20 August–10 October

In winter, rural and suburban localities are visited by Evening Grosbeaks, Pine Siskins, and American Tree Sparrows. Occasionally, in rural areas, Northern Shrikes, Pine Grosbeaks, and Common Redpolls appear in low trees and shrubs along roadsides, and (Northern) Horned Larks, Lapland Longspurs, and Snow Buntings in large flocks on open farmlands. The harbors at Buffalo and the Niagara River, which seldom freeze over, are worth field trips any time from November to April, as there are always good chances of identifying rare ducks and gulls in the wintering population of waterbirds and waterfowl.

Long Island, stretching about 125 miles eastward into the Atlantic from New York Harbor, has a coastline varied by bluffs, sand beaches and dunes, mud flats, and salt marshes. Among breeding birds of the salt marshes are Clapper Rails and both Sharp-tailed and Seaside Sparrows. The waters of several bays and canals, together with the offshore waters of Long Island Sound to the north and the open Atlantic to the east and south, attract large numbers of waterbirds and waterfowl in spring, fall, and winter.

Since Long Island is separated ecologically as well as geographically from mainland New York and to a large extent belongs within the metropolitan embrace of New York City, its bird-finding opportunities are presented along with those for New York City in the chapter, **New York City Area.**

Authorities for New York State and New York City Area

Dorothy W. Ackley, Arthur A. Allen, Irwin M. Alperin, G. Malcolm Andrews, Thomas G. Appel, Robert Augustine, Harold H. Axtell, Egbert Bagg IV, Guy Bartlett, Paul Bauer, John B. Belknap, Ned Boyajian, George W. Brack, John Bull, Geoffrey Carleton, Richard Chamberlain, Greenleaf T. Chase, Howard Cleaves, Allan D. Cruickshank, Emily Curtis, Harry N. Darrow, Robert W. Darrow, Thomas H. Davis, Jr., Scott Dearoff, Robert Deed, John C. Dye, Stephen W. Eaton, Eugene Eisenmann, Walter Elwood, Richard B. Fischer, Michael Gouchfeld, Clyde Gordon, Douglas Heilbrun, John Kieran, Allan S. Klonick, Roy Latham, John H. Mayer, Christopher K. McKeever, Gordon M. Meade, Harold D. Mitchell, John Morse, Eugene Mudge, William Norse, John C. Orth, Robert F. Perry, Theodore Pettit, Dennis Puleston, Merton Radway, Gilbert Raynor, Anita Raynsford, Charles H. Rogers, George A. Rose, Marge S. Rusk, Richard Ryan, James Savage, Francis G. Scheider, Frank R. Schetty, Mary Sheffield, Robert Sheffield, Christian G. Spies, Jr., Sally Hoyt Spofford, Rudolph H. Stone, Paul Stoutenburgh, Miriam C. Stryker, Edward D. Treacy, Guy Tudor, Paul S. Twitchell, William C. Vaughan, Otis Waterman, Audrey Wrede, Sam Yeaton.

References

Birds of New York State. By John Bull. Garden City, N.Y.: Doubleday/Natural History Press, 1974.

Enjoying Birds in Upstate New York: An Aid to Recognizing, Watching, Finding, and Attracting Birds in New York State North of Orange and Putnam Counties. By Olin Sewall Pettingill, Jr., and Sally Hoyt Spofford. 3d ed. Ithaca: Laboratory of Ornithology, Cornell University, 1971.

Birds of the Niagara Frontier Region. An Annotated Check-list. By Clark S. Beardslee and Harold D. Mitchell. Buffalo Society of Natural Sciences, Buffalo, 1965. Available from the Buffalo Museum of Science, Humboldt Park, Buffalo, N.Y. 14211.

ALBANY
John Boyd Thacher Park | Meadowdale Area

John Boyd Thacher Park (1,137 acres), southwest of Albany, encompasses the highest and most picturesque part of the Helderberg Escarpment on the northeastern edge of the Allegheny Plateau in eastern New York State. From the rim of the Escarpment, at approximately 700 feet elevation, one has a sweeping view of the mosaic of farms and woodlots in the Mohawk Valley, and of Albany and Schnectady in the distance. Except for the exposed cliffs, the slopes are heavily forested with hardwoods, mixed with stands of hemlock in the shaded areas especially near the talus at the base of the cliffs. In the woods on the lower slopes, nesting birds include the Worm-eating and Golden-winged Warblers; on the higher slopes, the Hermit Thrush, Blackburnian Warbler, White-throated Sparrow, and other species of northern affinities. Red-tailed Hawks, Red-shouldered Hawks, Pileated Woodpeckers, and Ruffed Grouse nest throughout.

To reach the Park entrance, drive west from Albany on US 20 to just beyond Guilderland, turn left on State 146 to Altamount, right on State 156 for 3 miles, and left on State 157. A path, Old Indian Ladder Trail, descends from this entrance for 2 miles through good places for birds to the base of the cliffs and State 156. To reach the Park by ascending the cliffs on foot, drive left at Altamount on State 156 for 3 miles to where a small stream, Indian Ladder Creek, crosses under the road. Park just before the culvert by a minor road leading to the right. Traill's Flycatchers and Louisiana Waterthrushes nest in the woods near the Creek; American Bitterns, Common Snipes, and Sedge Wrens nest in the adjacent sedge-choked marsh. Old Indian Ladder Trail leaves the minor road on the right at a right angle, and within 0.5 mile forks right. Do not cross the stream.

The **Meadowdale Area,** consisting of open, rolling farmlands with orchards, woodlands, and marshes, lies just north of the Helderberg Escarpment and State 156 and is bounded roughly by Altamount on the west, Guilderland Center and State 146 on the north, and Voorheesville and the West Shore Railroad on the east. Two roads run north and south through the Area; railroad tracks cross the Voorheesville Marsh on the eastern end.

Bird finding is good throughout the Meadowdale Area. The following trips are merely suggestions:

1. Drive east from Altamount on State 156 for 3 miles to Indian Ladder Creek crossing (*see above*). Continue on State 156 for 0.75 mile, turn left on Meadowdale Road for 0.5 mile, and park by a bridge across Black Creek. Search the boggy area for American

Bitterns, ducks, Red-shoulderd Hawks, Ruffed Grouse, Pileated Woodpeckers, Traill's Flycatchers, and Swamp Sparrows, which nest in the vicinity.

2. Drive south from Guilderland Center on the Altamount-Voorheesville Road for 1.3 miles, turn right on Hennessey Road. The top of the first hill is a fine lookout for migrating hawks. Continue on Hennessey Road across the railroad tracks, park the car, and walk east on the tracks above the Voorheesville marshes for 1.0 mile. A sluggish stream parallels the tracks for about 0.25 mile. Beyond are thickets and then open water on the right and ponds and pastures on the left. Marsh Wrens are abundant in the second half-mile. Other birds nesting in the area include the Green Heron, Least Bittern, American Bittern, Northern Harrier, Virginia Rail, Sora, Common Gallinule, Common Snipe, and Swamp Sparrow. Continue on Hennessey Road, turn right on Tygert Road, and right again on State 156.

BATAVIA

Iroquois National Wildlife Refuge | Oak Orchard Game Management Area | Towanda Game Management Area

The **Iroquois National Wildlife Refuge** (10,784 acres) in western New York State was established principally for migrating waterfowl and consists of marshland, swamp woodland, and wet meadows—flooded each spring by Oak Orchard Creek—and croplands and pastures. Canada Geese, Mallards, American Black Ducks, and Common Pintails stop here by the thousands in spring migration from late March through April. Shorebirds appear in April. Nesting ducks include those listed above plus Green-winged and Blue-winged Teal, Northern Shovelers, Wood Ducks, and Hooded Mergansers. There is a Great Blue Heron colony in the wetter section; Ruffed Grouse and Ring-necked Pheasants are common in drier areas. To reach headquarters, drive north and west from Batavia on State 63 for 13 miles.

Adjacent to the Refuge on the east and separated from it by Knowlesville Road is the New York State **Oak Orchard Game Man-**

agement Area (2,500 acres). To reach headquarters, drive north from Batavia on State 63 to Wheatville and turn right on Knowlesville Road for 2 miles. One may enter the Area, except during the hunting and trapping season, by obtaining permission from the New York State Department of Environmental Conservation (Scottsville, New York 14546), but the best bird finding is often from adjacent roads.

Drive north from headquarters on Knowlesville Road to Oak Orchard Creek crossing and look for Wood Ducks, Traill's Flycatchers, Prothonotary Warblers, and Northern Waterthrushes. Inspect the large marsh north of the Creek for Pied-billed Grebes, American and Least Bitterns, American Black Ducks, Common Gallinules, Black Terns, and Marsh Wrens, and the deciduous woods bordering the Creek for Scarlet Tanagers, Rose-breasted Grosbeaks, and many breeding warblers—for example, the Golden-winged, Cerulean, and Hooded. Continue north on Knowlesville Road, then right on Podunk Road for 2 miles to its end, and left into a parking area overlooking Stafford's Pond. From the time the ice goes out (usually by late March) until late April, thousands of Canada Geese and, sometimes, a few Snow Geese and Whistling Swans gather here. Later in spring, shorebirds appear on the mud flats around the Pond.

Bordering Iroquois Refuge on the southwest and separated from it by State 77 is the New York State **Towanda Game Management Area.** Drive north and west from Batavia on State 63 to Alabama and turn left on State 77. The best bird finding is along this road.

BINGHAMTON

Glenwood Cemetery | Broome County Airport | Chenango Valley State Park | Newark Valley State Land

The brush and conifers in the old **Glenwood Cemetery** (10 acres) make it one of the best places for the warbler migration in this south-central New York State city. Red Crossbills have been here in May, as well as Yellow-bellied Flycatchers, thrushes, kinglets, vireos, and Scarlet Tanagers. To reach the Cemetery, drive north on State 12, turn left on Prospect Street, and right on Glenwood Avenue.

The fields around **Broome County Airport** are good for open-country birds—Killdeers, (Prairie) Horned Larks, Eastern Meadowlarks, and Savannah Sparrows in summer; occasionally Upland Sandpipers, Lesser Golden Plovers, and Water Pipits in fall migration. Drive north from Binghamton on I 81 to Glen Castle, turn left and follow the signs.

A woods, partly deciduous and partly reforested with conifers, and two lakes make bird finding good throughout the year in **Chenango Valley State Park** (902 acres). To reach the Park, drive north from Binghamton on State 7 for 8 miles and turn left on State 369. Within the Park, two of the better walks are: (1) North from the former fish hatchery on a road along the Chenango River; (2) North from Cabin 12 in the cabin colony along a woodland trail. Watch for Blue-gray Gnatcatchers in the oaks. In winter, Mallards, American Black Ducks, Belted Kingfishers, Pileated Woodpeckers, White-breasted and Red-breasted Nuthatches, Northern Cardinals, Purple Finches, Pine Grosbeaks, Pine Siskins, American Goldfinches, and American Tree Sparrows are present. There may also be Winter Wrens, White-throated Sparrows, and Song Sparrows. The spring migration of warblers and thrushes is impressive. In fall, geese and ducks frequent the two lakes. In summer, the Wood Thrush, Veery, Black-throated Green Warbler, Northern Waterthrush, Canada Warbler, Scarlet Tanager, Rose-breasted Grosbeak, and Swamp Sparrow are residents.

The **Newark Valley State Land** (1,000 acres), reforested with deciduous and coniferous trees, often has Pine Grosbeaks in winter, and warblers and thrushes in spring migration. To reach the tract, drive west from Binghamton on State 17 for 9 miles, north on State 26 for 5 miles, west on State 38B for 2.8 miles, south for 2.3 miles to Oakley Corners (no settlement), and left for 0.5 mile to the parking area.

BUFFALO

South Harbor | Tifft Street | North Harbor | Bird Island | Forest Lawn Cemetery | Delaware Park | Niagara Loop | North Shore of Lake Erie | Beaver Meadow Refuge

This great city sprawling along the eastern end of Lake Erie and the east bank of the Niagara River offers splendid opportunities for the bird finder at any season of the year. Within easy reach are the fine beaches of both Lake Erie and Lake Ontario, for migrating shorebirds; the Niagara River, which almost never freezes, for transient and wintering waterbirds and waterfowl; and scattered marshes, fields, and woodlands for a diversity of nesting birds.

For the bird finder who prefers to watch birds from his car, Buffalo has several places. The first is **South Harbor.** Drive south from downtown Buffalo on State 5 (Fuhrmann Boulevard) over the Skyway Bridge and, immediately after crossing the bridge, turn right and proceed past the dock area to scan the waters of South Harbor for waterfowl, and the breakwalls for gulls any time of the year, Double-crested Cormorants in spring and fall, and Snowy Owls in winter.

Return to State 5 and continue south, stopping in the parking spaces to look over South Harbor. As soon as the Harbor is free of ice, usually before the end of March, ducks appear and build up to hundreds, sometimes thousands, of individuals. In late fall, before the Harbor freezes over, a few ducks are sometimes present. Thoughout winter look over the gull aggregations, mostly of Herring and Ring-billed, for Greater Black-backed, Glaucous, or Iceland Gulls. During migration, inspect the flocks of Bonaparte's Gulls for an occasional Little or Franklin's Gull, a very rare Black-legged Kittiwake, or an equally rare Laughing, Black-headed, or Sabine's Gull.

Near the end of South Harbor turn left on **Tifft Street,** an old willow-bordered road that crosses about 100 acres of willows and marsh surrounded by factories and railyards. In summer this is the only vegetation seen for several miles. Drive about a half-mile, turn south on a minor road, and park. Walk east on Tifft Street a few hundred feet to a viaduct, a good vantage point for inspecting the mud flats to the north for gulls and shorebirds. Then walk east and north on a road under the viaduct to a marsh, a breeding place for Black Terns, Virginia Rails, Soras, Common Gallinules, Marsh Wrens, and Swamp Sparrows.

To reach **North Harbor,** drive north on I.190 to Porter Avenue, west on Porter Avenue to its end, and turn left on Amvet Boulevard to LaSalle Park near the Water Works. From here one may scan the

Ship Canal and Harbor, the inner and outer breakwalls, and the Niagara River for migrating or wintering gulls and waterfowl and for shorebirds in spring, summer, and fall. On the sand spit at the west end of the outer breakwall Common Terns nest from late May to mid-July.

For **Bird Island,** drive north from downtown Buffalo on State 266 (Niagara Street) for 2.3 miles, turn left on West Ferry Street, cross the Ship Canal, and turn right onto the Island. On the shore of the Niagara River, beyond the disposal plant, is a cove where, from November to April, when the River is ice-free, one can see a few Common Goldeneyes, Buffleheads, Oldsquaws, and Common and Red-breasted Mergansers. From April to early May and from September to December, there are often hundreds of Bonaparte's Gulls along the Niagara River, and sometimes such unusual gulls as those mentioned for South Harbor.

Good places for transient landbirds are **Forest Lawn Cemetery** (2,670 acres) and **Delaware Park** (350 acres), both about 2 miles from downtown Buffalo on State 384 (Delaware Avenue). From mid-April to late May and from the last week of August to October, wave after wave of kinglets, vireos, and warblers pass among the shrubs and trees of Delaware Park. In April and October, various species of wild ducks, Herring Gulls, Bonaparte's Gulls, and such unexpected birds as Common Loons, Red-throated Loons, Red-necked Grebes, and Horned Grebes often join the tame Mallards on the lake in the Park. In winter a feeding station on the south side of this lake attracts woodpeckers, chickadees, nuthatches, Brown Creepers, and fringillids. At this same time of year a few Common Screech Owls, Great Horned Owls, Long-eared Owls, and Saw-whet Owls occasionally seek daytime shelter in the dense cover at the end of the Cemetery.

Of the trips afield, the **Niagara Loop,** a drive of approximately 60 miles combining bird finding with sightseeing, is especially good for transient and wintering waterfowl. From Buffalo take the Peace Bridge to Canada and turn right on Niagara River Boulevard, which follows the River closely to Niagara Falls. From the parking areas beside the road there are excellent opportunities in fall, winter, and spring for seeing loons, grebes, gulls, terns, and many species of waterfowl including Oldsquaws. From mid-March to mid-April, Whistling Swans usually stop on this part of the River during their northward flight.

In the Niagara Gorge below the Falls, look over the transient and wintering gull aggregations for 'white-winged' species, also for Black-legged Kittiwakes and Franklin's, Sabine's, and Little Gulls. From the Maid-of-the-Mist landing, during fall and winter before the ice bridge forms, one may have close views of Red-necked and Horned Grebes and waterfowl, particularly all three scoters and all three mergansers. In early fall, phalaropes are occasionally present. Sometimes there are Harlequin Ducks in the Gorge, but they are more often in the rapids above the Falls.

Cross the River on Rainbow Bridge to the American side, proceed east along the River on Robert Moses Parkway to I 190, and cross the bridge to Grand Island.

Grand Island, 27 miles in circumference, has roads all around the shore and through the interior. Leave I 190 at the first exit, N20, and drive east on East River Road. Just beyond Wood Creek, turn left into a parking lot beside the mouth of the Creek. The 'Sunken Island' area offshore, sometimes covered with water, is a good place for migrating waterfowl. A footbridge crosses Burntship Creek to Buckthorn Island State Park and paths lead over a mile to a causeway, a vantage point for waterfowl and shorebirds. Return to I 190, cross it, and take West River Parkway to Beaver Island State Park, the extreme southern point of Grand Island and excellent for viewing waterfowl. About two-thirds of the way south on West River Parkway, turn east on Love Road for about a mile to the well-marked entrance to the Buffalo Ornithological Society's 46-acre wildlife refuge, an attractive area, partly wooded, with an artificial pond.

To complete the Loop, drive north from Beaver Island State Park on South Parkway to I 190 and turn south on it to Buffalo.

Buffalo ornithologists consider the **North Shore of Lake Erie** in Canada the best place for transient shorebirds. Migration is at its height for various species from early May through the first week in June and from the last week of July through August. Cross the Niagara River on the Peace Bridge and turn left on Route 3C. Watch the lawns between the highway and the shore for gull and shorebird concentrations. After passing Old Fort Erie and before reaching the first house, park the car on the left and inspect the shore here and to the west for shorebirds, gulls, and terns. Continue westward on Route 3C, making trips to the shore on intersecting roads that go, in the following order, to Erie Beach, Crescent Beach (Kraft Road), Buffalo Road, Windmill Point, Yacht Harbor, Crystal Beach (east end), Point Albino (special permission needed in summer), Sherkston, Sugar Loaf, Morgan's Point, and Rockhouse Point (Camp Kvusta road), often the best shorebird area.

At Erie Beach, the remains of an amusement park, now largely overgrown with weeds and shrubs but with some large oaks and other trees, is an excellent place for transient landbirds, particularly from late August to early October. Trips in May are often rewarding. Opposite Rockhouse Point lies Mohawk Island (4 acres) with a colony of Herring Gulls, Ring-billed Gulls, and Common Terns. Nesting of the gulls is well under way in late May, terns usually in late June and early July. To reach Mohawk Island, continue to Dunnville, cross the Grand River, turn south to Port Maitland, and hire a powerboat.

Beaver Meadow Refuge (240 acres), about 35 miles southeast of Buffalo, is owned by the Buffalo Audubon Society. The tract includes a beaver pond and a rich, wet woodland of hemlock, birch, and many other trees well scattered. Wood Ducks and occasionally Hooded Mergansers breed here. Other nesting species include the Red-shouldered Hawk, Ruffed Grouse, Veery, Northern Waterthrush, Mourning, Hooded, and Canada Warblers, and Scarlet Tanager. To reach the Refuge, take State 78 southeast from I 90 to Java Village and drive east on Welch Road for 2.5 miles.

IRVING

Cattaraugus Creek | William P. Alexander Sanctuary

In spring and early summer there is no better place for variety of birds than Cattaraugus Creek Valley in western New York State. Nor is there a better place to observe hawk flights in spring than the mouth of **Cattaraugus Creek** and the high land just behind it. Leave I 90 at Exit 58, turn right on State 5, cross the Creek, turn left and again left beyond the underpass, and take the road to the Creek's mouth on the northeast side, where spectacular flights of Sharp-shinned and Cooper's Hawks sometimes take place. Return to State 5, drive west for about a mile, and turn south on Alleghany Road, which leads to higher ground and a fine vantage point for flights of Broad-winged Hawks.

Flights are heaviest when there are southwest winds. The first movement, dominated by Red-tailed and Red-shouldered Hawks, with fewer Northern Harriers, Cooper's Hawks, Bald Eagles, and Peregrine Falcons, reaches a peak at about 20 March. The second movement, with Broad-winged Hawks in largest numbers, reaches a peak about the end of April. A final, somewhat anticlimactic movement, dominated by Sharp-shinned Hawks, occurs in May with a peak between the 5th and 20th.

Farther up Cattaraugus Creek, southeast of Irving, outside the Indian Reservation and just across the road from Gowanda State Homeopathic Hospital, is a forested area of deciduous trees and shrubs interspersed with hemlock that is attractive in late spring and summer to such deciduous-forest birds as the Red-shouldered Hawk, Yellow-billed Cuckoo, Barred Owl, Great Crested Flycatcher, Blue Jay, Wood Thrush, Veery, Yellow-throated Vireo, Red-eyed Vireo, Ovenbird, Hooded Warbler, American Redstart, and Rose-breasted Grosbeak, as well as such coniferous-forest birds as the Magnolia, Black-throated Green, Blackburnian, Mourning, and Canada Warblers, and Slate-colored Junco. To reach this area, drive southeast from Irving toward Gowanda on State 438, which follows the Creek.

Close by this forest and in the Zoar Valley is the **William P. Alexander Sanctuary** (130 acres), owned by the Nature Sanctuary Society of Western New York. The entrance, between Gowanda and Springville, is marked by a sign. Birds regularly nesting here include the Great Horned Owl, Pileated Woodpecker, Solitary Vireo, Black-and-white, Golden-winged, Blue-winged, Black-throated Blue, Black-throated Green, Blackburnian, Chestnut-sided, and Canada Warblers, American Redstart, and Slate-colored Junco.

ITHACA

Sapsucker Woods Sanctuary | Tompkins County Airport | Stewart Park | Michigan Hollow Michigan Creek | North Spencer Marsh | Connecticut Hill Game Management Area | Taughannock State Park

The campus of Cornell University overlooks Ithaca from East Hill at the southern end of Cayuga Lake, one of the Finger Lakes in west-central upstate New York. Northeast of the campus is **Sapsucker Woods Sanctuary** (180 acres), a cool woodland of maples, birches, and beeches, with some hemlocks and white pines, and the setting for the Cornell Laboratory of Ornithology. To reach it, drive north from Ithaca on State 13, turn right on Warren Road, left on Hanshaw Road, and left again on Sapsucker Woods Road, which crosses the Sanctuary.

The Laboratory, on a 10-acre pond at the north end of the Sanctuary, is open to the public. From its observatory windows one may view, in comfort at any season, waterbirds and waterfowl, and other birds at the feeders. In winter the pond is kept partly open by an agitator and supplied daily with great quantities of grain, and thus attracts many ducks, chiefly Mallards and American Black Ducks, but occasionally Common Pintails, American Wigeons, and Hooded Mergansers. Nesting in Sapsucker Woods are such birds as Red-shouldered Hawks, Ruffed Grouse, Great Horned Owls, Pileated Woodpeckers, Yellow-bellied Sapsuckers, Brown Creepers, Northern Waterthrushes, Canada Warblers, and Scarlet Tanagers.

Each spring both American Woodcock and Common Snipes perform courtship flights over fields north of the **Tompkins County Airport.** Nesting in these same fields, later in the season, are several species of sparrows—Savannah, Grasshopper, Henslow's, Vesper, and Field. To reach the Airport, drive north from Ithaca on State 13 and turn left on Warren road.

Stewart Park, a city park at the southern end of Cayuga Lake, is partly forested with sycamores, willows, maples, and a thick undergrowth. Near the Lake is a small pond, the Louis Agassiz Fuertes Waterfowl Sanctuary. Yellow-billed Cuckoos, Blue-gray Gnatcatchers, Tufted Titmice, Cerulean Warblers, Northern Cardinals, Rose-breasted Grosbeaks, and Indigo Buntings are present in

BROWN CREEPER

the Park each spring. Pileated Woodpeckers are found regularly, and in some years Great Horned Owls nest close to the path. Over the weed beds offshore, during fall and in winter when the Lake remains open, small rafts of ducks gather—Redheads, Ring-necked Ducks, Canvasbacks, Greater Scaups, Buffleheads, and mergansers, together with most of the common surface-feeding ducks. By mid-August, if the water is low, shorebirds feed on the sand bars and mud flats near shore. To reach the Park, drive north on State 13 a short distance and watch for the turnoff sign.

South of Ithaca, **Michigan Hollow,** a wooded swamp on the divide between the Susquehanna River and Cayuga Lake drainage systems, cradles some beaver ponds, the headwaters of **Michigan Creek,** which meanders southward through **North Spencer Marsh.** In Michigan Hollow many deciduous-forest species breed together with such northern coniferous-forest species as the Yellow-bellied Sapsucker, Brown Creeper, Winter Wren, Hermit Thrush, Solitary Vireo, Nashville Warbler, Magnolia Warbler, Black-throated Blue Warbler, Blackburnian Warbler, Northern Waterthrush, Mourning Warbler, and Slate-colored Junco. Wood Ducks nest in the vicinity of the beaver ponds along Michigan Creek. Nearly all breeding species of birds common to marshes in upstate New York are represented in North Spencer Marsh.

To reach this area, drive south from Ithaca on State 96 for 15 miles to North Spencer. The Marsh is on the east side of this road. At North Spencer, turn left on Michigan Hollow Road and drive north for 5 miles along the east side of the Marsh and Creek and through Michigan Hollow to the beaver ponds.

Also south of Ithaca is **Connecticut Hill Game Management Area,** a tract of abandoned farmlands and mixed deciduous-coniferous forest managed for deer and Ruffed Grouse. The conifers at high elevations, nearly 2,100 feet in places, attract winter finches—Purple Finches, Pine Grosbeaks, Common Redpolls, and Pine Siskins. To reach the Area, drive south from Ithaca on State 13 to Newfield. Several roads north and south of this village lead west to Connecticut Hill. The best, in May or June, is Carter Creek Road, 4 miles south of Newfield, which follows Carter Creek past headquarters.

Look and listen for Wood and Hermit Thrushes, Black-throated Blue, Black-throated Green, Chestnut-sided, and Canada Warblers, Louisiana Waterthrushes, and Slate-colored Juncos. Beyond headquarters a Ruffed Grouse or Wild Turkey may appear along the side of the road. The only local breeding area of Prairie Warblers is along the edge of the spruce and pine plantation on the west side of Connecticut Hill. Red-breasted Nuthatches and Myrtle Warblers nest in the conifers; and, in the adjacent brushy fields, Brown Thrashers, Yellow-breasted Chats, and Rufous-sided Towhees nest. Here, too, American Woodcock perform their flight songs in spring.

About 10 miles north of Ithaca on State 89, **Taughannock State Park** (535 acres) includes a section of Cayuga Lake shore and a deep gorge at the head of which is Taughannock Falls, with a drop of 215 feet. The perpetual shadows and moisture on the south side of the gorge attract such northern coniferous-forest species as Winter Wrens, Canada Warblers, and Slate-colored Juncos, and, oddly enough, such more southern species as Rough-winged Swallows and Louisiana Waterthrushes.

Reference

Birding in the Cayuga Lake Basin. Edited by Mildred C. Comar, Douglas P. Kibbe, and Dorothy W. McIlroy. Ithaca: Laboratory of Ornithology, Cornell University, 1974.

KINGSTON
Slide Mountain

Slide Mountain (4,204 feet), the highest peak in the Catskills, has mixed hardwoods on the lower slopes, and hardwoods mingled with conifers above, but without a timber line. Breeding species typical of the more northern forests include the Ruffed Grouse, Pileated Woodpecker, Yellow-bellied Sapsucker, Yellow-bellied Flycatcher, both Red-breasted and White-breasted Nuthatches, and all the thrushes, except the Wood Thrush. There are warblers, the Northern Parula, Black-throated Blue, Myrtle, Black-throated Green, Blackburnian, Blackpoll, Mourning, and Canada being the most common. Near the summit, Winter Wrens, Slate-colored Juncos, and White-throated Sparrows nest.

Slide Mountain is an easy climb on a well-marked trail. There is a lean-to on the summit. To reach Slide Mountain from Kingston, drive west from I 87 at Exit 19 on State 28 to Big Indian, turn south on Oliverea Road, and proceed for 12 miles to Winisook Lodge, where the trail begins.

LAKE PLACID and the ADIRONDACKS
Chubb River Swamp | Mt. McIntyre | Heart Lake | Marcy Dam and Marcy Lake | Mt. Golden | Avalanche Pass and Avalanche Lake | Whiteface Mountain | Elk Lake

An excellent place for an assortment of birds characteristic of the northern coniferous forest in the Adirondacks is the **Chubb River Swamp.** Species regular in the breeding season include the Black-backed Three-toed Woodpecker, Yellow-bellied Flycatcher, Olive-sided Flycatcher, Boreal Chickadee, Red-breasted Nuthatch, Brown Creeper, Winter Wren, Hermit Thrush, Swainson's Thrush, Golden-crowned Kinglet, Northern Parula Warbler, Myrtle Warbler, Blackburnian Warbler, Slate-colored Junco, and White-throated Sparrow. Species seen occasionally are the Northern Goshawk, Spruce Grouse, Northern Three-toed Woodpecker, Gray Jay, Tennessee Warbler, Bay-breasted Warbler, Rusty Blackbird, Pine

Siskin, Red Crossbill, White-winged Crossbill, and Lincoln's Sparrow.

To reach the Swamp, drive southwest from the Lake Placid railroad station on the Averyville Road for 1.7 miles to the Chubb River, a tumbling stream. Park the car and walk upstream on a wood road through the first swamp and up a slight grade to the beginning of the second, nearly a mile from the car.

A climb up **Mt. McIntyre** (5,112 feet) in late spring and summer shows the changes in birdlife from the mixed stands of conifers and hardwoods of the lower slopes—habitat of Pileated Woodpeckers, Red-breasted Nuthatches, Hermit Thrushes, Swainson's Thrushes, Nashville Warblers, Magnolia Warblers, and Purple Finches—to the stunted spruces and firs near the timber line—habitat of Yellow-bellied Flycatchers, Boreal Chickadees, Winter Wrens, Golden-crowned Kinglets, Blackpoll Warblers, and (Bicknell's) Gray-cheeked Thrushes. The Gray-cheeked Thrushes sing during June and early July in early morning and in late afternoon until shortly after sunset.

To reach the trail up Mt. McIntyre, drive south from Lake Placid on State 73 to Elba, then right for 5 miles to Adirondack Lodge on Heart Lake (2,179 feet elevation). Park here and take the trail for 3.2 miles to the summit.

Some of the common summer-resident birds in the typical Adirondack forest adjacent to **Heart Lake** (*see above*) include the Least Flycatcher, Red-breasted Nuthatch, Wood Thrush, Veery, numerous warblers such as the Black-and-white, Tennessee, Nashville, and Black-throated Blue, and the Purple Finch. Bird finders who spend the night at the Lodge or at one of the campsites may hear a Great Horned Owl, Barred Owl, or Whip-poor-will.

From Heart Lake, also, an easy 2-mile trail leads south to **Marcy Dam** and **Marcy Lake** (2,366 feet elevation). In late spring and early summer along this trail one should see or hear the Yellow-bellied Sapsucker, Olive-sided Flycatcher, Solitary Vireo, Canada Warbler, Scarlet Tanager, and Rose-breasted Grosbeak. Around Marcy Lake one may find breeding Yellow-bellied Flycatchers, Philadelphia Vireos, and Mourning Warblers.

From Marcy Lake a rugged 3-mile trail goes up **Mt. Golden** (4,714 feet). At Arnold Lake, along the way, Black-backed Three-

NORTHERN GOSHAWK

toed Woodpeckers may be nesting and, farther on, one may hear a (Bicknell's) Gray-cheeked Thrush.

Another less-rugged trail from Marcy Lake leads for 2 miles to **Avalanche Pass** and **Avalanche Lake** (2,863 feet elevation). The partly wooded stretch just before reaching the Lake is promising for summer-resident Yellow-bellied Flycatchers, Solitary Vireos, Philadelphia Vireos, and Chestnut-sided, Blackpoll, and Canada Warblers.

An easy way to the higher reaches of the Adirondacks and their birds is to drive north from Lake Placid on State 86 to Wilmington and take a toll road that winds for 8 miles upward over a series of switchbacks nearly to the summit of **Whiteface Mountain** (4,868 feet). The road, Memorial Highway, is open from about 15 April to 1 November.

Elk Lake is surrounded by a forest of conifers and hardwoods against a backdrop of mountains rising to over 4,000 feet. In summer, Common Loons, American Black Ducks, and Hooded and Common Mergansers frequent the Lake. Sharp-shinned and Broad-winged Hawks nest in the woods. Among species breeding regularly are the Ruffed Grouse, Great Horned Owl, Barred Owl, Pileated

Woodpecker, Yellow-bellied Sapsucker, Traill's Flycatcher, Olive-sided Flycatcher, Tree Swallow, Boreal Chickadee, Red-breasted Nuthatch, Brown Creeper, Winter Wren, Hermit Thrush, Swainson's Thrush, Golden-crowned Kinglet, Solitary Vireo, probably twenty species of warblers, Slate-colored Junco, and White-throated Sparrow. The lucky bird finder may see a Black-backed Three-toed Woodpecker, Rusty Blackbird, Pine Siskin, Red Crossbill, or White-winged Crossbill.

Although Elk Lake is not far from Lake Placid, the route east on State 73 and south on I 87 is circuitous. Leave I 87 at Exit 29, North Hudson, go west on Blue Ridge Road for about 5 miles, and turn right to the Lake at the end of the road. Elk Lake Lodge is the only development on this privately-owned property. No camping is permitted. A trail leads from the Lodge area to the summit of Mt. Marcy.

Return to Blue Ridge Road, follow it west for about 10 miles to the Boreas River, and park on the far side of the bridge, on the right near an overgrown woods road. Inspect the barn near the road for Barn and Cliff Swallows, then walk along the woods road for about a mile, on the lookout for woodpeckers, Veeries, and warblers. Watch for the Indigo Buntings. On returning to Blue Ridge Road, walk south to Cheney Pond along a trail that can sometimes be rewarding. Continue west on Blue Ridge Road, turn right on State 28N to Newcomb, and stop at the Finch-Pruyn Company's picnic area, a good vantage point for Bobolinks, Savannah Sparrows, and swallows including the Purple Martin.

OSWEGO
Mouth of the Oswego River

At the **Mouth of the Oswego River** on eastern Lake Ontario, the harbor protected by two breakwaters and the River itself are wintering grounds for thousands of ducks and gulls. The best views of the harbor are from a hill beside the United States Naval Reserve Station on the west and Old Fort Ontario on the east. In the harbor regularly are several thousand scaups, Common Goldeneyes, and Common and Red-breasted Mergansers, plus lesser numbers of

Mallards, American Black Ducks, Redheads, and Canvasbacks. Also present are White-winged Scoters and possibly a few Surf Scoters or Black Scoters, Ring-necked Ducks, Ruddy Ducks, Hooded Mergansers, and, in some years, King Eiders. Red-necked and Horned Grebes appear fairly often. Buffleheads and Oldsquaws frequent the eastern side of the harbor; Pied-billed Grebes and half-hardy ducks stay on the western side, where the discharge from a power plant keeps the water open. Occasionally, a Snowy Owl appears on a breakwater.

The breakwaters and the falls at Oswego, and south on the River, are gathering places for gulls—Greater Black-backed, Herring, and Ring-billed with smaller numbers of Glaucous and Iceland. Common Pintails and Wood Ducks appear on the River. State 48, south from Oswego, runs close to the River and the dams at Minetto and Fulton.

PULASKI
Sandy Pond Area | Selkirk Shores State Park | Derby Hill

The **Sandy Pond Area,** bordering eastern Lake Ontario, has sand dunes, a pond with an inlet from the Lake, cattail marshes, brush, woods, and fields. To reach it, drive west from Pulaski on State 13 for 3 miles, north on State 3 to Sandy Pond Corners, and left on a road that swings around the marshes at the southwest corner of the Pond. Park, walk west 0.25 mile to the Lake shore, and then north along the sand spit to the inlet.

From middle to late May, many shorebirds—Black-bellied Plovers, Ruddy Turnstones, Dunlins, Red Knots, Sanderlings—as well as thousands of gulls and terns, frequent this area, especially the inlet. In late summer one may see such uncommon migrants as Whimbrels, Baird's and Western Sandpipers, and Northern Phalaropes.

Return by a path between the Pond and the dunes. Watch for the 'Warbler Pool,' a brush-flanked basin with a path along its north side. In mid-September, with northwest winds from 15 to 20 miles an hour, sixty species of landbirds, including from sixteen to twenty species of warblers, have rested here. In summer, search the marsh for breeding birds—bitterns, Mallards, American Black Ducks, Blue-winged Teal, rails, Common Gallinules, and Marsh and Sedge Wrens; watch the fields for Upland Sandpipers, Bobolinks, Eastern Meadowlarks, and Savannah, Grasshopper, Henslow's, and Vesper Sparrows.

South of Sandy Pond is **Selkirk Shores State Park** (647 acres). From Pulaski, drive west on State 13, turn south on State 3 for 0.5 mile, and right on a road marked 'Pine Grove.' After passing through a white pine grove where Pine Warblers nest, this road turns to parallel the brush- and tree-covered dunes between Lake Ontario and the Salmon River marshes. From the top of the dunes in spring, one has an eye-level view of the transient landbirds which sometimes gather in the trees and bushes in great numbers.

Return to State 3, continue south for 1.0 mile to Park headquarters, and obtain a map of the trails. The two main trails along the Lake are the best for bird finding. In fall, transient sparrows gather in the fields behind headquarters and in the brushy marsh south of the parking lot.

Continue south on State 3, turn west on State 104B, and west again on the first minor road, Sage Creek Drive, which leads up a hill, **Derby Hill,** one of the best places on the Lake for watching migrating hawks in spring. From the top of the hill is a view north and northwest over Lake Ontario. When the winds are south or southwest from 15 to 20 miles an hour, hawks, swallows, and blackbirds pass from west to northeast in impressive numbers. A change in the weather will stop the flight. This lookout is also excellent for migrating waterfowl in fall, especially during a northwest wind.

ROCHESTER

Durand-Eastman Park | Sea Breeze | Mouth of the Genesee River | Island Cottage Woods | Braddock Heights Sand Spit | Braddock Bay State Park | Shore Acres and Salmon Creek | Highland Park | Reed Road Swamp | Monroe County Airport

Rochester's location at the mouth of the Genesee River on the south shore of Lake Ontario is excellent for birds—waterbirds, waterfowl, shorebirds, and marsh-dwelling birds, as well as open-field and woodland species.

Northeast of the city is **Durand-Eastman Park** (794 acres); its rolling terrain covered with native deciduous trees plus extensive plantings of conifers and other introduced trees and shrubs, three ponds with an adjoining marshland, and a mile-long stretch of sand-and-rock beach provide a diversity of habitats and good bird finding at any season.

Loons, grebes, and diving ducks feed and rest offshore on Lake Ontario during fall and spring. As many as twenty-two varieties of warblers pass through the Park in migration. Hawks appear during April, moving easterly toward Irondequoit Bay. During winter the crabapple orchard south of the Park zoo is the best place for small birds such as American Robins, Cedar Waxwings, Purple Finches, both crossbills, and 'northern' grosbeaks.

To reach the Park, drive north from I 490 on State 47, the Outer Loop, and, near its terminus at Lake Ontario, watch directional signs to the Park, which is west of State 47.

Sea Breeze, at the northern end of State 47 and adjacent to Irondequoit Bay outlet, is a good vantage point for large waterfowl and gull concentrations. Over ten thousand Red-breasted Mergansers may show up here when the ice leaves in late winter.

At the **Mouth of the Genesee River,** west of Durand-Eastman Park, are two wide concrete piers extending into Lake Ontario. Broad, sandy beaches stretch east and west of the piers. One can reach East Pier (Summerville Pier) by driving west through Durand-Eastman Park on Lake Shore Boulevard, and continuing to its end near the Coast Guard Station. For West Pier (Charlotte Pier), drive south from East Pier on St. Paul Boulevard, turn right on Bridge Street, cross the Genesee River, and go north on Lake Avenue. From downtown Rochester, go north on State Street (which becomes Lake Avenue).

Both beaches and piers are good lookouts for waterbirds and waterfowl from mid-August through April.

The Lake Ontario shore, west of the Genesee River, is excellent for shorebirds in May, August, and September and offers vantage points for ducks, gulls, and terns.

From West Pier drive west on Beach Avenue to the Rochester

Gas and Electric Russell Station Steam Plant. On the south side of the road is a meandering stream with mud flats attractive to shorebirds, particularly in August and September. On the Lake side, the turbulent water near the pumping station outlets lures many ducks, gulls, and terns.

From Lake Ontario Parkway, which runs west from Lake Avenue, there are several places well worth investigating.

Island Cottage Woods. Go north from the Parkway at Long Pond. The wet, partly forested area between Buck and Long Ponds has many transient flycatchers, kinglets, thrushes, warblers, sparrows, and other birds in spring and fall.

Braddock Heights Sand Spit. Drive north from the Parkway on East Manitou Road to Braddock Heights. From the northwest side of this community, walk out on a dike (East Spit) in Braddock Bay. In the proper season the outlying waters, neighboring marshes, and sand spits are most productive. Canada Geese, Brant, Redheads, Canvasbacks, and Common Goldeneyes, as well as many shorebirds may be expected.

Braddock Bay State Park. Turn west from East Manitou Road at a directional sign to the Park (837 acres) and the main parking area from which there is an unobstructed view of Braddock Bay and the Lake. This is a fine vantage point for impressive hawk flights. From the last of March through mid-May, thousands of diurnal predators and other migrants fly within view. In late April as many as 4,000 Broad-winged Hawks may pass this point in a few hours.

Shore Acres and **Salmon Creek.** About 6 miles west of Braddock Bay, drive south from the Parkway on West Manitou Road to Salmon Creek, a good area for birds. Cross the Creek and drive north on Payne Beach Road to the Lake. Turn right on a cottage road to a parking place at the edge of Shore Acres Pond, well worth investigating for shorebirds in August and early September.

Highland Park (108 acres), with its arboretum and sheltered ravines, is superb for warbler migrations in late April, early May, and September, and for finches and the occasional thrush in winter. To reach the Park, drive south from the city center on South Avenue for about 3 miles and turn left on Reservoir Avenue.

Reed Road Swamp, owned and managed by the Genesee Ornithological Society, comprises a moist deciduous woods with heavy undergrowth. A highway through the Swamp permits easy access to its rich breeding avifauna, which includes the Ruffed Grouse, Yellow-billed and Black-billed Cuckoos, Pileated Woodpecker, Red-bellied Woodpecker, Red-headed Woodpecker, Great Crested Flycatcher, Veery, Blue-gray Gnatcatcher, Scarlet Tanager, Rose-breasted Grosbeak, and many species of warblers, as well as the Turkey Vulture, Great Horned Owl, Barred Owl, and Saw-whet Owl. To reach the Swamp, drive south from Rochester on State 383, turn right on Ballantyne Road, and left on Reed Road, which passes down a hill for about a half-mile and leads into the Swamp. A path from the west side of Reed Road penetrates the remote sections.

In nesting season the fields surrounding the **Monroe County Airport** are attractive to such open-country birds as Upland Sandpipers, Short-eared Owls, (Prairie) Horned Larks, Bobolinks, and Grasshopper Sparrows. Drive south from city center on State 383, turn right on Weidner Road, and continue until the grassy edges of the Airport are on the right.

ROME

Rome Sand Plains | Verona Beach State Park

The **Rome Sand Plains,** in central New York State, with ridges and dunes interspersed with ravines, harbor a variety of landbirds. The following trip in June and early July should be productive: Drive west on State 49 to 1.1 miles beyond its intersection with State 46, turn right on Lauther Road for about 0.5 mile to Hogback Road, park the car, and walk into the woods on a sandy road on the left. Ovenbirds and Hermit Thrushes nest here; Red-breasted Nuthatches, Brown Creepers, and Black-and-white and Myrtle Warblers are regular.

Drive east on Hogback Road for 0.4 mile and walk left on a woods road that runs about a mile to Oswego Road. The last third is along a ravine, cut by a brook, in habitat for Black-throated Green, Blackburnian, and Canada Warblers and the Northern Waterthrush. Return to Hogback Road and continue east across some dunes to Oswego Road. Turn left on Oswego Road and drive to its junction with Lauther Road on the left and Humiston Road on the right. Proceed right on Humiston Road for 0.7 mile to railroad tracks, park, and walk along the tracks, which pass through a bog. Nashville and Chestnut-sided Warblers, Common Yellowthroats, Rufous-sided Towhees, and White-throated Sparrows nest around this bog.

The beaches of **Verona Beach State Park** (726 acres) on the east end of Oneida Lake, especially the stretch north of the Barge Canal, constitute a fine area for fall transient shorebirds. Among twenty-five species recorded here are the Ruddy Turnstone, Greater and Lesser Yellowlegs, Red Knot, Short-billed Dowitcher, White-rumped and Baird's Sandpipers, and Sanderling. Species of land-birds nesting in the Park, in the area east of State 13, include Chestnut-sided and Canada Warblers, Rufous-sided Towhees, and White-throated Sparrows. To reach the Park, drive west from Rome on State 49 and south on State 13; or drive north on State 13 from Exit 34 on I 90.

SALAMANCA

Allegany State Park | Rock City Park | Allenberg Bog Refuge

In southwestern New York State, **Allegany State Park** (65,000 acres), with elevations reaching 2,500 feet, embraces vast forests of hemlock, beech, maple, and birch on shaded slopes and oak in the sunnier places. The following 60-mile loop trip through the Park and vicinity, scenic at any time of year, is most rewarding for birds in June and early July. Watch for the Northern Goshawk, Sharp-shinned Hawk, Red-shouldered Hawk, and Broad-winged Hawk, all of which breed in the area.

From State 17 in Salamanca, on the north side of the Park, take ASP 1, which soon enters the Park and goes up a hill for about a mile to an observation point overlooking the Allegheny River Valley. From here one may see hawks riding updrafts above the adjacent ridges. The road then passes through an oak woodland, a breeding habitat for the Hermit Thrush, Black-and-white Warbler, and Slate-colored Junco. Continue on ASP 1, over the top of a hill, past a fire tower where there is another good view of the Park, to the Administration Building. Stop here for a map of the Park and information on the best places to see Wild Turkeys. Look over the nearby lake for ducks.

Continue on ASP 1 for 2 miles to the South Administration Building, where a fine stand of hemlock with a rocky stream harbors summer-resident Red-breasted Nuthatches and Winter Wrens. Proceed on ASP 1 to ASP 3, follow it to the left through deciduous woods to Pennsylvania 46, and turn left to Bradford, Pennsylvania. Here, take Pennsylvania 346, Pennsylvania 646, and New York 16 to Rock City, New York.

Almost a half-mile east of Rock City is privately-owned **Rock City Park,** where one may listen to the chorus of Hermit and Swainson's Thrushes. Other birds in the immediate vicinity include at least six warblers, the Magnolia, Black-throated Blue, Black-throated Green, Blackburnian, Hooded, and Canada.

Take State 16 north to Olean and, at the first traffic light, at the bottom of the hill just before crossing the Allegheny River, turn left on a road that follows the south bank of the River until it joins State 17. This road traverses bottomland forest and open, shrubby areas attractive to Tufted Titmice, Blue-winged Warblers, Mourning Warblers, and Northern Cardinals. Return to Salamanca on State 17.

The **Allenberg Bog Refuge** (318 acres), owned by the Buffalo Audubon Society, lies northwest of Salamanca between the villages of Napoli and New Albion. To reach the Refuge, drive north on State 353 to Little Valley, turn left on State 242 to Napoli, and then north on the road to New Albion.

The area includes a sphagnum bog and many acres of rhododendron. Among nesting birds are Broad-winged Hawks, Wild Turkeys, Brown Creepers, Winter Wrens, Wood, Hermit and Swainson's Thrushes, Veeries, and Northern Waterthrushes. Because the trails in the Bog are obscure, a map with detailed directions, available from the Buffalo Museum of Science, Humboldt Park, Buffalo, is helpful.

SCHENECTADY

Karner Pine Barrens | Niskayuna Wide Waters | Vischer Ferry Game Management Area | Stony Creek Reservoir | Hudson River | Tomhannock Reservoir

Southeast of this city on the Mohawk River, urban development threatens the **Karner Pine Barrens,** about three square miles of sandy soil with low scrub oaks and scattered pitch pines. At present, ridges of scrub, sometimes separated by wet ravines, attract Northern Harriers, Brown Thrashers, Pine and Prairie Warblers, Rufous-sided Towhees, and Chipping Sparrows from May to September. To reach the Barrens, drive southeast from Schenectady on State 5, and turn right on State 155, which bisects the area. Investigate any of the winding roads on the right or left.

The **Niskayuna Wide Waters** are extensive marshes along the Mohawk River southeast of the city. Nesting here regularly are American Bitterns, Virginia Rails, Common Gallinules, Marsh Wrens, and Red-winged Blackbirds, as well as a few Least Bitterns, Mallards, American Black and Wood Ducks, and American Coots. Late summer brings Great Blue Herons, Great Egrets, and hordes of Tree and Bank Swallows. In October and April, Mallards, American Black Ducks, Common Pintails, Blue-winged Teal, and American Wigeons feed here.

Leave I 87 at Exit 6, drive west on State 7 for 0.2 mile, north on Forts Ferry Road for 2 miles, and then left on the River Road for 4.2 miles, stopping frequently to investigate the marshes on both sides of the road and to follow one of the several roads and trails that lead from the River Road across an old railroad right-of-way to the marshes and the River. Cross Lisha Kill, and when River Road turns left, continue straight on Lock 7 Road for 1.0 mile, park at the turnaround, and scan the River above and below the dam.

Return to River Road, continue west, and take a left fork on Rosendale Road for 0.8 mile. Park on the left behind a firehouse and at the entrance to Lisha Kill Natural Area (100 acres), open to the public from about 1 March to mid-April. A 0.5-mile trail leads to the main ravine through stands of large hemlocks and ravines with mixed woods; a 1.5-mile trail, with some climbing, circles the Preserve, where one may expect to see Brown Creepers and a variety of warblers and finches.

Across the Mohawk River from the Niskayuna Wide Waters is **Vischer Ferry Game Management Area** with extensive marshes and swamps and several shallow ponds. Mallards, American Black and Wood Ducks, Blue-winged Teal, and Hooded Mergansers nest; birds of prey and landbirds are common. To reach the Area, drive east from I 87 at Exit 8 for 0.5 mile, turn south on Dunsback Road to its end, west on Clam Steam Road, and make a right swing onto River View Road and across I 87. Investigate the roadside ponds and marshes, and park in the Game Management Area, 1.5 miles on the left. Walk along the roads and paths extending to the River and along the old Erie Canal. Continue west on River View Road for 2 miles. On the left, inspect Ferry Road, which dead-ends at the River. Return to River View Road, go west for 1.75 miles, then left opposite Sugar Hill Road to Lock 7. Park in the area overlooking the dam and powerhouse. West of this turnoff, River View Road intersects State 146.

Stony Creek Reservoir, relatively shallow with swamp woodland, is a good place for waterbirds and waterfowl, especially Ring-necked Ducks in the migration seasons. To reach it, drive north from Vischer Ferry Game Management Area on Van Vranken Road and park at the end. Drive east on Crescent Road for 1.0 mile, north on Moe Road for 1.0 mile, and west on Englemore Road across the Reservoir. Turn left at the end of Englemore Road and return to Vischer Ferry.

The **Hudson River,** from below the Federal Dam at Troy north to the mouth of the Mohawk River at Cohoes and the Hoosick River at Stillwater, is fine for ducks in winter. The birds—Mallards, American Black Ducks, Common Goldeneyes, and Common Mergansers predominating—congregate on the open water and ice at the Dam.

The following trip, with turnoffs to locks and dams, should be worthwhile: From the junction of US 4 and State 7 in Troy, drive north on State 4 for 0.5 mile, turn left, cross the River, and go right on George Street for 0.75 mile, stopping to look over the River from the small park below the Dam. Turn left on Tibbits Avenue to the traffic light, right on Cohoes Avenue for 1.0 mile, and right on State 32 to Stillwater. Investigate the entrance roads to all locks and dams along the way. Cross the River at Mechanicville and return to Troy by driving south on River Road.

Northeast of Troy, the **Tomhannock Reservoir** (5 miles long) is a gathering area for loons, grebes, and diving ducks in spring and fall. State 7 east of Troy crosses its southeastern tip. Watch the marsh edge on the far left bank for surface-feeding ducks. Turn left on a road that follows the northeast shore.

SENECA FALLS
Montezuma National Wildlife Refuge

One of the outstanding places in the Finger Lakes region of upstate New York for waterbirds and shorebirds as well as waterfowl and landbirds is **Montezuma National Wildlife Refuge,** with headquarters on US 20 about 5 miles northeast of Seneca Falls, near the Free Bridge over the Seneca River. Of its 6,432 acres, about 2,750 are open-water marsh with pondweeds, waterweeds, water smartweeds, and patches of bulrushes; 500 acres of cattail marsh; 1,000 of swamp timber; 1,450 of brushy lands and fallow fields; and 300 for grain production. About 10 miles of roads in the Refuge are open to bird finders, except during nesting season, when some of them are closed.

The breeding bird population includes Pied-billed Grebes, American and Least Bitterns, Northern Harriers, Canada Geese, Mallards, American Black Ducks, a few Gadwalls, Blue-winged Teal, a few Northern Shovelers, Redheads, and Ruddy Ducks, Virginia Rails, Soras, Common Gallinules, American Coots, Black Terns, and Marsh Wrens; along the dikes, Sedge Wrens and Swamp Sparrows; in the swamp timber, Wood Ducks, sometimes a few Prothonotary Warblers, Golden-winged Warblers, Cerulean Warblers, and Mourning Warblers.

Waterfowl concentrations reach their peaks in the first two weeks of April and the last two weeks of October. As many as 25,000 Canada Geese, with a few Snow Geese, often appear in one season. About 100,000 ducks have stopped by at one time, the most abundant being the Mallard, American Black Duck, Common Pintail, American Wigeon, Redhead, Ring-necked Duck, and Canvasback. A few Whistling Swans sometimes drop in in early spring, soon after the ice melts. Shorebirds visit the shallows and mud flats in late May and during August and September. Great Egrets, together with several Little Blue Herons and Snowy Egrets, may linger in the marsh during August.

SYRACUSE
Oakwood Cemetery | Morningside Cemetery | Onondaga Lake | Oneida Lake | Whiskey Hollow Road | Beaver Lake Three Rivers Game Management Area | Camillus Valley

Two of the best places in this central New York State city for transient landbirds in spring and fall are **Oakwood Cemetery** and **Morningside Cemetery.** Both have park-like stands of oak. Blackbirds and many sparrows appear in March; Brown Creepers, Hermit Thrushes, and kinglets in early April, warblers in mid-May and early September, and many sparrows in October. Weedy fields and a ravine in the southern part of Morningside Cemetery are especially good for sparrows. To reach the Cemeteries, drive south on South Salina Street, east on Colvin Street, and north on Comstock Avenue.

Despite industrial development, many shorebirds still frequent the mud flats and shallows on the west shore of **Onondaga Lake** from late July to late September. Pectoral, Least, and Semipalmated Sandpipers come almost any time during this period; the greater variety, including Ruddy Turnstones and White-rumped, Baird's, and Stilt Sandpipers are here from mid-August to mid-September.

For a tour of several vantage points, drive north on North Salina Street, turn left on Hiawatha Boulevard, and right on State 48A, following the Lake Shore. At the Solway Process trestle over State 48A, park and scan the mud flats and small pool to the southeast. Then walk along a road paralleling a waste canal. Drive on 1.2 miles to where the highway dips down to Nine Mile Creek, park at the brink of the descent, walk along a road leading to a 60-foot bluff overlooking the Lake, and scan the beach and mud flats below. In the winter, (Northern) Horned Larks, Lapland Longspurs, and Snow Buntings are common on the 'waste beds' between State 48A and the Lake, also a favorite hunting ground of Rough-legged Hawks, American Kestrels, Short-eared Owls, and, occasionally, Snowy Owls and Northern Shrikes.

During migration in the fall and early spring before the ice has gone out completely, loons, grebes, Canada Geese, ducks, and sometimes Whistling Swans gather on the southwest shore of **Oneida Lake** from Brewerton to Bridgeport. To reach this section, drive north on I 81 to the Brewerton Exit and east along the Lake on Bartell, Muskrat Bay, Beach, and Lake Shore Roads to State 31 near Bridgeport.

Whiskey Hollow Road, west of Baldwinsville, runs for about 1.5 miles through a narrow, steep-sided gorge with a contrast in flora between the north and south-facing slopes. Golden-winged, Blue-winged, Cerulean, Blackburnian, and Mourning Warblers all summer here. To reach the area, drive north on State 48 (Exit 39 from I 90) almost to Baldwinsville, turn left on State 31 for about 2 miles—until it makes a sharp turn left—and leave it here, continuing straight ahead across Dead Creek. In migration seasons, stop by the Creek to look for ducks and shorebirds. After crossing the Creek, go left on the first road, West Dead Creek Road, for 2.5 miles, and then right on Whiskey Hollow Road, stopping at every turnoff to look and listen.

In spring and fall, **Beaver Lake** is important for waterbirds and waterfowl—Common Loons, Red-necked, Horned, and Pied-billed Grebes, Canada Geese with occasional Snow Geese, possibly twenty-four species of ducks, and American Coots. To reach the Lake, drive north on State 48 to Baldwinsville, west on State 370 for 4 miles, and, just beyond East Mud Lake Road, right on Vann Road. Park overlooking the Lake.

Certain places in the **Three Rivers Game Management Area** are outstanding for fall-migrating sparrows. To reach the Area, drive north on State 48 to Baldwinsville, east on State 31 to Phillips, and north on Phillips Street, which becomes Sixty Road, to its junction with Kellogg and Potter Roads. In late August, September, and October the Area 'boils' with sparrows—Savannah, Grasshopper, Henslow's, Vesper, and Song in the fields; Chipping, Field, White-crowned, White-throated, Lincoln's, and Swamp in the alder-willow-buttonbush swamps.

The **Camillus Valley,** west of Syracuse, is excellent for summer-resident warblers. The following trip might be rewarding: Drive west from Syracuse on State 5 to Camillus, south on State 321 along the northwestern side of the Valley for about 1.5 miles and park in a highway storage area. Walk back along the road for Black-and-white, Golden-winged, Cerulean, Mourning, Hooded, and Canada Warblers and American Redstarts; also for Red-bellied Woodpeckers,

Great Crested, Traill's, and Least Flycatchers, Baltimore Orioles, Scarlet Tanagers, and Rose-breasted Grosbeaks.

This place is also good for migrating warblers in the spring. Continue on State 321 until it turns right beside a small, sharp rise called Nose Hill, park the car, and look for a Rough-winged Swallow colony on the Hill, and then walk on 0.25 mile to a small marsh. Golden-winged and Cerulean Warblers are regularly reported here, as well as Red-tailed Hawks, Great Horned Owls, and Pileated Woodpeckers. Turn south on State 174 to a culvert under a one-line railroad track. Walk west along the track for a treetop view of breeding warblers—Black-and-white, Black-throated Green, Cerulean, and Blackburnian.

UTICA

Frankfort Gorge | Ilion Gorge

In central New York State, both **Frankfort Gorge** and **Ilion Gorge,** steep-walled and wooded, are worth exploring for summer-resident birds. Stop at the frequent turnouts for parking. Look for the Louisiana Waterthrush along streams and Eastern Phoebes under bridges.

To reach the Gorges, drive southeast from Utica on State 5S. Turn right on State 171 for Frankfort Gorge and on State 51 for Ilion Gorge. At the head of Ilion Gorge, just beyond Cedarville, turn right on Swamp Road, park, and listen for Yellow-bellied Sapsuckers, Northern Waterthrushes, and White-throated Sparrows in the white cedar swamp.

WATERTOWN

Pillar Point | Perch River Marsh

During spring and fall, loons, grebes, diving ducks, and gulls gather along the shore of **Pillar Point,** a peninsula in eastern Lake Ontario with level farmland, some woods, rocky shores, and marsh and mud flats at Sherwin's Bay on its east side. In late fall there are thousands of scaups in Black River Bay on the south shore. In late fall and winter, hardy ducks remain where the Lake is not frozen. Landbirds in winter include the Rough-legged Hawk, Northern Harrier, an occasional Snowy Owl, (Northern) Horned Lark, Water Pipit, Lapland Longspur, and Snow Bunting. To reach the Point from Watertown, drive west on State 12F for 8 miles, turn right on State 180 to Dexter, and turn left on Lakeview Drive, following 'Pillar Point' signs to South Shore Drive, which continues along the shore.

In the **Perch River Marsh,** a game management area maintained by the New York State Department of Environmental Conservation, a low dam known as the 'Dike' has created 3,000 acres of shallow marshland with cattails and other marsh vegetation. There are two access roads: (1) Drive north on State 12 for 7 miles, then right on Parish Road for 0.75 mile, and left into a parking area. (2) Drive north on State 12 for 8 miles—to beyond the bridges—and right on Vaadi Road for one mile to the Marsh.

Below the Dike in early spring are Canada Geese and a few species of surface-feeding ducks; above the Dike are diving ducks, chiefly scaups and Buffleheads. Nesting in or near the Marsh are Pied-billed Grebes, American Bitterns, Mallards, American Black Ducks, Blue-winged Teal, Virginia Rails, Soras, Common Gallinules, American Coots, Common Snipes, and Black Terns. In late summer, herons and shorebirds frequent the scraped-out depression below the Dike. In winter, Rough-legged Hawks and Snowy Owls sometimes hunt here.

New York City Area

ROBERT S. ARBIB, JR.

PIPING PLOVER

Few large cities are so well endowed by nature and man for the study of birds as New York. Concealed in the urban sprawl are parks, marshes, forests, harbors, beaches, bays, backyards, and even garbage dumps and sewer outlets. Nowhere in the city is one more than a mile from a leafy square, a patch of shrubbery, or a waterfront.

New York City and its nearby counties have produced records for over 410 species, and the list still grows. There are many reasons for this. The climate is variable, with temperatures ranging from subzero to 100 degrees in one year. The city lies athwart the Atlantic flyway, and its shorelines, rivers, and ridges serve as flightlines for migrating birds. Its borders are water-rimmed, with its eastern arm far out in the Atlantic, bringing a constant stream of pelagic birds, as well as strays from faraway places. And lastly, New York City lies within the breeding range of many birds. In three hours by car from Times Square, the bird finder may reach the breeding areas of as many as 160 species. Some of the breeding areas are the most northern for certain southern species, and some are the most southern for certain northern species.

The variety of habitats for birds is remarkable. Some are man-made, others man-altered. Some places are almost primeval in their unspoiled beauty, and a few places in certain seasons give the illusion at least of being untouched by man. But the climax forests, fresh-water marshes, and other habitats are almost gone. Nearly all such distinctive natural environments have merged into the artificial habitat of the city.

Long Island, south of the two glacial moraine ridges that transect it from east to west, is a sandy outwash plain, flat and featureless, and originally covered either with open fields and meadows, or, on the poorer soils, with oak-pine scrub and laurel. Parts that are almost pure sand support red cedar, wild cherry, gray birch, and bayberry; often these sandy areas run down to salt marshes. Along the richer margins of the streams and in their shallow valleys a variety of deciduous trees abound—red maple, alder, willow, birch, and shadbush. North of the ridges the land is more rugged, with rolling hills, cloaked with oak and laurel, and some rather steep valleys, forested with oak, maple, beech, tulip tree, and other hardwoods.

In New York City itself, these differences exist wherever the land is unchanged, but the far more prevalent habitat is the landscaped park with both native and exotic trees and shrubs. North of the city, the wooded areas, like northern Long Island, show a predominance of maple and beech, with some birch, basswood, and white oak; in the valleys there are oak, hickory, dogwood, laurel, and willow. Throughout, suburban development means a mixture of the native and the imported: streets lined with Norway or silver maple, sycamore, or white oak; gardens with flowering trees and shrubs of a hundred different species.

Although eastern Long Island still has much farmland, in recent years a building boom has reduced this habitat substantially. Along the southern shore stretches a sand-barrier beach, with dunes, tidal flats, and miles of phragmites and spartina.

In summary, there are woodlands, evergreen groves, beaches,

tidal estuaries, salt marshes, lakes, and streams; and habitats created by man that are no less inviting to birds—fields and farms, parks and cemeteries, orchards and estates, reservoirs with encircling pines, and, to the joy of every suburbanite, the thicket in the lot next door, the glade behind the garage, and the feeder at the kitchen window.

Listed below are some of the species breeding regularly in such upland habitats as woodlands, woodland edges, groves, thickets, landscaped areas, and dooryards.

Mourning Dove
Yellow-billed Cuckoo
Black-billed Cuckoo
Common Screech Owl
Yellow-shafted Flicker
Downy Woodpecker
Eastern Kingbird
Great Crested Flycatcher
Eastern Pewee
Barn Swallow
Blue Jay
Black-capped Chickadee
Tufted Titmouse
White-breasted Nuthatch
House Wren
Carolina Wren
Northern Mockingbird
Gray Catbird
Brown Thrasher
Wood Thrush
Red-eyed Vireo
Yellow Warbler
Chestnut-sided Warbler
Ovenbird
Common Yellowthroat
American Redstart
Baltimore Oriole
Common Grackle
Scarlet Tanager
Northern Cardinal
Rose-breasted Grosbeak
House Finch
American Goldfinch
Rufous-sided Towhee
Chipping Sparrow
Song Sparrow

Birds migrating northward over the New York City region seem to fly in a broad mass, most heavily concentrated along the two shores of the Hudson River but with little reference to the more minor topographic features. In normal years, spring transients noticeably thin out eastward on Long Island, being seen infrequently east of Freeport on the south shore and Smithtown on the north. Spring migration begins in late February, increases with every favorable weather front, reaches its peak between 25 April and 22 May, and then tapers off through the first week in June.

In fall, conversely, birds seem more dependent on local topographical marks, and there are many distinct sub-flightlines. Some follow the two shores of the Hudson River, some the Connecticut shore of Long Island Sound; some, such as hawks and smaller birds, the northeast to southwest ridge lines in mainland New York and New Jersey; and others, the outer beach lines of Long Island westward to Staten Island and New Jersey. Winds and weather influence the concentration of birds: a cold front with its attendant northwest winds will mass birds all along the coastal strip. Fall migrations begin with the shorebirds in early July and are heaviest for landbirds during the last two weeks of August and until mid-October, tapering off through early November, with a concurrent build-up of waterfowl and wintering bird populations attending every advancing cold front.

In late spring a northward movement of southern herons begins, reaching a peak in late August. September may bring exotic rarities in the wake of an unusual tropical hurricane. Winter visitants—e.g., the Northern Goshawk, Rough-legged Hawk, Snowy Owl, Northern Shrike, and (Ipswich) Savannah Sparrow—may be expected between 1 November and 1 April, almost anywhere, but most often along the coast.

Metropolitan New York City includes the five city boroughs (Manhattan, Brooklyn, Queens, The Bronx, and Richmond or Staten Island); the remaining part of Long Island east of Brooklyn and Queens; five northern counties (Westchester, Putnam, Dutchess, Rockland, and Orange) along the Hudson River north of the city; and nearby counties in Connecticut and New Jersey. From the hundreds of fine places for bird finding within the New York State boundaries of metropolitan New York City, thirty-five have been selected because they are especially rewarding, accessible, and not likely to be covered with concrete in the foreseeable future. (*See* New Jersey and Connecticut chapters for locations adjacent to the New York State boundaries.)

The bird finder unfamiliar with New York City should have a generous supply of recent road maps, possibly street maps, and 'Rapid Transit' maps that are available at most newsstands. Where roads and tracts are marked 'private,' he should always ask permission for entry to look for birds there. In the places described, directions for reaching them are sometimes by public transportation (e.g., by IRT,

IND, and BMT subways), sometimes by both public transportation and private car, and sometimes by car only, usually from the nearest superhighway.

The Linnaean Society of New York (American Museum of Natural History, Central Park West at 79th Street) and the National Audubon Society (950 Third Avenue) sponsor the New York Rare Bird Alert, a telephone-recording service for bird finders. Upon dialing (212) 832-6523, one hears an up-to-six-minute recording that details recent sightings of interest in the area of Metropolitan New York City.

References

Birds of the New York [City] Area. By John Bull. New York: Harper & Row, 1964. Reprint, New York: Dover Publications, 1975.

Enjoying Birds around New York City. An Aid to Recognizing, Watching, Finding, and Attracting Birds in New York City, Long Island, the Upstate Counties of Westchester, Putnam, Dutchess, Rockland, and Orange, and Nearby Points in New Jersey and Connecticut. By Robert S. Arbib, Jr., Olin Sewall Pettingill, Jr., and Sally Hoyt Spofford. Boston: Houghton Mifflin, 1966. Available from the Laboratory of Ornithology, Cornell University, Ithaca, N.Y. 14853.

MANHATTAN
Central Park | Fort Tryon Park | Inwood Hill Park | Upper Bay

Manhattan offers two basic habitats—parks and waterfronts. In the parks for much of the year are only such permanent-resident birds as woodpeckers, jays, and chickadees, plus the introduced Rock Doves, European Starlings, and House Sparrows. The summer-resident list is longer, and, during spring and fall migrations, the parks, squares, churchyards, even the smallest green plots, come alive with birds. Manhattan can be a haven for migrating birds; it can also be a menace. On certain foggy nights in autumn, when a sudden lowering of the cloud ceiling and following winds combine, the tallest buildings take a toll of hundreds of landbirds. The waterfronts along the Hudson and East Rivers are worth watching at any season.

The best place for birds in Manhattan is **Central Park** (862 acres) extending from 59th to 110th Streets between Fifth Avenue and Central Park West (Eighth Avenue). A rectangular, rolling tract of lawns, woodlots, gardens, and shrub plantings, broken by outcroppings of Manhattan's pre-Cambrian rocks, with several small bodies of water and a growing clutter of buildings and roads, the Park provides sanctuary for small migrating birds caught at daybreak over the stone vastness of the city. Some thirty-eight species of warblers have been recorded here, among the most abundant being the Black-and-white, Northern Parula, Yellow, Magnolia, Myrtle, and American Redstart; nine species of flycatchers, all the eastern thrushes and vireos, and most of the finches and sparrows. Regular spring transients with southern affinities are the Blue-gray Gnatcatcher, Kentucky Warbler, and Orchard Oriole. Because of its light undergrowth, the Park is one of the best areas for observing in spring such secretive birds as the Mourning Warbler and Lincoln's Sparrow. The Park list has reached 250 species, many of them extralimital, with the best one-day list of 101 species, including 29 warblers.

A favorite spot during migration is the Ramble, on the west-central side between 72nd and 79th Streets, a neglected hilly area where paths wind through groves of locust, willow, maple, birch, oak, and sycamore beside a lake. From autumn to spring the Reservoir at 90th Street is a haven for loons, grebes, waterfowl, and gulls, including quite frequently a 'white-winged' species.

To reach the various places in the Park, take the Madison Avenue bus north to the desired street and walk west one block, or take the IND 8th Avenue subway.

Fort Tryon Park (61 acres), between Henry Hudson Parkway and Broadway, north of 192nd Street, is largely natural woodland, excellent for warblers and other transients in season. Its breeding-bird list is more like that of the northern counties.

Inwood Hill Park, on the northwestern tip of Manhattan at West 207th Street, also has a natural woodland attractive to transients, and, in addition, the highest point on Manhattan Island (405 feet), a good vantage point for fall flights of hawks, swifts, jays, and thrushes.

Transportation. North-bound from midtown: for Fort Tryon Park,

Madison Avenue bus No. 4 or IND 8th Avenue subway to 190th or Dyckmann Street. For Inwood Hill Park, same bus or subway to 207th Street.

From the Staten Island Ferry across **Upper Bay** from the southern tip of Manhattan to Staten Island, one may see, in April and May, flocks of Bonaparte's Gulls and possibly a rare Little Gull among them. At all times of the year, other gulls, including the common Herring and Ring-billed, are usually in sight. Wilson's Storm Petrels formerly appeared here in late summer, and may appear again.

Transportation. South-bound from midtown: IRT Lexington Avenue or Broadway–7th Avenue subway, or East Side or West Side bus, to Ferry terminal at the Battery.

BROOKLYN, KINGS COUNTY
Prospect Park

The best area in Brooklyn is **Prospect Park** and the adjacent Brooklyn Botanic Garden (together about 600 acres), west of Flatbush Avenue and south of Grand Army Plaza. Wooded hills, evergreen groves, ponds, winding brooks, meadows, and landscaped areas, along the Harbor Hill glacial moraine ridge, serve as a migration flightline for hawks and smaller landbirds in spring and fall. For a walk that will include the most rewarding places in the Park, stroll north from the Flatbush Avenue bus stop through the Vale of Kashmir and the rose garden; cross East Drive; and follow the ridge line southwestward, around Boulder Bridge to the Ravine, thence around Swan Boat Pond and Marsh; skirt the Quaker Cemetery; watch for transients on Lookout Hill and the Point of the Lake; and then return to Three-Arch Bridge, the Bandstand Pond, and the wooded area just north.

Transportation. From Manhattan: BMT Brighton Beach subway express to Prospect Park Station, or IRT subway local to Grand Army Plaza. By car, from Manhattan Bridge, straight out Flatbush Avenue.

QUEENS
Jamaica Bay Wildlife Refuge | Jacob Riis Park | Forest Park

The **Jamaica Bay Wildlife Refuge,** now part of Gateway National Park, is a must for anyone looking for birds in the New York City area. Application for a permit—required to enter—may be obtained in the interpretive building at the Park entrance upon one's first visit.

The Refuge, a man-made sanctuary with ponds, dikes, plantings, salt marsh, and mud flats, is worth a visit at any season but is most exciting from July to October, when concentrations of herons, waterfowl, gulls, terns, shorebirds, and transient landbirds are at their peak. During spring, from early April to early June, it is often equally rewarding. In summer, breeding species include the Green and Little Blue Herons, Cattle, Great, and Snowy Egrets, Louisiana Heron, Glossy Ibis, Common Gallinule, American Coot, and a variety of waterfowl. Since the early 1950s a total of 312 species have been noted at the Refuge.

Cross Bay Boulevard bisects the Refuge from north to south. East Pond (fresh water) lies east of the Boulevard. In fall, stand on the shore at the southwestern corner of East Pond and look along the southern edge for shorebirds. The best time for shorebirds here and at West Pond is high tide, which crests at the Refuge about an hour later than at Sandy Hook, as announced in local newspapers.

At West Pond (also fresh), west of the Boulevard, Lesser Golden Plovers appear regularly, Baird's Sandpipers sometimes, and the rare Buff-breasted Sandpipers regularly in the turf edging the water. Watch the northeastern corner for possible, but rare, phalaropes. In fall the Refuge teems with herons, gulls, terns, swallows, and broods of local waterfowl; the plantings swarm with transient landbirds. In winter the site is excellent for Short-eared Owls, (Ipswich) Savannah Sparrows, and Lapland Longspurs, and the ponds are excellent for waterfowl. The bays nearby support large flocks of Brant.

Transportation. From Manhattan: IND Rockaway subway line to Broad Channel, then walk north about a mile to the entrance. By car: from Kennedy Airport, west on State 27 about 2 miles, then south on Cross Bay Boulevard to a prominent sign on the right at the entrance, 1.4 miles south of the first bridge.

Not far from the Jamaica Bay Refuge is **Jacob Riis Park,** on the south shore of Rockaway Peninsula. Bird finding here is fair in winter, good in spring, and excellent in autumn, especially after a drop in temperature with northwest winds. Explore all the paths in the central mall, walk west around the north side of the golf course, and search the area around the old army post farther west. Then leave the car at the western end of Neponset Avenue, walk through the fence, and search the nearby bushy fields.

In fall, Riis Park is the best place for rarities such as the Western Kingbird, Blue Grosbeak, and Lark and Clay-colored Sparrows, and fine for landbirds including the Loggerhead Shrike, Philadelphia Vireo, Orange-crowned, Connecticut, and Mourning Warblers, and many flycatchers, jays, thrushes, tanagers, and finches, and, in late fall, winter finches. Watch for hawks streaking westward in the fall.

Transportation. From Manhattan: IRT Flatbush Avenue subway to end, or No. 935 Bus to Fort Tilden. By car: from Manhattan Bridge, Flatbush Avenue to its end over Marine Parkway Toll Bridge. Leave car at the extreme western end of the Park's public parking field.

Forest Park (538 acres), in east-central Queens, is one of the city's 'hottest' spots for landbirds in spring migration, in large part because of its extensive mature oak forest, with a heavy undergrowth of greenbriar and blackberry thickets. After a warm front in May, the Park may abound in flycatchers, thrushes, vireos, warblers, and others. Three observers recorded nearly a hundred species on one May day. Best bird finding is north of Forest Park Drive, which runs roughly east and west and is closed to vehicular traffic during the day, from Myrtle Avenue east to Metropolitan Avenue, especially toward the eastern end. During May, a morning visit to Forest Park combined with an afternoon visit to Jamaica Bay Wildlife Refuge (only 20 minutes away by car) will often produce a list of over a hundred species.

Transportation. From Manhattan: IND Jamaica subway to Union Turnpike station, then walk west to the Park. By car: from Long Island Expressway (I 495), turn south on Woodhaven Boulevard to Metropolitan Avenue, then east to Forest Park Drive, then right—into the Park. For Cross Bay Boulevard to Jamaica Bay Refuge, return to Woodhaven Boulevard and drive straight south.

THE BRONX
Van Cortlandt Park | Pelham Bay Park

Despite its industrial clutter, the Bronx features some of the best parks in the area.

Van Cortlandt Park (1,132 acres) lies east of Broadway between 240th and 263rd Streets. For the best places, walk east from the IRT station at 242nd Street, across playing fields to a railroad track, turn north, cross rowboat pond, and enter a small wooded marsh bordering a golf course. Swallows, Fish Crows, and Rusty Blackbirds are here in early spring; Wood Ducks, Traill's Flycatchers, Marsh Wrens, Warbling Vireos, and, possibly, the Green Heron, Virginia Rail, and Sora in nesting season. The wooded slopes on the eastern side of the Park are pine-oak scrub, excellent for spring migration and good in fall. The Sycamore Swamp in the northeastern corner is fine for spring warblers, including rarer species such as Golden-winged and Cerulean.

Transportation. From Manhattan: IRT 7th Avenue–Broadway subway to 242nd Street. By car: Henry Hudson Parkway north to Exit 17 (Broadway).

Pelham Bay Park, on Long Island Sound in the northeast corner of the Bronx, is a sprawling area of bay front, creeks, meadows, salt marshes, woodland, and landscaped grounds. Extremely good at any season, it is best in winter, when in the waters off Orchard Beach, Twin Islands, and the Lagoon—all in the northeast corner—are American Black Ducks, American Wigeons, Canvasbacks, Common Goldeneyes, Buffleheads, Red-breasted Mergansers, and other waterfowl, plus Horned Grebes, and the occasional loon. In midwinter, Great Cormorants perch on the rocks of Twin Islands, and Purple Sandpipers feed at the water's edge.

Transportation. Northeast from Manhattan: IRT Lexington Avenue subway, Pelham Bay Local, to terminus. By car: Pelham Bay Park Exit from Hutchinson River Parkway; drive east and north to the large parking area just west of Orchard Beach.

GREAT CORMORANT

STATEN ISLAND, RICHMOND COUNTY
Oakwood Beach | Great Kills Park

Although Staten Island has received less attention from bird watchers than the other boroughs, and there are fewer records for this area, its southeastern shore, with a series of parks and vantage points, should be a trap for birds funneling down Long Island in autumn and pouring across from New Jersey in spring. Intensive urban development has followed the completion of a superhighway that cuts across the northern part, and two bridges that tie it to New Jersey and Brooklyn.

Oakwood Beach and **Great Kills Park,** half-way down the southeastern shore, are best for birds from spring through fall. In winter there are many species of sea ducks, as well as loons, grebes, many gulls including the Greater Black-backed, and Purple Sandpipers. Short-eared Owls are in view between the marina and Cookes Point at the southern end of the Park. In summer one may see terns, Black Skimmers, and occasionally jaegers and Wilson's Storm Petrels from Oakwood Beach. In spring and fall the Beach, just north of the Park, has a good variety of shorebirds and a fair flight of hawks and other transient landbirds.

Transportation. From Manhattan: Staten Island Ferry from the Battery to St. George, Staten Island Rapid Transit (SIRT) to Oakwood Heights, then walk east on Guyon Avenue to Oakwood Beach, and south to the Park. By car: south from I 278 on Hyland Boulevard, left on Guyon Avenue to the Beach, then south to the Park.

LONG ISLAND EAST OF QUEENS

Lawrence Beach | Lawrence Marsh | Point Lookout | Jones Beach State Park | John F. Kennedy Memoral Wildlife Sanctuary | Gilgo Beach | Robert Moses State Park and Fire Island | Moriches Inlet | Moriches Bay | Shinnecock Bay and Inlet | Mecox Bay | Montauk Peninsula | Mill Neck Smithtown Area | Orient Point | Orient Beach State Park | Atlantic Ocean

Long Island east of Queens has lost much of its semirural character. Farmlands have disappeared, and with them habitats for birds. Despite this, the Nassau County area is good for spring migration—better here than farther east, and exciting everywhere in summer and winter. In fall, following a cold front with northwest winds and a sudden drop in temperature, bird finding is often sensational. At such a time, birds caught over the water at dawn beat back and drop exhausted in the shrubbery of the south-shore parks. Any mile of Long Island's southern barrier beach can be rewarding. The same beach can be productive in spring, when an early-morning fog follows a night flight.

Just north across Reynolds Channel from Atlantic Beach are **Lawrence Beach** and **Lawrence Marsh,** a noted spring migration funnel. Here are landscaped estates surrounded by creeks, salt marshes, and tidal flats with nesting egrets, herons, ibises, bitterns, American Black Ducks, Northern Harriers, Clapper Rails, American Woodcock, Traill's Flycatchers, Marsh Wrens, and Seaside and Sharp-tailed Sparrows. The wetlands are feeding and resting grounds for herons, waterfowl, and shorebirds, and are noted for rare visitants in all seasons. Oak groves bordering the suburban streets are stopovers for migrating landbirds. Even the smallest patch of woodland harbors transients in May.

Beach and Marsh are reached by car from the end of Rockaway Boulevard (Meadow Lane), and left on Causeway Road to Sage Pond, where Least Bitterns, Common Gallinules, and Virginia Rails have bred. At Causeway Road and Pond Crossing (Pond X), look for a narrow lane called Mallow Way, which leads to a dike with an excellent view of the salt marsh. Six species of rails have been seen here; four of them probably breed.

Continue on Causeway Road, turn right on Barrett Road and again right on the driveway to the yacht basin, on the southern edge of the golf course, which gives access to another salt marsh and Bannister Creek, both best for waterfowl in colder months.

Warning: The entire Lawrence Beach area is infested with ticks in spring and summer.

In recent winters **Point Lookout,** a residential community at the eastern end of Long Beach Strip of Jones Inlet, which has a fast-moving current, rock jetties, and shoals, has been a good place for the Great Cormorant, Harlequin Ducks, eiders, alcids, and rare gulls. In spring and fall, distant flats at low tide are shorebird havens.

Transportation. By Long Island Railroad (LIRR) to Long Beach and bus to Point Lookout. By car, east from Atlantic Beach. Parking limited to Lido Boulevard, or in private drives if permission is requested politely. Winter parking at Hempstead Beach, just west of the village.

Jones Beach State Park (2,413 acres), east of Jones Inlet, includes 8 miles of barrier beach, dunes, fields, lawns, and plantings backed by creeks, bays, and salt marshes— all excellent habitats for birds. The beach has colonies of sea birds in summer and is a vantage point for pelagic species—shearwaters, storm petrels, and jaegers offshore. In fall, migrating landbirds are everywhere. On a day in late September, with a northwest wind following the passage of a cold front, thousands of swallows 'boil' through the air, and hawks, flickers, warblers, sparrows, and dozens of other species of landbirds crowd every cover.

Worthwhile locations, starting at Jones Beach Inlet, are:

Short Beach, for ducks and other sea birds. Both eiders, Little and Black-headed Gulls, Black-legged Kittiwakes, and Dovekies are seen occasionally in late fall and winter. The Purple Sandpiper is regular in numbers on the long jetty at the mouth of the Inlet; Dunlins and Sanderlings are also present in winter.

Leave the car in *West End Parking Field 2* and walk west to the jetty at Jones Inlet. Search the pines east of *West End Parking Field 1* for Barn, Long-eared, and Saw-whet Owls, and among the dunes for Snowy and Short-eared Owls in winter. Watch adjacent lawns and dunes for (Northern) Horned Larks and Snow Buntings.

In the vicinity of *West End Parking Field 2* are nesting Piping Plovers and nesting colonies of Common and Little Terns and Black Skimmers. Drive slowly and do not enter the area during nesting season.

On the channel side, opposite *Parking Field 4,* is the Fishing Station and a service garage, landscaped with sycamores, evergreens, and beach plums, a fine place for migrating landbirds in fall. Also search the pines along the northern edge of *Parking Fields 4 and 5* and western end of Zach's Bay. In winter, look for redpolls, siskins, and crossbills in the evergreens.

The entire beach strip is a flightline in fall for migrating predators—Northern Harriers, Ospreys, and falcons, all headed westward. During migration, search around buildings and the central water tower for dead and injured birds; watch the wires, light poles, and dune tops for hawks, shrikes, and other rarities.

The strip is reached by car by driving south from State 27 on Meadowbrook Parkway to Ocean Parkway, and east or west along the beach.

East of Jones Beach State Park lies the **John F. Kennedy Memorial Wildlife Sanctuary.** The permit, required for entry, is obtainable by writing to the Town Clerk, Oyster Bay, Long Island, New York 11771. The Sanctuary's center, a brackish pond about 3,500 feet long, surrounded by phragmites and spartina and, on the north side, by dunes, is excellent for fall-migrating and wintering waterbirds and waterfowl. During fall the pond margins attract many shorebirds—Stilt Sandpiper, both dowitchers, and occasionally Hudsonian and Marbled Godwits and American Avocet.

Also in fall, landbirds throng among the bayberry, poison ivy, pitch pine, and wild cherry on the dunes. From an observation tower one may watch migrating hawks, swallows, and small landbirds at this same season.

During breeding season, nine wading-bird species frequent the Sanctuary, including Louisiana and Little Blue Herons, Yellow-crowned Night Herons, and Glossy Ibises. Willets breed in the salt marshes along the Sanctuary's northern perimeter, where Clapper Rails and Seaside and Sharp-tailed Sparrows abound.

The Sanctuary is reached by car by driving east from Jones Beach on Ocean Parkway, entering the Tobay Beach Parking Area, and from its southwest corner taking an access road to the Sanctuary parking area. Walk west.

About 2 miles east of Jones Beach is **Gilgo Beach,** a cottage settlement and well-known migration trap in fall. Flycatchers, thrushes, vireos, warblers, and occasionally a Lark or Lincoln's Sparrow lurk around the cottages. The beach stretches eastward for miles, with dunes to the south and salt marsh and tidal flats to the north. Look for Rough-legged Hawks and Snowy and Short-eared Owls in winter; gull, tern, and skimmer colonies in summer.

To reach Gilgo Beach by car, drive east on Ocean Parkway, park in Gilgo Beach village parking lot, and walk east to a former Coast Guard Station, a brick structure with lawns and pine groves.

Oak Beach, near the east end of the barrier beach and facing Fire Island Inlet, is excellent for sea birds in winter. This is reached by car by driving east from Gilgo Beach on Ocean Parkway. Enter the Oak Beach community parking area, and take an access road west to a series of jetties along the Inlet.

Continue east on Ocean Parkway and cross the bridge to **Robert Moses State Park** (1,000 acres), at the western end of **Fire Island.**

This recently developed Park offers the most spectacular concentrations of fall transients on Long Island. Especially impressive are mid-September to early October flights of falcons, flickers, swallows, and landbirds, but favorable weather conditions—the passage of a cold front with northwest winds—are essential. Over 100 Merlins have passed over in a single day, over 1,000 American Kestrels, and over 10,000 Yellow-shafted Flickers. The best vantage points to witness the flights are at the extreme east end of the parkway, just beyond *Parking Field 5,* and the northeast corner of *Parking Field 2.*

For landbirds, walk east from the Park about a half-mile to Fire Island Lighthouse, searching all thickets and around the Lighthouse and other buildings.

At the Park, look for landbirds near the northeast corner of the pitch-putt golf course and in the vicinity of the boat basins. From mid-May to late September pelagic species including shearwaters, Wilson's Storm Petrels, and Parasitic Jaegers are often seen from the beach when there are onshore winds, especially during storms. In late fall, watch for Northern Gannets and Black-legged Kittiwakes; during the winter months, for loons, grebes, and sea ducks. When there are irruptions of winter finches, large numbers of Common Redpolls, Pine Siskins, and crossbills have been found wherever there are pines, especially near the water tower at the base of the bridge.

To reach the Park by car, drive via Ocean Parkway (*see above*), or south from State 27A on Robert Moses Causeway.

An outer beach, cut by narrow **Moriches Inlet,** protects the sand bars and broad tidal flats of wide shallow **Moriches Bay** from the sea. In this area, famous for its colonies of terns and skimmers in summer, shorebirds and storm-borne exotics in spring and fall, and waterfowl in winter, bird finding is best from a rowboat. Look for shorebirds and waterbirds at low tide on and around West and East Inlet Islands, the West 0.25 mile north of the Inlet, the East about 0.5 mile to the northeast; expect twenty or more species on a good day, among them larger shorebirds such as the Whimbrel, Willet, Hudsonian and Marbled Godwits, and many smaller species. Breeding birds include the American Oystercatcher, Piping Plover, Common, Roseate, and Little Terns, and Black Skimmer. Moriches Inlet is especially good for unusual terns—Gull-billed, Royal, and Caspian.

Reached by car to East Moriches on State 27A. Rowboats, with or without motors, are available at the southern end of Atlantic Avenue. For the Inlet, drive south from State 27A through the town of Westhampton Beach, cross the bridge to the Beach, and turn west on Dune Road to its end at Suffolk County Park. East Inlet Island is due north of the far end of the large parking lot and can be reached at low tide by wading knee-deep.

Farther east is **Shinnecock Bay and Inlet,** a marine bay sheltered by a barrier beach with access to the sea through the narrow Inlet. The sand bars, a few hundred yards from the Inlet and best approached by rowboat, are sometimes very good for shorebirds and southern strays. Greater Black-backed and Herring Gulls nest on islands at the Inlet; Piping Plovers and Little Terns nest on Shinnecock Beach, along with American Oystercatchers, Spotted Sandpipers, and (Prairie) Horned Larks. In winter the beach is excellent for Snowy and Short-eared Owls during flight years; the Purple Sandpiper is regular at the jetties guarding the Inlet. Also watch for Glaucous, Iceland, Black-headed, and Little Gulls.

To reach Bay and Inlet by car, drive south from State 27A at Canoe Place on Lynn Avenue to Dune Road, west to Tiana Beach or east to the Inlet and jetties. On the average the west side of the Inlet offers better bird finding, and is easier to reach.

Farther east, flocks of Canada Geese winter on **Mecox Bay** or in the rye fields nearby. Whistling Swans appear regularly in late fall and winter. A good variety of marine waterbirds winter here. In late summer and fall, look for shorebirds, terns, and storm-blown vagrants at the flats closest to the ocean beach. In September, search the potato fields for flocks of Lesser Golden Plovers and, sometimes, Buff-breasted Sandpipers.

To reach Mecox Bay by car, go south from State 27A at the intersection of State 27 and 27A on Flying Point Road to end; or south from State 27 at Hayground on Mecox Road, south on Job's Lane, and west on Dune Road.

State 27 ends at **Montauk Peninsula,** a famous bird-watching spot on the easternmost tip of Long Island. In late fall and winter it is the best place for Common and Red-throated Loons, Red-necked and Horned Grebes, Northern Gannets, Great Cormorants, Harlequin Ducks, eiders, scoters, Red-breasted Mergansers, 'white-winged' gulls, Black-legged Kittiwakes, Bonaparte's and other gulls, and particularly alcids. Every part of the 14 miles from Amagansett east is rewarding at this season; several segments are exceptional. In fall, Montauk Point State Park is an exciting area for landbirds, one of the best places on Long Island to find western strays. From early June through early October, several boats leave Montauk Harbor daily for Cox's Ledge, the most prolific area for pelagic birds available from Long Island (*see* **Atlantic Ocean** *below*).

In winter, park at Montauk Point and search the pine groves for landbirds. Scan the Atlantic from the bluffs beside the lighthouse or the Park restaurant, and then walk northwestward to Shagwong Point, another good vantage point for eiders, Harlequin Ducks, and possibly alcids. Of the alcids only Razorbills and Dovekies are regular in small numbers, but both murres and Black Guillemots have been seen. At times, thousands of scoters of all three species are viewed off the Point. The Red-breasted Merganser is probably the most abundant duck. From 15 August to 15 November, migrating shearwaters, Wilson's Storm Petrels, Parasitic Jaegers, and terns sometimes skirt the Point. On the moors in winter, half-hardy birds find shelter in the stunted oak woods and dense thickets.

East Lake Drive, on the east side of Lake Montauk, leads to the highest hills on the Peninsula; east of the hills are Oyster Pond, a haven for ducks when ice-free, and Big Reed Pond surrounded by marshes. Watch the wires for American Kestrels and Northern Shrikes; the hills for Red-tailed and Rough-legged Hawks; the lowland marshes for Northern Harriers and Short-eared Owls; and the dunes for Snowy Owls. From the jetties at the ends of East or West Lake Drives, look for loons, Great Cormorants, Harlequin Ducks, Purple Sandpipers, and gulls. Large numbers of terns, including many Roseates, are also here during late summer and fall.

Other areas worth investigating on the Peninsula are: *Fort Pond,* north of Montauk Village; sand dunes and pine woods in Hither Hills State Park (1,755 acres), 4 miles west of Montauk; and *Lazy Point Road,* north from State 27 between Hither Hills Park and Napeague Harbor, to *Goff Point,* where large flocks of Common Goldeneyes, Buffleheads, Oldsquaws, and Red-breasted Mergansers winter.

At **Mill Neck,** an estate community west of Oyster Bay, rolling hills slope to a river valley that drains north into Mill Neck Bay. The Bay has a typical salt-creek habitat with herons, waterfowl, shorebirds, rails, and gulls in season. Beaver Lake, south of the dam on Cleft Road, has waterfowl in colder seasons. At the southern end of the lake, south of the railroad embankment, is a marshy estuary with nesting records for the Least Bittern, Wood Duck, Virginia Rail, and Belted Kingfisher. South of this estuary is Shu Swamp Sanctuary, a wet woodland with oak, beech, maple, and tulip tree, and exciting in spring migration but especially fine in April for transient Eastern Phoebes, swallows, Palm Warblers, and Purple Finches. Nesting birds include the Wood Duck, Red-shouldered and Broad-winged Hawks, Yellow-throated Vireo, and Louisiana Waterthrush.

Transportation. From Manhattan: Long Island Railroad to Mill Neck. Walk west along the railroad embankment, keeping alert for trains; or take encircling roads. *For the Sanctuary,* walk south on Mill Neck Road for 200 yards to the entrance on the west side of the road. By car: *for the dam,* turn north from State 25A on State 106 to Oyster Bay, turn left on West Main Street, and left again on Cleft Road. *For the station,* turn north from State 25A on Wolver Hollow Road, right on Chicken Valley Road, right on Glen Cove Road, and left on Mill Neck Road. Parking is permitted only at the station.

The **Smithtown Area,** on the Nissequogue River south of Smithtown Bay, is worth investigating. Two ponds in the headwaters of the River have ducks—Gadwall, Redhead, Ring-necked Duck, Ruddy Duck, and Hooded Merganser—regularly in fall: the first, on a private estate just south of State 25; the second, Stump Pond, farther south in a wild tract with marshes and woodlands. Other ponds that are private may eventually become a nature reserve.

Landing Road, north from State 25 just east of Smithtown Branch, crosses the River on a white bridge that overlooks cattail marshes bordered by a wooded hillside. In fall and early winter this area holds the Eastern Phoebe, Winter and Carolina Wrens, Myrtle Warbler, Gray Catbird, American Robin, and Golden-crowned Kinglet, as well as wintering hawks, blackbirds, finches, and sparrows. Farther east, River Road, north from the intersection of State 25 and 25A, follows the River to Smithtown Bay. Watch for 'bay' ducks on the water and Brown Creepers, Eastern Bluebirds, and Cedar Waxwings in the woods. Where River Road forks, go left to Short Beach for wintering sea birds, or right to Long Beach for waterfowl on Long Island Sound.

State 25 ends at the northern fluke of Long Island, **Orient Point,** an excellent area for sea birds in winter and migrating birds in fall. From the causeway, east of Marion, look north over Dan Pond in summer for herons, Northern Harriers, Ospreys, gulls, and terns; and south to Orient Harbor for summering but non-breeding loons, cormorants, and even White-winged Scoters. The salt marsh, east of the causeway, has numerous breeding species, including Clapper Rails.

The extreme end of Orient Point overlooks Plum Gut. In winter, watch for loons, grebes, flocks of Greater Scaups, Common Goldeneyes, Oldsquaws, an occasional Harlequin Duck, Common and King Eiders, all three scoters, and Red-breasted Mergansers. Search for Purple Sandpipers on the rocks and be alert for almost any species of gull.

In **Orient Beach State Park** (357 acres), just west of the Point, during winter, look for American Black Ducks, Greater Scaups, Common Goldeneyes, and Oldsquaws on the water, the occasional Snowy Owl or Short-eared Owl on the meadows, (Northern) Horned Larks on the beaches, and Snow Buntings at the parking field. In summer, look for breeding Ospreys, Piping Plovers, Common and Little Terns, and (Prairie) Horned Larks.

Two miles west of the Point, turn south from State 25 on Narrow River Road. Ospreys nest in the marshland. In winter, look for Rough-legged Hawks and Northern Harriers, American Kestrels, Snowy and Short-eared Owls, (Northern) Horned Larks, Eastern Meadowlarks, Lapland Longspurs, and other open-country species. Return to State 25 by turning right on Village Lane.

Perhaps the most fascinating and certainly the most unpredictable habitat for birds is the **Atlantic Ocean.** Fishing boats are available from Sheepshead Bay, Freeport, Captree State Park, Montauk, and many other south-shore ports. Usually the farther east the better. The best seasons are late May through late June, and late August through September. Pelagic bird watching is best after three days of winds from the southwest, south, and southeast, except in winter, when a 'nor'easter' is best.

There are two types of boat: The first—the 'open' boat or party boat—will take anyone until capacity is reached, and goes where the captain wants to. Boats fishing for flounder or bottom fish will not go far offshore; those fishing for mackerel or bluefish may go 20–35 miles out—and this is worthwhile. The second type—the charter boat—must be booked in advance and will go where the passengers wish.

As the boat leaves the harbor, start to attract gulls by 'chumming'—spreading a trail of minced menhaden, day-old bread, or puffed cereal soaked in fish oil. A large gull following may attract shearwaters, storm petrels, jaegers, other pelagic species farther out.

Be prepared for the pelagic trip with pills for seasickness and plenty of warm, waterproof clothing at any season, and the warmest possible in winter.

Highly recommended are two trips: to Cox's Ledge and to Hudson Canyon. Cox's Ledge lies about 40 miles east-southeast of Montauk Point. From mid-May through late October, open boats visit here daily from Montauk Harbor and Point Judith, Rhode Island. Hudson Canyon, about 90 miles southeast of New York City, is included in fishing trips to the edge of the continental shelf north or south of the true 'canyon.' Open boats visit here from Sheepshead Bay, Brooklyn, and various New Jersey ports on a sporadic schedule. Most open boats grant groups of bird finders (non-fishermen) a discount; inquire before embarking. A fishing trip to Cox's Ledge takes 12 hours; Hudson Canyon, about 26 hours. Charter trips can be arranged. For information concerning open boats to either Cox's Ledge or Hudson Canyon, consult the fishing advertisements in the sports section of local newspapers.

On either of these trips one has reasonably good prospects of seeing many sea birds, including the species listed below (with annotations by Thomas H. Davis, Jr., from numerous trips).

Northern Fulmar. Regular in small numbers, especially during June; also a few in late fall.

Cory's Shearwater. In varying numbers, midsummer to early November.

Greater Shearwater. In varying numbers, usually common, mid-May through October.

Sooty Shearwater. Common mid-May through June, occasionally later.

Manx Shearwater. One or more individuals, June and occasionally later.

Audubon's Shearwater. Occasional at Hudson Canyon, late summer.

Leach's Storm Petrel. Small numbers at Hudson Canyon, June; rare elsewhere to October.

WILSON'S STORM PETREL

Wilson's Storm Petrel. Abundant, late May to mid-September, a few to mid-October far offshore.

Northern Gannet. Present throughout year; hundreds often in November and April, dozens in winter, small numbers regularly to early June, and stragglers through summer.

Northern Phalarope. Uncommon in spring and fall.

Red Phalarope. Uncommon in spring and fall, large numbers occasionally in late fall.

Parasitic Jaeger. Usually within sight of land, less common farther offshore.

Long-tailed Jaeger. A few each year, especially in June; less common in early September.

Great Skua. Increasing in numbers, a few regularly on June trips, still casually otherwise.

Black-legged Kittiwake. Common from late fall through winter, occasionally in summer.

Sabine's Gull. Occasional in late summer, not every year.

Dovekie. Small numbers in November.
Atlantic Puffin. Twice in November at Hudson Canyon; not expected much closer to shore.

WESTCHESTER, PUTNAM, and DUTCHESS COUNTIES

Playland Lake–Manursing Island | Croton Point Park | Shrub Oak Memorial Park | Cruger's Island

Directly north of the Bronx and bordered on the east by Long Island Sound and Connecticut, on the north by Columbia County, and on the west by the Hudson River lie (from south to north) Westchester, Putnam, and Dutchess Counties. Birdlife is similar in all three, except that Westchester has a waterfront on Long Island Sound, attractive to waterfowl in winter, and Dutchess is more rugged and has higher hills—up to 2,300 feet in the eastern third—that harbor certain bird species of northern affinities.

Playland Lake–Manursing Island in Rye extends from Oakland Beach to North Manursing Island. Between the two sites is the best bird finding in the Sound shore area, with sandy beaches, shingle, and bluff, and behind them Playland Lake, with its surrounding marshlands, meadows, woods, tidal flats, and bays.

Playland Lake is a 'rowboat lake' in summer, but after Labor Day the ducks return. On the Lake one October morning, 12,000 Greater Scaups were counted. Well into December, flocks of Canada Geese, American Black Ducks, Canvasbacks, Greater Scaups, Common Goldeneyes, and Buffleheads rest here, along with numerous mergansers and American Coots. Offshore are more waterbirds and waterfowl—Horned Grebes, Oldsquaws, a few scoters, and an occasional loon. In winter, Great Blue Herons feed along the shores of the islands in the Lake; and the birch groves and thickets, east of the Lake, shelter Mourning Doves, American Robins, Hermit Thrushes, and Red-winged Blackbirds. Wintering finches and sparrows are often abundant. Watch for Red-tailed and Rough-legged Hawks and Northern Harriers in winter and Ospreys in spring. Laughing Gulls are common in spring and fall; Bonaparte's Gulls, in spring.

By car: Exit from I 95 in Rye to Playland Parkway. For Manursing

Way, turn left from Playland Parkway (just before the Park entrance) on Forest Avenue for about a mile, then left to Kirby Lane.

In spring and fall the woodlands north and south of the Lake become 'migration traps' for flycatchers, thrushes, vireos, warblers, and other landbirds. Breeding birds include the Belted Kingfisher, Traill's Flycatcher, Carolina Wren, Veery, White-eyed Vireo, and the abundant House Finch. Manursing Way goes east from Forest Avenue. Follow it to the end and, after investigating the evergreen groves in winter, turn left on Kirby Lane for another vantage point for herons, waterfowl, and shorebirds.

Croton Point Park, on the Hudson just south of Peekskill, is the best place inland for Red-necked Grebes and a good place for waterfowl, owls, 'white-winged' gulls, and half-hardy landbirds. For owls, search the evergreens on the southern slope of the Point; for sparrows in fall, the bushes around the garbage dump; for hawks, the woodland; for gulls, the garbage dump or fields near the Park entrance. This Park should be a vantage point for seeing birds using the Hudson River flyway, but bird watchers seem to have neglected it during migration season.

By car: Drive west from US 9 on Croton Point Avenue.

Shrub Oak Memorial Park, northwest of Shrub Oak at the base of Piano Mountain, has second-growth woodland with a wooded swamp and lake. The landbird migration in spring is good on the southern slope. Worm-eating, Golden-winged, Blue-winged, and Hooded Warblers breed. About a hundred yards west of the lake, a trail leads to the summit of Piano Mountain, an autumn lookout for hawks.

To reach the Park by car, head west from Peekskill on US 6 to Shrub Oak and turn north on Barger Road.

Cruger's Island, on the Hudson north of Rhinebeck, has large marshy bays on the north and south. Nesting birds are: Least Bittern in the north bay, Virginia Rail, both cuckoos, Traill's Flycatcher, Brown Creeper, Marsh Wren, Blue-gray Gnatcatcher, and several warblers. This is a good place for an occasional Bald Eagle from late February to early March, and for migrating waterfowl in April.

By car: North from Rhinebeck on US 9 for about 2 miles, left on State 9G for about 3 miles, and then left to the Island on a minor road marked 'Whalebone Inn.' When the road is impassable (often in spring), park on the right of the sheds near the entrance and walk.

ROCKLAND COUNTY and ORANGE COUNTY
Bear Mountain–Harriman State Park | West Point Military Reservation

These two Counties—opposite Putnam and Westchester Counties—on the west side of the Hudson River are perhaps the least populated and most rural in the Greater New York region; yet houses, factories, and shopping centers are fast replacing the farms, orchards, and woodlands of the rolling, hilly land.

All driving directions are from US 9W, which follows the west side of the Hudson River from New York City to Albany.

Bear Mountain–Harriman State Park (42,000 acres on the Rockland–Orange County line) is a natural area with hardwood forests, rhododendron bogs, and spring-fed lakes, productive at any season. In summer, Turkey Vultures, Broad-winged Hawks, Pileated Woodpeckers, and warblers such as the Worm-eating, Black-throated Blue, Black-throated Green, Blackburnian, Hooded, Canada, and Louisiana Waterthrush nest.

One entrance is just off US 9W between Nyack and Newburgh. At Park headquarters, across a pedestrian bridge east of Bear Mountain Inn, are a trailside museum, zoo, nature trail, and maps showing hiking trails.

Seven Lakes Drive, west from Bear Mountain Inn, can be rewarding. It passes the entrance to Memorial Drive to the summit of Bear Mountain (1,306 feet), an excellent outlook for hawks in fall; rambles through woodland and beside lakes, passing down the west side of Lake Tiorati, a resting place for waterfowl in late autumn; crosses State 210 and continues south as Stony Brook Drive to join State 17 near Ramapo, a village 2 miles south of Ramapo Torne (1,120 feet), the highest hill in the area and largely unexplored for birds.

In the southern part of the **West Point Military Reservation,** a narrow, twisting road called Mine Road (or Forest of Dean Road) crosses the Reservation and is open to the public. Just north of Bear Mountain Circle, on US 9W at Fort Montgomery, turn west on Mine Road. At its intersection with State 293, turn right and return to US 9W. There are several places along Mine Road worth a stop to look for birds. About a half-mile from Bear Mountain Circle, stop at the Long Mountain Trail crossing. In spring or summer a walk west along the Long Mountain Trail to Weyant's Pond should yield a variety of birds in a short time. The Trail winds through fields and woodlands, crosses Popolopen Creek over a bridge that is a vantage point for marsh birds and, possibly, Bank Swallows nesting in an exposed sand bank. Farther on, the Trail crosses an old mining road—a short walk on this road in either direction may yield Blue-gray Gnatcatchers—and eventually reaches Weyant's Pond.

Along Mine Road beyond the Trail crossing, watch for Golden-winged Warblers and perhaps a Golden-winged × Blue-winged hybrid; on the farms, overgrown with alder, willow, and birch, look for nesting White-eyed Vireos, Yellow-breasted Chats, Indigo Buntings, and American Goldfinches. The American Woodcock performs here in April. In winter, look for Common Redpolls and Pine Siskins. Farther north on the steep-walled, wooded slopes, breeding species include the Red-tailed Hawk, Red-shouldered Hawk, Broad-winged Hawk, Ruffed Grouse, and Pileated Woodpecker. Mine Road next passes along the north shore of Stillwell Lake, where loons, transient waterfowl, and possibly a Bald Eagle appear in winter.

North Carolina

AMERICAN OYSTERCATCHERS

North Carolina's long chain of sand islands fronting on the Atlantic, the wide sounds and vast salt marshes lying behind them, and the lakes and trackless swamplands near the coast comprise one of the finest wilderness areas remaining on the eastern seaboard. It is not surprising, therefore, that bird finders in North Carolina focus their attention on species enjoying this remoteness: on American Oystercatchers, Wilson's Plovers, and Black Skimmers that nest on the beaches and sand islands, on the hordes of Whistling Swans and Snow Geese that pass the winter on the protected waters, on the many herons and egrets that colonize in situations varying from island thickets to mainland swamps. The birds of the high Appalachians of western North Carolina are no less deserving of attention, for they include a fascinating array of species with distinctly northern affinities.

Nearly half of North Carolina is part of the Coastal Plain, a flat to slightly rolling country with maximum elevation of about 500 feet, extending from the coast westward to the Fall Line. The sand islands or banks forming a narrow barrier between the sea and the sounds possess beaches and shifting sand dunes with relatively little vegetation, save sea oats and other grasses and, occasionally, scattered clumps of bayberry and inkberry. The sounds back of the banks are characteristically shallow, the water being fresh at the river mouths and becoming increasingly saline toward the sea inlets. Marshes fringe the sounds and the low islands in the sounds. Some are salt marshes containing an abundance of cord grasses; others are brackish or fresh-water marshes, with cattails, bulrushes, and quillreeds. American Oystercatchers, though never abundant, are conspicuous birds of the outer beaches, a pair or two usually nesting near the many inlets; but far more common are Wilson's Plovers, which scratch out their nesting hollows almost anywhere on the shell-littered beaches and deposit their eggs, usually in May or June. In addition to these two shorebirds, other species nesting regularly on the beaches or in the salt or brackish marshes are the following:

Clapper Rail
Willet
Laughing Gull
Gull-billed Tern
Common Tern
Little Tern
Royal Tern
Black Skimmer
Marsh Wren
Boat-tailed Grackle
Seaside Sparrow

The area for 30 to 80 miles inland from the sounds is the so-called tidewater belt of the Coastal Plain; the terrain is low, with many wooded swamps—locally referred to as 'pocosins' or 'dismals'—containing such trees as bald cypresses, white cedars, and red maples. One of the best known, ornithologically, is the Great Dismal Swamp on the Virginia border (*see under* **Suffolk,** Virginia); another, the swamp bordering Great Lake in Croatan National Forest (*see under* **Morehead City**), is much less well known.

Prothonotary and Northern Parula Warblers are common in the tidewater swamps, and a few Swainson's Warblers may be expected.

Not infrequently the swamps contain wading-bird colonies of varying size, with some or all of the following species as occupants: Great Blue Heron, Little Blue Heron, Great Egret, Snowy Egret, Louisiana Heron, Yellow-crowned Night Heron, and Black-crowned Night Heron. In the larger swamps, such as that bordering Great Lake, are colonies of Double-crested Cormorants, and a few pairs of American Anhingas may reside.

On some lower levels of the Coastal Plain are great sedge-covered savannas and, along rivers and streams, extensive bottomlands that produce hardwoods such as sycamores, black birches, ashes, tupelos, and various oaks. On the drier, sandy uplands at the Fall Line, which runs roughly southward between Raleigh and Fayetteville from Roanoke Rapids to Hamlet, are second-growth forests of longleaf and loblolly pines, with scattered stands of gums and oaks.

Extending west from the Fall Line to the Blue Ridge Mountains is the Piedmont Plateau, an area of rolling, red-clay hills rising from 500 to 2,500 feet. As it is the most densely populated part of the state, its wilderness has been greatly reduced; nonetheless there are large tracts supporting red and white oaks, hickory, and aspen; others supporting beech, maple, and tulip tree; still others supporting shortleaf and Jersey pines. Nearly all the forests are second growth.

Among landbirds breeding in North Carolina almost exclusively east of the Blue Ridge are Fish Crows, in suitable tree growth near the coast; Painted Buntings, in thickets near the coast from Beaufort southward; Chuck-will's-widows, in bottomland woods; Red-cockaded Woodpeckers, Brown-headed Nuthatches, and Pine Warblers, in pine woods; Bachman's Sparrows, in open pine woods and brushy fields; Yellow-throated Warblers, in pine woods or in nearby swampy places.

The Blue Ridge Mountains, an irregular chain crossing the state in a northeast-southwest direction, rise sharply above the Piedmont Plateau, then slope westward to a plateau cut into small mountain ranges and deep valleys. Bordering the plateau farther west are the Great Smoky Mountains, which form part of the North Carolina-Tennessee boundary. The Blue Ridge and the Great Smoky Mountains comprise the highest, most impressive, mountain mass east of the Mississippi. Over forty peaks have altitudes exceeding 6,000 feet. One, Mt. Mitchell in the Black Mountains—a northwest spur of the Blue Ridge—reaches 6,684 feet, highest point in eastern United States. Dozens of other peaks have elevations ranging from 4,000 to 6,000 feet.

The forests of the mountain region up to 4,000 feet contain red oak, white oak, basswood, sugar maple, beech, tulip tree, and hickory, with stands of pine on the dry, more exposed areas. From 4,000 feet up to 5,000 feet the hardwoods become mixed with hemlock, spruce, and fir, and above 5,000 feet the forests are principally spruce and fir. Except in the Great Smoky Mountains and atop the highest peaks, most of the original timber has been destroyed and replaced by second growth. Breeding birds characteristic of the North Carolina mountains are listed below. Species marked with an asterisk are confined mainly to the coniferous forests of the highest peaks.

Ruffed Grouse
* Saw-whet Owl
Yellow-bellied Sapsucker
* Olive-sided Flycatcher (*Great Smokies*)
Northern Raven
* Black-capped Chickadee
* Red-breasted Nuthatch
* Brown Creeper
* Winter Wren
Veery
* Golden-crowned Kinglet
Solitary Vireo
Black-throated Blue Warbler
Black-throated Green Warbler
Blackburnian Warbler
Chestnut-sided Warbler
Canada Warbler
Scarlet Tanager
Rose-breasted Grosbeak
* Pine Siskin
* Red Crossbill
Slate-colored Junco

In most deciduous forests of the Coastal Plain, Piedmont Plateau, lower slopes of the mountains, and mountain valleys, and in most of the fields, wet meadows, brushy lands, orchards, and dooryards of nearby farmlands, the following species breed regularly:

DECIDUOUS FORESTS

Turkey Vulture
Black Vulture (*except in mountains*)
Red-shouldered Hawk
Yellow-billed Cuckoo
Black-billed Cuckoo (*chiefly in mountains*)
Common Screech Owl

Barred Owl
Whip-poor-will
Yellow-shafted Flicker
Pileated Woodpecker
Red-bellied Woodpecker
Hairy Woodpecker
Downy Woodpecker
Great Crested Flycatcher
Acadian Flycatcher
Eastern Pewee
Blue Jay
Carolina Chickadee
Tufted Titmouse
White-breasted Nuthatch
Wood Thrush
Blue-gray Gnatcatcher
Yellow-throated Vireo
Red-eyed Vireo
Warbling Vireo (*chiefly in mountains*)
Prothonotary Warbler (*east of mountains*)
Northern Parula Warbler
Ovenbird (*uncommon near coast*)
Louisiana Waterthrush
Kentucky Warbler
Hooded Warbler
American Redstart (*east of mountains*)
Summer Tanager

FARMLANDS

Common Bobwhite
Mourning Dove
Red-headed Woodpecker
Eastern Kingbird
Eastern Phoebe
House Wren
Bewick's Wren (*chiefly in mountains*)
Carolina Wren
Northern Mockingbird
Gray Catbird
Brown Thrasher
Eastern Bluebird
Loggerhead Shrike
White-eyed Vireo
Yellow Warbler (*except near coast*)
Prairie Warbler
Common Yellowthroat
Yellow-breasted Chat
Eastern Meadowlark
Orchard Oriole
Common Grackle
Northern Cardinal
Blue Grosbeak (*east of mountains*)
Indigo Bunting
American Goldfinch
Rufous-sided Towhee
Grasshopper Sparrow
Chipping Sparrow
Field Sparrow
Song Sparrow

Since there are no great river valleys traversing the state in a north-south direction, most landbirds tend to filter through in a somewhat dispersed fashion. Waterbirds, waterfowl, and shorebirds, on the other hand, pass along the coast in tremendous numbers. Such sites as the chain of sand islands east of the sounds (e.g., Hatteras Island, east of Pamlico Sound) and Wrightsville Beach, 10 miles east of Wilmington, are excellent vantage points from which to observe migratory movements. The following dates give some indication of when to expect the main flights:

Waterfowl: 15 February–1 April; 25 October–15 December
Shorebirds: 20 April–20 May; 10 August–15 October
Landbirds: 25 March–5 May; 10 September–5 November

In the tidewater belt of the Coastal Plain are a dozen or more natural lakes that, together with the coastal sounds, are favorite winter resorts for hordes of waterfowl (November through February). To the wide surface of Mattamuskeet, largest of the lakes, come Whistling Swans; to the submerged sand bars and mud flats of Pamlico Sound come Brant and Snow Geese; to nearly all these waterways, where there is isolation and food, come Canada Geese and ducks. The resulting waterfowl concentrations are undoubtedly North Carolina's paramount attraction to bird finders. For a wide variety of birds in winter, the Wilmington area has few rivals in North Carolina. Christmas bird counts commonly yield from 150 to 160 species.

Authorities

Edna Lanier Appleberry, Thomas D. Burleigh, Willie G. Cahoon, J. H. Carter III, B. Rhett Chamberlain, Elizabeth Barnhill Clarkson, Lockhart Gaddy, R. J. Hader, William L. Hamnett, Thomas L. Quay, Ben F. Royal, Phillips Russell, James L. Stephens, Jr., Henry M. Stevenson, Marcus B. Simpson, Jr., Paul W. Sturm, David L. Wray.

ANSONVILLE
Lockhart Gaddy's Wild Goose Refuge

Near this south-central town, 20 miles south of Albermarle, is **Lockhart Gaddy's Wild Goose Refuge.** Drive south from Ansonville on US 52 for 1.5 miles, turn left at a directional sign, and proceed a half-mile to the entrance.

The Refuge has a 10-acre pond, where, from October to mid-March, many hundreds of Canada Geese and smaller numbers of Mallards, American Black Ducks, Common Pintails, and Ring-necked Ducks pass the winter. Remaining on the Refuge all summer are a few decoy Canada Geese.

The Refuge started in 1934 when Lockhart Gaddy built a one-acre pond for fishing purposes. In the fall of that year, nine Canada Geese were induced by six decoy Canada Geese to alight on the pond, and here they remained all winter. Thereafter, when food supply and pond were alike enlarged, the number increased dramatically each season.

ASHEVILLE

Mt. Mitchell | Mt. Mitchell State Park | Craggy Gardens | Mt. Pisgah Area | Devil's Courthouse | Tanasee Bald | Richland Balsam

Lofty **Mt. Mitchell,** 33 miles northeast of Asheville and famous for its supremacy of elevation in eastern United States, is reached by driving north and eastward on the Blue Ridge Parkway for 27 miles and turning off left on State 128, which climbs 4.5 miles to a parking area near the summit of the mountain. Though Mt. Mitchell's elevation of 6,684 feet is unspectacular from afar—it is one of many closely associated peaks rising in a somewhat leisurely manner—the marvelous view from its higher slopes and the charm of the coniferous forest at its summit make up for any shortcomings in outward appearance.

Mt. Mitchell was once densely forested from base to summit, hardwoods predominating below an elevation of approximately 4,500 feet, conifers above; but lumbering and repeated fires laid waste all but a fringe of conifers at the summit. Ornithologically only the summit, which now comprises **Mt. Mitchell State Park** (1,469 acres), is of primary interest. Here, red spruces mixed with firs are of sufficient height and density to hold a small breeding population containing the following species of northern affinities: Ruffed Grouse, Saw-whet Owl (may be heard in the evening from the parking area), Black-capped Chickadee, Red-breasted Nuthatch, Brown

SAW-WHET OWL

Creeper, Winter Wren, Gray Catbird, American Robin, Golden-crowned Kinglet, Solitary Vireo, Black-throated Green Warbler, Blackburnian Warbler, Canada Warbler (more common below the summit), Slate-colored Junco.

In certain cutover areas on the lower slopes that have produced second-growth trees, additional northern species breed: Yellow-bellied Sapsucker (uncommon), Eastern Pewee, Veery, Cedar Waxwing, Black-throated Blue Warbler, Chestnut-sided Warbler, Common Yellowthroat, Rose-breasted Grosbeak. One or more Northern Ravens usually appear about the summit, and probably at least one pair nests on the mountain.

Although June and July are best for observing summer residents on the summit, since they are in full song at that time, more species can be seen beginning in late July, when the population is increased by Eastern Pewees, Carolina Wrens, Pine Warblers, Ovenbirds,

Hooded Warblers, American Redstarts, and other species that wander up the mountain from their nesting habitats below. From late August to early November the following northern transients appear commonly: Hermit Thrush, Swainson's Thrush, Ruby-crowned Kinglet, Black-and-white Warbler, Tennessee Warbler, Magnolia Warbler.

The Park road up Mt. Mitchell is closed nightly after eight o'clock. If one wishes to hear Saw-whet Owls, he must either stay overnight in the Park (there is a campground near the parking area) or make arrangements with Park officials to be let out after closing time.

The Blue Ridge Parkway, in its winding course from Virginia south and westward past Mt. Mitchell and Asheville to the Great Smoky Mountains, goes over high elevations where the birdlife is much the same as on the slopes of Mt. Mitchell and the Great Smokies (*see under* **Gatlinburg, Tennessee**). Five stops along the Parkway are particularly recommended. Mileages are from Oteen, a community directly east of Asheville, on US 70.

Craggy Gardens (5,497 feet elevation), on the Parkway 18 miles north of Oteen, features rhododendron 'gardens' or 'balds' overlooking hardwood forests below and superb scenery for miles beyond. June, when the rhododendrons are in bloom, is also the best time for finding mountain birds. From the visitor center, where Parkway naturalists are available for consultation, well-maintained trails lead through the gardens and forests and over adjacent peaks.

The following few breeding species show the variety of birds one may expect: Northern Raven, Gray Catbird, American Robin, Veery, Black-throated Blue Warbler, Chestnut-sided Warbler, Canada Warbler, Rufous-sided Towhee, Slate-colored Junco. Should the bird finder be at Craggy Gardens in September and October, just after a cold front, he may see large flights of south-bound hawks.

The **Mt. Pisgah Area** (4,900 feet), a tourist center on the Parkway 26 miles south of Oteen, is adjacent to rhododendron balds, oak forests, and spruce-fir woods through which foot trails run (map available at motel). High-country birds breeding here include the Ruffed Grouse, Cedar Waxwing, Rose-breasted Grosbeak, and many other species identical to those at Craggy Gardens and Mt. Mitchell. An ornithological feature after sunset in early spring is the flight-singing of American Woodcock, readily heard from the parking lot.

South and westward from the Mt. Pisgah Area, the Parkway penetrates spruce-fir forests at high elevations with many overlooks and well-marked nature trails leading away. The most rewarding, because they offer opportunities for Saw-whet Owls and other species of the upper heights of the Blue Ridge Mountains, are: **Devil's Courthouse** (5,700 feet), 45 miles from Oteen; **Tanasee Bald** (5,622 feet), 46 miles from Oteen; and **Richland Balsam** (6,540 feet), highest point on the entire Blue Ridge Parkway, 53 miles from Oteen.

ELIZABETH CITY
Pea Island National Wildlife Refuge

On North Carolina's Outer Banks, bounded on one side by the Atlantic and on the other by Pamlico Sound, the **Pea Island National Wildlife Refuge,** a part of Cape Hatteras National Seashore, occupies 5,915 acres of Hatteras Island from Oregon Inlet south almost to the town of Rodanthe. (The Refuge retains the name Pea Island because its northern two-thirds was once an island with that name, separated from Hatteras Island by an inlet later filled in by ocean storms.) Except for a ridge of low dunes, the sandy terrain is level, with a broad beach on the ocean front, salt marshes extending into Pamlico Sound, and fresh-water marshes and pools, created by dikes, in the interior. Trees are virtually absent, but there are shrubs around the pools. Predominating plant growth consists of grasses, although bulrushes are prevalent in the salt marshes, and such aquatic plants as smartweeds, pondweeds, and cattails have become established in the fresh-water marshes.

The Pea Island Refuge is famous as a wintering area for Snow Geese. Through December, January, and usually February, perhaps half the entire population from northeastern North America stay, along with Canada Geese and Brant, in the waters offshore in Pamlico Sound.

Shorebird migrations in May and October are consistently impressive, the following species being relatively abundant: Semipalmated Plover, Black-bellied Plover, Ruddy Turnstone, Greater Yellowlegs, Lesser Yellowlegs, Red Knot, Least Sandpiper, Dunlin, Semipalmated Sandpiper, Sanderling. A few shorebirds are always present at other times, with the possible exception of early summer.

Despite environmental limitations, a fine variety of birds nests in the Refuge and its vicinity. The salt marshes are occupied by Clapper Rails, Marsh Wrens, and Seaside Sparrows, and the marshes created by fresh-water impoundments have proved attractive to pairs of Mallards, American Black Ducks, and (surprisingly) Gadwalls, as well as Great Blue Herons, Green Herons, Little Blue Herons, Great Egrets, Snowy Egrets, Louisiana Herons, both Black-crowned and Yellow-crowned Night Herons, Least Bitterns, a few Glossy Ibises, and Red-winged Blackbirds. In the grasses, above the high-water mark, Eastern Meadowlarks are common; many Tree and Barn Swallows find suitable nesting sites; Cattle Egrets appear beside the Refuge roads; and Boat-tailed Grackles show up nearly everywhere.

Other birds in suitable habitats either on the Refuge or on small islands in Pamlico Sound are Wilson's Plovers, Killdeers, Willets, Laughing Gulls (nesting abundantly on islands in the Sound), Gull-billed Terns, Forster's Terns, Common Terns, Little Terns, Royal Terns, Black Skimmers, and a few Common Nighthawks.

State 12 coming south from Whalebone Junction on Bodie Island—northernmost of the Outer Banks—and crossing Oregon Inlet by bridge, traverses the entire length of the Refuge. To reach Whalebone Junction, and thence the Refuge, from Elizabeth City, take US 158 east and south. (Other routes to Whalebone Junction are US 64 east from Plymouth and US 264 east and north from Washington, North Carolina.) South from Whalebone Junction, on State 12, stop at the visitor center of Cape Hatteras National Seashore for information about the natural history of the Outer Banks.

Before and after crossing the bridge over Oregon Inlet, park the car and scan the watery expanse and adjacent flats for waterfowl, waterbirds, and shorebirds. Except during late spring and summer months, possibilities for waterbirds and waterfowl will include Common Loons, Horned Grebes, Buffleheads, White-winged and Surf Scoters, and Red-breasted Mergansers. From the bridge south for the next 13 miles, State 12 passes through the Refuge. About 6 miles from the bridge, walk east on a dike overlooking an impoundment where there are likely to be wintering or transient American Wigeons, Northern Shovelers, Ring-necked Ducks, and Lesser Scaups along with many American Coots. In spring and summer, Gray Catbirds, Common Yellowthroats, Rufous-sided Towhees, and Song Sparrows are numerous in the shrubby thickets along the dike.

Two miles farther on, inquire at Refuge headquarters whether there are any notable bird species or aggregations presently to be seen. Continue south of the Refuge on State 12 and, at Rodanthe, walk out on the fishing pier, which juts into the Atlantic well above the surf. Its tip is a fine vantage point for viewing the passage of sea birds, including Northern Gannets and Double-crested Cormorants, during winter and migration seasons.

MOREHEAD CITY
Great Lake

Between Morehead City and its sister city, Beaufort, US 70 crosses the Newport River, part of the Intracoastal Waterway, on a bridge and sand-filled causeway. South of the bridge is a large sandy island formed by dredging, and here many pairs of Little Terns and Black Skimmers nest, together with a few scattered pairs of Gull-billed Terns, Common Terns, and Wilson's Plovers. The nesting season extends from May to mid-July and sometimes longer. A road to the island goes south from State 70 just west of the center of Morehead City.

Great Lake in Croatan National Forest (306,300 acres), northwest of Morehead City, is a picturesque spot surrounded by age-old cypress, tupelo, and other virgin timber, good habitat for both Prothonotary and Swainson's Warblers. On the south side of the Lake, in isolated trees, is a colony of Double-crested Cormorants. A few American Anhingas and Ospreys nest in the same area. To reach Great Lake, drive west and northwest from Morehead City on US 70 for 17 miles to Havelock, then turn left and proceed on a Forest Service road for 8 miles to a hunt club on Ellis Lake. Cross Ellis Lake by boat—usually for hire at the hunt club—and walk 1.0 mile west to Great Lake. The shore must be explored on foot. Though Great Lake is difficult to reach, the results are well worth the effort, since it is undoubtedly one of the choice spots for birds in North Carolina.

RALEIGH
Lake Raleigh | Lake Wheeler | Swift Creek Lowgrounds | William B. Umstead State Park

This capital city, 362 feet above sea level, is on the Fall Line, where the Piedmont Plateau and the Coastal Plain meet. Within a 15-mile radius of the Capitol are small bodies of water, open fields, and woodlands—all good for bird finding at any season.

Two reservoirs, **Lake Raleigh** and **Lake Wheeler,** and their immediate surroundings southwest of the city, offer particularly fine opportunities. About 80 acres in size, Lake Raleigh is bordered variously by fields, willow thickets, pine woods, and hardwood swamps. Take US 64B west from the city, turn left on Avent Ferry Road for 0.5 mile, and then left on Lake Raleigh Road for 0.5 mile. Park the car near the first sharp turn to the left and walk along the same road, following the northeast shore. During the spring and summer, look for open-habitat birds back from the Lake, such as Eastern Meadowlarks, Orchard Orioles, Blue Grosbeaks, and Indigo Buntings. From October to April, look for waterfowl on the Lake. Although many may be in view here, a better vantage point is from the south shore, reached by returning to Avent Ferry Road, continuing 0.7 mile to Trailwood Drive (State 1348), and turning left to Tanager Street (State 1449). Turn left on Tanager Street, and immediately left again on an unmarked road for 0.7 mile to a pine woods on the right; park and walk 500 feet farther, then go left on a trail through the woods to a bluff overlooking the Lake. Among waterfowl in view should be Mallards, American Black Ducks, Gadwalls, Green-winged and Blue-winged Teal, American Wigeons, Wood Ducks, Ring-necked Ducks, Lesser Scaups, Ruddy Ducks, and Hooded Mergansers.

Larger Lake Wheeler (340 acres) is reached from the city by driving south from US 64B on South Saunders Street for 2 blocks, turning right on Lake Wheeler Road (State 1371) for 4.6 miles, and right again on State 1379 for 2.4 miles to a causeway over the upper end of the Lake. From the causeway and the adjacent high banks, one may see the same duck species as on Lake Raleigh, though in greater numbers. At the same time, Horned and Pied-billed Grebes and American Coots are often very common, and several species of gulls and terns are frequent visitants. In late summer and early fall, when the water level is usually low enough to expose mud flats, fifteen or more species of shorebirds and sometimes Water Pipits may be in view.

Excellent for transient warblers during April and the first of May, and for nesting landbirds later, are the **Swift Creek Lowgrounds,** south of the city. Starting at Memorial Auditorium, drive south on US 401 for 4.3 miles, then left on Old Stage Road (State 1006) for 2.2 miles to the bridge crossing Swift Creek. Park 1,000 feet beyond the bridge and begin looking for transient warblers and other birds along the road and in the hardwood swamps on either side. Prothonotary Warblers breed in the swamps, as do other parulids, together with Red-shouldered Hawks, woodpeckers (including the Pileated), and many flycatchers, thrushes, and vireos.

William B. Umstead State Park (5,200 acres), northwest of Raleigh, embraces the most extensive forest in the region. From belt road I 64 north of the city, drive northwest on US 70 for 6 miles, turn left into the Park and go on for 1.6 miles to a dead-end parking area between two lakes. From here trails go around the lakes and through the woods, which are mainly second-growth pines and hardwoods with scattered brushy clearings.

Wild Turkeys are the distinctive ornithological feature. Among fifty other breeding species are the Prairie Warbler, Blue Grosbeak, Bachman's Sparrow, and Field Sparrow in brushy clearings; elsewhere in suitably wooded upland and/or lowland habitats are the following: Whip-poor-will, Acadian Flycatcher, Carolina Chickadee, Brown-headed Nuthatch, Wood Thrush, Yellow-throated and Solitary Vireos, Northern Parula Warbler, Yellow-throated Warbler, Louisiana Waterthrush, Kentucky Warbler, Hooded Warbler, Scarlet and Summer Tanagers.

SOUTHERN PINES
Weymouth Woods Sandhills Nature Preserve

An area representing the pine-clad sandhills in the drier uplands of the Coastal Plain at the Fall Line is the **Weymouth Woods Sandhills Nature Preserve** (413 acres) in south-central North Carolina. From Southern Pines, drive southeast on Indiana Avenue to State 2074, then left for 0.75 mile to the entrance on the left, and proceed to the interpretive center.

In the surrounding forest of longleaf and loblolly pines, look for Red-cockaded Woodpeckers and Brown-headed Nuthatches in any season and for Solitary Vireos and Pine Warblers in late spring and summer. Take one of the trails leading down to and circling the beaver ponds. Their swampy edges and the adjacent pine-hardwoods forest will yield summer-resident Acadian Flycatchers, Prothonotary Warblers, Ovenbirds, Kentucky Warblers, Hooded Warblers, and American Redstarts, as well as many transient warblers in late summer and fall. Watch for a pair of permanent-resident Red-tailed Hawks.

WASHINGTON
Mattamuskeet National Wildlife Refuge | Swanquarter National Wildlife Refuge

The **Mattamuskeet National Wildlife Refuge,** on the North Carolina coast, is a must for the bird finder. Here, in winter, vast numbers of Canada Geese and Whistling Swans, together with throngs of ducks and American Coots, gather from their northern breeding grounds to form waterfowl concentrations as impressive as any in southeastern United States. And the remarkable thing about these concentrations is their accessibility.

The Refuge covers altogether 50,179 acres, 40,000 of which comprise the prominent physiographical feature, Lake Mattamuskeet, largest fresh-water body in North Carolina. Though about 18 miles long and 6 miles wide in extreme dimensions, the Lake is little more than 3 feet deep anywhere. Because it is so shallow, its bottom is easily disturbed by wave action, causing the waters to be perpetually muddy and preventing aquatic plants from taking root. Thus, for waterfowl, the Lake is merely a resting area. In order to feed, the birds either go to several Refuge impoundments with marshy shores where there are emergent or submerged plants, or make flight excursions elsewhere. A remarkable spectacle results in early morning when great flocks of birds rise from the Lake, where they pass the night in the safety of open water, and leave for their feeding grounds. In late afternoon when the flocks return to settle on the Lake, the spectacle is no less remarkable.

The peculiar development of the Refuge accounts for its unusual appearance and facilities. About 1914, with the idea of creating a 'New Holland,' futile attempts were made to drain Mattamuskeet so that the bottom soil could be used as farmland. Drainage canals and a pumping station, installed at enormous expense, failed. Later, in 1934, the Federal Government acquired the land as a waterfowl sanctuary, except for two areas set aside for managed hunting along the south side of the Lake. The pumping station was converted into a combined headquarters for Refuge personnel and lodge for visitors. The lodge has since been closed.

To reach the Refuge, drive east from Washington on US 264 to New Holland on the south shore of Lake Mattamuskeet, where new headquarters lies next to the old pumping station, recognizable by a 120-foot smokestack. At headquarters obtain a checklist of Refuge birds and instructions about travel by car on the Refuge.

The most satisfactory time to visit the Refuge is in fall (except during hunting season), winter, and early spring. The Whistling Swans usually arrive in November and remain until February. They stay near the southern and eastern shores of Mattamuskeet, and one can sometimes view them from the highway, perhaps 500 at a time. The Canada Geese arrive at about the same time but stay longer, usually until the middle of March. Although they are in sight from the shore, one can see them better by driving onto the canal 'peninsulas'—banks of the former canals now densely overgrown with trees and shrubs. A walk along the peninsulas in early morning when the geese are taking off for distant feeding grounds, or in evening when they are returning, will be rewarded by countless flock patterns silhouetted against the sky and by a tumult of sonorous voices.

The ducks, everywhere present from fall to spring, are for the most part Mallards, American Black Ducks, Gadwalls, Common Pintails, Green-winged and Blue-winged Teal, and American Wigeons—all surface feeders.

Mattamuskeet does not attract many wintering or transient water-birds or shorebirds. Summer birds are comparatively few in number of species. Broods of Wood Ducks occasionally appear along the canals. Ospreys nest in tall trees or stubs near the edge of the Lake. Great Blue Herons, Green Herons, Little Blue Herons, Great Egrets, and Least Bitterns occur in the marshy borders. Pileated Woodpeckers and Prothonotary Warblers are common in the cypresses and tupelos on the northern border of the Lake; Brown-headed Nuthatches, Pine Warblers, and Prairie Warblers frequently occupy the stretches of open pine woods back from the shore; Gray Catbirds, Brown Thrashers, White-eyed Vireos, and Rufous-sided Towhees are very numerous in the thickets and low trees on the canal peninsulas.

Over-wintering diving ducks are the feature of the **Swanquarter National Wildlife Refuge,** 15,501 acres of woods and salt-water marsh bordering Pamlico Sound not far southwest of the Mattamuskeet Refuge and administered by it. Although Whistling Swans winter in the area, their numbers are smaller than at Mattamuskeet. Redheads, Lesser Scaups, Buffleheads, and Ruddy Ducks are the prominent waterfowl.

The entrance drive to the Refuge leaves US 264 about 12 miles west of New Holland and leads south 2 miles to the patrolman's headquarters on Rose Bay.

WILMINGTON

Greenfield Park | Fort Fisher | Airlee Gardens | Wrightsville Beach | Orton Plantation Gardens | Orton Pond Dam | Battery Island

Within a 20-mile radius of this river-port city in southeastern North Carolina, one has quick access to ocean fronts, marshes, swamplands, hardwood forests, pinelands, and open countrysides.

The best bird finding in Wilmington is in **Greenfield Park** (250 acres), a short distance south from downtown on US 421. Half the acreage within this area comprises Greenfield Lake, a sanctuary and winter haven for a small but varied number of waterbirds and waterfowl: Pied-billed Grebes, Mallards, American Black Ducks, Gadwalls, American Wigeons, Common Pintails, Ring-necked Ducks, Lesser Scaups, Ruddy Ducks, and American Coots. All are easily viewed from a 5-mile drive that circles the Lake.

Standing in the water around the edge of the Lake are moss-draped cypresses, nesting habitat for Wood Ducks and Prothonotary Warblers in spring and summer. Across the drive from the pond is a forest of pines and hardwoods with shrubby borders, and here, in late spring and summer, appear such breeding birds as Great Crested Flycatchers, Acadian Flycatchers, Tufted Titmice, Brown-headed Nuthatches, Wood Thrushes, White-eyed Vireos, Yellow-throated Vireos, Northern Parula Warblers, Yellow-throated Warblers, Pine Warblers, Prairie Warblers, Hooded Warblers, Orchard Orioles, Northern Cardinals, Summer Tanagers, Common Grackles, Rufous-sided Towhees, and Bachman's Sparrows. In winter, Red-breasted Nuthatches, Winter Wrens, Hermit Thrushes, Golden-crowned Kinglets, Ruby-crowned Kinglets, Solitary Vireos, Myrtle Warblers, White-throated Sparrows, Fox Sparrows, and Song Sparrows appear regularly.

On the coast south of Wilmington is **Fort Fisher,** a state-owned historical site, reached by continuing south from Greenfield Park on State 421 through Carolina Beach to its end at the tip of a peninsula. Here is an ocean beach, a small protected bay bordered by a salt marsh, and, inland, a forest of live oaks. In winter, the small bay has both Common and Red-throated Loons, many Horned Grebes, and various diving ducks; and a stone jetty extending far southward from the peninsula often attracts American Osytercatchers, Ruddy Turnstones, and an occasional Purple Sandpiper. The mud flats east of the jetty bring in Marbled Godwits and many other shorebirds. In early fall the forest of live oaks abounds in transient small landbirds.

Eight miles east of Wilmington on US 76 are the landscaped **Airlee Gàrdens,** a plantation where Painted Buntings reside from mid-April to carly fall.

From Airlee Gardens, US 74 runs east along Wrightsville Sound

TUFTED TITMOUSE

to Wrightsville Station and crosses a bridge over the Intracoastal Waterway, a causeway over Wrightsville Sound, and a bridge over Banks Channel to **Wrightsville Beach.** While crossing the causeway, watch for shorebirds on the flats to the left. After crossing the Banks Channel bridge, turn left on Lumina Avenue and drive to the gate of the causeway leading out to the sewage-disposal plant. Park here and explore the mud flats and salt marshes in the vicinity.

Drive back to the bridge and continue south on Waynick Boulevard to the parking circle at the end, watching for birds along the way. The entire area can yield a good list of birds in any season: in winter, several kinds of diving ducks, as well as loons, Northern Gannets, Double-crested Cormorants, Greater Black-backed Gulls, Herring Gulls, Ring-billed Gulls, Bonaparte's Gulls, Forster's Terns, and Caspian Terns; in spring (15 April to 15 June) and fall (1 August to 1 October), many shorebirds; from early spring to fall, on or near the shore, American Oystercatchers, Black-bellied Plovers, Wilson's Plovers, Laughing Gulls, Little Terns, and Black Skimmers (abundant).

For another worthwhile tour, go south from Wilmington on US 17 and, immediately after crossing the Brunswick River Bridge, turn left on State 133 (River Road, marked by a sign 'To Orton Gardens'), leading directly to **Orton Plantation Gardens.** Here, in spring and summer, are Painted Buntings and many other breeding passerines. A diked riverside marsh attracts herons and egrets all year and, after being flooded in fall, brings in surface-feeding ducks and a few Fulvous Tree Ducks for the winter.

On leaving Orton Gardens, turn left on State 133 and continue to **Orton Pond Dam.** Stop here and listen for Pileated Woodpeckers in the adjoining deep woods. In spring and early summer, watch for Ospreys nesting on dead tree stumps near the Dam.

From Orton Pond Dam, continue south on River Road to Southport. Directly opposite the waterfront and about three-quarters of a mile offshore is **Battery Island,** site of a wading-bird colony each year from April to August. Although the low sandy island is about two-thirds of a mile long and about half that distance wide, the colony occupies only a dense growth of shrubby cedars and yaupons running parallel with the northern shoreline. The entire colony consists mainly of Little Blue Herons, Cattle Egrets, Great Egrets, Snowy Egrets, Louisiana Herons, Black-crowned Night Herons, Glossy Ibises, and White Ibises. Fish Crows and Boat-tailed Grackles nest in the same shrubby growth, and Mourning Doves on the ground. Where there are sandy flats above the high-tide line, American Oystercatchers, Willets, Gull-billed Terns, and Little Terns also nest. The bird finder wishing to visit the Island can hire a small boat in Southport.

Ohio

COMMON BOBWHITES

In pioneer days one of the finest deciduous forests on the continent covered Ohio. Oaks, hickories, maples, beeches, walnuts, and elms grew in lavish abundance. With the retreat of these forests before the onrush of agricultural activities came the retreat of the associated birdlife. Today the Common Bobwhite and other birds of open farmlands are predominating species in the state, where once forest species such as the Red-tailed Hawk, Whip-poor-will, Pileated Woodpecker, Wood Thrush, Yellow-throated Vireo, Cerulean Warbler, and Ovenbird were widespread.

The bird finder should not get the impression that all deciduous-forest birds have gone from Ohio, or that his opportunities for bird finding will now be limited to open fields, pastures, and dooryards. Woodland birds are plentiful in the state, for nearly every farm has its woodlot, nearly every community its shaded and shrub-bordered cemetery and park; and in the northeastern corner and southeastern quarter of the state there are many large forested tracts. In addition, there are the waterbird, waterfowl, and shorebird aggregations on the southern shore of Lake Erie and on numerous inland lakes and reservoirs.

Much of the eastern and southern terrain is hilly, with gorges and valleys: actually a continuation of the western slope of the Allegheny Plateau, though it is much less spectacular. On the other hand, near Lake Erie and in the western part of the state the terrain is prevailingly level to rolling, and generally low. The southern two-thirds of the state drains into the Ohio River, the remainder into Lake Erie.

The forests of the state are predominantly deciduous. Beech and maple prevail, with an admixture of wild cherry, white oak, and red oak. In drier situations, forests of oak and hickory are typical. Swamp forests consisting of ash, silver maple, and sometimes pin oak and swamp white oak occupy a large area in the northwest corner. In unglaciated southeastern Ohio the forest cover consists of mixed hardwoods, such as tulip tree, beech, white oak, black birch, and yellow buckeye. Conifers are decidedly local. Hemlock grows in the gorges of eastern Ohio and white pine in a few places in the northeastern part of the state. Pitch, scrub, and shortleaf pines are locally abundant in the southeastern sector. In several spots in the northeastern corner, forest, climate, and drainage conditions attract a few nesting birds characteristic of northern coniferous forests—e.g., the Black-throated Green Warbler, Northern Waterthrush, and Purple Finch.

Some of the birds breeding regularly in the deciduous woods and farmlands (fields, brushy lands, wet meadows, orchards, and dooryards) of the state are the following:

DECIDUOUS WOODS

Red-shouldered Hawk
Yellow-billed Cuckoo
Black-billed Cuckoo
Common Screech Owl
Barred Owl
Whip-poor-will
Yellow-shafted Flicker
Hairy Woodpecker
Downy Woodpecker
Great Crested Flycatcher

Acadian Flycatcher
Eastern Pewee
Blue Jay
Black-capped Chickadee (*chiefly northern Ohio*)
Tufted Titmouse
White-breasted Nuthatch
Wood Thrush
Blue-gray Gnatcatcher
Red-eyed Vireo
Warbling Vireo
Cerulean Warbler
Ovenbird
American Redstart
Baltimore Oriole
Scarlet Tanager (*uncommon*)
Rose-breasted Grosbeak (*chiefly northern Ohio*)

FARMLANDS

Common Bobwhite
Mourning Dove
Red-headed Woodpecker
Eastern Kingbird
(Prairie) Horned Lark
Barn Swallow
House Wren
Carolina Wren
Northern Mockingbird
Gray Catbird
Brown Thrasher
Eastern Bluebird
Yellow Warbler
Common Yellowthroat
Bobolink (*except southern Ohio*)
Eastern Meadowlark
Common Grackle
Northern Cardinal
Indigo Bunting
American Goldfinch
Rufous-sided Towhee
Savannah Sparrow (*chiefly northern Ohio*)
Grasshopper Sparrow
Henslow's Sparrow
Vesper Sparrow
Chipping Sparrow
Field Sparrow
Song Sparrow

Central and southern Ohio are far enough south to be within the breeding ranges of certain species characteristic of southern deciduous forests and brushy lands. Among these species the following are more or less regular as breeding birds, increasing in numbers toward the southern boundary of the state:

Turkey Vulture
Black Vulture (*southern Ohio*)
Carolina Chickadee
Bewick's Wren
White-eyed Vireo

Prothonotary Warbler
Worm-eating Warbler
Yellow-throated Warbler
(*southern Ohio*)
Louisiana Waterthrush
Kentucky Warbler
Yellow-breasted Chat
Hooded Warbler
Summer Tanager
Bachman's Sparrow

An outstanding ornithological attraction in Ohio is an area called the 'Oak Openings,' a few miles west of Toledo in the northwest corner of the state, where such diversified habitats as open oak woods, swampy woodlands, bogs, wet prairies, and dry prairies exist in remarkably close proximity. Within a square mile it is possible to find an extraordinary combination of landbirds; some of them are here at the northernmost or southernmost limits of their ranges. No bird finder should miss visiting this unique spot.

Along Lake Erie there are extensive cattail marshes bordered by grasses and sedges and, on higher ground, forests of oak, basswood, and maple. Often there are open, sluggish ponds within the marshes. Most of the marshes have a typical bittern-rail association. The Lake Erie shoreline is in some places rocky, in other places sandy, backed by sand dunes; in both situations, transient shorebirds frequently appear. East Harbor State Park (*see under* **Port Clinton**) and the Ottawa National Wildlife Refuge (*see under* **Toledo**) have particularly good resting and feeding places for waterfowl. Nearby are several mud flats, sand bars, and points of land—the best observation points in Ohio for a wide variety of waterbirds, waterfowl, and shorebirds.

Inland from Lake Erie are numerous lakes and marshes. Most of the bodies of water of ornithological interest are artificial—e.g., Mosquito Creek Reservoir near Warren, and Pymatuning Reservoir (*see under* **Linesville, Pennsylvania**) in the northeastern part of the state, Buckeye Lake near Hebron in the central part, and Grand Lake St. Marys (*see under* **St. Marys**) in the western part. Transient waterfowl gather in fair numbers on these bodies of water in spring and fall, many remaining through the winter. Shorebirds stop here too—more often in fall, when the water level is low.

Landbirds generally migrate through Ohio in a widely dispersed manner. Nevertheless, thrushes, kinglets, vireos, and warblers often move in large 'waves,' especially along the river valleys that trend in a north-south direction. On the Lake Erie shore in spring, north-bound small landbirds frequently 'bunch up' on peninsulas and near-shore islands before taking off across the Lake. Here, at the same time, north-bound hawks become numerous. On South Bass Island (*see under* **Put-in-Bay**) the migration of both small landbirds and hawks is often spectacular.

The following timetable applies chiefly to the main flights in northern Ohio:

Waterfowl: 1 March–10 April; 15 October–1 December
Shorebirds: 1 May–1 June; 1 August–5 October
Landbirds: 15 April–15 May; 1 September–25 October

Among migratory species of small landbirds that winter regularly in Ohio are the (Northern) Horned Lark, Brown Creeper, Golden-crowned Kinglet, Eastern Meadowlark, Red-winged Blackbird, Slate-colored Junco, American Tree Sparrow, and Song Sparrow. Evening Grosbeaks, Purple Finches, and Pine Siskins appear sporadically. Lapland Longspurs and Snow Buntings show up frequently in northern Ohio near Lake Erie; many Winter Wrens, Myrtle Warblers, American Goldfinches, and White-throated Sparrows regularly pass the winter in southern Ohio.

Authorities

Donald J. Borror, Louis W. Campbell, Vera Carrothers, Floyd B. Chapman, F. W. Fais, Arthur P. Harper, Edward Hutchins, Richard B. Klein, Harry S. Knighton, Harold Mayfield, Karl H. Maslowski, Vincent P. McLaughlin, Carl M. Newhous, Edward S. Thomas, Milton B. Trautman, Walter A. Tucker, Harold E. Wallin, Arthur B. Williams.

Reference

A Guide to Ohio Outdoor Education Areas. Columbus: Ohio Department of Natural Resources, 1970.

CINCINNATI

Cincinnati Nature Center | Newtown Gravel Pits | Spring Grove Cemetery | Sharon Woods County Park
California Nature Preserve | Miami–Whitewater Forest County Park | Shawnee Lookout County Park | Little Miami River Anthony Meldahl Dam

NORTHERN CARDINAL

Cincinnati, in the southwestern corner of Ohio on a bend of the Ohio River, has no claims to spectacular bird aggregations, and yet it has habitats sufficiently diversified to provide many hours of pleasant and profitable bird finding at any season of year. In Christmas bird counts for the past twenty-five years, this city has led the nation in the number of Northern Cardinals reported. It might well be, as one naturalist suggests, that Cincinnati is the cardinal capital of the world.

The **Cincinnati Nature Center** (750 acres), in a delightful rural setting 20 miles from downtown Cincinnati, concentrates on outdoor education and is a pleasure for any bird finder to visit. The interpretive building and 6.5 miles of trails through hardwood forest, along ravines, past a 5-acre lake and several small ponds, and across old fields with brushy edges, give a real sampling of the different habitats and the different birds. In addition, the Center, being a clearing house for local information on birds, has personnel ready to advise the bird finder and direct him to any 'hot' spots. The Center is open to the public from Monday through Friday throughout the year. Although weekends are reserved for members and their guests, the bird finder unable to visit the Center during the week may obtain permission for a weekend visit by phoning or writing the Cincinnati Nature Center, 4949 Tealtown Road, Milford, Ohio 45150.

To reach the Center from downtown Cincinnati, go east on US 50, turn right on State 125, and then left (east) on State 32. Just after passing the ring road, I 275, turn left on Gleneste–Williamsville Road to a 'T,' then right for about two blocks, and left again on Tealtown Road for about 2 miles to the entrance on the left. Without going into the city, one can exit from I 275 east on State 32 and continue as above.

Bird finding is excellent at the Center throughout the year. Evening Grosbeaks, Purple Finches, Slate-colored Juncos, and White-throated Sparrows are common in winter. An abundant mast crop usually brings Red-headed Woodpeckers. The lake and ponds attract a very limited number of waterfowl in migration. Between late February and late April, American Woodcock give their flight songs above the fields.

During the warbler migration in early May some twenty-two species are regularly recorded. Red-tailed, Red-shouldered, and Broad-winged Hawks, American Kestrels, Common Screech Owls, Great Horned Owls, and Barred Owls nest in the big woods. Other birds nesting on the property include the Pileated and Red-bellied Woodpeckers, Acadian Flycatcher, Carolina Wren, Brown Thrasher, Eastern Bluebird, and such warblers as the Blue-winged, Yellow, Cerulean, Yellow-throated, Common Yellowthroat, Kentucky, and Yellow-breasted Chat. Both the Scarlet and Summer Tanagers and Rufous-sided Towhees are abundant, and Northern Cardinals are positively thick.

Ask at the Nature Center about conditions and directions for the nearby **Newtown Gravel Pits.** This is an on-and-off situation, with bird finding sometimes excellent and sometimes very poor.

Although all city and adjacent county parks are good places to observe transient and nesting landbirds, **Spring Grove Cemetery,** about 6 miles from downtown, is one of the best places for year-round bird finding. Drive north on I 75, exit on Mitchell Avenue, go north to Spring Grove Avenue, and left for 0.5 mile to the entrance on the left. At the gate, ask permission to look for birds.

Several small sections are still quite natural and wild, but most of the area presents a park-like appearance, with diversified plantings of ornamental shrubs interspersed with native stands of beech, oak, and maple. There are several small lakes that invite such birds as Pied-billed Grebes, Black-crowned Night Herons, Wood Ducks, and American Coots during spring migration in April and May. Many transient warblers and other small landbirds tarry here during the last two weeks of April and throughout May. Among warblers that remain to nest, the Yellow, Cerulean, Kentucky, and Common Yellowthroat are common. Other birds presumably nesting are the Yellow-shafted Flicker, Hairy and Downy Woodpeckers, White-breasted Nuthatch, Carolina Wren, Cedar Waxwing, Scarlet Tanager, Chipping Sparrow, Field Sparrow, and Song Sparrow. In winter the Cemetery is one of the best places to see such erratics as Evening Grosbeaks, Purple Finches, Pine Siskins, and Red and White-winged Crossbills.

Fourteen miles north of downtown is **Sharon Woods County Park** (740 acres), with a large artificial pond, open fields, shrubby areas, and a climax woodland of oak, beech, and maple. Here bird finding, though best in April, May, September, and October, gives gratifying results at any time of year. Transient Pied-billed Grebes, American Wigeons, Blue-winged Teal, Ring-necked Ducks, Lesser Scaups, and American Coots often occupy the pond; in the meadows, nesting Killdeers, (Prairie) Horned Larks, and Eastern Meadowlarks are common. The Acadian Flycatcher, Carolina Chickadee, Wood Thrush, Blue-gray Gnatcatcher, Red-eyed Vireo, and Warbling Vireo are among many breeding species. Transient warblers, common in both spring and fall, are the Tennessee, Nashville, Cape May, Black-throated Green, Black-throated Blue, and Canada. Common also are transient Hermit, Swainson's, and Gray-cheeked Thrushes.

To reach Sharon Woods, drive north on I 75, turn east on I 275 to US 42, then south to the entrance on the left.

A 60-acre woodland, principally of beech, oak, and maple, through which a small creek winds, the **California Nature Preserve** is excellent for summer-resident birds such as the Scarlet Tanager, Summer Tanager, and Louisiana Waterthrush. It is also one of the few spots within the city limits where one can expect to find Great Horned Owls, Pileated Woodpeckers, Acadian Flycatchers, Worm-eating Warblers, and Cerulean Warblers nesting regularly. To reach the Preserve, drive east from Cincinnati on US 52 to the entrance about a mile beyond the Little Miami River bridge.

Miami–Whitewater Forest County Park (2,031 acres) includes bits of fine beech-maple forest and several small lakes that attract a fine group of nesting birds: Red-tailed Hawk, Great Horned Owl, the most common woodpeckers including the Pileated, and a great variety of warblers—Black-and-white, Cerulean, Ovenbird, Louisiana Waterthrush, Kentucky, and Hooded. The Park is also a good spot to hear the flight song of the American Woodcock.

To reach the Park, take I 74 west, and then go north on Dry Forks Road for about 2 miles to the entrance.

Shawnee Lookout County Park (1,010 acres), at the mouth of the Great Miami River west of the city, is one of the better territories for transient shorebirds and ducks as well as landbirds. The Park, covered with old second-growth timber interspersed with thickets and meadows, occupies some river bottomland. In late winter and early spring, when high water floods the bottomlands, ducks concentrate in fair numbers, among them both surface-feeding and diving species such as Mallards, American Black Ducks, Gadwalls, Common Pintails, Green-winged and Blue-winged Teal, American Wigeons, Common Goldeneyes, Buffleheads, and mergansers. The Park is one of the few places in the county where the Prothonotary Warbler nests regularly, and the only place where, in recent times, Black Vultures and American Redstarts have been found nesting. Many species of songbirds appear either as nesting birds or transients. Myrtle Warblers and American Goldfinches commonly winter here.

To reach Shawnee Lookout Park, take I 74 west from downtown, go south on I 275 to Kilby Road, turn south to US 50, then west; or go west from downtown on US 50 to Elizabethton, and take Lawrenceburg Road east and south to the Park entrance.

In the extensive bottomlands in Indiana, across the Great Miami River from Shawnee Lookout Park, is an old oxbow, an especially fine place for shorebirds in both spring and fall. About twenty species show up here, including the Semipalmated Plover, Lesser Golden Plover, Solitary Sandpiper, Greater and Lesser Yellowlegs, White-rumped Sandpiper, Baird's Sandpiper, Short-billed Dowitcher, Stilt Sandpiper, and Buff-breasted Sandpiper. Among waterbirds that come in as transients are Horned Grebes, Double-crested Cormorants, Great Egrets, Black-crowned Night Herons, Bonaparte's Gulls, Common Terns, and Black Terns.

Consult the Cincinnati Nature Center for the most up-to-date directions to the oxbow. Most of the land around the oxbow is privately owned; thus the bird finder must stay on the road or close to the water's edge.

East of Cincinnati, the mouth of the **Little Miami River** offers approximately the same waterfowl and waterbirds as those already listed. From downtown Cincinnati, go east on US 52. Just before reaching the west bank of the Little Miami, turn right toward the Ohio River to a parking lot above a marina. From this lot it is possible to walk to the mouth of the Little Miami, where, in winter when it is largely frozen over, one may find a fair concentration of ducks, especially Mallards and American Blacks with a scattering of Common Goldeneyes and Common Mergansers. Swallows circle above the area during spring and fall migrations; Killdeers nest along the gravel drives; (Prairie) Horned Larks and Eastern Meadowlarks in nearby fields; and Traill's Flycatchers in the scrubby willow growths. In spring a wide variety of vireos and warblers move through the bottomlands: Yellow-throated and Philadelphia Vireos, Magnolia, Cape May, Bay-breasted, Blackpoll, and Canada Warblers.

From early November through mid-April the **Anthony Meldahl Dam,** across the Ohio River, can be a good place for bird finding. Drive east from Cincinnati on US 52 for about 30 miles to just beyond the town of Moscow, and watch for the turn to the Dam on the right. There is a parking lot and an observation platform at the Dam. A spotting scope is almost essential, since the viewer is still a considerable distance from the River itself. Among birds commonly sighted are Common Loons, Horned Grebes, Pied-billed Grebes, Canada and Snow Geese, surface-feeding and diving ducks, and Herring and Ring-billed Gulls.

Below the Dam on the Kentucky side is a large gravel bar often frequented from late August to October by a variety of shorebirds, including Semipalmated Plovers, Ruddy Turnstones, and Pectoral Sandpipers. Ospreys sometimes appear overhead. To reach this area from central Cincinnati, drive south on I 75 across the Ohio River and then east at the first exit on State 8 for about 30 miles. Park along the road and follow fishermen's paths down to the gravel bar.

CLEVELAND

North Chagrin, Rocky River, and Hinckley Metroparks | Gordon Park | White City Park | Edgewater Park | Shaker Lakes

Forming a circle of green around this great city are more than 18,000 acres in ten reservations owned by the Cleveland Metroparks System. Thus preserved are a diversity of natural habitats—rivers, lakes, swamps, rocky cliffs, tree-clad hills, and floodplain woodlands—with a corresponding variety of birds in any season. Considered outstanding for birds are the **North Chagrin, Rocky River,** and **Hinckley Metroparks.**

North Chagrin Metropark (1,719 acres) is a woodland composed for the most part of beech, sugar maple, and hemlock along the Chagrin River, 16 miles east of central Cleveland. Drive east on I 90 to just beyond the city limits and go south on State 91 to the entrance on the left. Trails lead to all the significant features, including Sunset Ponds, where an aerating system keeps some of the water open all winter. Mallards and American Black Ducks consequently over-winter, feeding on grain crops planted especially for them.

Some permanent-resident birds in North Chagrin are the Ruffed Grouse, Barred Owl, Pileated Woodpecker, Red-bellied Woodpecker, Hairy Woodpecker, Downy Woodpecker, Black-capped Chickadee, Tufted Titmouse, White-breasted Nuthatch, and Northern Cardinal. Summer residents include the Great Crested Flycatcher, Eastern Pewee, Wood Thrush, Yellow-throated Vireo, Red-eyed Vireo, Black-throated Green Warbler, Cerulean Warbler, Ovenbird, Louisiana Waterthrush, Hooded Warbler, American Redstart, Baltimore Oriole, Scarlet Tanager, and Rose-breasted Grosbeak. Many migrating thrushes, warblers, and fringillids pass through in the spring and fall. Evening Grosbeaks and Pine Siskins are often among the common winter visitants: Purple Finches,

Slate-colored Juncos, American Tree Sparrows, and White-throated Sparrows. A few Canada Geese are resident, nesting on a small island in Sunset Pond.

Rocky River Metropark (5,667 acres), running north and south along the meandering Rocky River on the western limits of Cleveland and Lakewood, has a typical northern Ohio floodplain forest of cottonwood, sycamore, black walnut, boxelder, Ohio buckeye, and black maple. The upper valley has fine groves of beech and sugar maple. Drive west from central Cleveland on US 20, crossing Rocky River; turn south on State 252 (Columbia Road); then east on Cedar Point Road to the parking lot for the interpretive center, 0.25 mile north, where information is available.

The area is especially good during migration for a few surface-feeding ducks and many small landbirds. Among breeding species, many of them the same as in North Chagrin Metropark, are the Green Heron, Killdeer, Spotted Sandpiper, Rough-winged Swallow, Carolina Wren, Yellow Warbler, Common Yellowthroat, American Goldfinch, Rufous-sided Towhee, and Song Sparrow.

Hinckley Metropark (1,924 acres), 20 miles south from central Cleveland, embraces Hinckley Lake, hillside woods of beech, maple, oak, and chestnut, and spectacular ledges of conglomerate rock. To reach the Metropark, follow I 71 south, exit east on State 303, then turn south and east on Bellus Road to a park road on the right that leads past the dam at the north end of Hinckley Lake. In early spring and summer, scan the Lake for a variety of surface-feeding ducks, including Wood Ducks, some of which stay through summer. In late summer, look for Great Egrets and other wading birds.

The Metropark is notable as a haven for Turkey Vultures. On an early morning in April as many as fifty to a hundred of these birds may be seen taking off from the high ledges where they roost at night, to soar in the air at one time. Where there are shrubby places near the Lake, listen and search for summer-resident Black-billed Cuckoos, Traill's Flycatchers, and Blue-winged Warblers.

In fall, winter, and early spring, many waterbirds and waterfowl gather off the Lake Erie shore behind the breakwaters. Particularly good views are afforded from **Gordon Park** (112 acres) and **White City Park** (12 acres) on the east shore, and **Edgewater Park** (17 acres) on the west shore. *For Gordon Park,* drive east on I 90, exit north on East 72nd Street and go to Lake Shore Boulevard, which immediately crosses the Park close to shore. *For White City Park,* go east on I 90, exit north on 136th Street, turn east on State 283 (Lake Shore Boulevard), turn left on a narrow road just before the intersection with East 140th Street, and park northeast of the Easterly Sewage Treatment Plant. *For Edgewater Park,* drive west on US 20 and take the Edgewater Park Exit at West 70th Street.

As long as there is open water, views from any one of these vantage points should yield Horned Grebes and diving ducks: Lesser Scaups, Common Goldeneyes, Buffleheads, Ruddy Ducks, Red-breasted Mergansers, sometimes a few Redheads and Canvasbacks, and occasionally an Oldsquaw or two. One can count on an abundance of gulls—especially Herring, Ring-billed, and Bonaparte's—a few Greater Black-backed, and the possibility of a Glaucous or a Black-legged Kittiwake.

Shaker Lakes are two small bodies of water called Lower Shaker Lake and Upper or Horseshoe Lake, in a 275-acre public park east of Cleveland, between Shaker Heights and Cleveland Heights. A few transient grebes, wading birds, and ducks occasionally visit the Lakes. During spring migration in April and early May, transient thrushes, vireos, and many warblers are prevalent.

To reach Shaker Lakes from central Cleveland, drive east on US 322 about 7 miles, turn right on Coventry Road, and left on North Park Boulevard to Lower Shaker Lake; continue on North Park Boulevard to Horseshoe Lake, 0.5 mile east of the intersection with Lee Road. The best starting point for a walk is the Shaker Lakes Regional Nature Center, at 2600 South Park Boulevard, on Lower Shaker Lake. Trails radiate out from here to the best bird-finding areas in the Park.

COLUMBUS

Greenlawn Cemetery | Scioto River | Highbanks, Sharon Woods, Blendon Woods, Blacklick Woods, and Darby Creek Metropolitan Parks

Situated on a level to gently rolling plain very near the geographical center of Ohio, this capital city has in its vicinity ponds and streams,

LONG-EARED OWL

several unspoiled woodlands, and large reservoirs—all offering opportunities for bird finding.

An outstanding place for transient landbirds is **Greenlawn Cemetery,** on the southwestern edge of the city. This area contains, within one-quarter square mile, native oaks, introduced conifers and shrubs, lawns, one small pond in the center, and wet fields in the southeastern section. In March and April, owls, particularly the Long-eared and Saw-whet, sometimes stay in the conifers during the day. At the peak of migration in early May, one can list from seventy to a hundred bird species, including nearly all warblers known to Ohio and such elusive birds as the Yellow-bellied Flycatcher, Philadelphia Vireo, and Lincoln's Sparrow.

To reach the Cemetery, take I 71 west and south from the center of the city, and exit on Greenlawn Avenue (west), which leads directly to the Cemetery.

Transient waterbirds and waterfowl can be seen in March, early April, October, and November on the **Scioto River** from Scioto Boulevard between Greenlawn Avenue and Mound Street. Pied-billed Grebes, Redheads, Canvasbacks, Lesser Scaups, Common Mergansers, Red-breasted Mergansers, Herring Gulls, and Ring-billed Gulls are among birds that may be expected. Although few birds winter along this stretch of the River, Lesser Scaups have been seen in the fast-moving water just below Greenlawn Dam.

To reach that part of the River, exit from I 71 on Greenlawn Avenue (east), cross the River, and turn left on Scioto Boulevard; or turn left on a dead-end road just before crossing the River.

Owned by the Metropolitan Park District of Columbus and Franklin County are 7,500 acres that include five natural-area parks, primarily forests with such features as ravines, streams, wetlands, small bodies of water, and open spaces. All have nature trails. Perhaps the finest of the five is **Highbanks Metropolitan Park,** 9466 Columbus Pike, 3 miles north of Worthington, and reached by driving north from outer-belt I 270 on US 23 for about 2 miles. Named for the massive shale bluffs overlooking the Olentangy River, Highbanks (1,050 acres) has upland forest of beech and maple in the many deep ravines. Nature trails lead from old fields to bottomlands, providing ready access to representative habitats.

Permanent-resident birds include the Common Screech Owl, Great Horned Owl, Barred Owl, Pileated Woodpecker, Red-bellied Woodpecker, Carolina Chickadee, Tufted Titmouse, and Carolina Wren; summer residents, the Green Heron, Wood Duck, Red-tailed Hawk, Acadian Flycatcher, Bewick's Wren, Wood Thrush, Blue-gray Gnatcatcher, Yellow-throated Vireo, Red-eyed Vireo, Blue-winged Warbler, Cerulean Warbler, Kentucky Warbler, Ovenbird, Louisiana Waterthrush, and Scarlet Tanager. Among transients are the Hermit Thrush, Swainson's Thrush, Gray-cheeked Thrush, Veery, Solitary Vireo, about twenty species of warblers—e.g., the Black-throated Blue and Black-throated Green, Blackburnian, Palm, Northern Waterthrush, Mourning, and Canada—and

Rose-breasted Grosbeak. Winter visitants are the Purple Finch, Pine Siskin, Slate-colored Junco, American Tree Sparrow, and White-throated Sparrow.

Among birds nesting along the forest edges and open country with grassy fields and brushy areas are the Common Bobwhite, Eastern Bluebird, Bobolink, Indigo Bunting, Grasshopper Sparrow, Henslow's Sparrow, and Field Sparrow. During migrations and in winter, Highbanks has habitats attractive to Golden-crowned Kinglets, Myrtle Warblers, White-crowned Sparrows, and Swamp Sparrows.

North of Columbus, **Sharon Woods Metropolitan Park,** 6911 Cleveland Avenue, reached by exiting north from I 270 on Cleveland Avenue, has 760 acres of level terrain with prairie-like old fields interrupted by steep forested ravines. Along the Park's 2 miles of nature trails the rewards for the bird finder compare favorably with those in Highbanks.

Northeast of Columbus, **Blendon Woods Metropolitan Park** (4265 East State Route 161), reached by exiting east from I 270 on State 161 for 1.0 mile, has 646 acres of deep wooded ravines, wooded uplands, and old fields. Its major ornithological attraction is 11-acre Walden Pond, the finest place in central Ohio for waterfowl during colder months. The birds are induced to stay the winter with tons of grain scattered annually around and in the Pond, kept free of ice by an aerating system. So that visitors may enjoy, as well as fully appreciate, the avian aggregation even in severest weather, the Park has heated observation shelters equipped with spotting scopes, and a naturalist on duty to answer questions.

The waterfowl usually in greatest evidence are Canada Geese, Mallards, American Black Ducks, Gadwalls, Common Pintails, Blue-winged Teal, American Wigeons, and Lesser Scaups. Among others are Green-winged Teal, Northern Shovelers, Redheads, Canvasbacks, Common Goldeneyes, Buffleheads, and Hooded Mergansers. Before and after the coldest months, there are usually Pied-billed Grebes, occasionally Horned Grebes, and American Coots. Ospreys sometimes show up around the Pond, and a few wading birds, such as Great Blue Herons, along the shore.

East of Columbus, **Blacklick Woods Metropolitan Park,** 6975 East Livingston Avenue, reached from I 70 by exiting north on

Bryce Road and turning right on East Livingston Avenue, encompasses 632 acres. Included is a 56-acre tract of primeval forest consisting of beech and sugar maple mixed with white ash, black cherry, hickory, white oak, and red elm on the uplands, silver maple, pin oak, bur oak, and swamp white oak in wet areas. Birdlife is about equivalent to that of Highbanks in variety but is more abundant.

About 6 miles west of Columbus, **Darby Creek Metropolitan Park,** 1775 Darby Creek Drive, reached by driving west on US 40 and then south for 3 miles to Darby Creek Drive, has 369 acres of oak-hickory woodland, open meadows, and bottomland woodland bordering Big Darby Creek. During migration in late April and early May, the bluffs overlooking the Creek are fine vantage points for watching warblers at treetop level. An abundance of flowering shrubs and trees attracts many hummingbirds in late spring and early summer, as do meadow flowers later in the season.

DAYTON

Aullwood Audubon Center | Englewood Reserve

Both the **Aullwood Audubon Center** and the **Englewood Reserve,** northwest of this city in southwestern Ohio, are excellent for bird finding throughout the year. To reach them, drive north from Dayton on I 75, exit west on I 70, take the next exit (State 48) north a short distance, and then turn right (east) on US 40 across a huge dam. At the end of the dam's guard rail, turn right into the Center, or left into the Reserve.

The Aullwood Center (70 acres), owned and operated by the National Audubon Society, has a diverse landscape of woodlands, a wooded swamp, a pond and stream, old fields, a grassy meadow, and a 10-acre restored grass prairie. Mature stands of oak and hickory, mixed with basswood, sugar maple, and tulip tree comprise the upland woods; sycamore and cottonwood line the stream; white ash dominates the swamp and, along with honey locust, black walnut, and red cedar, invades the adjacent fields. In this variety of habitats, about 140 bird species are recorded each year.

Breeding birds include the Common Bobwhite, Great Horned

Owl, Red-bellied and Red-headed Woodpeckers, Acadian and Traill's Flycatchers, Carolina Chickadee, Wood Thrush, White-eyed and Red-eyed Vireos, Blue-winged and Kentucky Warblers, Common Yellowthroat, Yellow-breasted Chat, Orchard and Baltimore Orioles, Indigo Bunting, and Field Sparrow. From mid-March to mid-April the American Woodcock gives its crepuscular aerial and vocal displays above the fields. The prairie in winter attracts many fringillids, especially the American Tree Sparrow, White-throated Sparrow, Song Sparrow, and an occasional Swamp Sparrow.

The Englewood Reserve (1,000 acres), managed by the Dayton-Montgomery County Park District, includes a part of the valley and floodplain of the Stillwater River, where there are marshy areas, impounded waters, and woodlands. A pine plantation frequently has Long-eared Owls and a Saw-whet Owl or two in the winter. Highly recommended is a walk in the Reserve at the height of the spring warbler migration, which peaks at about 7/8 May.

HEBRON

Buckeye Lake | Hebron Fish Hatchery

In 1833 the 'Great Swamp,' 25 miles east of Columbus and just south of Hebron, was flooded to form a reservoir called **Buckeye Lake** and designed to regulate the water supply for a series of canals that were to connect the Ohio River with Lake Erie. The Lake became a haven for waterbirds and waterfowl in a section of the country that then lacked large bodies of water. When, not long thereafter, the National Road was built across Ohio just north of Buckeye Lake, the surrounding country was soon deforested to provide farmlands. In later years the swampy area bordering the Lake was drained to make room for summer resorts.

Buckeye Lake as seen today is a shallow body of water 7.5 by 1.5 miles in maximum diameters, with a water surface of approximately 4,200 acres in a part of the state thickly populated during vacation season. Buckeye Lake State Park on the north shore, reached from I 70 by exiting south on State 79, is a resort center. Encircling the Lake are state highways from which roads lead off to the shore. Boats can be hired in many places.

At the shallow, eastern end of the Lake remains a small section of a vast cattail marsh, where one may find nesting the Least Bittern, Virginia Rail, Common Gallinule, Marsh Wren, and Red-winged Blackbird. In shrubs on the shore are Traill's Flycatchers and Yellow Warblers. Boats may be hired at Thornport (on State 13 south from I 70) to investigate this marsh, which extends west beyond Honey Point.

On the north side of the Lake are three patches of woodland, two of which—Lakeside Woods and Bound's Woods (Jack's Neck Woods,)—are adjacent to the Lake. The best bird finding is in Bound's Woods, one mile east of Buckeye Lake State Park and conveniently reached by car or on foot. Although its 200 acres of white ash and red maple in the lowlands, and beech, sugar maple, and white oak on higher ground, have been lumbered and grazed, the following are among the species still nesting: Red-shouldered Hawk, Barred Owl, Red-bellied Woodpecker, Great Crested Flycatcher, Eastern Pewee, Tufted Titmouse, White-breasted Nuthatch, Wood Thrush, Red-eyed Vireo, Ovenbird, Yellow-breasted Chat, and American Redstart. Great Blue Herons nest in a small colony. In late April and early May, Bound's Woods is a good place to look for small transient landbirds.

A good representation of waterfowl shows up on the Lake during migration, from late February to April and during October and November. The population usually includes Mallards, American Black Ducks, American Wigeons, Common Pintails, Green-winged and Blue-winged Teal, Northern Shovelers, Wood Ducks, Redheads, Ring-necked Ducks, Canvasbacks, Lesser Scaups, Common Goldeneyes, Ruddy Ducks, and Red-breasted Mergansers. Early in fall, Herring Gulls, Ring-billed Gulls, Bonaparte's Gulls, Common Terns, and Black Terns appear. Late October and early November is the time for Canada Geese. In November, Common Loons pass over the Lake, sometimes dropping down to rest and feed. Since the waterfowl tend to stay at the deep western end of the Lake, Seller's Point, about 2.5 miles southwest of the State Park and just south off State 79 on State 360, is one of the best vantage points.

A satisfactory place for shorebirds is the **Hebron Fish Hatchery,** northwest of Buckeye Lake. From I 70, exit south on State 37 for about a mile, turn east (left) on State 79 for about another mile; immediately after crossing a canal, turn north (left) along the canal about 0.5 mile to the Fish Hatchery. Some of the shorebirds frequenting the edges of the ponds, from late August to early October, when the water is low, are the Semipalmated Plover, Killdeer, Spotted Sandpiper, Solitary Sandpiper, Greater and Lesser Yellowlegs, Pectoral Sandpiper, Least Sandpiper, and Semipalmated Sandpiper. At the same time of year, back of the pools, a swampy woods penetrated by foot trails attracts large numbers of small transient landbirds.

HILLSBORO
Fort Hill State Memorial

One of the best preserved of Ohio's prehistoric Indian hilltop enclosures, built long before the discovery of America, is in the **Fort Hill State Memorial** (1,197 acres) in the southwestern part of the state. The picturesque setting of the enclosure, the remarkably diversified habitats in its vicinity, and the presence of many southern bird species make the Memorial one of the outstanding spots for bird finding.

The ancient enclosure sits on the summit of Fort Hill, 1,283 feet above sea level and 423 feet above Baker Fork, a stream on the north and west that has cut a deep, scenic gorge through shale and limestone. The higher, exposed slopes of Fort Hill are covered with open oak woods, and the more shaded north and east slopes, with densely growing sugar maples and tulip trees. Flowering shrubs and wildflowers are everywhere abundant, providing a succession of blooms from early April to fall. On the Memorial property, which embraces Fort Hill and part of the gorge, are abandoned fields reverting to brushy thickets and saplings; beyond the property stretch meadows and cultivated fields.

When migration is under way in April and early May, no fewer than fourteen warblers species may be found in the wooded environments. Breeding birds are as diversified as the habitats. In the woods nest the Red-tailed Hawk, Barred Owl, Great Horned Owl, Yellow-billed Cuckoo, Whip-poor-will, Red-bellied Woodpecker, Acadian Flycatcher, Great Crested Flycatcher, Carolina Chickadee, Tufted Titmouse, Wood Thrush, Blue-gray Gnatcatcher, Yellow-throated Vireo, Black-and-white Warbler, Worm-eating Warbler, and Scarlet Tanager.

The Louisiana Waterthrush and the Kentucky Warbler—and, more rarely, the Northern Parula Warbler—may be spotted in the dark, wooded gorge and in the ravine entering the gorge; the Yellow-throated Warbler appears in the edges of woods near streams along which sycamores grow. The Summer Tanager is a characteristic bird of the open woods and groves. Both the Carolina and the Bewick's Wrens are fairly common around the Memorial buildings. The Eastern Phoebe and the Rough-winged Swallow have nests on the cliffs of the gorge.

In the old fields at the south end of the Memorial, where there are tangles of weeds, shrubs, and vines, Field Sparrows are numerous, and there are often several pairs of Prairie Warblers. Other

BEWICK'S WREN

species inhabiting the brush areas near woodland edges, fence rows, and streams are the Yellow-breasted Chat, Indigo Bunting, Rufous-sided Towhee, and Bachman's Sparrow. The Turkey and Black Vultures are conspicuous permanent residents of the Memorial, nesting in hollow trees, in crannies in the cliffs of the gorge, or on the ground.

The Fort Hill State Memorial may be reached from Hillsboro by driving southeast on State 124 for 16 miles, and turning north on State 41 for 3 miles to the entrance on the left.

LEBANON
Fort Ancient State Memorial

The **Fort Ancient State Memorial** (696 acres), in southwestern Ohio, embraces a prehistoric Indian enclosure on a bluff rising about 270 feet above the Little Miami River. Age-old walls, varying from 4 to 23 feet in height, circle the crest of the bluff, to form an enclosure of approximately 100 acres. From the walls of the enclosure, deep, forested ravines, each with its own stream, lead down to the River and valley below. Predominating trees on the north and east slopes are beech and maple, on the south and west, oak and hickory. Brushy areas are numerous. Within the enclosure are open groves and stretches of grass.

In a section of the state where soil cultivation is extensive and forests are few, the Memorial draws bird finders interested in woodland species. Moreover, it is noteworthy ornithologically because in its avifauna are breeding species near the northern limits of their ranges—as, e.g., the Black Vulture, Summer Tanager, Bewick's Wren, White-eyed Vireo, Kentucky Warbler, and Yellow-breasted Chat. Other species known to nest include the Great Crested Flycatcher, Acadian Flycatcher, Wood Thrush, Blue-gray Gnatcatcher, Blue-winged Warbler, Cerulean Warbler, Scarlet Tanager, and Rufous-sided Towhee.

The Memorial may be reached from I 71, east of Lebanon, by exiting east on State 350, which soon enters the Memorial grounds and the northern end of the enclosure.

LOGAN
Sugar Grove Region | Hocking Hills State Parks

Lying west and south of this city in southeast-central Ohio is the 10-square-mile **Sugar Grove Region,** one of the most scenic areas in Ohio. Here are deep, narrow gorges with cliffs of massive coarse-grained sandstone, cool ravines, fascinating caves, lovely waterfalls, and magnificent forests. Oak and hickory predominate on the dry slopes, shortleaf pine on the cliff edges, and hemlock, Canada yew, and yellow birch in the moist shady ravines.

The rewards for bird finding are many in any season. Among birds nesting are the Turkey Vulture, Black Vulture, Red-tailed Hawk, Broad-winged Hawk, Ruffed Grouse, Black-billed Cuckoo, Great Horned Owl, Barred Owl, Whip-poor-will, Pileated Woodpecker, Acadian Flycatcher, Cliff Swallow, Bewick's Wren, White-eyed Vireo, Yellow-throated Vireo, Black-and-white Warbler, Pine Warbler, Prairie Warbler, Louisiana Waterthrush, Kentucky Warbler, Yellow-breasted Chat, Hooded Warbler, and Bachman's Sparrow.

Because of the protection afforded by the ravines and the excellent forest cover, the winter population of birds is impressively heavy. Christmas bird counts usually record between fifty and sixty species, nearly all landbirds. The migration of small landbirds is probably most exciting in early May. In late April and early May, and again in September, when small landbirds are migrating, one may watch the movements of kinglets, vireos, and warblers from the edges of the ravines at treetop level.

Six sites in the Sugar Grove Region outstanding for their natural features are state parks with the collective name **Hocking Hills State Parks.** All six have foot trails along which one may see or hear birds representative of the Region. State 374, reached by driving north from Logan on US 33 for 7 miles, turning left on State 180, and left again on State 374, passes the entrance to five of the Parks in the following order: Cantwell Cliffs (60 acres), Rock House (160 acres), Conkles Hollow (559 acres), Old Man's Cave (150 acres), and Cedar Falls (40 acres). The entrance to Ash Cave (80 acres) is on the north side of State 56, reached by following State 374 to its terminus on State 56 and turning right. To return to Logan, backtrack on State

56 and State 93, turn left to US 33, and then right. The entire loop tour is about 50 miles.

PAINESVILLE
Headlands Beach State Park | Mentor Marsh

Northwest of Painesville in northeastern Ohio are two adjacent places where bird finding is productive the year round: **Headlands Beach State Park** and **Mentor Marsh.** The Park (120 acres), on Lake Erie at the mouth of the Grand River, has beaches for shorebirds on the north, and a breakwater and Fairport Harbor for waterfowl on the east. Southwest of the Park and a half-mile inland from the Lake stretches 860-acre Mentor Marsh through which Black Brook meanders. In 1971 the State of Ohio designated the Marsh an 'Interpretive Nature Preserve.' The Cleveland Museum of Natural History now administers a program of field trips, classes, and research from the Marsh House on Corduroy Road, which bisects the Marsh.

The unspoiled ponds and adjoining marshy areas are ideal for birds. In late March and April, transient waterfowl and waterbirds regularly include Mallards, American Black Ducks, Green-winged and Blue-winged Teal, Wood Ducks, Ring-necked Ducks, Lesser Scaups, Buffleheads, and Ruddy Ducks, with a few Horned and Pied-billed Grebes. By mid-April the marshy areas yield Great Blue Herons, Green Herons, Black-crowned Night Herons, American and Least Bitterns, Virginia Rails, Soras, Common Gallinules, and American Coots. Later in the month all the swallows represented in Ohio are present, darting about over the water surfaces in search of food. During the first two weeks in May, warblers and other landbirds throng in the shrubs and small trees on the north side of the Marsh.

To reach both areas, exit north from I 90 on State 44, and go through Painesville. For the Marsh, exit west on State 283 for about a half-mile, then turn right on Corduroy Road to the Marsh House, where a map and current information are available. For the Park, continue on State 44 to its end at the Lake.

PORT CLINTON
East Harbor State Park

About 6 miles east of Port Clinton on the southwestern shore of Lake Erie, and bordered by State 163 and 269, is **East Harbor State Park** (1,613 acres); the entrance is on State 269. Much of the Park consists of a 2.5-mile sand beach, marsh, and open water. At several nearby points, boats may be rented for exploring the marsh, where, in May and June, one may find Pied-billed Grebes, Black-crowned Night Herons, American Bitterns, Least Bitterns, Mallards, American Black Ducks, King Rails, Virginia Rails, Soras, Black Terns, Marsh Wrens, and Red-winged Blackbirds. At least twenty species of transient waterfowl gather in the marsh and adjacent harbor waters from late March to April and from late September through October. American Wigeons, Common Pintails, Ring-necked Ducks, Canvasbacks, and Lesser Scaups are unusually numerous among the transients.

PORTSMOUTH
Shawnee State Forest | Ohio Brush Creek

Well worth visiting for a variety of summer-resident birds is **Shawnee State Forest,** west of Portsmouth near Ohio's southern boundary. The Forest embraces 59,000 acres of ruggedly picturesque country with sharp-ridged hills and many creeks running through ravines and hollows. Conifers, chiefly white, red, and shortleaf pines, cloak the drier ridges and slopes; elsewhere the predominant tree cover consists largely of oaks and hickories. Broad-winged Hawks are numerous. Pine Warblers are common in the conifers, and Yellow-throated Warblers, in the deciduous woods along the creeks. Among other parulids are the Black-and-white, Black-throated Green, Prairie, and Hooded Warblers.

From Portsmouth, drive west and south on State 52 for 7 miles to the village of Friendship, turn right on State 125 to the Boy Scout Camp OYO, turn right again on Forest Road 1, which, in the next mile, passes through stands of red and white pines where Pine Warblers may be heard frequently. Backtrack to State 52 and continue south for 3 miles, then turn right on Forest Road 1, which follows up Pond Run. Stop along the way and look for Yellow-throated Warblers in the tops of the sycamores.

Whip-poor-wills are common summer residents in Shawnee State Forest. West of the Forest, in the woods along **Ohio Brush Creek,** are also Chuck-will's-widows at one of the northernmost points in the breeding range of the species. During any late evening from early May to July, one may have the unusual experience of hearing both species simultaneously giving their repetitive chants. At times the Whip-poor-wills, being more plentiful, tend to drown out the fewer Chuck-will's-widows. To reach the best listening post, drive west from Friendship on State 125 for 23 miles to the village of Blue Creek; continue an additional 5 miles on State 125, then turn left on Waggoner Riffle Road, continue for 5 miles, and park on the berm overlooking the Creek.

PUT-IN-BAY

South Bass Island | Kelleys Island | Starve Island

Among the twenty-one islands in the shallow western end of Lake Erie, **South Bass Island,** with its fine harbor and village of Put-in-Bay, is probably the best known. Four miles north of the nearest point on the mainland, the Island's surface, 3.5 by 1.5 miles in greatest length and width, is slightly rolling and low, the highest elevation only 69 feet above lake level. Except for a few gravel and sand beaches, most of the shore consists of rocky cliffs. Cultivated vineyards, neglected fields, and small stretches of maple-hackberry woods have entirely replaced the Island's original forest cover.

Outstanding ornithological attractions on South Bass Island are the hawk and small-landbird migrations. The hawk migration is impressive only in spring, from 10 April to 1 May, when Sharp-shinned Hawks, Cooper's Hawks, Red-tailed Hawks, Red-shouldered Hawks, and Broad-winged Hawks pass over, occasionally in large numbers. The small-landbird migration, though very spotty in both spring and fall, is best between 1 May and 20 May, 20 September and 10 October, during the first three hours of daylight on cloudy, windy, or rainy days. At such times one may see hundreds of vireos, warblers, and other passerines flying overhead or dropping down into the vegetation.

Strangely enough, the direction of the flights in both spring and fall depends on the direction of the wind. If the wind is from the north, the birds fly northward; if the wind is from the south, they fly southward. Lighthouse Point, on the extreme southern tip of the Island, is the best place to watch the flights. Ferry boats from the mainland village of Catawba Island dock a few hundred yards east of Lighthouse Point.

Kelleys Island, lying southeast of South Bass Island and 3.5 miles north of the mainland at Marblehead Point, comprises an area of 2,888 acres and has 18 miles of rocky shoreline. Of special interest to the bird finder are two cattail marshes—Kelleys Pond near the village on the south shore, and Carp Pond near the west shore. Here are typical bittern-rail associations.

Herring Gulls nest in a colony on a limestone islet, **Starve Island,** directly off the southeast shore of South Bass Island. The surface of Starve Island, only five feet above lake level, is about 100 by 30 yards in greatest length and width.

There are passenger and auto ferries to South Bass and Kelleys Islands from Sandusky, and to South Bass Island from Port Clinton and Catawba Island. There is also air service to South Bass Island from Port Clinton. Starve Island can be reached only by private or chartered boat.

ST. MARYS
Grand Lake St. Marys

Probably the best area for year-round bird finding in west-central Ohio is **Grand Lake St. Marys** (13,500 acres), formed over a century ago by the construction of two low earthen dams. The Lake—sometimes called Lake St. Marys—is surrounded by highways. State 29, west of St. Marys, passes close to the northern shore; State 703, south of Celina, the west shore; State 219, the south shore; and State 364, the east shore. Numerous gravel roads lead from the highways to the water's edge.

The south side of the Lake is particularly productive. In nesting season, from the last week of May to mid-July, look for Pied-billed Grebes, American Bitterns, Least Bitterns, Northern Harriers, King Rails, Virginia Rails, Soras, Common Gallinules, American Coots, Marsh Wrens, and Red-winged Blackbirds in the cattail marshes; Prothonotary Warblers among the stumps and dead trees along the shore; and Great Blue Herons and Black-crowned Night Herons, which have colonies in the tall trees back from the shore. Mallards, Blue-winged Teal, and Wood Ducks are regular breeding birds.

At the east side of the Lake, on State 364, is a State Fish Hatchery with many ponds varying in size from an eighth of an acre to 7 acres. The larger ponds support cattail edges where marsh birds are common. Gravel roads around the ponds make it possible to use a car for bird finding, provided permission is first obtained at the Hatchery office.

Impressive numbers of geese and ducks stop at Grand Lake during spring migration, beginning in February; gulls and terns appear in April, shorebirds in May. Equally impressive numbers of transient landbirds gather in the woods along the shore in April and early May. From late August to mid-October, when the larger ponds at the Fish Hatchery are usually being drained, the soft muddy bottoms exposed attract many shorebirds, including the Semipalmated Plover, an occasional Upland Sandpiper, Spotted Sandpiper, Solitary Sandpiper, Greater and Lesser Yellowlegs, Pectoral Sandpiper, Least Sandpiper, and Semipalmated Sandpiper.

Reference

Birds of the Lake St. Marys Area. An Annotated Check List and Migration Dates. By Clarence F. Clark and James P. Sipe. Columbus: Ohio Department of Natural Resources, Division of Wildlife, 1970.

TOLEDO

Oak Openings Metropolitan Park | Maumee River Rapids | Ottawa National Wildlife Refuge

Most of the land around this city at the extreme western end of Lake Erie is level, black, and fertile—ideal for farming. Before the area was cleared and drained it was known as the Black Swamp. Yet, oddly enough, a few miles west of this city there is an area covering many square miles that is strikingly different from all the surrounding country. This is the famous Oak Openings. Here there is yellow, sandy soil whose surface is dry in some places, and either level or heaped into dunes; wet in other places, and either level or hollowed out. The vegetation, varying according to the surface, has produced a number of different habitats: woods with black and white oaks, widely spaced, growing on dunes; swamp forest consisting of aspens, pin oaks, maples, and tupelos, growing in wet hollows; bogs with alders and willows, growing densely in wet areas at the bases of dunes; wet prairies with luxuriant growths of grasses and occasional shrubs, growing in areas that are flooded in spring; bare dunes and dry prairies sparsely covered with grasses and low shrubs.

Altogether 3,200 acres of the Oak Openings are within **Oak Openings Metropolitan Park.** To reach it from central Toledo, drive west on State 2, the Airport Highway. After going 18 miles, under the Ohio Turnpike and past the Airport, turn left on Wilkins Road to the main entrance on the right. Most of the Park trails start here at the Shelter House. Beginning in June, the mosquitoes become abundant and aggressive. More than one bird finder has walked hopefully into Oak Openings and then been forced to retreat in disorder because he was not suitably dressed to withstand their attacks.

The Park habitats are so intermixed and closely adjoined that they attract a correspondingly varied birdlife to a relatively small area. In a walk through the Park in late spring and early summer, one may observe Great Crested Flycatchers, Blue-gray Gnatcatchers (uncommon), Yellow-throated Vireos (uncommon), and Scarlet Tanagers *in the oak woods; in the swamp forests,* Acadian Flycatchers, Wood Thrushes, Veeries, American Redstarts, and Rose-breasted Grosbeaks; *in the bogs,* Golden-winged Warblers, Blue-winged Warblers (uncommon), Chestnut-sided Warblers (uncommon), Mourning Warblers (uncommon), Yellow-breasted Chats, and Rufous-sided Towhees; *in the wet prairies,* Common Yellowthroats and Henslow's Sparrows; *on the bare dunes and prairies,* (Prairie) Horned Larks, Vesper Sparrows, Lark Sparrows, and Field Sparrows. Some of these habitats are so close together—all can be found within a square mile—that one may hear a Lark Sparrow and a Veery, or a Scarlet Tanager and a Henslow's Sparrow, singing at the same time.

The birdlife of Oak Openings is remarkable in other respects. Several species nesting commonly here—e.g., the Veery and Golden-winged Warbler—either do not nest commonly elsewhere in the vicinity or are absent. A few species such as the Acadian Flycatcher and Yellow-breasted Chat, though nesting here, are near the northern limits of their ranges. Brewster's and Lawrence's Warblers, hybrids resulting from interbreeding of the Golden-winged and Blue-winged Warblers, appear with unusual frequency. The Brewster's Warbler is much more common than the Lawrence's Warbler, which is the recessive hybrid.

In winter the **Maumee River Rapids,** between the communities of Maumee and Grand Rapids southwest of Toledo, provide almost the only open water in the vicinity, and thus attract great numbers of ducks after Lake Erie freezes over. In August and September, when the water is low, shorebirds often congregate along the exposed River bottoms. US 24 southwest from the city runs very close to the Rapids on the north shore.

Just north of Toledo, reached from I 75, is the Erie State Game Area in Michigan, an excellent place for waterbirds, waterfowl, and transient landbirds (*see under* **Erie, Michigan**).

East of Toledo, some of the finest remaining cattail marshes that once bordered much of western Lake Erie are within the **Ottawa National Wildlife Refuge** (5,518 acres). To reach the Refuge, drive east from the city on State 2 for 18 miles; after passing the intersection with State 590 from the right, take the next left to Refuge headquarters. The stellar ornithological attraction is the assemblage of Whistling Swans in early spring. As soon as the shallows and open water are clear of ice—about the first of March—these great birds begin stopping off on their northward flight; by mid-April, most have continued on. Other waterfowl that show up with the Whistling Swans, or soon thereafter, are vast numbers of Canada Geese, Mallards, and American Black Ducks and smaller numbers of Common Pintails, Green-winged and Blue-winged Teal, American Wigeons, Northern Shovelers, Redheads, Lesser Scaups, Common Goldeneyes, and Common Mergansers.

In late summer and early fall, Great Blue Herons, Black-crowned

Night Herons, and several Great Egrets are among conspicuous birds. By mid-October the waterfowl population starts increasing rapidly until nearly all species reach, and sometimes exceed, the same number of individuals as in spring—except for Whistling Swans, which are poorly represented. Many Canada Geese and ducks linger into December, as long as the water stays open.

Reference

The Birds of the Toledo Area. By Lou Campbell. Toledo: The Blade, 1968.

WARREN
Mosquito Creek Reservoir Wildlife Area

The **Mosquito Creek Reservoir Wildlife Area** (11,857 acres), north of this city, encompasses 10-mile-long Mosquito Creek Reservoir covering 7,600 acres. This may be reached from the city by taking State 5 north for 6 miles, then turning left on State 46 to Mecca, or, without going into Warren, from I 80 directly north on State 11 for 15 miles, then west on State 88 to Mecca. From Mecca, State 88 immediately crosses the Reservoir to the west side, which, north and south of the highway, has the best vantage points.

In spring and fall, many waterfowl gather just as they do at Meander Reservoir (*see under* **Youngstown**). When the water level is low in late summer and early fall, exposing extensive mud flats, Mosquito Creek Reservoir is probably the best place in northeastern Ohio for shorebirds. More common species usually include the Semipalmated Plover, Black-bellied Plover, Greater and Lesser Yellowlegs, Pectoral Sandpiper, Least Sandpiper, and Semipalmated Sandpiper. Often present, though few in number, are the Whimbrel, Baird's Sandpiper, Dunlin, Short-billed Dowitcher, Stilt Sandpiper, White-rumped Sandpiper, and Northern Phalarope.

YOUNGSTOWN
Mill Creek Park | Meander Reservoir | Berlin Reservoir

Though situated in what is predominantly the rolling farmland of northeastern Ohio, this city of steel mills is fortunate in having **Mill Creek Park** (2,389 acres), which extends southwest along Mill Creek for 13 miles to the outskirts. The Park's natural beauty, with its deep gorge, cliffs, and tributary ravines, is further enhanced by the impoundment of three small lakes. Despite picnic grounds, ball fields, a golf course, and bridle paths, there are several hundred acres of unspoiled deep woods containing stands of beech, maple, oak, and hemlock, some of them virgin. The best of the woods—Boardman Woods and Flats—is in the southern end.

From mid-April through May, and throughout September, there are throngs of transient landbirds in the Park; sometimes they pass through the gorge and ravines and linger near the lakes. In late March and April, American Woodcock perform their crepuscular displays above the open, undisturbed localities where, later in the season, Henslow's Sparrows and Field Sparrows nest. Birds commonly nesting in the woods and their brushy edges include the Red-shouldered Hawk, Common Screech Owl, Barred Owl, Yellow-billed Cuckoo, Pileated Woodpecker, Red-bellied Woodpecker, Acadian Flycatcher, Traill's Flycatcher, Eastern Pewee, Carolina Wren, Wood Thrush, Veery, Yellow-throated Vireo, Blue-winged Warbler, Cerulean Warbler, Louisiana Waterthrush, Yellow-breasted Chat, Baltimore Oriole, Scarlet Tanager, and Rose-breasted Grosbeak.

The main entrances to the Park, which connect with drives along both sides of Mill Creek, are at the Mahoning Avenue Bridge, Glenwood Avenue at Falls Avenue, and Canfield Road (US 62). Bird finders should visit the Old Mill, a grist mill converted into a museum and interpretive center, for information on the most productive areas for birds, and specific directions for reaching Boardman Woods and Flats. The Old Mill, just off Canfield Road where it crosses Mill Creek, is the starting point of several well-marked foot trails through the woods overlooking the gorge.

Besides Mill Creek Park for bird finding, Youngstown has within a radius of 20 miles several reservoirs, built to supply water for the steel mills, that are havens for waterfowl from the time when the ice breaks up in the spring—about the end of February—to mid-April, and from early October until the water freezes over—usually about mid-December. All are sanctuaries the year round.

The nearest of the large impoundments is **Meander Reservoir** (1,720 acres), crossed by State 18, 10 miles west of Youngstown.

Canada Geese abound, and sometimes flocks of Whistling Swans and a few Snow Geese appear. Among many ducks are Mallards, American Black Ducks, Common Pintails, Green-winged and Blue-winged Teal, Lesser Scaups, Common Goldeneyes, and Ruddy Ducks. Starting in late July, when the water level drops, shorebirds come en masse and stay until mid-October. In September, impressive numbers of wading birds, including Little Blue Herons and Great Egrets, put in an appearance.

Another impoundment, **Berlin Reservoir** (4,700 acres), with similar opportunities for bird finding, is southwest of Youngstown, reached by driving south for 5 miles on State 7, then turning west on US 224, which crosses the Reservoir after 15 miles. Still another is the Mosquito Creek Reservoir north of the city (*see under* **Warren**).

Pennsylvania

HAWKS OVER KITTATINNY RIDGE

From northeastern North America—Labrador, Newfoundland, the Maritime Provinces, and New England—hawks by the thousands funnel into eastern Pennsylvania, sailing along the western face of Kittatinny Ridge. They come when the autumn colors are rich and the westerly winds are brisk. From high points such as Hawk Mountain they can be seen parading by: fleet accipiters and falcons, slow buteos and eagles, each making use of the favoring updrafts. A thousand feet or more below them stretches the floor of the valley, a mosaic of pastures, harvested fields, and woodlots. Truly there are few avian spectacles in the eastern United States that can excel the hawk movements along the Kittatinny.

Just as the hawk migration dominates the ornithological scene in Pennsylvania, so does the Appalachian Mountain system dominate the state's topography. Mountain ridges traverse the eastern and central portions of the state from northeast to southwest, highest point being Mt. Davis (3,213 feet) in the southwest. Southeast of the mountains, a strip of the Coastal Plain comprises the valleys of the Delaware, Schuylkill, and Susquehanna Rivers, and west and northwest the high Allegheny Plateau slopes gradually westward to the Ohio boundary.

In former years the summits of the ridges, where rocky, were cloaked with pitch pine, but summits that were less rocky and covered with more soil had chestnut, chestnut oak, and other hardwoods predominantly. Moist, cool ravines often contained yellow birch, black birch, red maple, and a luxuriant growth of rhododendron and mountain laurel. Most forests over the Allegheny Plateau once consisted of white pine and hemlock interspersed with red maple, sugar maple, beech, and yellow birch, but today these have been extirpated except in a few ravines and valleys (e.g., Cook Forest State Park in northwestern Pennsylvania; *see under* **Brookville**). The forests, where they exist, usually consist of a secondary deciduous growth. Birdlife of forest areas above 2,000 feet and, in hemlock ravines, down to 1,000 feet shows a number of species characteristic of more northern forest and thicket associations. In the following list of species known to nest, those marked with an asterisk are confined mainly to higher areas in the northern part of the state.

* Northern Goshawk
Yellow-bellied Sapsucker
* Red-breasted Nuthatch
Brown Creeper
Winter Wren
Hermit Thrush
Swainson's Thrush
* Golden-crowned Kinglet
Solitary Vireo
Nashville Warbler
Northern Parula Warbler
Magnolia Warbler
Black-throated Blue Warbler
* Myrtle Warbler
Black-throated Green Warbler
Blackburnian Warbler
Northern Waterthrush
Mourning Warbler
Canada Warbler
Purple Finch
Slate-colored Junco
* White-throated Sparrow

In southeastern Pennsylvania the heavily settled Coastal Plain is gently undulating. Hemlock, yellow birch, and red maple grow sparsely along the banks of the rivers and on a few steep hills; woodlots of swamp white oak, swamp chestnut oak, sweet gum, magnolia, and red cedar are interspersed among settlements and farming lands. In the southwestern part of the state, also heavily settled, the west-sloping Allegheny Plateau, cut by the Ohio River Valley, is extremely hilly, with wide intervening bottomlands. The remaining forests consist of sycamore and willow along rivers and streams, red maple, beech, black birch, and black and red oaks on the bottomlands and steep hillsides, hemlock in the ravines. In southern Pennsylvania, breeding birds characteristic of more southern associations reside commonly in the deciduous woods and thickets; among them are the following:

Red-bellied Woodpecker
Acadian Flycatcher
Carolina Chickadee
Blue-gray Gnatcatcher
Worm-eating Warbler
Cerulean Warbler
Kentucky Warbler
Louisiana Waterthrush
Yellow-breasted Chat
Hooded Warbler

In most of the mountain valleys through the center of the state there are farmlands and small settlements. In the extreme northwestern part of the state, the Allegheny Plateau slopes gradually downward to a narrow plain extending for 40 miles along Lake Erie between New York and Ohio. Here is some of the richest farming country in the state.

Wherever the bird enthusiast travels in Pennsylvania, he will likely find the following birds breeding regularly in the deciduous forests of lower elevations and on the farmlands (fields, wet meadows, brushy lands, orchards, and dooryards):

DECIDUOUS WOODS

Turkey Vulture
Red-shouldered Hawk
Broad-winged Hawk
Ruffed Grouse
Yellow-billed Cuckoo
Black-billed Cuckoo
Barred Owl
Whip-poor-will
Yellow-shafted Flicker
Hairy Woodpecker
Downy Woodpecker
Great Crested Flycatcher

Least Flycatcher
Eastern Pewee
Blue Jay
Black-capped Chickadee
Tufted Titmouse
White-breasted Nuthatch
Wood Thrush
Yellow-throated Vireo
Red-eyed Vireo
Warbling Vireo
Ovenbird
American Redstart
Baltimore Oiole
Scarlet Tanager
Rose-breasted Grosbeak

FARMLANDS

Common Bobwhite
Red-headed Woodpecker (*local*)
Eastern Kingbird
Eastern Phoebe
(Prairie) Horned Lark
Tree Swallow (*local*)
American Goldfinch
Rufous-sided Towhee
Savannah Sparrow
Grasshopper Sparrow
Vesper Sparrow
Chipping Sparrow
Field Sparrow
Song Sparrow
Common Yellowthroat
Bobolink
Eastern Meadowlark
Common Grackle
Northern Cardinal
Indigo Bunting
House Finch (*eastern Pennsylvania*)
Barn Swallow
House Wren
Carolina Wren
Northern Mockingbird
Gray Catbird
Brown Thrasher
Eastern Bluebird
Yellow Warbler
Chestnut-sided Warbler

Lakes and marshes are relatively small and few, but wherever they exist there is a greater variety of birds. Conneaut Lake in northwestern Pennsylvania, only 2.5 miles long and less than a mile wide, is the state's largest natural lake; but several reservoirs—notably, Wallenpaupack in the northeast, Lake Ontelaunee in the southeast, and Pymatuning on the northwestern boundary—have greater surfaces. Recent studies demonstrate their value as feeding and resting grounds for waterfowl and shorebirds. The upland marshes or swamps are either of the cattail-sedge type, usually of small extent, or of the tamarack-sphagnum type, in a few cases quite large. Those of the latter type are particularly attractive ornithologically since they encourage the nesting of several northern species. The lowland marshes, such as those along the Delaware River, are coastal in character: small counterparts of the big marshes in New Jersey and Delaware.

The Delaware, Susquehanna, and Ohio River Valleys are important migration routes for large numbers of landbirds. In the last week of April and the first two weeks of May, waves of warblers stop in parks and woodlands. Transient waterfowl and shorebirds appear commonly on the lower Delaware and Susquehanna Rivers, on the large reservoirs, and at Presque Isle on Lake Erie, even though there seems to be no main flyway across the state. Undoubtedly the greatest feature of migration is the fall movement of hawks along the easternmost ridges of the Appalachians from Delaware Water Gap to Waynesboro. Peak flights may be expected within the following dates:

Waterfowl: 1 March–10 April; 15 October–1 December
Shorebirds: 1 May–1 June; 1 August–5 October
Landbirds: 15 April–15 May; 1 September–25 October

During winter, Evening Grosbeaks, Purple Finches, Slate-colored Juncos, American Tree Sparrows, and White-throated Sparrows frequent suburban localities almost anywhere in the state. In the southern parts, where winters are milder, many passerines—e.g., Hermit Thrushes, Myrtle Warblers, White-crowned Sparrows, Fox Sparrows, and Swamp Sparrows—appear in sheltered situations.

Authorities

Jacob B. Abbott, C. U. Atkinson, Herbert H. Beck, Maurice Broun, Joseph M. Cadbury, David B. Freeland, William C. Grimm, George E. Grube, Helen Harrington, Hal H. Harrison, Donald S. Heintzelman, Quintin Kramer, Robert C. Leberman, Leo A. Luttringer, Jr., Harold B. Morrin, Earle L. Poole, Edith Portman, Barton L. Sharp, Phillips B. Street, Jean Stull, George B. Thorp, Grace E. White, Merrill Wood.

References

Pennsylvania Birds. An Annotated List. By Earle L. Poole. Philadelphia: Delaware Valley Ornithological Club, 1964. Available from Philadelphia Academy of Natural Sciences, 19th Street and the Parkway, Philadelphia, Pa. 19103.

Birds of Pennsylvania. When and Where To Find Them. By Merrill Wood. Rev. ed. University Park: Pennsylvania State University, Agricultural Experiment Station, 1973.

Where To Find Birds in Western Pennsylvania. Edited by David B. Freeland. Pittsburgh: Audubon Society of Western Pennsylvania, 1975.

ALLENTOWN
Bake Oven Knob | Lehigh Furnace Gap | Bear Rocks

On the Kittatinny Ridge, about 20 miles north of Allentown, are three good hawk lookouts: at Bake Oven Knob, Lehigh Furnace Gap, and Bear Rocks. Since 'as the crow flies' none are far from Hawk Mountain (*see under* **Hamburg**), the species observed, the times they appear, and the weather conditions governing flights are about the same as at that famous lookout—and they are not so crowded by hawk watchers!

To reach **Bake Oven Knob,** probably the best of the three, drive north from Allentown on State 309 to exactly 2 miles beyond its junction with State 143, go right onto a paved road (County 39056) for 2.1 miles, and then turn left on an unmarked road that goes between a white house and white outbuildings. Continue to two parking areas at the top of the Ridge. Walk east along the Appalachian Trail for about a third of a mile. Shortly after passing a field with large rocks, and climbing a steep grade, note an old cement foundation. The South Lookout is in view here, about 150 feet east of the Trail. When the winds are from the east or south, observers use the South Lookout, high above the Great Valley. When the winds are west and the hawks fly along the north side of the mountains, observers move north along the Trail a short distance to the North Lookout.

Lehigh Furnace Gap is good for watching hawks only when the winds are from the south or the east. To reach the Gap from Bake Oven Knob, return to County 39056, drive east for 2.4 miles, turn left on an obscure road (partly hidden by a red brick house) for 1.1 miles to a T, and left on another dirt road to the top of the Ridge, marked by a series of rocky outcrops and a power line. Any one of the outcrops is suitable for hawk watching when the winds are favorable.

Both the hawk watching and scenery are superb at **Bear Rocks,** an open area on the Ridge a scant 1.5 miles southwest of Bake Oven Knob. To reach Bear Rocks, drive north from Allentown on State 309 to the parking areas near Bake Oven Knob. (See above directions to Bake Oven Knob.) Leave the car and walk southwest on the Appalachian Trail for about 1.5 miles to the lookout, a huge pile of boulders on the north side of the Trail. *Warning:* Watch for snakes. There are copperheads in this area.

BROOKVILLE
Cook Forest State Park

Invariably delightful to western Pennsylvania bird enthusiasts is a trip to **Cook Forest State Park** (7,822 acres), containing Pennsylvania's largest remaining primeval woods. Here giant white pines and hemlocks, sometimes as large as five feet in diameter, rise in stately columns to form a thick, green canopy nearly a hundred feet above the forest floor. Only subdued light reaches the scant undergrowth on the floor. No less impressive than the pines and hemlocks are a few scattered trees of other kinds, the most noticeable being yellow birch, beech, red maple, and black birch. In the more moist, rocky places there are alder and rhododendron thickets. The avifauna of these woods, as would be expected, shows a tinge of the northern coniferous association. Among nesting birds are the Winter Wren, Hermit Thrush, Solitary Vireo, Magnolia Warbler, Black-throated Blue Warbler, Black-throated Green Warbler (the most abundant parulid), Blackburnian Warbler, Northern Waterthrush, Canada Warbler, and Slate-colored Junco.

From Brookville, just south of Exit 13 on I 80, drive northwest on State 36 for about 15 miles to Cooksburg just south of the Forest. The tract extends from southeast to northwest for about 6 miles and is traversed by Thom's Run, a small stream along the western slope, where the best primeval stand of timber is located. The upper part of the Rhododendron Trail, the lower part of the Joyce Kilmer Trail, and the main part of the Longfellow Trail pass through the forest.

CARLISLE
Sterrets Gap | Waggoners Gap

A good and easily accessible observation post for fall hawk flights is **Sterrets Gap,** on State 34 just 8 miles north of the Carlisle toll gate on the Pennsylvania Turnpike. The observation post is atop a mountain about 75 yards west of the highway. Another fine lookout is **Waggoners Gap,** on State 74 about 12 miles northwest of Carlisle. Just as the highway crosses into Perry County, look for a radio tower on one side of the road and an abandoned building with a parking lot on the other. When the winds are from the south or the east, stay near the tower; when the winds are from the west or the north, watch from a rock pile behind the building. The species to be seen, the times in fall when they can be expected, and the weather conditions governing their appearance are identical with those at Hawk Mountain (*see under* **Hamburg**).

CHAMBERSBURG
Wilson College Campus | Caledonia State Park | Mountain Lake | Tuscarora Mountain

On **Wilson College Campus,** at Philadelphia and College Avenues, is an artificial body of water called Wolf Lake, the small islands of which provide sandy shores and swampy areas that are fine for birds. Ten miles east of Chambersburg on US 30 lies **Caledonia State Park** (1,444 acres), in heavily wooded mountain country at 1,500 feet elevation, where the Wood Thrush, Ovenbird, Yellow-breasted Chat, Scarlet Tanager, and Rufous-sided Towhee are breeding birds. Just east of the Park, a turn north on State 233 brings one to a reservoir where transient waterbirds and waterfowl tarry.

Northwest of Chambersburg, on State 75 in Fannetsburg, **Mountain Lake,** an artificial impoundment about 2 miles long, is attractive to large numbers of loons, grebes, ducks, and American Coots in March and early April. Whistling Swans appear early in March; Canada Geese between 15 and 20 March. To reach Mountain Lake, drive west from Chambersburg on US 30 to Fort Loudon and north on State 75.

The Pulpit on the summit of **Tuscarora Mountain,** west of Chambersburg, is a favorite hawk lookout from mid-September to mid-November. To reach the Pulpit, drive west from Chambersburg on US 30 for about 20 miles to an inn on the summit of Tuscarora Mountain, park, and follow a well-marked trail from the back of the inn.

COLUMBIA
Conejohela Flats

The Susquehanna River is consistently attractive to many of the large transient birds. This is particularly true in southeastern Pennsylvania, south of the Columbia-Wrightsville Bridge, where the River, backed up by the Safe Harbor Dam, widens, surrounding a group of low marshy islands and a number of mud and coal-silt bars known as the **Conejohela Flats.** Many transient waterfowl gather on the River near the Flats in October/November and March/April.

The two northern islands are high enough to support birches and sycamores, where, in early summer, Black-crowned Night Herons have nesting colonies and, later in summer, various other wading birds—Great Blue Herons, Little Blue Herons, Cattle, Great, and Snowy Egrets, and Glossy Ibises—have roosts for the night after foraging during the day on the Flats. The other islands are marshy with sedges and cattails and, on higher ground, with low birches, mallows, and willows. Mallards, American Black Ducks, and Wood Ducks summer on the Flats. Traill's Flycatchers nest in good numbers.

The barren mud- and coal-silt bars come into their own from about 20 April to 20 May and from 15 August to 15 September, when they attract hordes of shorebirds, but more in fall than in spring. In late summer and fall also one may find both Least and American Bitterns, waterfowl, rails, gulls, and terns, both Marsh and Sedge Wrens, Water Pipits, and large flocks of Bobolinks. In winter there are almost always Short-eared Owls and Snow Buntings, occasionally Northern Harriers and Lapland Longspurs, and in certain winters a Snowy Owl.

The great spring event on the River is the arrival of the Whistling Swans. They begin to come in as soon as the ice opens up enough to allow small pools—in January, February, or March, depending on the weather—and build up to greatest numbers (possibly over three thousand) by mid-March. After that they vanish almost over night. They return in fall from October to December but never in such great numbers.

For the best view of the Flats, drive south from Columbia on State 441 for about 4 miles to a historical marker on the north edge of Washington Boro. Look over the area from here with binoculars or spotting scope and from other high points along State 441 for the next mile. To visit the Flats, drive west from Columbia across the River on State 462 and turn south on State 624 to Long Level, where there are row boats for hire. In good weather the Flats are about a 20-minute row from Long Level. Anyone using a motorboat should beware of the many submerged stumps and sand bars that make travel on the River hazardous.

DOWNINGTOWN
Marsh Creek State Park

From the Downingtown Interchange on the Pennsylvania Turnpike, about 30 miles west of Philadelphia, drive north on State 100 for 1.2 miles to the village of Eagle. Go left in front of the Eagle Tavern for 0.1 mile, and left again onto Lyndell Road to **Marsh Creek State Park.** A finger of the 535-acre reservoir will soon show up on the right, and Park headquarters shortly thereafter on the left. This impoundment is good for waterfowl from November through April, except on rare occasions when it freezes over. Concentrations of Canada Geese are sometimes spectacular. A winter flock of 1,000 is not uncommon, and as many as 3,000 have been noted on the water or in nearby fields. A wide variety of other waterfowl as well as waterbirds is usually present in small numbers, most regular being the Pied-billed Grebe, Mallard, Common Pintail, American Wigeon,

Ring-necked Duck, Greater and Lesser Scaups, Bufflehead, Ruddy Duck, Common Merganser, and American Coot.

ERIE

Presque Isle State Park

Jutting into Lake Erie near the western outskirts of this city is a somewhat curved peninsula, about 7 miles long, called Presque Isle. This strip of land and Erie Bay, which it encloses, are a part of **Presque Isle State Park** (3,203 acres), on some maps referred to as Pennsylvania State Park at Erie. The great variety of habitats and their corresponding variety of birds make Presque Isle an exceptionally worthwhile area for the bird finder.

Gull Point, a bird sanctuary at the eastern end, has a beach, sand bars, and shallow, sandy-bottomed pools, attractive to transient shorebirds in May, late August, and September. Species appearing here with greatest regularity are the Semipalmated Plover, Black-bellied Plover, Ruddy Turnstone, Baird's Sandpiper, Least Sandpiper, Dunlin, Semipalmated Sandpiper, and Sanderling. The Killdeer and Spotted Sandpiper are summer residents, and Common Terns sometimes have a nesting colony.

The interior of the peninsula consists of parallel ridges, in reality sand hills. Those near the beach fronting Lake Erie are covered with coarse grasses, shrubs, and aspens; those farther inland, with a variety of deciduous trees, a few hemlocks, and thick underbrush. Among nesting birds are the Barred Owl, Pileated and Red-headed Woodpeckers, Acadian Flycatcher, Eastern Pewee, Veery, Red-eyed and Warbling Vireos, Chestnut-sided Warbler, American Redstart, Northern Cardinal, Rose-breasted Grosbeak, Rufous-sided Towhee, and Field Sparrow.

Scattered in the interior are shallow, marsh-bordered lagoons where surface-feeding ducks—Gadwalls, Common Pintails, Green-winged and Blue-winged Teal, American Wigeons, and Northern Shovelers—and Hooded Mergansers appear in late March and in April, again in October and November. Mallards, American Black Ducks, and occasionally Wood Ducks are present from spring through fall. Transient shorebirds such as Solitary Sandpipers,

Greater and Lesser Yellowlegs, Pectoral Sandpipers, and Short-billed Dowitchers feed and tarry near the edges; during the summer, Least and American Bitterns, Virginia Rails, Soras, and both Marsh and Sedge Wrens nest farther back in the cattails and sedges; Green Herons, still farther back in tall shrubs and low trees.

Capacious Erie Bay attracts waterfowl and waterbirds in early spring and late fall. Notable among the waterfowl are many Whistling Swans and diving ducks—Canvasbacks, Greater and Lesser Scaups, Common Goldeneyes, Buffleheads, Oldsquaws, White-winged Scoters, Common and Red-breasted Mergansers—many of which stay through the winter if the water remains open. Transient waterbirds include Herring and Ring-billed Gulls in any season, Horned and Pied-billed Grebes and Bonaparte's Gulls in spring and fall, and Greater Black-backed Gulls in winter. During early summer a few Common Gallinules and Black Terns nest in the cattail marshes bordering the west side of the Bay.

An exciting time to visit Presque Isle is in late April and May, when the peninsula hosts vast numbers of migrating passerine species from flycatchers to sparrows prior to their taking off northward across Lake Erie. Although fall migration in September is less impressive in numbers of individuals, it is nonetheless impressive in the variety of species.

To reach the Park, drive west from Erie on State 5 and turn north on State 832 directly to the Park where it intersects paved Park roads. From I 90, south of the city, exit on I 79 north to its terminus on 25th Street, then turn left to State 832 north. After entering the Park, stop at the administration building for a copy of *Finding Birds on Presque Isle* by Jean Stull, with its fine map and specific directions to the best spots for birds; and ask permission to enter the Gull Point Bird Sanctuary.

GETTYSBURG
Gettysburg National Military Park

When entering the **Gettysburg National Military Park** (2,394 acres), obtain a map at the gate. Then locate Devil's Den and Big Round Top. Devil's Den, on Sickles Avenue west of Warren Avenue, features a small brook and pond, with a rocky cliff on one side and a woodlot, for the most part of oak, on the other. This is a favorite wintering ground for such fringillids as Slate-colored Juncos, American Tree Sparrows, White-throated Sparrows, and Song Sparrows. During the last week of April and the first week of May, great numbers of warblers and other small landbirds stop here on their way north. Later, Red-bellied Woodpeckers, Tufted Titmice, Cerulean Warblers, Yellow-breasted Chats, and Hooded Warblers are among the regular breeding birds.

At Big Round Top, about one mile distant, one may look down, except in winter, on both Turkey and Black Vultures as they continually patrol the surrounding fields and forests. While driving through the Park in late spring and summer, watch for Upland Sandpipers in open fields where they nest.

HAMBURG

Rattling Run | Hawk Mountain Sanctuary

At Port Clinton, 4 miles north of Hamburg, US 61 crosses **Rattling Run,** a fine mountain stream bordered by a deciduous woodland mixed with mature stands of hemlock. Beneath the heavy forest canopy there is an undergrowth of rhododendron and laurel. Along a path that follows the creek for at least 2 miles, one is likely to find, in late May and June, such breeding species as the Worm-eating and Black-throated Green Warblers, Louisiana Waterthrush, and the Hooded and Canada Warblers.

Farther north of Hamburg on a spur of Kittatinny Ridge—the eastern chain of the Appalachian system—is **Hawk Mountain Sanctuary** (2,000 acres), the world's first preserve for diurnal predators. For the bird finder a visit here in fall will be a never-to-be-forgotten thrill.

Whereas Kittatinny Ridge is generally broad so that hawks separate in their line of flight and thus are not readily followed, at Hawk Mountain the Ridge becomes suddenly narrow and high (1,200 to 2,000 feet), bringing the birds both closer together and closer to the ground. From observation points on the crest of the Ridge, where the highest is North Lookout (1,521 feet), as many as 32,155 hawks have been counted in a single season (1974). Usually they fly low, scudding over the tops of the trees. Sometimes one can look down upon them as they pass the crest; sometimes they ride the wind directly over the crest, so close that field glasses are unnecessary.

The hawk migration gets under way as early as 20 August, when the vanguard—mostly Bald Eagles and Ospreys—appears. In the third week of September comes one of the most exciting parts of the migration—an immense, and often sudden, flight of Broad-winged Hawks. On a single September day, observers counted 11,392 individuals. Ospreys often dominate the scene about 20 September. By 7 October, most of the Broad-winged Hawks have passed, and many Sharp-shinned Hawks and a small number of Cooper's Hawks take their place. Perhaps the most exciting part comes from mid-October to mid-November, when one may see as many as twelve to fourteen species of hawks, among them many Northern Goshawks and Red-tailed Hawks. At this time, too, one may see hundreds of waterfowl and throngs of small birds—from Snow Geese to Golden-crowned Kinglets—all within a few hours.

The bird finder must bear in mind that all flights are subject to weather conditions in New England and to local winds. Study the weather maps. Two or three days after a climatic disturbance in New England, flights will appear at Hawk Mountain, provided there are strong westerly winds.

To reach Hawk Mountain Sanctuary, drive north from Hamburg on State 61 to Molino, thence right on State 895 to Drehersville; cross the Little Schuylkill River and go up the blacktop road to Sanctuary headquarters in an old stone house. Park here and walk along a well-marked, 0.75-mile trail to the bare edge of a great rock-strewn promontory overlooking a wide expanse of farming country and forested hills. The hawk flyway is within a few yards of this promontory, the main observation point. Headquarters is open daily all year long, and Saturday evenings in fall.

For other good hawk lookouts in Pennsylvania, *see under* **Allentown, Carlisle,** and **Chambersburg.**

LANCASTER
Middle Creek Wildlife Management Area

The **Middle Creek Wildlife Management Area** in southeastern Pennsylvania, owned and operated by the Pennsylvania Game Commission, occupies more than 5,000 acres including a 400-acre shallow-water lake and 70 acres of water in a series of ponds and potholes. An extensive deciduous forest, small pine forests, open fields (some crop-planted), and brushy lands are among the diverse habitats. A resident population of several hundred Canada Geese is augmented from fall to spring by thousands more from the north. Transient and wintering ducks abound. A few Rough-legged Hawks are present each winter, and Ruffed Grouse are common residents. In March and early April, American Woodcock perform their aerial displays above fallow fields.

To reach the Management Area from Lancaster, drive north on State 501 to Brickerville, turn right on US 322 to the village of Clay, then turn north and follow directional signs. At the visitor center, ask about roads and foot trails leading to the best vantage points for birds.

LIGONIER
Powdermill Nature Reserve

The **Powdermill Nature Reserve** (1,800 acres), a research station of Carnegie Museum, lies southeast of Ligonier along the western slope of Laurel Hill in beautiful Ligonier Valley east of Pittsburgh. A secondary hardwood forest covers the tract except for some old fields, grassy areas, strip-mined land, a few small ponds, and a small cattail marsh.

During a period of nineteen years, researchers noted 219 species of birds on the Reserve, and the majority were forest species. The list of breeding species includes the Green Heron, Mallard, Red-shouldered Hawk, Broad-winged Hawk, Ruffed Grouse, Wild Turkey, American Woodcock, Barred Owl, Pileated and Red-bellied Woodpeckers, Acadian and Least Flycatchers, Carolina Wren,

Wood Thrush, Blue-gray Gnatcatcher, White-eyed and Solitary Vireos, Golden-winged Warbler, Northern Parula Warbler, Cerulean Warbler, Prairie Warbler, Louisiana Waterthrush, Kentucky Warbler, Mourning Warbler, Hooded Warbler, Yellow-breasted Chat, Scarlet Tanager, Indigo Bunting, American Goldfinch, and Field Sparrow.

To reach the Reserve, drive east from Ligonier on US 30 for 2 miles and right on State 381 for 6 miles. Just beyond the Pennsylvania State Agricultural Experiment Farm, turn left into a parking area beside the headquarters building. A small field museum, 0.25 mile farther south, is open on weekends from April through October.

LINESVILLE
Pymatuning Reservoir | Pymatuning State Game Refuge

Pymatuning Reservoir, in northwestern Pennsylvania and northeastern Ohio, originally a scythe-shaped body of water and marsh that comprised about 10,400 acres (15 miles long by about 2 miles wide at the widest point), has now become, because of damming, two reservoirs covering 17,000 acres, with a shoreline of about 70 miles. The **Pymatuning State Game Refuge** occupies 3,670 acres of water, marsh, and land on the Upper Reservoir. The Reservoir itself is shallow (seldom more than a few feet deep) and has several mud flats and twenty-one low islands densely covered with trees, shrubs, vines, cattails, and sedges. The surrounding land is broken up into small lagoons and marshy bayous, bordered by trees and shrubs. All in all, both Refuge and Reservoir have the requisites for a successful resting, feeding, and nesting area for waterfowl, marsh birds, and shorebirds, as well as landbirds commonly associated with moist areas. The abundance of birdlife is ample proof.

Waterfowl breeding commonly are the Canada Goose (large resident population), Mallard, American Black Duck, Gadwall, Blue-winged Teal, Northern Shoveler, Wood Duck, and Hooded Merganser. Nesting is under way in May. Nearly all species of waterfowl of northeastern United States show up here at one time or another during their migrations. The peak of the duck flights in spring occurs between 20 March and 15 April, in fall between 10 October and 20 November.

Among marsh birds nesting regularly, if not commonly, are the Pied-billed Grebe, American Bittern, Least Bittern, Virginia Rail, Sora, Common Gallinule, American Coot, Black Tern, Marsh Wren, Sedge Wren, Red-winged Blackbird, and Swamp Sparrow. There is a small Black-crowned Night Heron colony within the Refuge and a large Great Blue Heron colony nearby. Green Herons are common summer residents. During summer, Great Egrets and other southern wading birds visit the Refuge, frequently arriving in June. Some transient shorebirds appearing commonly on the mud flats in May and September are the Semipalmated Plover, Solitary Sandpiper, Greater and Lesser Yellowlegs, Pectoral Sandpiper, Dunlin, Short-billed Dowitcher, Least Sandpiper, and Semipalmated Sandpiper.

In woods and open areas there is a diversity of nesting birds,

LEAST BITTERN

among them the American Woodcock, Upland Sandpiper, Pileated Woodpecker, Brown Creeper, Veery, Blue-winged Warbler, Northern Waterthrush, Canada Warbler, and Henslow's Sparrow. At least two pairs of Bald Eagles nest in the area. The Reservoir, Refuge headquarters, and a waterfowl museum are directly south of Linesville on an unnumbered road.

MEADVILLE
Erie National Wildlife Refuge

The **Erie National Wildlife Refuge** (5,012 acres) in northwestern Pennsylvania was established to provide one more stopping place for migrating waterfowl and one more haven for resident birds. The area includes upland hardwood forests and swamp woodlands, marshes, and abandoned fields and pastures. Impoundment of two creeks has added more marsh and open water for the birds. And gravel roads and dikes have provided human visitors with easy access to bird-finding areas. Personnel and visitors have listed 224 bird species on the Refuge.

Some waterfowl nesting commonly are the Canada Goose, Mallard, American Black Duck, Blue-winged Teal, Wood Duck, and Hooded Merganser. Among species breeding in the wooded uplands are the Ruffed Grouse, American Woodcock, Veery, and Scarlet Tanager. Black Terns are common summer residents in the marshes, as are Pied-billed Grebes, Marsh and Sedge Wrens, and Swamp Sparrows. In winter, (Northern) Horned Larks and Snow Buntings show up on snowy fields.

To reach headquarters, drive east from Meadville on State 27 for 12 miles to Mt. Hope, then south on State 173 for 2 miles.

MEDIA
Tyler Arboretum

An attractive spot southwest of Philadelphia for bird finding in any season is the **Tyler Arboretum.** From Media, take US 1 west for 2 miles, turn right on State 352 at Lima; then turn right on Forge

Road, and immediately right again on Painter Road. The Arboretum, a 675-acre tract of woods and fields on the left, features a notable collection of trees, shrubs, and plants. A map showing the nearly 20 miles of trails through fields and woodlands and along streams is available at the office on weekdays and on Sunday afternoons.

Some 170 species of birds have been found here. Among the eighty species that nested are the Green Heron, Wood Duck, American Woodcock, Great Horned Owl, Red-bellied Woodpecker, Acadian Flycatcher, Eastern Bluebird, Blue-gray Gnatcatcher, White-eyed, Yellow-throated, and Warbling Vireos, Worm-eating, Blue-winged, Cerulean, Chestnut-sided, and Kentucky Warblers. Twenty-two species of transient warblers may be expected each spring, 1–20 May, and fall, late August through October. Regular winter visitants include the Yellow-bellied Sapsucker, Red-breasted Nuthatch, Brown Creeper, Winter Wren, Hermit Thrush, Golden-crowned Kinglet, Purple Finch, Slate-colored Junco, and White-throated Sparrow.

PHILADELPHIA

Rittenhouse Square | Wissahickon Section of Fairmount Park | Pennypack Park | Tinicum National Environmental Center | Pennsbury Manor | Mill Grove

This huge metropolis, sprawling in the valley of the Delaware River, has several parks and undeveloped areas where bird finding is always rewarding.

Rittenhouse Square, in the heart of Philadelphia at 18th and Walnut Streets, is an oasis of green in a cityscape of pavements and sky-reaching edifices. Though a mere city block in size, this park attracts a surprising number of landbirds during both spring (April/early May) and fall (September) migrations. Over a hundred species have been recorded in recent years. Of warblers and fringillids, probably Common Yellowthroats and White-throated Sparrows are the most numerous. Other birds of regular appearance are the Yellow-bellied Sapsucker, White-breasted Nuthatch, Brown Creeper, House Wren, Gray Catbird, Brown Thrasher, American Robin, and Hermit Thrush. Most of the birds are seen either in the silver maples or in shrubbery bordering the walks and grassy plots. Early morning, after arrival of a 'wave' during the previous night and before the disturbing daytime activities of the city are under way, is the best time to look for birds here.

Fairmount Park, within the city limits, occupies 3,845 acres bordering the Schuylkill River and its tributary, Wissahickon Creek, for 10 miles. The **Wissahickon Section of Fairmount Park,** north of the Schuylkill River, is an unspoiled wooded ravine extending from the mouth of Wissahickon Creek northwestward to the city limits of Chestnut Hill. Four areas in, or adjacent to, this section of Fairmount Park are worth investigating: Wissahickon Ravine, which members of the Delaware Valley Ornithological Club find excellent for migrating and nesting landbirds; Cresheim Valley, a branch of the Wissahickon Section, extending northeastward along Cresheim Creek; Carpenter's Woods; and the Morris Arboretum.

For the most part the vegetation of both the Wissahickon Ravine and Cresheim Valley consists of fine stands of hemlock (especially on the slopes along the creeks), white oak, red oak, tulip tree, mulberry, and sycamore. Birches are fairly abundant, and white pine, beech, white ash, and sugar maple are scattered throughout. Commonest shrubs are spicebush and witch hazel. Some bird species reported nesting are the following: Broad-winged Hawk, Common Screech Owl, Great Crested Flycatcher, Acadian Flycatcher, Tufted Titmouse, Wood Thrush, Blue-gray Gnatcatcher, Kentucky Warbler (the commonest warbler), Baltimore Oriole, Scarlet Tanager, Northern Cardinal, and Indigo Bunting. In summer, Carolina Wrens, Louisiana Waterthrushes, and Yellow-breasted Chats appear from time to time and probably nest. In the meadows bordering Wissahickon Creek, northeast of the Park limits, Upland Sandpipers and (Prairie) Horned Larks find conditions suitable for nesting.

To reach various parts of the Wissahickon Ravine, drive northwest from US 1 on US 422, which passes close to the northeast side of the Park, and turn left on one of the roads leading southwest. For Cresheim Valley, turn left from US 422 on Cresheim Valley Road. For Carpenter's Woods, turn northwest from US 1 on Wissahickon Avenue. The Woods is about 2 miles distant on the right, between Sedgwick Street and Mount Pleasant Road. For Morris Arboretum, turn north from US 422 on the Bethlehem Pike for 0.5 mile and then left on Stenton Avenue for 0.75 mile to the entrance.

Thousands of flycatchers, kinglets, vireos, parulids, and fringillids pass through the Wissahickon Section in spring and fall. Usually the peaks of the warbler migration are reached in the first week of May and in the second and third weeks of September. For a rewarding trip, involving an entire day, park the car at the intersection of US 422 and Cresheim Valley Road and walk down Cresheim Creek to Wissahickon Creek and follow Wissahickon Creek northwest to the Morris Arboretum. Return to the car by bus.

Pennypack Park, in the northeast part of the city, extends from a point near the Delaware River northwestward about 5 miles to Pine Road. Pennypack Creek runs through the area, which is primarily deciduous woodland surrounded by farmland. The Park is excellent for migrating landbirds; warblers, especially, are abundant from late April to late May and from mid-September to early October. In fall, Connecticut Warblers feed here on giant ragweed. Birds nesting in the northern section of the Park, near Pine Road, include the Orchard Oriole, Baltimore Oriole, and Grasshopper Sparrow. In winter a search through the tangles of honeysuckle may reveal 'whitewash,' denoting the presence of Saw-whet Owls, which appear here quite regularly. US 1 and 13 both cross the Park.

West of the International Airport on the southwest side of the city is a vast fresh-water marsh and flatland known to Philadelphia bird finders as 'Tinicum.' State 291, the Industrial Highway, and I 95 bisect the marsh, which stretches from the Airport along the Delaware River to Essington. **Tinicum National Environmental Center** occupies more than 300 acres of the marsh northwest of the Airport, between the Airport and Darby Creek.

Bird finding is good at any season in the Center. Winter brings Rough-legged and Red-tailed Hawks to feed on marsh rodents; American Black Ducks, Greater Scaups, and Ruddy Ducks to rest and feed offshore in the River; and (Northern) Horned Larks to occupy the flatlands. Spring and fall bring enormous flocks of Common Pintails and many Green-winged and Blue-winged Teal, together with numerous other duck species, to feed on the wet flatlands. Early fall regularly brings great numbers of shorebirds, including such species as the Lesser Golden Plover, Ruddy Turnstone, White-rumped Sandpiper, and Stilt Sandpiper. In late spring and summer, Virginia Rails and Common Gallinules nest in the marsh, and a few Traill's Flycatchers, in the bordering shrubs. At the same time, Yellow Warblers commonly frequent the willows along Darby Creek and nest in low shrubs nearby.

To reach the Center from the Airport area, go northwest on Island Avenue, turn left on Lindbergh Boulevard, and proceed to the entrance on 86th Street.

For anyone who enjoys combining history with bird finding there are two delightful places: **Pennsbury Manor,** the restored home of William Penn, northeast of the city near Tullytown; and **Mill Grove,** the first home of John James Audubon in America, northwest of the city in the vicinity of Valley Forge.

One good place for ducks near Philadelphia is the area around Pennsbury Manor. Drive northeast from the city on I 95, take the exit to Bristol and go south, then turn left on US 13, exit at Tullytown, and from there drive north on the Bristol Pike a short distance, bear right on Bordentown Road for 2 miles, and right again on Ford Mill Road along Scott's Creek to the Manor. To the left of Ford Mill Road is Scott's Creek, on the right are fresh-water ponds formed by excavations.

Ducks of nearly all eastern species gather on Scott's Creek and the ponds before and after the hunting season and remain throughout winter. If the surfaces freeze over, the birds move to nearby Delaware River. Pennsbury Manor is itself worth a visit for its historical interest. Moreover, its grounds and surrounding farmlands and woodlands support an avifauna that is typical of the lowlands of the Delaware River Valley.

Mill Grove, always lovely, is at its very best in May, when flowering shrubs are in bloom and small transient landbirds are passing through. To reach it, take I 76, the Schuylkill Expressway, to US 202 south (near the exit from Pennsylvania Turnpike) and shortly thereafter turn north on State 363. Continue past Valley Forge State Park and across the Schuylkill River, and turn left on Audubon Road to the main entrance at Pawling Road, and the parking area beyond. The old stone house, beautifully restored, rests in a formal setting of lawns, trees, and shrubs overlooking Perkiomen Creek. Stop at the house (now headquarters) for a map of the 120 acres, showing the more than 6 miles of trails that wind through deciduous woods, brushy areas, and meadows, and past a nursery with plantings of trees and shrubs attractive to birds, a small pond, and two creeks.

The Ring-necked Pheasant, Downy Woodpecker, Tufted Titmouse, White-breasted Nuthatch, and Northern Cardinal are year-round residents. Birds common in summer include the Mourning Dove, Chimney Swift, Yellow-shafted Flicker, Great Crested Flycatcher, Eastern Phoebe, Barn Swallow, House Wren, Gray Catbird, Brown Thrasher, Wood Thrush, Eastern Bluebird, Louisiana Waterthrush, Baltimore Oriole, and Song Sparrow.

PITTSBURGH

Frick Park | Riverview Park | Squaw Run Park | Todd Wildlife Sanctuary | Raccoon Creek | Moraine State Park | Lake Oneida

Surrounding this great city, where the Monongahela and Allegheny Rivers join to form the Ohio, are high plateaus and wooded ridges, shaded ravines and broad, open valleys, parks and artificial ponds—features attractive to a variety of birds.

Within the city limits, **Frick Park** (380 acres), north of US 30 just east of the Squirrel Hill Tunnel and 4 miles from the Golden Triangle, is an excellent area for birds. The main entrance is just off Forbes Street. Here, among wooded hillsides and valleys, are good food and cover for both transient and resident birds. During the course of the year, over a hundred species appear. The Hermit Thrush, Fox Sparrow, White-throated Sparrow, and most of the northern warblers pass through in spring and fall. Migration peaks for the Hermit Thrush come in the third week in April and the first two weeks of October; for the White-throated Sparrow, in the last two weeks of April and in October; for the Fox Sparrow, the first two weeks of April and the last two weeks of October; and for most warblers, the first week of May and in September. Breeding birds include the Acadian Flycatcher, Carolina Chickadee, Kentucky Warbler, and Rose-breasted Grosbeak.

Riverview Park (350 acres), in the northwestern part of the city, west of US 19, is another excellent area for birds. The terrain, cut by deep valleys and ravines, is for the most part forested with oak of several species, mixed with ash and sugar maple. In the few open areas, such as along trails, shrubs grow extensively. Many birds pass through the Park in their spring and fall migrations, peaks of abundance coinciding with those for Frick Park.

Among species regularly breeding are the following: Black-billed Cuckoo, Yellow-shafted Flicker, Hairy Woodpecker, Downy Woodpecker, Great Crested Flycatcher, Eastern Pewee, Tufted Titmouse, White-breasted Nuthatch, House Wren, Wood Thrush, Blue-gray Gnatcatcher, Red-eyed Vireo, Ovenbird, Kentucky Warbler, Yellow-breasted Chat, Baltimore Oriole, Scarlet Tanager, Northern Cardinal, Rose-breasted Grosbeak, Indigo Bunting, American Goldfinch, Rufous-sided Towhee, and Field Sparrow.

Squaw Run Park, in the suburban community of Fox Chapel 5 miles northeast of Pittsburgh, has excellent bird-finding habitat, including a warbler migration 'pocket' that can be spectacular. As many as 1,000 warblers of thirty species have been found in a single day in May in the tall white oak trees along Trillium Trail in the Park; September migrations are only slightly less rewarding. Typical summer residents along the Trail are the Acadian Flycatcher, Veery, Louisiana Waterthrush, Kentucky Warbler, and Hooded Warbler.

To reach Squaw Run Park, drive northeast from the city on State 28 to the Fox Chapel Exit, go left (north, away from the Allegheny River) on Fox Chapel Road for about a mile; turn left on Squaw Run Road and proceed 1.2 miles to the parking lot at the base of Trillium Trail. To reach the best warbler area, climb the Trail to a point just above a small pond, then go left along the side of the ridge, watching the tops of the tall oaks.

Still farther northeast of Pittsburgh is the **Todd Wildlife Sanctuary** (160 acres), owned by the Audubon Society of Western Pennsylvania. This is a delightful woodland of huge maples and hemlocks bordering a stream known as Watson's Run. Birds nesting regularly include the Ruffed Grouse, Acadian Flycatcher, Magnolia Warbler, Black-throated Green Warbler, Chestnut-sided Warbler, Louisiana Waterthrush, and Canada Warbler. To reach the Sanctuary, drive north from the city on State 28 to State 356, turn left (northwest) for about a mile and turn right on Monroe Road; at a golf course, take the right fork onto Kepple Road. The Sanctuary lies on the right about 2 miles from the fork.

West of the city is **Raccoon Creek,** flowing north into the Ohio southwest of Monaca. In its vicinity are some of the best places for birds in the Pittsburgh area. The country near the mouth of the Creek consists of steep hills where oaks, birches, white ashes, and tangles of grapevines and bittersweet grow abundantly. Raccoon Creek itself winds for many miles through a great variety of country. Near the Creek are thickets of wild roses, grasses, and sedges. Here, and on the bordering, wooded hillsides, are excellent bird habitats. During winter, some of the species observed are the Red-tailed Hawk, American Kestrel, Ruffed Grouse, Red-bellied Woodpecker, Yellow-bellied Sapsucker, Tufted Titmouse, White-breasted Nuthatch, Brown Creeper, Winter Wren, Carolina Wren, Slate-colored Junco, and American Tree Sparrow. Of many points along Raccoon Creek from which to start trips for birds, the following are recommended:

Go north from Pittsburgh on State 60 or 51 to Monaca, turn south on State 18, and east on (1) State 151 to the Creek crossing at Brocktown; (2) US 30 to junction of Raccoon Creek and Big and Little Traverse Creeks at Pattons Point, then east on US 30, turning south on State 931 to Murdockville; (3) US 22 to the Creek crossing at Bavington.

Moraine State Park, 40 miles north of Pittsburgh near Butler, encompasses Lake Arthur in an ancient glacial basin. The 3,225-acre body of water, formed in 1970 by the damming of Muddy Creek, is a prime stop-over point for migrating waterfowl in spring and fall. The surrounding parkland offers a wide diversity of habitats, including cattail marsh, lowland swamp, hardwood forest, and prairie-type landscape; over a hundred bird species appear here in mid-May.

From mid-March through April, hundreds of Whistling Swans show up, along with large flocks of ducks, among them chiefly the Mallard, Green-winged and Blue-winged Teal, American Wigeon, Redhead, Ring-necked Duck, Canvasback, Lesser Scaup, and Red-breasted Merganser, with smaller numbers of such species as Gadwalls and Ruddy Ducks. October through mid-November is best for fall waterfowl. The variety of birds nesting in the Park includes Virginia Rails, Soras, Sedge Wrens, Grasshopper Sparrows, and a great many Henslow's Sparrows in the fields, once censused at 90 singing males. Large flocks of Bonaparte's Gulls, some Caspian Terns, and huge flocks of migrating swallows appear over the surface of Lake Arthur in April and early May.

To reach Moraine State Park, follow I 79 north from Pittsburgh to US 422, then go east to the Day Use Area Exit. At Park headquarters, obtain maps and the permission necessary to visit the State Game Commisssion's Propagation Area marsh at Lake Arthur's eastern end.

A visit to **Lake Oneida,** east of Moraine State Park, can be rewarding from late July through September, when, because of the low water level at this time, there are shallows and mud flats attractive to many wading birds and shorebirds. Lake Oneida may be reached by continuing east on US 422 to the intersection of State 38 and 68, then following State 38 north for about 5 miles, until the Lake appears on the right. The northern end has the most extensive mud flats.

READING
Lake Ontelaunee | Hay Creek

Lake Ontelaunee, a 1,080-acre reservoir formed by impounding the waters of Maiden Creek, is a stopping-off place for transient waterfowl, and for wintering waterfowl as long as it remains open. In late March, April, October, and November it is possible to see at least twenty-one species of waterfowl in one day. The 3,600-acre land area surrounding the Lake is an auxiliary game refuge, where trespassing is forbidden during October, November, and December; still, from roads that pass around a large part of the Lake, one can identify most of the birds with a telescope or high-powered binoculars. In addition to waterfowl, transient shorebirds are present in May, August, and September; the following are common: Semipalmated Plover, Common Snipe, Solitary Sandpiper, Greater and Lesser Yellowlegs, Pectoral Sandpiper, Least Sandpiper, and Semipalmated Sandpiper.

To reach the Lake, drive north from the city on US 222 for about 7 miles to the village of Maiden Creek and turn left on State 73.

Southeast of Reading, a rich coniferous and deciduous woodland borders **Hay Creek,** where a variety of species nests in late May, June, and early July. Among them are the Acadian Flycatcher, Least Flycatcher, Wood Thrush, Veery, White-eyed Vireo, Yellow-throated Vireo, Black-and-white Warbler, Worm-eating Warbler,

PRAIRIE WARBLER

Blue-winged Warbler, Cerulean Warbler, Chestnut-sided Warbler, Louisiana Waterthrush, Hooded Warbler, Canada Warbler, American Redstart, Scarlet Tanager, and Rose-breasted Grosbeak.

From Exit 3 on I 176, south of Reading, drive east on State 724 to Birdsboro and turn right on State 82, which parallels the Creek.

STATE COLLEGE

Bear Meadows | Barrens

Around State College in central Pennsylvania are a few mountains reaching elevations approximating 2,200 feet. All are forested and inhabited by such large birds as the Ruffed Grouse, Wild Turkey, Great Horned Owl, Pileated Woodpecker, and Northern Raven. Near Pennsylvania State University, on the eastern edge of town, are a pond and cattail marsh, experimental farm plots, and bushy fields—all habitats for birds common in this part of the state.

The best place for a diversity of breeding warblers is **Bear Meadows,** a sphagnum bog covered with a dense rhododendron-hemlock forest. To reach the Meadows, drive east of State College on US 322 for 4 miles—1.0 mile beyond Boalsburg—to a sign indicating the direction to Bear Meadows, 5 miles distant, on an unimproved road to the right.

West of State College lie the **Barrens,** a scrub-oak pitch-pine tract about 2 by 10 miles in greatest extents, with open grassy areas used by American Woodcock as singing fields from late March to mid-May. Other birds occupying the Barrens are Ruffed Grouse, Wild Turkeys, Prairie Warblers, and Rufous-sided Towhees. To reach the Barrens, drive west on US 322 for 3 miles to a service station, turn left on an unimproved road, continuing straight ahead past a road turning left, then taking the next left fork, which soon passes through the middle of the Barrens, close to the American Woodcock singing fields. The birds give their flight songs soon after sundown and before sunrise.

STROUDSBURG

La Bar's Nursery | Delaware Water Gap | Delaware Water Gap National Recreation Area | Pocono Environmental Education Center | Beltzville Lake State Park | Wild Creek Reservoir | Penn Forest Reservoir | Long Pond

The Pocono Mountains, about 100 miles north of Philadelphia and easily accessible from the Northern Extension of the Pennsylvania Turnpike, comprise the area nearest that city where birds typical of more northern associations nest. In these Mountains, covered with mixed conifer and hardwood forests, cut by ravines, and dotted with lakes, bogs, and marshy meadows, are many habitats for Red-breasted Nuthatches, Brown Creepers, Hermit Thrushes, Veeries, Cedar Waxwings, Solitary Vireos, Black-and-white Warblers, Magnolia Warblers, Black-throated Blue Warblers, Black-throated Green Warblers, Blackburnian Warblers, Chestnut-sided Warblers, Ovenbirds, Northern and Louisiana Waterthrushes, Canada Warblers, American Redstarts, Purple Finches, Slate-colored Juncos, and White-throated Sparrows. Although usually rare, a few Saw-whet Owls, Traill's Flycatchers, Golden-crowned Kinglets, Golden-winged Warblers, Nashville Warblers, and Myrtle Warblers breed here.

Bird finding is best from mid-May to mid-July, when the birds are in full song, but the fall warbler migration in September and the fall hawk flights along the ridges are excellent, and a winter trip to the Poconos may produce a great many northern finches during 'invasion' years.

La Bar's Nursery, south of I 80 in Stroudsburg at the west end of Bryant Street, is a productive area for transients in May and worth a visit simply to view the blooming azaleas and rhododendrons. House Finches breed here in numbers.

A few miles south of Stroudsburg on State 611 is the spectacular **Delaware Water Gap.** During May and June, a walk from the overlook parking lot south along State 611 for a mile or so should yield singing Winter Wrens, Black-and-white Warblers, Worm-eating Warblers, Northern Parula Warblers, Yellow Warblers, Chestnut-sided Warblers, Louisiana Waterthrushes, Hooded Warblers, American Redstarts, Scarlet Tanagers, and Indigo Buntings. Rough-winged Swallows nest in crevices in the wall between the road and the river below, and hawks are often overhead availing themselves of the updrafts.

From I 80 south of Stroudsburg, take State 402 north 4 miles to Marshalls Creek, turn right on US 209, and soon enter the **Delaware Water Gap National Recreation Area,** which includes a diversity of habitats: the Delaware River, the floodplain bordering it on the Pennsylvania side, the steep hillsides on both sides of the River, and the rolling uplands beyond, both open and wooded. Over 250 species of birds have thus far been identified within the Area, according to the checklist of the **Pocono Environmental Education Center,** which is reached by turning left 7 miles beyond the blinker light at Bushkill. Visitors are welcome to stop for assistance in finding their way to the best areas for birds.

The following circle tour from Stroudsburg passes some of the best bird-finding areas. Drive southwest from Stroudsburg on US 209 to the Pennsylvania Turnpike at the Mahoning Valley Interchange. Turn right here on Harrity Road and immediately right again on Pohopoco Drive to headquarters of **Beltzville Lake State Park.** Beltzville Lake is a large impoundment created by the damming of Pohopoco Creek. Although the Lake offers little bird finding because of disturbance by powerboats, the recesses away from the main expanse, as well as the fields and woodlands along its shores, are worth exploring. Continue on Pohopoco Drive for 5 miles and watch for the entrance gate to **Wild Creek Reservoir** on the left. Go in and drive up to the overlook at the dam. There may be transient waterfowl in season, and Prairie Warblers may be singing nearby in May and June.

Upon returning to the gate, turn left for 0.4 mile to an intersection, then turn left again and go 2.1 miles to a vista on the left overlooking the upper end of the Reservoir. The next 0.6 mile is through a Norway spruce plantation, excellent for transient warblers in May and September. Cape May and Bay-breasted Warblers seem particularly attracted to this area and sometimes appear in remarkable numbers. Magnolia Warblers are regular, and transient thrushes frequent. Pine Grosbeaks have been reported here in winter on several occasions.

Less than a mile beyond is the overlook at **Penn Forest Reservoir,** a huge protected impoundment with 7 miles of shoreline. The Reservoir draws a variety of waterfowl, including unexpected species when a storm or cold front passes through during migration. Golden-winged and Pine Warblers breed in the red pines between the overlook and the north end of the Reservoir.

Four miles beyond the Reservoir the road dead-ends at State 903. Turn right for 11 miles, and turn right again on State 115 for 2.8 miles to a secondary road on the left at the Pocono Raceway. Take this road for 2 miles to a bridge. To the south of the bridge stretches **Long Pond,** a wild, meandering stream, wide in spots, and highly recommended for exploration by canoe. Look for Traill's Flycatchers, Sedge Wrens, Common Yellowthroats, and Swamp Sparrows in the grasses and shrubs bordering it. Great Blue Herons, American Bitterns, Mallards, American Black Ducks, and Wood Ducks regularly summer here, and Northern Harriers occasionally. To the north of the bridge, the woods has Black-throated Blue Warblers, Northern Waterthrushes, and Canada Warblers, but perhaps the three commonest warblers are the Magnolia, Black-throated Green, and Blackburnian.

Just beyond the bridge, a local road to the left leads through some mature woodland worth exploring for breeding species such as the Myrtle, Black-throated Green, and Blackburnian Warblers, Slate-colored Juncos, and White-throated Sparrows. The road intersects State 940 at Pocono Lake. To return to Stroudsburg, drive east on State 940, looking for birds all the way, to I 380, turn south to I 80, and then east.

Reference

Birds of the Pocono Mountains, 1955–1975. By Phillips B. Street. In *Cassinia,* no. 55 (1974–1975). Available from Academy of Natural Sciences of Philadelphia, 19th Street and the Parkway, Philadelphia, Pa. 19103.

South Carolina

DENNIS M. FORSYTHE AND SIDNEY A. GAUTHREAUX, JR.

AMERICAN ANHINGAS

South Carolina is rich in ornithological history and in the number of bird species within its borders. The first known list of South Carolina birds came in 1562 from Captain Jean Ribault, leader of the Huguenot colony at Port Royal. Over two centuries later, in 1772, Mark Catesby began systematic studies, and from his specimens, descriptions, and drawings the Swedish naturalist, Linnaeus, authored the scientific names for many of our more common North American species, with South Carolina as the type locality.

The first major studies of Carolina birdlife were undertaken early in the nineteenth century by John James Audubon and his coworker, the Reverend John Bachman, and culminated early in the twentieth with the enormous contributions of Arthur T. Wayne. Numerous observers have continued this ornithological tradition to the present, with the result that 77 species have been described from South Carolina and more than 470 species have been recorded as appearing in the state.

South Carolina is a small state—thirty-ninth in size among the contiguous forty-eight—consisting of about 30,000 square miles. But within its roughly triangular borders is a wide diversity of habitats. The broad base of the triangle rests on the Atlantic Ocean, with a coastline of over 281 miles, interrupted by such large indentations as Winyah Bay, Bulls Bay, Charleston Harbor, and St. Helena and Port Royal Sounds. The apex of the triangle, the northwest corner of the state, rests in the Blue Ridge Mountains, where Sassafras Mountain attains 3,548 feet, highest point in the state.

Three general physiographical divisions are discernible from east to west: the Coastal Plain, the Piedmont Region, and the Mountain Region.

The *Coastal Plain,* the largest natural division, is probably the most visited and the most productive area for the bird finder. Approximately 100 miles wide, it stretches from the Atlantic Ocean inland to the Fall Line, whose most prominent feature is a series of sand dunes, the 'sandhills,' from 3 to 40 miles wide, running southwestward across the state from Cheraw to Augusta, Georgia. The Coastal Plain is, in turn, divided into the Lower Coastal Plain, from the Atlantic to about US 301, and the Upper Coastal Plain, from US 301 to the Fall Line. A series of long, narrow, barrier islands form the eastern boundary of the Coastal Plain. These islands, with their sand beaches, sea oats-covered dunes, and live oak-cabbage palmetto-wax myrtle forest, have come under increasing pressure for development in recent years. Between the barrier islands and the mainland are the farmed and well-populated 'sea islands' and expanses of cord grass salt marsh, oyster banks, and mud flats, transected by the Intracoastal Waterway and numerous tidal creeks. Birds nesting exclusively on the coastal islands, beaches, and marshes are the following:

Brown Pelican
Clapper Rail
American Oystercatcher
Wilson's Plover
Willet
Laughing Gull
Gull-billed Tern
Forster's Tern

Little Tern
Royal Tern
Sandwich Tern
Caspian Tern (*rarely*)
Black Skimmer
Boat-tailed Grackle
Seaside Sparrow

Common on the inland Coastal Plain, though not entirely restricted to it, are wooded swamps and pinelands. The swamps are characterized by such trees as bald cypress, tupelo gum, red maple, and Walter's or spruce pine. Extensive brakes of swamp cane and tangles of vines are abundant. Examples of this habitat include I'On Swamp (*see under* **Charleston**), Four Hole Swamp, and tracts along the Congaree and Savannah Rivers. Breeding birds typical of such habitat are the Acadian Flycatchers, Red-eyed Vireos, Prothonotary Warblers, Swainson's Warblers (uncommon), Northern Parula Warblers, and (Wayne's) Black-throated Green Warblers.

The pinelands feature mixed and pure stands of loblolly and pitch pines, with an understory of blueberry, bracken fern, and various grasses. Breeding birds characteristic of these pine forests include Red-cockaded Woodpeckers, Brown-headed Nuthatches, Pine Warblers, and Bachman's Sparrows.

The *Piedmont Region,* the next largest division, extends from the Fall Line northwest to the Blue Ridge Mountains. This rolling hill country with its red clay soil is the most populous and industrialized region of the state. Once heavily forested, the tracts remaining consist largely of second-growth oak, hickory, and pine on higher ground, black and sweet gums, maple, and willow in the river valleys.

The smallest division, the *Mountain Region,* includes the western portions of three counties on the Georgia and North Carolina borders. Besides Sassafras Mountain, at least two other peaks exceed 3,000 feet. On their forested slopes, oak, hickory, and tulip tree predominate. In the valleys and ravines the forests have sweet birch, holly, hemlock, and white pine, with a dense understory of rhododendron and mountain laurel. Breeding birds characteristic of the Mountain Region include:

Ruffed Grouse
Northern Raven
House Wren
Solitary Vireo

Swainson's Warbler (*rhododendron thickets*)
Black-throated Blue Warbler
Black-throated Green Warbler
Chestnut-sided Warbler
Ovenbird
Canada Warbler (*high elevations only*)
American Redstart
Scarlet Tanager
Rose-breasted Grosbeak
Song Sparrow

The following breeding birds are characteristic of deciduous woods below 3,000 feet and all adjoining farmlands (fields, wet meadows, brushy lands, orchards, and dooryards):

DECIDUOUS WOODS

Turkey Vulture
Black Vulture
Red-shouldered Hawk
Yellow-billed Cuckoo
Common Screech Owl
Barred Owl
Chuck-will's-widow
Whip-poor-will (*except coastal region*)
Yellow-shafted Flicker
Pileated Woodpecker
Red-bellied Woodpecker
Hairy Woodpecker
Downy Woodpecker
Eastern Kingbird
Great Crested Flycatcher
Acadian Flycatcher
Eastern Pewee
Blue Jay
Common Crow (*except near coast*)
Fish Crow (*coastal region only*)
Carolina Chickadee
Tufted Titmouse
White-breasted Nuthatch
Wood Thrush
Blue-gray Gnatcatcher
Yellow-throated Vireo
Red-eyed Vireo
Northern Parula Warbler
Yellow-throated Warbler
Louisiana Waterthrush
Kentucky Warbler
Hooded Warbler
Summer Tanager

FARMLANDS

Common Bobwhite
Mourning Dove
Red-headed Woodpecker
Eastern Kingbird
Carolina Wren
Northern Mockingbird
Brown Thrasher
American Robin (*above Fall Line*)
Eastern Bluebird

Loggerhead Shrike
White-eyed Vireo
Yellow Warbler (*above Fall Line*)
Prairie Warbler
Common Yellowthroat
Yellow-breasted Chat
Eastern Meadowlark
Common Grackle
Northern Cardinal
Blue Grosbeak
Indigo Bunting
Painted Bunting (*Coastal Plain*)
Rufous-sided Towhee
Grasshopper Sparrow (*above Fall Line*)
Bachman's Sparrow
Chipping Sparrow
Field Sparrow

During migration, large numbers of waterfowl, waterbirds, and shorebirds pass along the coast. No similar landbird numbers, however, pass through the state, as there are no topographical features in South Carolina to channel their movements. Good spring and fall landbird migrations take place in the Piedmont and Mountain Regions, but the spring migration along the coast is relatively poor. The shorebird migration is also poorer along the coast during spring as compared with fall. The following general dates give some indication of when to expect the main flights:

Waterfowl: 15 February–1 April; 25 October–15 December
Shorebirds: 20 April–20 May; 10 August–15 October
Landbirds: 25 March–5 May on coast; 15 April–30 May in Piedmont; 10 September–5 November

During winter, one can observe between fifty and ninety bird species in the Mountain and Piedmont Regions and the Upper Coastal Plain. But along the coast where one sees the largest number and variety of birds, counts of over 140 species are possible at such places as Myrtle Beach, Charleston, and Hilton Head. Accounting for these high figures is the large number of permanently residing species, joined by many winter visitants such as Common and Red-throated Loons, Horned Grebes, Northern Gannets, waterfowl, shorebirds, gulls, terns, Tree Swallows, American Robins, kinglets, Palm Warblers, Rusty Blackbirds, Evening Grosbeaks, Purple Finches, and Savannah, Sharp-tailed, and White-throated Sparrows. Winter concentrations of both geese and ducks are impressive at the

Santee National Wildlife Refuge (*see under* **Summerton**), and ducks at the Savannah River and Cape Romain National Wildlife Refuges (*see under* **Hardeeville** and **Charleston**).

Authorities

Harry E. LeGrand, Jr., Willie M. Morrison, William N. Neely, H. Douglas Pratt, Alexander Sprunt, Jr.

CHARLESTON

Charleston Waterfront | Fort Sumter National Monument | Drum Island | Pitt Street Causeway | Sullivans Island | Folly Island | Magnolia Gardens and Nursery | Bear Swamp | Francis Marion National Forest | Cape Romain National Wildlife Refuge

About midway on the South Carolina coast, Charleston enjoys a wide variety of bird habitats in its vicinity. The main part of the city, on a peninsula in Charleston Harbor between the mouths of the Cooper and Ashley Rivers, is accessible by US 17 (the 'coastal route') coming in from the north over the Cooper River Bridge and from the south over the Ashley River Bridge, and by I 26 coming down the peninsula from the west and connecting with US 17.

The **Charleston Waterfront** offers close views of waterbirds, waterfowl, and a few shorebirds. To reach this area, drive south on US 17 from its intersection with I 26 and, just before it goes over the Ashley River Bridge, turn left on Lockwood Drive. After about two blocks, the City Marina appears on the right, a salt water pond on the left, and the United States Coast Guard Station straight ahead. Stop here and scan the open water, salt marshes, and, at low tide, the mud flats and sand bars.

Present all year are Brown Pelicans, Double-crested Cormorants, Little Blue Herons, Snowy Egrets, Louisiana Herons, Clapper Rails, American Oystercatchers, Laughing Gulls, Royal Terns, and Marsh Wrens. From November through March the variety is increased by such species as Horned and Pied-billed Grebes, Lesser Scaups, Red-breasted Mergansers, Black-bellied Plovers, and Herring, Ring-billed, and Bonaparte's Gulls. To see more birds along the Waterfront, drive around the inland side of the Coast Guard Station to Murray Boulevard and then East Battery, both of which border Charleston Harbor and have pulloffs for viewing.

Fort Sumter National Monument, 5 miles out in Charleston Harbor and in view from Murray Boulevard and East Battery, is an artificial rocky island that attracts Ruddy Turnstones and an occasional Purple Sandpiper in winter. Surf Scoters, Black Scoters and sometimes a few Oldsquaws are probable winter visitants offshore.

Drum Island, a 100-acre spoil area on which rests the middle span of the Cooper River Bridge, has dense stands of baccharis and red mulberry at the north and south ends where nest many thousands of wading birds: Great Blue Herons, Little Blue Herons, Great Egrets, Snowy Egrets, Louisiana Herons, Black-crowned and Yellow-crowned Night Herons, and Glossy and White Ibises. Willets, Laughing Gulls, Little Terns, and occasionally Black-necked Stilts appear on the barer portion of the Island.

Both Fort Sumter and Drum Island are accessible only by boat. Harbor cruises out of Charleston, from the dock at the Fort Sumter Hotel on the Battery, the City Marina, and the adjacent Charleston Inn, stop at Fort Sumter and draw near Drum Island, providing good chances for seeing its birds. From April to November, party fishing boats from the City Marina and Charleston Inn go 15 to 70 miles out to sea, offering a good opportunity to see pelagic birds such as shearwaters and Wilson's Storm Petrels.

The **Pitt Street Causeway** in Mt. Pleasant is an excellent vantage point for birds in any season. From I 26, take US 17 north across the Cooper River Bridge; turn right on Business 17 through Mt. Pleasant and across Shem Creek to State 703; at the next traffic signal, turn right on McCants Drive and go about 12 blocks; then turn left on Pitt Street, which leads to the Causeway overlooking mud flats, creeks, and salt marsh.

Least Bitterns, American Oystercatchers, Wilson's Plovers, Little Terns, and Seaside Sparrows are some of the birds one may expect in late spring and summer; from October through March one should look for shorebirds—e.g., Semipalmated Plovers, Black-bellied Plovers, Whimbrels (spring and fall only), Willets, Dunlins, Short-billed Dowitchers, Western Sandpipers, Marbled Godwits—and other species such as Northern Harriers, Ospreys, Black Skimmers,

Fish Crows, Boat-tailed Grackles, and Savannah and Sharp-tailed Sparrows.

Sullivans Island, a barrier island, is worth a trip in any season. To reach it from Charleston, proceed as above to State 703 and stay on it across the Intracoastal Waterway to Sullivans Island; leave State 703 where it turns left, and continue toward the front beach for one block, then turn right on Middle Street to Fort Moultrie on the west end of the Island. During summer, Little Terns and Wilson's Plovers nest on the beach in front of the Fort. In spring and fall, large flocks of swallows stay temporarily around the wax-myrtle thickets. In winter, inspect the rock groins in front of the Fort for Ruddy Turnstones and the occasional Purple Sandpipers. Off the east end of Sullivans Island is Breach Inlet, crossed by State 703 to the Isle of Palms. In winter, look this over for Greater Black-backed and other gulls, as well as for possible 'bay' ducks such as Oldsquaws and scoters.

For an equally rewarding coastal area, visit **Folly Island,** another barrier island about 12 miles southeast of Charleston. Take US 17 south across the Ashley River Bridge about 0.5 mile, then bear left on State 171 to its end on Folly Beach, a summer resort 9 miles distant. On the way, State 171 crosses Wappoo Cut to James Island, a typical sea island with suburbs, agricultural lands, and partially wooded areas, as well as extensive salt marshes and mud flats.

About 3 miles after State 171 crosses Wappoo Cut, turn left on Fort Johnson Road to the College of Charleston Marine Resources Laboratory, 4 miles distant. Here, with a fine view of Charleston Harbor, together with representative salt marshes and mud flats, one may see pelicans, cormorants, wading birds, bay ducks, Clapper Rails, shorebirds, gulls, terns, Boat-tailed Grackles, and Marsh Wrens. In the willows and other woody growth behind the Laboratory, look for transient and wintering landbirds that will include American Robins, Myrtle Warblers, Palm Warblers, Common Yellowthroats, and White-throated Sparrows.

Return to State 171 and continue. On reaching the end at Folly Beach, one may turn right and drive as far as the road goes (about 3 miles), then walk about 200 yards over the dunes to Stono Inlet. The sand bar, Bird Key, at its entrance, is a breeding site for Wilson's Plovers, Laughing Gulls, Gull-billed Terns, Royal Terns,

BOAT-TAILED GRACKLE

Sandwich Terns, and Black Skimmers. At Folly Beach one may also turn left and drive about 2 miles toward Morris Island Inlet. Although the road ends at the United States Coast Guard Station (open from dawn to dusk), one may sign in here and walk the rest of the way to the Inlet. The sand bars are especially attractive in winter to resting gulls and terns. Daily winter sights offshore include Common and Red-throated Loons, sometimes Northern Gannets in passage far out, cormorants, scoters, and Red-breasted Mergansers. In grassy areas along the dunes, from mid-November to late March, watch for (Ipswich) Savannah Sparrows. Folly Beach itself, from Stono Inlet to Morris Island Inlet, is attractive to many shorebirds—Red Knots, 'peeps,' and Sanderlings—during their migrations and in winter when human disturbances are minimal.

Magnolia Gardens and Nursery (350 acres), truly excellent for waterbirds, waterfowl, and small landbirds, is a privately owned plantation but open to the public for a fee. To reach the Gardens, go south on US 17 over the Ashley River Bridge, then northwest on State 61 for 12 miles. At the entrance, turn into the Gardens and go to the office for information on birds present at the time. Behind, and about 100 yards to the left of the office, is fresh-water Marsh Pond (130 acres), ringed by cattails, where American Anhingas, Great Blue Herons, Black-crowned and Yellow-crowned Night Herons, American Bitterns, Glossy and White Ibises, Ospreys, and

Common Gallinules are present all year and Least Bitterns in summer. Throughout winter, the bird population on and around Marsh Pond is greatly augmented by about fifteen species of ducks, often including a few Fulvous Tree Ducks, together with American Coots, and occasionally one or more Bald Eagles.

In the semi-wooded and agricultural lands are habitats in which one may find summer-resident Wood Thrushes, Orchard Orioles, Blue Grosbeaks, Indigo Buntings, and Painted Buntings as well as Northern Parula Warblers, Yellow-throated Warblers, and Pine Warblers the year round. In the cypress backwaters of the Gardens, Wood Ducks, Prothonotary Warblers, and Hooded Warblers nest, and many surface-feeding ducks pass the winter.

To look for summer-resident Swainson's Warblers, backtrack about 2 miles on State 61 and turn left on unmarked Bee's Ferry Road, which connects State 61 with US 17 to Charleston. After traversing open fields for about 2 miles, the Road goes through a dense swamp, **Bear Swamp,** for the remaining 3 miles to US 17. Several pairs of Swainson's Warblers breed in Bear Swamp along with Barred Owls, Great Crested Flycatchers, Acadian Flycatchers, Eastern Pewees, White-eyed Vireos, Prothonotary Warblers, Northern Parula Warblers, and Hooded Warblers. Listen and look for Swainson's Warblers in dense thickets. *Caution:* Beware of poisonous snakes.

The **Francis Marion National Forest** north of Charleston, covering more than 240,000 acres, is rich in diversity of habitats for landbirds and consequent variety of species. The following tours are suggested. Before taking them, obtain maps and further information available from the District Ranger, Forest Service, United States Department of Agriculture, Box 298, Moncks Corner, South Carolina 29461.

Drive north on US 17 over the Cooper River Bridge for about 10 miles, turn left on State 41, crossing the Wando River (good for waterbirds, waterfowl, and shorebirds at low tide) at 5 miles, and passing the town of Cainhoy at 7 miles to the junction of Brick Church Road (Forest Service Road 188) on the left and Hoover Road (FS 183) on the right. Take Hoover Road, which goes through a woods of loblolly pine for the next 3 miles. Observable from the Road are several clans of Red-cockaded Woodpeckers, as well as Red-bellied

Woodpeckers, Brown-headed Nuthatches, Eastern Bluebirds, and Pine Warblers. Look for Wild Turkeys, which tend to be more active during March through May at dawn and dusk. Summer-resident Chuck-will's-widows call regularly at night.

Although Bachman's Sparrows may be found along the Hoover Road, a better area is easily reached by returning to State 41, turning right and going on for 2 miles, then right again on a dirt road (FS 229) for about 0.5 mile. Stop at this point and search the pine woods with bushy undergrowth. Bachman's Sparrows reside here the year round but are more easily found during April through June, when they reveal themselves by singing.

For a second trip, take US 17 about 9 miles north from its junction with State 41, passing the turnoff for Moore's Landing on the right at 5 miles, and turn off left on unmarked I'On Swamp Road—the only gravel road entering US 17 from the left after the Moore's Landing turnoff. For the first 1.5 miles the Road passes through a hardwood-pine woods; then it crosses two wooden bridges over the I'On Swamp, wooded primarily with cypress and tupelo. Among birds residing here are Pileated Woodpeckers and (Wayne's) Black-throated Green Warblers. On the left about a mile farther on, the Road passes a field overgrown with broom sedge and brush, attractive to Sedge Wrens and fringillids in winter. Anywhere along the Road in late spring and summer, always watch for both the Swallow-tailed and Mississippi Kites.

The **Cape Romain National Wildlife Refuge** stretches for 15 miles along the Atlantic coast northeast of Charleston. A primary haven for wildlife on the eastern seaboard, this vast area of 60,000 acres east of the Intracoastal Waterway includes salt marshes with a maze of creeks and channels, Bulls Bay with mud flats, oyster banks, and sand bars, and three barrier islands—Cape, Raccoon, and Bulls—with broad, shell-strewn beaches fronting on the ocean.

Ornithological interest focuses on Bulls Island, famed through the years for its wealth of birdlife in a remote setting of great beauty and charm. Situated south of Bulls Bay in the southern part of the Refuge, 6 miles long by 2 miles wide, and higher than the other islands, Bulls Island enjoys a rich forest cover, primarily of live oaks, magnolias, and loblolly pines—many of them huge and ages old—shading an understory of palmetto, myrtle, holly, bay, and yaupon laced by vines of many varieties. Among the common breeding birds one may see or hear from the roads and trails are the Chuck-will's-widow, Pileated and Red-bellied Woodpeckers, Great Crested Flycatcher, White-eyed and Red-eyed Vireos, Northern Parula Warbler, Yellow-throated Warbler, Pine Warbler, Summer Tanager, and Painted Bunting. Here and there on the Island are fresh-water ponds that are attractive to Pied-billed Grebes, Wood Ducks, and Common Gallinules.

Bulls Island merges westward with salt marshes, year-round habitat for Clapper Rails, Marsh Wrens, and Boat-tailed Grackles; and foraging areas for hosts of wading birds, such as herons, egrets, Wood Storks, and both Glossy and White Ibises. Northeastward toward Bulls Bay, the Island tapers to a long sand spit. Here and offshore in Bulls Bay on mud flats, oyster banks, and sand bars the shorebirds congregate: American Oystercatchers, Wilson's Plovers, and Willets in any month, since they breed on the Refuge and stay throughout winter; Semipalmated Plovers, Black-bellied Plovers, Ruddy Turnstones, Whimbrels, Greater and Lesser Yellowlegs, Red Knots, Least Sandpipers, Dunlins, Short-billed Dowitchers, Semipalmated Sandpipers, Western Sandpipers, and Sanderlings during their migrations in May, August, and September, less commonly in winter, when they may be joined occasionally by one or more Marbled Godwits.

Present in any season are Brown Pelicans, Laughing Gulls, Gull-billed Terns, Forster's Terns, Little Terns, Royal Terns, Sandwich Terns, Caspian Terns, and Black Skimmers. All nest in the Refuge, usually on treeless islands, except Forster's Terns, which prefer the salt marshes.

During winter the bird population is swelled by Common Loons, Horned Grebes, Double-crested Cormorants, vast numbers of waterfowl (Canada Geese, Mallards, American Black Ducks, Gadwalls, Common Pintails, Green-winged and Blue-winged Teal, American Wigeons, Northern Shovelers, Ring-necked Ducks, Canvasbacks, Greater and Lesser Scaups, Buffleheads, Surf Scoters, Ruddy Ducks, Hooded Mergansers, and Red-breasted Mergansers), American Coots, Herring Gulls, Ring-billed Gulls, and usually a few Red-throated Loons, Greater Black-backed Gulls, and Bonaparte's Gulls.

Refuge headquarters is at Moore's Landing, reached from

Charleston by driving north on US 17 for 20 miles and then south on SeeWee Road for 5 miles. Boats are accessible at the Landing for trips to Bulls Island, 3 miles distant.

CLEMSON
Lake Hartwell

Lake Hartwell, largest of the reservoirs in northwestern South Carolina, formed by impounding the Savannah River, has two good vantage points near Clemson for transient and wintering waterbirds and waterfowl.

1. From the intersection of US 123 and US 76 in Clemson, drive south on US 76 for 2.2 miles, right on State 22 for 1.0 mile, then left on State 139 for 1.3 miles to the Lake. Here along the causeway one can scan the Lake to the north and south and to the west near a small island and shoreline.

2. From I 85, south of Clemson, drive north on State 24 for 6 miles toward Townville; right on State 184 for 3.4 miles, crossing over an arm of Lake Hartwell; turn right again on State 21 and go 3 miles to its end on Oconee Point for scanning the Lake.

Among waterbirds one may expect to see from both vantage points during colder months are Pied-billed Grebes, American Coots, and Ring-billed Gulls and probably a few Common Loons, Horned Grebes, surface-feeding and diving ducks, and Herring Gulls. During migration, particularly spring migration in March and early April, some of the birds that stop off during their northward passage are Blue-winged Teal, Lesser Scaups, Red-breasted Mergansers, Ospreys, Bonaparte's Gulls, and usually a few Forster's, Common, Caspian, and Black Terns.

HARDEEVILLE
Savannah National Wildlife Refuge

South of Hardeeville, in the southernmost corner of the state, US 17 bisects the **Savannah National Wildlife Refuge** before and after crossing the Savannah River into Georgia. More than half the Refuge, or 7,229 acres, is in South Carolina; the remaining 5,450 acres are in Georgia. Headquarters buildings are in the South Carolina section, in view from US 17.

When established in 1927, the Refuge contained many rice fields, abandoned early in the century. Beginning in 1935, about 3,000 acres were reclaimed and improved by a system of dikes with mechanisms to control the water level for the cultivation of crops as food for migrating and wintering waterfowl. The rest of the Refuge, about 10,000 acres, is comprised of marshes and swamps in which cypresses, magnolias, tupelos, and live oaks thrive—ideal habitats for a rich variety of birds.

The Refuge is consequently excellent for bird finding all year long. Turnoffs along the highway through the Refuge provide good vantage points for seeing birds in the fields, marshes, and swamps, but there are better opportunities along some of the dikes in the Refuge. Inquire at headquarters about dikes that are accessible by car or on foot at the specific time of visit.

At any time of the year, look for American Anhingas, herons, egrets, and White Ibises, as well as Wood Ducks—the only waterfowl breeding in the Refuge. Watch the sky for Red-tailed or Red-shouldered Hawks and, except during spring and summer, scan the fields for a Northern Harrier circling low. In spring and summer, watch the flooded fields for Least Bitterns, King Rails (common), Purple Gallinules (also common), and Common Gallinules. In August, when the water level is lowered, look over the fields for Wood Storks along with White Ibises, feeding on stranded fish.

From September throughout winter, when the fields are again flooded, listen for, and try for a glimpse of, Virginia Rails and Soras, both of which are numerous. Later in November and from then until late March, some thousands of wintering waterfowl—especially Mallards, American Black Ducks, Gadwalls, Common Pintails, Green-winged Teal, American Wigeons, Northern Shovelers—are in view from the highway.

HARTSVILLE
Kalmia Gardens

For an excellent representation of breeding landbirds common to the woods and woodland edges of northeastern South Carolina, visit **Kalmia Gardens** (60 acres), 3 miles west of Hartsville on an extension of Carolina Avenue. This arboretum, bordering Black Creek, features biotic communities with trees, shrubs, and flowering plants typical of the state's interior. Paths lead to all of them from the parking lot. Especially rewarding is the path along the edge of a heavily wooded swamp, flanked with shrubbery and open woods. During late spring and early summer, expect to see or hear: six woodpeckers, the Yellow-shafted Flicker, Pileated, Red-bellied, Red-headed, Hairy, and Downy; three flycatchers, the Great Crested, Acadian, and Eastern Pewee; the Wood Thrush; three vireos, the White-eyed, Yellow-throated, and Red-eyed; and at least nine warblers, the Prothonotary, Northern Parula, Yellow, Yellow-throated, Prairie, Kentucky, Common Yellowthroat, Hooded, and American Redstart. The same path is equally rewarding during migration in late March and early April and again in late August through September, when many northern thrushes and warblers are passing through the area.

McBEE
Carolina Sandhills National Wildlife Refuge

The **Carolina Sandhills National Wildlife Refuge** in northeastern South Carolina embraces 46,000 acres of sandhill country along the Fall Line between the Coastal Plain and the Piedmont Plateau. When the acreage was acquired as a National Wildlife Refuge in 1939, much of it was submarginal cropland. In the years since, terrain that once seemed hopeless now supports extensive stands of loblolly pines and oaks, habitat suitable for Wild Turkeys, which were reintroduced, and for Red-cockaded Woodpeckers, Brown-headed Nuthatches, Yellow-throated Warblers, and Summer Tanagers. Streams, impounded here and there, formed small ponds for wintering waterfowl—chiefly Mallards and American Black Ducks—and resident Wood Ducks.

Refuge headquarters is on US 1, 4 miles northeast of McBee. From here one may tour the Refuge, following a route that offers many good vantage points. The hardwoods along the creeks and adjacent brushy edges are worth examining for the Yellow-billed Cuckoo, Chuck-will's-widow, Prairie Warbler, Hooded Warbler, Orchard Oriole, and Indigo Bunting. In open country, look especially for Red-headed Woodpeckers and Loggerhead Shrikes.

MYRTLE BEACH
Huntington Beach State Park

On US 17, south of this seaside resort and about halfway to Georgetown, turn off just south of Murrells Inlet and enter **Huntington Beach State Park** (2,500 acres). The entrance is on the east side of the highway opposite the entrance to Brookgreen Gardens. After passing through the gate and turning sharply left on the Park road, stop the car and explore the adjacent pine woods, a year-round habitat for Red-cockaded Woodpeckers, Brown-headed Nuthatches, and Pine Warblers. Continue eastward on the main road, stop when passing through a big stand of moss-draped live oaks, and look and listen for permanent-resident Pileated Woodpeckers and Yellow-throated Warblers and summer-resident Acadian Flycatchers, Northern Parula Warblers, (Wayne's) Black-throated Green Warblers, and Summer Tanagers.

Eventually the main road comes to a marshy area—probably the best spot for birds in the Park—and crosses it by causeway, dividing it into fresh-water marsh on the right and a salt-water marsh on the left. In the fresh-water marsh, Pied-billed Grebes and Common Gallinules are present all year, a few Least Bitterns in spring and summer, and many surface-feeding and diving ducks in winter. In the salt marsh one can count on many herons and egrets at any time and, though less conspicuous, Clapper Rails, as well. Black Skimmers commonly patrol the water channels, summer or winter, and nest in the vicinity. Shorebirds frequent the mud flats during their migrations in May, August, and September; some stay all winter.

Notable among the species are Semipalmated Plovers, Black-bellied Plovers, Whimbrels, Willets, Greater and Lesser Yellowlegs, Dunlins, Semipalmated Sandpipers, and Western Sandpipers.

Beyond the causeway the road forks. Take the left fork to a parking area on the beach. Along the way the road passes through a myrtle thicket, suitable habitat for summer-resident Indigo and Painted Buntings. By chance, one or more of the buntings may be perched on the utility wires overhead. From the parking area, walk north on the beach, watching ahead for summer-resident Little Terns and for such transient shorebirds as Piping Plovers, Ruddy Turnstones, Red Knots, and Sanderlings, while keeping an eye offshore for wintering or migrating Common Loons, Red-throated Loons, Horned Grebes, Northern Gannets, and one or more species of scoters. Search the beach grass for fringillids. Besides Song Sparrows, there should be Savannah Sparrows and possibly one of the Ipswich race may be among them.

PICKINS
Sassafras Mountain

The species of birds characteristic of the mountains in northwestern South Carolina are well represented on **Sassafras Mountain** (3,548 feet), the state's highest peak, almost on the North Carolina line. The best time to find the birds is in late spring, preferably mid-May to mid-June, when they are in full song.

Drive north from Pickens on US 178 for 19 miles to the community of Rocky Bottom, then take the paved road for 4.9 miles to the summit of Sassafras. Along the road as it meanders up through the forest cover—predominantly of hardwoods—stop often to listen or search for the Solitary Vireo, Black-and-white Warbler, Swainson's Warbler, Worm-eating Warbler, Black-throated Blue Warbler, Black-throated Green Warbler, Chestnut-sided Warbler, Ovenbird, and Canada Warbler. Both Black-throated Blue Warblers and Ovenbirds become increasingly numerous near the summit.

In case the Swainson's Warbler was missed, try searching for it, after descending the Mountain, in the ravines along US 178 for 3 miles north of Rocky Bottom. The Swainson's prefers the rhododendron–mountain laurel thickets.

SUMMERTON
Santee National Wildlife Refuge

In the upper Coastal Plain of south-central South Carolina, the **Santee National Wildlife Refuge** covers 74,353 acres, of which all but 3,400 acres constitute the waters of two great hydroelectric reservoirs, Lake Marion and Lake Moultrie. Managed for wintering waterfowl since its establishment in 1941, the Refuge harbors thousands of Canada Geese, a few Snow Geese, sundry ducks—primarily Mallards, American Black Ducks, Gadwalls, Common Pintails, Green-winged Teal, American Wigeons, Wood Ducks, and Ring-necked Ducks—and American Coots from late November to early March. The best place for observing large concentrations is near headquarters, just off US 301, 7 miles south of Summerton, where the birds feed on grains, sunflower seeds, and other crops planted especially for them.

Tennessee

GREAT EGRETS

Reelfoot Lake and the Great Smoky Mountains . . . if Tennessee had no haunts for birds other than these two wilderness areas, the state would be a mecca for bird finders.

Situated in Tennessee's northwestern corner, Reelfoot is the only large natural lake in the state; it measures 12 miles long and 4 miles wide at the widest point. Its character is extraordinary—in fact, nearly as extraordinary as the way it was created years ago by a series of earthquakes. From its shallow waters emerge majestic cypresses, thick mats of surface vegetation, and vast stretches of man-high grasses; about some of its shores grows a swamp forest, dense and almost impenetrable. Exploring Reelfoot will yield a variety of birds from bitterns, rails, and gallinules to Mississippi Kites and Prothonotary Warblers.

The Great Smokies, towering along the eastern boundary of the state, possess a character in some ways as extraordinary as that of Reelfoot. Though the mountains are high, with elevations exceeding 6,000 feet, their outlines are softened by a cover of rich forests and, not infrequently, by a veil of grayish mist. Beneath the canopy of the forests lies an understory of shrubs that bloom in riotous color. Birds are everywhere: Carolina Wrens, Wood Thrushes, and Hooded Warblers on the lower slopes; Winter Wrens, Veeries, and Canada Warblers on the higher. Since the Great Smokies are embraced by a National Park, fine roads and trails meander from shaded valleys to lofty summits, making easily accessible the best avian haunts. Delightful is the word for bird finding in the Great Smokies.

Tennessee has three major natural divisions: mountainous east Tennessee; middle Tennessee, with its bluegrass plain; and west Tennessee, a low plateau.

In east Tennessee, the Unaka, Unicoi, and Great Smoky Mountains—part of the Appalachian uplift—form a crestline along which runs the North Carolina boundary. The highest point, Clingmans Dome (6,643 feet), is also the highest point in the state. Paralleling the escarpment on the west is the Great Valley of east Tennessee, a southward extension of Virginia's Shenandoah Valley. Its fertile surface and numerous ridges of 300 to 800 feet elevation are drained by the Tennessee River, which flows southward into Alabama. In the Great Valley lie the cities of Knoxville and Chattanooga. West of the Great Valley rises the Cumberland Plateau, a broad highland with elevations averaging 2,000 feet, but reaching as high as 3,600 feet in the mountains near the Kentucky line.

Middle Tennessee, one of the richest agricultural areas in the state, is largely a plain. On the Cumberland River in the north-central part rests the city of Nashville. The plain, from 400 to 600 feet above sea level, is surrounded by the so-called Highland Rim, approximately 400 feet higher. The Rim merges with the Cumberland Plateau on the east and flanks the Tennessee River on the west, as this waterway returns from its Alabama circuit and flows north into Kentucky.

West Tennessee is mainly a plateau extending from the Tennessee River to the bluffs overlooking the Mississippi River bottomlands. Its gently rolling terrain has elevations ranging between 200 and 500 feet. The city of Memphis sprawls back from the Mississippi bluffs in the southwest corner of west Tennessee; in the northwest corner, Reelfoot Lake lies within the Mississippi bottomlands, an area typically studded with swamps, marshes, and small lakes.

Tennessee was originally covered with great forests, principally of hardwoods. Hickory, tulip tree, walnut, chestnut, basswood, and various kinds of oaks were relatively widespread; sweet gum and sycamore were common in west Tennessee; sugar maple, beech, and black cherry were among the chief trees growing in certain mountainous parts of east Tennessee. Very few of these forests remain today, having been replaced either by second-growth hardwoods, usually of the original type, or by farmlands (fields, wet meadows, brushy lands, orchards, and dooryards). Some of the bird species regularly breeding in the second-growth hardwood forests and farmlands of Tennessee are the following:

HARDWOOD FORESTS

Turkey Vulture
Black Vulture
Broad-winged Hawk
Yellow-billed Cuckoo
Black-billed Cuckoo (*mainly east Tennessee*)
Common Screech Owl
Barred Owl
Chuck-will's-widow
Whip-poor-will (*uncommon in west Tennessee*)
Yellow-shafted Flicker
Pileated Woodpecker
Red-bellied Woodpecker
Hairy Woodpecker
Downy Woodpecker
Great Crested Flycatcher
Acadian Flycatcher
Eastern Pewee
Blue Jay
Carolina Chickadee
Tufted Titmouse
White-breasted Nuthatch (*uncommon in west Tennessee*)
Wood Thrush
Blue-gray Gnatcatcher
Yellow-throated Vireo
Red-eyed Vireo
Warbling Vireo (*uncommon in east Tennessee*)
Prothonotary Warbler (*rare in east Tennessee*)
Northern Parula Warbler (*mainly west Tennessee*)
Yellow-throated Warbler (*rare in east Tennessee*)

HARDWOOD FORESTS (*cont.*)

Louisiana Waterthrush
Kentucky Warbler
Hooded Warbler
American Redstart
Baltimore Oriole
Scarlet Tanager (*mainly highlands*)
Summer Tanager

FARMLANDS

Common Bobwhite
Mourning Dove
Red-headed Woodpecker (*mainly west Tennessee*)
Eastern Kingbird
Eastern Phoebe
Barn Swallow
Bewick's Wren (*rare except in middle Tennessee*)
Carolina Wren
Northern Mockingbird
Gray Catbird
Brown Thrasher
Eastern Bluebird
Loggerhead Shrike
White-eyed Vireo
Yellow Warbler (*rare in west Tennessee*)
Prairie Warbler
Common Yellowthroat
Yellow-breasted Chat
Eastern Meadowlark
Orchard Oriole
Common Grackle
Northern Cardinal
Indigo Bunting
Dickcissel (*rare in east Tennessee*)
American Goldfinch (*uncommon in west Tennessee*)
Rufous-sided Towhee
Grasshopper Sparrow (*except east Tennessee*)
Chipping Sparrow
Field Sparrow
Song Sparrow (*except west Tennessee*)

Coniferous trees in Tennessee were, and still are, confined to very limited areas. Bald cypresses stand in the swamps of the river bottomlands in west Tennessee; both shortleaf and scrub pines grow on the Highland Rim, on the Cumberland Plateau, and in the Great Valley of east Tennessee; white and pitch pines appear at widely scattered spots in east Tennessee. Hemlocks thrive in the cool ravines and shaded slopes of the Cumberland Plateau and the mountains of east Tennessee; spruces and firs are abundant in the mountains of east Tennessee at elevations above 5,500 feet.

The limited distribution of conifers in the state necessarily limits the distribution of birds usually associated with them. There are, however, some stands of pine sufficiently extensive to attract Red-cockaded Woodpeckers, Brown-headed Nuthatches, and Pine Warblers. Also, the stands of spruce and fir high up in the mountains of east Tennessee are extensive enough to hold fairly large populations of Red-breasted Nuthatches, Brown Creepers, Golden-crowned Kinglets, and other birds characteristic of cool coniferous forests. (For a discussion of the birdlife inhabiting the spruce-and-fir region of Roan Mountain and the Great Smoky Mountains, *see under* **Roan Mountain** and **Gatlinburg.**)

Natural marshes and swamps are few, except in the river bottoms of west Tennessee. For this reason, such birds as Great Blue Herons and Wood Ducks usually bred only in this section of the state. But the later construction of a system of dams across the Tennessee River and some of its tributaries by the Tennessee Valley Authority created several large reservoirs (frequently referred to as lakes), and many of them provide nesting habitats for these birds. A good example is Kentucky Lake in the Tennessee National Wildlife Refuge (*see under* **Paris**). Thus more water-loving birds than heretofore show up in other sections of the state, and, in certain localities, such birds are now common.

There are at least three major paths of migration in Tennessee, all passing through the state in a north-south direction: the Great Valley of east Tennessee; the Tennessee River Valley between middle and west Tennessee; and—by far the most important—the Mississippi River Valley (often called the Mississippi flyway). Whereas in former years most waterbirds and waterfowl apparently followed the Mississippi Valley, large numbers now move north and south along the Tennessee River reservoirs, frequently stopping to rest and feed. Fort Loudoun Lake near Knoxville, Chickamauga Lake near Chattanooga, and parts of Kentucky Lake in the Tennessee National Wildlife Refuge attract many geese, ducks, gulls, and terns. The main bird flights through Tennessee may be expected within the following dates:

Waterfowl: 15 February–1 April; 25 October–15 December
Shorebirds: 20 April–20 May; 10 August–15 October
Landbirds: 25 March–5 May; 10 September–5 November

Tennessee is a wintering area for many northern birds. Reelfoot Lake has thousands of waterfowl, as do the reservoirs formed by impoundment of the Tennessee River and its tributaries. Hordes of European Starlings and blackbirds descend on the state, at dusk converging in clouds to roost in favorite sites. Bald Eagles are an ornithological feature each winter at Reelfoot Lake. Winter bird finding in the Knoxville, Nashville, and Memphis areas ordinarily yields between sixty-five and seventy-five species, the majority landbirds. While a good proportion are permanent residents, northern species such as the Yellow-bellied Sapsucker, Hermit Thrush, Myrtle Warbler, White-crowned Sparrow, White-throated Sparrow, Fox Sparrow, and Swamp Sparrow are well represented, and Red-breasted Nuthatches, Evening Grosbeaks, Purple Finches, and Pine Siskins appear sporadically.

Authorities

Fred J. Alsop III, Fred W. Behrend, Vandiver L. Childs, Ben B. Coffey, Jr., Wendell Crews, Eugene Cypert, Dan DeGroat, Kenneth H. Dubke, Albert F. Ganier, Katherine A. Goodpasture, Lee R. Herndon, Joseph C. Howell, Amelia R. Lasky, Don Manning, Chester R. Markley, Karl H. Maslowski, Harry C. Monk, Richard Nevius, Ruth Reed Nevius, Walter R. Spofford, Arthur Stupka, James T. Tanner, Bruce P. Tyler, Robert Sparks Walker.

BROWNSVILLE
Hatchie National Wildlife Refuge

The **Hatchie National Wildlife Refuge,** in south-central west Tennessee, borders the south side of the Hatchie River. About 80 per cent of the Refuge's 9,398 acres are bottomlands forested with oaks, sweet gum, and hickory. The remaining acreage consists of old agricultural lands, meandering streams, and small ponds, some of them stream cutoffs. Among common breeding landbirds are the Wild Turkey, Red-headed Woodpecker, Great Crested Flycatcher, Acadian Flycatcher, Carolina Chickadee, Carolina Wren, Brown Thrasher, Prothonotary Warbler, Swainson's Warbler, Cerulean Warbler, Hooded Warbler, American Redstart, Orchard Oriole, Summer Tanager, and Indigo Bunting.

The Refuge is south of Brownsville; both I 40 and State 76 pass through it parallel to each other. Access to the Refuge is from State

76; leave I 40 south either from Exit 21 (Brownsville) or Exit 20 (Stanton-Koko Road). Before visiting the Refuge, go to headquarters at 34 North Lafayette Street in Brownsville, north of I 40, 4 miles from Exit 21. Inquire here about roads leading from State 76 into the Refuge for the most productive bird finding. Also inquire about obtaining a boat for floating the Hatchie River bordering the Refuge—an ideal way to observe both the Prothonotary and Swainson's Warblers, Wood Ducks, and Hooded Mergansers, which nest on the Refuge, and a variety of wading birds such as the Green, Little Blue, and Yellow-crowned Night Herons.

CHATTANOOGA

Chickamauga Lake | Booker T. Washington State Park

Some of the best bird-finding spots in southeastern Tennessee lie along **Chickamauga Lake** and its feeder inlets. Formed by impounding the Tennessee River, the narrow Lake has a surface area of 35,400 acres extending from near Chattanooga northeast for 59 miles. The following loop tour leads to some of the best spots.

From the city, drive northeast on State 58 along the east side of Chickamauga Lake for 5 miles, turn left on Champion Road into **Booker T. Washington State Park** (350 acres), and proceed to the boat ramp. This is a fine vantage point for viewing a variety of transient and wintering waterbirds and waterfowl including Common Loons, Horned Grebes, Pied-billed Grebes, Ring-necked Ducks, Redheads, Canvasbacks, Lesser Scaups, Common Goldeneyes, Hooded Mergansers, and Red-breasted Mergansers. Herring Gulls, Ring-billed Gulls, and a few Bonaparte's Gulls are usually present.

Along the Park's foot trails through the least disturbed wooded and shrubby areas, search for such wintering birds as Brown Creepers, Winter Wrens, Hermit Thrushes, both Golden-crowned and Ruby-crowned Kinglets, Cedar Waxwings, Purple Finches, and White-throated Sparrows. Where there are stands of pine, look for Red-breasted Nuthatches, Evening Grosbeaks, and Pine Siskins. The Park in April and May is ideal for transient warblers: Black-and-white, Tennessee, Cape May, Chestnut-sided, Bay-breasted, Blackpoll, Kentucky, and about fifteen others.

Return to State 58, continue for 18 miles, and turn left on State 60. At 2.7 miles, the first water impoundment on the right—Mungers Pond (privately owned; easily viewed from the car)—has a small wintering population of Buffleheads and a few other ducks.

Continue on State 60 for about 5 miles to the landing for the Blythe Ferry, park, and walk up the bluff overlooking Chickamauga Lake. In winter a Bald Eagle perched in one of the high trees or stubs near the water is always a possibility, and Great Blue Herons standing in the shallows are almost a certainty.

Backtrack on State 60 for about a mile and turn left on the paved Blythe Ferry–Charleston Road; after a short distance, take the first dirt road to the left and continue to its end. If the gate on the left into the Hiwassee Hunting Area is open (closed during hunting season for waterfowl), enter, park, and walk toward the slough ahead. Throughout winter the commonest waterfowl are Canada Geese, Mallards, and American Black Ducks. Look for a Northern Harrier sailing low over the slough, or a Red-tailed Hawk resting on a stub. In April and early May, migrating Bobolinks stop off in the fields back from the slough; in early summer, Dickcissels and Grasshopper Sparrows nest in the fields, Blue Grosbeaks and Indigo Buntings in the adjacent hedgerows.

Go back and turn left on the Blythe Ferry–Charleston Road to State 58; turn left, continue a mile, then turn off on the paved Lower River Road along the Hiwassee River, stopping during the next 7 miles to scan the mud flats in view from the Road: first at Sugar Creek, then at Candies Creek, and finally at South Mouse Creek. In August and September, when the water level of these Hiwassee River inlets is low, almost any shorebird species regularly migrating south through eastern United States may be expected. Among the commonest are the Semipalmated Plover, Spotted Sandpiper, Solitary Sandpiper, Greater and Lesser Yellowlegs, Pectoral Sandpiper, Least Sandpiper, Short-billed Dowitcher, and Semipalmated Sandpiper. At the same time, wading birds are well represented by Great Blue Herons and such other species as the Little Blue Heron, Cattle Egret, Great Egret, and Snowy Egret.

One may return to Chattanooga by continuing on the Lower River Road to US 11 at Charleston and turning south, or entering I 75, which parallels US 11, at the nearest interchange.

DOVER
Paris Landing State Park | Cross Creeks National Wildlife Refuge

In extreme northwestern middle Tennessee are two places especially good for winter bird finding.

One is **Paris Landing State Park** (820 acres) on a harbor of Kentucky Lake, reached by driving west from Dover on US 79 for 16 miles. From the shore one can usually see Bonaparte's Gulls, along with the ever-present Herring and Ring-billed Gulls; also Horned Grebes as well as Pied-billed Grebes, and a few Common Loons.

The other place is the **Cross Creeks National Wildlife Refuge,** reached from Dover by driving east on State 49 for 2 miles, then turning left on a gravel road to headquarters. Extending for 10 miles on either side of Barkley Lake, an impoundment of the Cumberland River, the Refuge embraces 9,892 acres of bottomlands consisting of deciduous woods, cultivated fields, marshes, and ponds: ideal for a wide variety of birds.

Thanks to water impoundments and croplands created to attract waterfowl, thousands of geese and ducks pass the winter on the Refuge. Commonest species are Canada Geese (including a decoy flock), Mallards, American Black Ducks, Gadwalls, Common Pintails, American Wigeons, Northern Shovelers, and Wood Ducks (permanent residents). Often appearing are Redheads, Canvasbacks, Buffleheads, Ruddy Ducks, and Hooded, Common, and Red-breasted Mergansers.

A few Bald Eagles over-winter. Among small landbirds often appearing during winter are the Water Pipit, Purple Finch, Le Conte's Sparrow, American Tree Sparrow, and frequently the Evening Grosbeak.

ELIZABETHTON
Wilbur Lake | Watauga Lake

East of Elizabethton are **Wilbur Lake** and **Watauga Lake,** reservoirs created by damming the Watauga River for flood control and electric power. For good bird finding in their environs, take the following trips.

1. From the junction of Broad Street and Lynn Avenue in Elizabethton, drive east on State 19E for 0.6 mile, turn left on State 91 over the Watauga River via the Gilbert Peters Bridge at 0.7 mile and turn off right at 1.0 mile on a paved road, park on the shoulder, and explore the thickets and partially open areas on either side of the road. From late January to early March, listen in the evening for the flight songs of the American Woodcock. From about 25 April through 10 May and again from 15 September through 20 October, the area is especially rewarding for a variety of transient thrushes, warblers, finches, and sparrows.

Some of the birds nesting here are the Green Heron, Common Bobwhite, Brown Thrasher, Blue-gray Gnatcatcher, White-eyed Vireo, Common Yellowthroat, American Redstart, and Field Sparrow. Among the birds expected in winter are the Yellow-bellied Sapsucker, Brown Creeper, Hermit Thrush, Myrtle Warbler, Purple Finch, Slate-colored Junco, Fox Sparrow, and White-throated Sparrow.

Return to State 91, turn right, proceed about 4.1 miles, and turn off right, opposite a small store and service station, on a road with a directional sign to Watauga Dam. After crossing a concrete bridge over Stony Creek, turn right immediately and keep right. At 6 miles, cross the iron bridge over the Watauga River and follow the River to Wilbur Dam at 8.1 miles. From here the road parallels Wilbur Lake, narrow and horseshoe-shaped, for 0.6 mile, with several pulloffs and a picnic area extending to a bridge at 8.8 miles.

In any season, scan the Lake from the pulloffs and bridge for Wood Ducks, and from October to April look for a variety of other ducks, chiefly Mallards, American Black Ducks, Gadwalls, American Wigeons, Northern Shovelers, Redheads, Ring-necked Ducks, Canvasbacks, Lesser Scaups, Common Goldeneyes, Buffleheads, Hooded Mergansers, and Red-breasted Mergansers. Occasionally along with the ducks are a few Common Loons, Horned Grebes, and Pied-billed Grebes. Immediately before the bridge, turn right on a road leading up to Horseshoe Church, park, and explore the adjacent pine-hardwoods forest, where breeding birds include the Yellow-throated Warbler, Ovenbird, Louisiana Waterthrush, Hooded Warbler, and Summer Tanager.

Cross the bridge, bear left and leave the Lake, then drive up and over Iron Mountain, where the road ends at 11.2 miles with a view of Watauga Lake, a large body of water extending eastward. After 1 November, Wautauga Lake attracts rafts of ducks for short periods. Earlier in fall and in spring, the Lake also brings in some gulls—Herring, Ring-billed, and a few Bonaparte's—and terns, such as Forster's, Common, a few Caspian, and Black.

2. For bird finding on the south side of Watauga Lake, take a second trip south from Elizabethton on US 321 to Hampton, about 4 miles, then eastward, making two stops: at 8.6 miles, turn left off US 321 and drive to the shore of Watauga Lake; at 10 miles on US 321, stop at a parking area on the left and walk down to Watauga Point. Both places will afford good views of waterbirds and waterfowl on the Lake, and the walk down to the Lake, through woods and shrubby areas, should be productive.

Look for such year-round species as the Pileated Woodpecker, Carolina Chickadee, Carolina Wren, and Rufuous-sided Towhee. Winter visitants should include the Winter Wren, and on the beaches and sand bars perhaps a few Horned Larks and an occasional Water Pipit. During migration in both spring and fall, a great many passerine species tend to concentrate here briefly.

Farther along US 321 at 10.8 miles, just before crossing Little Stony Creek, are roadside tables near hemlocks in which Northern Parula Warblers are summer resident.

At 13.2 miles, turn left off US 321 onto State 67 and proceed to Butler. Here at 15.4 miles, turn right onto a dirt road and drive down toward an arm of Watauga Lake, park, and investigate the mud flats, fields, and streams bordered with shrubs and low trees. In April/May and August/September—provided that the water level is low—the mud flats bring many shorebirds, especially the Semipalmated Plover, Spotted Sandpiper, Solitary Sandpiper, Greater and Lesser Yellowlegs, Pectoral Sandpiper, Least Sandpiper, and Semipalmated Sandpiper. In shrubby places, look especially for Lincoln's Sparrows and Swamp Sparrows.

Return to State 67 and continue. At 19.5 miles, turn right immediately beyond a country store, cross Doe Creek on a concrete bridge and follow it to its confluence with Roan Creek at 23.6 miles. Park on the right and walk onto the bridge over Roan Creek, a good vantage point for this part of Watauga Lake. In August and September, when the water level is low, the shallows and mud flats are attractive to wading birds—Great Blue Herons, Little Blue Herons, Great Egrets, and Snowy Egrets—as well as to numerous shorebirds that may include the Dunlin, Short-billed Dowitcher, and Buff-breasted Sandpiper. A few gulls and terns sometimes feed and rest here. The woods and thickets on both sides of the Roan Creek Bridge and along the Lake provide excellent bird finding in any season. Early in summer the Whip-poor-will calls regularly at night.

GATLINBURG
Great Smoky Mountains National Park

The Great Smoky Mountains, astride the boundary between eastern Tennessee and western North Carolina, are the backbone of the **Great Smoky Mountains National Park,** embracing 800 square miles. From northeast to southwest for 71 miles, the crest of the Great Smokies zigzags through the Park, maintaining an altitude of at least 5,000 feet for 36 miles and possessing sixteen peaks higher than 6,000 feet.

Although most big mountain ranges have a preponderance of barren heights, crags, and escarpments, the Great Smokies are richly blanketed with trees and shrubs. There is no timber line. The peaks either are completely forested or have 'balds'—peculiar openings giving the illusion of an alpine region.

Above 6,000 feet the trees are largely spruce and fir; from 6,000 feet down to 4,500 feet, they become mixed with northern hardwoods, predominantly beech and yellow birch, together with hemlock, black cherry, sugar maple, and other trees. Below 4,500 feet in damp, sheltered places are 'cove' hardwoods consisting of the same species of hardwoods along with red oak and basswood among others—all attaining great size; on dry slopes and ridges the stands are primarily oaks—white, chestnut, red, and black—and pitch pine. The magnificence of these unbroken forests is probably unrivaled in the eastern United States. Within the confines of the Park alone, approximately 40 per cent of the timber is virgin. Shrubs, especially the great white rhododendron, form an extensive undergrowth. In June, when the rose-pink rhododendron, flame azalea, and mountain laurel are in bloom at higher elevations, the flower spectacle is at its best.

There are few environments more delightful for the bird finder. With altitudinal variations of 5,000 feet or more, one may journey from valleys where the avifauna is distinctly southern, to slopes and summits where the avifauna is akin to that of northern forests. Yet these vertical contrasts in avifauna are not sharply marked, since a number of southern species invade the higher slopes. Differing in this respect from the birdlife of most big mountain ranges, the Great Smokies offer the bird finder an element of perplexity as well as delight.

On the slopes of the Great Smokies the following species of northern affinities reside during nesting season; all of them breed.

Ruffed Grouse (*all altitudes*)
Saw-whet Owl (*uncommon*)
Yellow-bellied Sapsucker (*uncommon*)
Olive-sided Flycatcher (*uncommon*)
Northern Raven
Black-capped Chickadee (*common above 3,000 feet*)
Red-breasted Nuthatch
Brown Creeper
Winter Wren
Veery
Golden-crowned Kinglet
Solitary Vireo
Golden-winged Warbler (*usually below 5,000 feet*)
Black-throated Blue Warbler
Black-throated Green Warbler (*all altitudes*)
Blackburnian Warbler
Chestnut-sided Warbler
Canada Warbler
Scarlet Tanager
Rose-breasted Grosbeak
Pine Siskin (*irregular*)
Red Crossbill (*irregular*)
Slate-colored Junco (*common above 3,000 feet*)

These species and many others may be observed from late April through early July by entering the Park on US 441 from Gatlinburg, at an elevation of 1,300 feet, and with stops and side trips on foot, exploring habitats at increasingly higher elevations to the summit of Clingmans Dome at 6,643 feet—Tennessee's loftiest peak.

First, stop at the Sugarlands Visitor Center by turning right off US 441, 1.5 miles from Gatlinburg, and walk the self-guiding nature trail for a sampling of such breeding birds of the lower altitudes as the Wood Thrush, White-eyed Vireo, Yellow-throated Vireo, Black-and-white Warbler, Golden-winged Warbler, Ovenbird, Louisiana Waterthrush, Kentucky Warbler, Hooded Warbler, American Redstart, and Summer Tanager.

Then backtrack to US 441 and drive up into the Park. Stop at Alum Cave Parking Area, 9 miles from the Center at 3,800 feet and take the 2.5-mile foot trail to Alum Trail Bluffs at 4,900 feet. Moderately steep, this trail leads through northern hardwoods with a variety of breeding birds including the Winter Wren, Carolina Wren, Wood Thrush, Veery, Solitary Vireo, Red-eyed Vireo, Northern Parula Warbler, Black-throated Blue Warbler, Black-throated Green Warbler, Blackburnian Warbler, Canada Warbler, Scarlet Tanager, and Rose-breasted Grosbeak. Along the way the trail ascends through balds where Chestnut-sided Warblers are abundant.

Back on US 441, proceed to Newfound Gap, 13.5 miles from the Center; turn right off US 441 on the Clingmans Dome Road, which follows State Line Ridge for 9 miles, ascending from 5,040 feet at Newfound Gap to 6,311 feet, where it terminates at the Forney

BLACK-THROATED BLUE WARBLER

Ridge Parking Area. From here climb the paved trail for 0.5 mile to the observation tower on the summit of Clingmans Dome.

Newfound Gap roughly marks the beginning of forests dominated by spruce and fir, hence the first dependable opportunities to see or hear, at pulloffs, the Olive-sided Flycatcher, Red-breasted Nuthatch, Brown Creeper, Golden-crowned Kinglet, Pine Siskin, and Red Crossbill. The nature trail at Collins Gap (5,729 feet), midway between Indian Gap and the Forney Ridge Parking Area on the Clingmans Dome Road, is especially productive. To hear the Saw-whet Owl, listen on a clear, quiet, moonlit spring evening from almost anywhere along the Clingmans Dome Road where there are dense stands of spruce and fir.

Reference

Notes on the Great Smoky Mountains National Park. By Arthur Stupka. Knoxville: University of Tennessee Press, 1963.

JAMESTOWN
Pickett State Park and Forest

Some of the most rugged highlands of the Cumberland Plateau, with elevations exceeding 2,000 feet, are included in the 11,742 acres of **Pickett State Park and Forest** on the Kentucky line in east Tennessee. The terrain is cut by deep, steep-walled ravines, or gulfs, largely forested with second-growth deciduous timber and some fine stands of pine. Hemlock, often bannered with *Usnea*, grows in the deepest gulfs near streams. A few Red-cockaded Woodpeckers are permanent resident. Among other breeding species are the Broad-winged Hawk, American Kestrel, Barred Owl, Pileated Woodpecker, Red-bellied Woodpecker, Acadian Flycatcher, White-breasted Nuthatch, Wood Thrush, Northern Parula Warbler, Louisiana Waterthrush, Kentucky Warbler, Hooded Warbler, Scarlet Tanager, and Summer Tanager.

To reach Pickett State Park and Forest from Jamestown, drive north 2 miles on US 127, east on State 154 for 14 miles to the entrance, and then follow directional signs to headquarters. Inquire here about trails to the best habitats for birds.

KINGSPORT
Bays Mountain Park

This city in northeastern Tennessee owns and operates **Bays Mountain Park** on a mountain top (elevation 1,840–2,440 feet) 6 miles southwest of the business district. To reach the Park and stop for bird finding along the way, take State 93 south and, at 1.5 miles south of the Holston River Bridge, turn right on Reservoir Road. Just before crossing Horse Creek (first bridge), park opposite the golf course on the left and walk into the grassy fields on the right (north). During late spring and summer, look for Eastern Meadowlarks in the field, Traill's Flycatchers in the willows along Horse Creek, and Green Herons, Least and American Bitterns, Wood Ducks, Virginia Rails, and Soras on the three cattail-ringed ponds and the wet grassy areas west of them.

Continue on Reservoir Road for about 5 miles to the Park entrance, well marked by directional signs. Within the 1,300-acre Park is a large reservoir, attractive in winter to waterbirds and waterfowl including Pied-billed Grebes, Mallards, American Black Ducks, Gadwalls, Common Pintails, Green-winged and Blue-winged Teal, American Wigeons, Northern Shovelers, Redheads, Canvasbacks, Lesser Scaups, Common Goldeneyes, and Buffleheads. Wood Ducks are permanent residents. Many miles of trails radiate from the nature center, providing easy access to both the shore of the reservoir and the woodlands consisting primarily of oak and hickory, with hemlock on the lower slopes to pines on the ridges.

Among many nesting birds are the Yellow-billed Cuckoo, Pileated and Red-bellied Woodpeckers, Great Crested Flycatcher, Eastern Pewee, Carolina Chickadee, Tufted Titmouse, White-breasted Nuthatch, Blue-gray Gnatcatcher, White-eyed Vireo, Yellow-throated Vireo, Red-eyed Vireo, Black-and-white Warbler, Worm-eating Warbler, Pine Warbler, Ovenbird, Kentucky Warbler, Hooded Warbler, and both Scarlet and Summer Tanagers. Among species usually appearing in winter are the Red-breasted Nuthatch, Brown Creeper, Hermit Thrush, Golden-crowned Kinglet, Evening Grosbeak, and Pine Siskin.

KNOXVILLE

Fort Loudoun Lake | University of Tennessee Plant Science Farm | Sharp's Ridge Park

The focal point for bird finding in the Knoxville region of eastern Tennessee is **Fort Loudoun Lake,** a reservoir with a surface area of 14,560 acres, formed by the flooding of the Tennessee River from the city southwest for 55 miles. For finding a good representation of bird species, take the following loop tour.

From the intersection of I 40 and I 75 in the center of Knoxville drive west on I 40 for about 3.5 miles to the Bearden Interchange, exit, and go south to the first traffic light, then turn left on Northshore Drive. After about 1.25 miles Northshore Drive merges with Lyons View Drive, and 0.5 mile beyond, Northshore Drive goes right and Lyons View Drive left. Continue on Lyons View Drive for about 0.5 mile, then turn left across a causeway to the Fourth Creek Sewage Treatment Plant. Here in April/May and August/September, the temporary puddles along the road and the larger ones near Fort Loudoun Lake, all on the grounds of the Plant, attract shorebirds, particularly Semipalmated Plovers, Common Snipes, Spotted Sandpipers, Solitary Sandpipers, Lesser Yellowlegs, and Pectoral, Least, and Semipalmated Sandpipers.

Backtrack to Northshore Drive, turn left, and continue west for 8 miles to Concord Park Road. Along much of the way, where the Drive parallels the shore of Fort Loudoun Lake, stop at convenient places and scan the open water for waterbirds and waterfowl. In April/May and August/September, look for transient Forster's, Common, and Black Terns. From late September into April, look for Common Loons, Horned Grebes, Pied-billed Grebes, Great Blue Herons along the shore, Mallards, American Black Ducks, Gadwalls, Common Pintails, Green-winged and Blue-winged Teal, American Wigeons, Northern Shovelers, Redheads, Ring-necked Ducks, Canvasbacks, Lesser Scaups, Buffleheads, and American Coots.

Upon reaching Concord Park Road, turn left into the parking lot and boat-launching area, and again scan the Lake for transient or wintering waterbirds and waterfowl. Herring Gulls, Ring-billed Gulls, and occasionally Bonaparte's Gulls often rest on the exposed tip of land opposite the boat ramp. Return to Northshore Drive, go left for 0.6 mile to a bridge crossing an inlet; park at the west end, walk onto the span, and look over the Lake for wintering diving ducks, including Common Goldeneyes. Continue on Northshore Drive for 0.7 mile to the next bridge over another inlet; cross it and park on the left at its far end.

Continue on Northshore Drive for about a half-mile; if the gate is open, turn left into the Concord Park campground. A loop drive through it offers a fine vantage point for nearly all the wintering waterbirds and waterfowl previously mentioned and, in addition, the possibility of seeing Greater Scaups and mergansers. A winter walk through the campground should produce Yellow-bellied Sapsuckers, Hermit Thrushes, Golden-crowned Kinglets, Myrtle Warblers, and White-throated Sparrows.

Return to Northshore Drive and continue west for about 1.5 miles; turn right onto Harvey Road and follow it for 1.5 miles, then turn right onto Boyd Station Road, which immediately goes under a railroad overpass. For 1.0 mile Boyd Station Road parallels the railroad on the right and grassy fields on the left. In midspring and late summer, look over the fields, especially the broad patches of crimson clover, for migrating Bobolinks. During summer, listen for Grasshopper Sparrows and keep an eye on the fence rows and utility wires for Blue Grosbeaks. Watch for Red-tailed Hawks and American Kestrels.

After 1.0 mile Boyd Station Road becomes Virtue Road and intersects Turkey Creek Road, 0.5 mile distant. Continue on Virtue Road for about 1.5 miles, stopping at the bridge over the creek. On the right is a good spot for wintering sparrows, including the White-crowned. Less than a mile ahead, Virtue Road intersects US 70 (Kingston Pike). A right turn on it will lead back to Knoxville, 15 miles away.

For a shorter trip that will yield most of the birds on the loop tour, and other species besides, visit the **University of Tennessee Plant Science Farm,** south of Knoxville. This experimental agricultural facility may be reached from the city by taking I 40 west and exiting south on US 129 (Alcoa Highway). Three miles after crossing Fort Loudoun Lake, watch for the sign marking the entrance on the right. Drive in, taking the left fork past the manager's house, continuing straight ahead, and parking near the pumphouse and chain barrier.

The Farm embraces 212 acres of mostly flat river bottomland only a few feet higher than Fort Loudoun Lake, which surrounds it on three sides. Walk the roads that crisscross the Farm, *always keeping off the planted field plots.* The entire area is excellent for the common variety of shorebirds, as well as numerous rarities, in late spring and late summer. From late September through April, scan the drainage ditches and the Lake itself for waterfowl. Surface-feeding ducks and sometimes Snow Geese forage on the fields near the Lake, and on the Lake are frequent rafts of diving ducks. Search all low-lying wet places and shortgrass fields for transient or wintering Common Snipes, Horned Larks, Water Pipits, Savannah Sparrows, and the possibility of Upland Sandpipers and Buff-breasted Sandpipers. In April and September, migrating swallows, representing all eastern United States species swarm over the area. In late summer and fall there is always a chance of seeing an Osprey or two patrolling the Lake.

Sharp's Ridge Park (79.2 acres), on forested Sharp's Ridge in the northwestern part of the city, is excellent for transient small landbirds in spring, especially the last ten days of April and the first ten of May. In early morning one may record five species of vireos, over twenty species of warblers, and at least forty other species.

To reach the Park, take I 75 north from the center of Knoxville for 2 miles, exit east onto I 640 and follow it for 2.2 miles; exit south and turn left on Broadway for about 0.5 mile; turn right onto Ludlow Street, cross the railroad, and continue to Freemason Street; turn right a few hundred feet, then right again onto Sharp's Ridge Road to the crest (1,300 feet), bearing left at the fork in the gap. On reaching the crest, leave the car and look for birds west along the road, which runs the length of the Ridge to its west end.

LEXINGTON
Perryville Bridge

Twenty-one miles east of Lexington on State 20 is a large colony of Cliff Swallows under the eastern approaches to the **Perryville**

Bridge, which crosses the permanently flooded Tennessee River. Formerly the Cliff Swallows occupied low cliffs overhanging the River, but when the water level was raised by artificial means and the cliffs were inundated, the birds were forced to resort to man-made structures. Smaller colonies are on each side of the Savannah Bridge (US 64), which crosses the Tennessee River about 50 miles farther south, just west of Savannah. From mid-May, when nest-building begins, to mid-July, these colonies are centers of bustling activity.

MEMPHIS

Riverside Park | Meeman-Shelby State Forest Park

In this large metropolis on the east bank of the Mississippi in the southwestern corner of Tennessee, the place most highly recommended for bird finding is **Riverside Park** (375 acres). On the southwestern edge of the city, where it extends along the Tennessee Chute of the Mississippi for 1.5 miles, the Park has many acres of hardwoods—chiefly oaks, hickories, sweet gums, and tulip trees—interlaced with drives and paths. Here and there are grassy areas, shrubby thickets, and picnic grounds. Bird finding is excellent in April and early May, when the small-landbird migration is at its height.

Characteristic breeding birds include the Yellow-crowned Night Heron (small colony), Yellow-billed Cuckoo, Barred Owl, Red-bellied Woodpecker, Great Crested Flycatcher, Eastern Pewee, Carolina Chickadee, Tufted Titmouse, Carolina Wren, Northern Mockingbird, Gray Catbird, Brown Thrasher, Wood Thrush, White-eyed Vireo, Yellow-throated Vireo, Northern Parula Warbler, Yellow-throated Warbler, Kentucky Warbler, Yellow-breasted Chat, Hooded Warbler, Baltimore Oriole, Orchard Oriole, Summer Tanager, Northern Cardinal, Indigo Bunting, and Rufous-sided Towhee. The Painted Bunting is always a possibility, though it is more likely on the southern outskirts of the city along country byways bordered by shrubs, scattered trees, and utility wires and poles, which it uses for singing perches. The Park is accessible from I 55, which traverses its eastern side, by exiting either at South Parkway or Mallory Avenue.

Meeman-Shelby State Forest Park (12,512 acres), 12 miles northwest of the city, is mostly on the Mississippi loess bluffs, deeply cut by ravines. The bottomlands, from the bluffs to the River, 1 to 3 miles away, are heavily forested with water locust, pecan, hickory, red maple, hackberry, sweet gum, tupelo, cottonwood, and cypress. Although the Park and its adjacent bottomlands hold many species of birds identical with those in Riverside Park, they offer in addition the Mississippi Kite, Broad-winged Hawk, Chuck-will's-widow, Red-headed Woodpecker, Acadian Flycatcher, Blue-gray Gnatcatcher, Prothonotary Warbler, Swainson's Warbler, Cerulean Warbler, Louisiana Waterthrush, and American Redstart. Careful searching should yield a few Wild Turkeys.

The Park may be reached from US 51 on the north edge of the city by turning off left (northwest) on North Watkins Street and following directional signs to the Park via Benjestown Road. Two roads enter the Park. The first, Grassy Lake Road, eventually winds down a bluff to the bottomlands. The second, Bluff Road, goes to the Park office, where maps are available. From here, take Riddick Road, which becomes a one-way scenic route, and stop at available pulloffs to explore the slopes and ravines of the bluff for birds.

NASHVILLE

Radnor Lake Natural Area | Warner Parks | Centennial Park | Shelby Park

The best locality for bird finding is the state-owned **Radnor Lake Natural Area** (770 acres), 7.5 miles south of Nashville's center, reached by going south on US 31 (Franklin Pike). About 4.5 miles from the city limit, turn right onto Otter Creek Road and proceed 1.5 miles to Radnor Lake. For 0.5 mile Otter Creek Road borders the Lake, affording excellent views of the open water and shore.

A reservoir covering 61.3 acres, surrounded by woods with cattails and other emergent aquatic plants near shore, Radnor Lake attracts waterbirds, waterfowl, and shorebirds during their migrations in spring and fall. Shorebirds are less frequent in spring because of the high water level. A small variety of waterbirds and waterfowl pass the winter: commonly, Pied-billed Grebes, Mallards, American Black Ducks, Ring-necked Ducks, Lesser Scaups, Buffleheads, and

American Coots; less commonly, Horned Grebes, Canada Geese, Common Pintails, Northern Shovelers, Canvasbacks, and Common Goldeneyes. Green Herons and Wood Ducks reside the year round. Great Blue Herons, Little Blue Herons, and Great Egrets, together with Spotted Sandpipers, Solitary Sandpipers, and other shorebirds begin appearing in summer and are prevalent by September.

Back from Radnor Lake are old fields bordered with brush, marshy areas, and steep hills and narrow ravines largely forested with mature hardwoods—a variety of habitats that hold a corresponding variety of breeding birds, including the following: Turkey Vulture, Black Vulture, Red-tailed Hawk, Yellow-billed Cuckoo, Common Screech Owl, Great Horned Owl, Barred Owl, Pileated Woodpecker, Red-bellied Woodpecker, Great Crested Flycatcher, Acadian Flycatcher, Eastern Pewee, Carolina Chickadee, Tufted Titmouse, Carolina Wren, Wood Thrush, Blue-gray Gnatcatcher, Yellow-throated Vireo, Prothonotary Warbler, Blue-winged Warbler, Cerulean Warbler, Yellow-throated Warbler, Louisiana Waterthrush, Kentucky Warbler, Hooded Warbler, Baltimore Oriole, and Summer Tanager. The Winter Wren, Hermit Thrush, and Ruby-crowned Kinglet, Myrtle Warbler, Slate-colored Junco, White-throated Sparrow, and Swamp Sparrow are among common birds in winter. Marked trails, open from 9:00 A.M. until sundown, provide access to the different habitats.

The city-owned **Warner Parks** (Percy Warner Park and adjoining Edwin Warner Park), comprising together 2,141 acres, are the next best locality for bird finding. Lying 9 miles southwest of the center of Nashville, they may be reached by driving west on US 70S (West End Avenue) to Belle Meade Boulevard, turning left (south), and proceeding 2.8 miles to the entrance. Paved drives, bridle paths, and foot trails traverse the Parks.

The rugged, hilly terrain of Warner Parks is covered to a great extent with mature hardwoods, in which the bird species are almost identical with those in the forest surrounding Radnor Lake; but it also has many grassy areas and shrubby thickets that are particularly attractive to such birds as the following: Bewick's Wren, Northern Mockingbird, White-eyed Vireo, Yellow-breasted Chat, Orchard Oriole, Northern Cardinal, Indigo Bunting, Rufous-sided Towhee, Field Sparrow. Eastern Bluebirds are common summer residents, since nesting boxes suitable for them have been placed in open areas. In early May and mid-September the woods often throng with transient small landbirds, viewed conveniently along the bridle paths and foot trails.

Two other localities offer good bird-finding opportunities: **Centennial Park** (134 acres), in the western suburbs on US 70S (West End Avenue) between 25th and 28th Avenues; and **Shelby Park** (361 acres), on the Cumberland River at the city's eastern border. Exit from I 24 on Shelby Avenue and drive east to its end. Small hardwood groves in both places attract small breeding populations and large numbers of transients.

Reference

Birds of the Nashville Area. Compiled by Henry E. Parmer. 3d ed. Nashville: Tennessee Ornithological Society (Box 1301, Nashville, Tenn. 37203), 1975.

NATCHEZ TRACE STATE PARK AND FOREST

Midway between Nashville and Memphis and readily accessible from I 40 cutting through it, is **Natchez Trace State Park and Forest,** a vast, largely forested tract of 44,977 acres. Oaks, hickories, sweet gums, and other hardwoods, together with some stands of loblolly pine, comprise much of the tree cover. Four lakes lie within the area.

In late spring and summer the Whip-poor-will is common as well as the Chuck-will's-widow. Other residents common at the same season include the Wood Thrush, White-eyed Vireo, Red-eyed Vireo, Black-and-white Warbler, Prairie Warbler, Kentucky Warbler, Yellow-breasted Chat, Summer Tanager, Rufous-sided Towhee, and Field Sparrow. Breeding Eastern Phoebes, Pine Warblers, and Scarlet Tanagers can be found.

From Exit 33 of I 40 a paved road with numerous pulloffs goes north and south through the Forest. Recommended for bird finding is the road south from I 40. At 1.7 miles, go east for about 3 miles to Cub Creek Lake, where the surrounding woods has a particularly good variety of birds.

PARIS
Tennessee National Wildlife Refuge

The **Tennessee National Wildlife Refuge** on Kentucky Lake in northwestern Tennessee comprises 51,347 acres of land and water, divided into three separate units, namely, from north to south, Big Sandy, Duck River, and Busseltown. Refuge headquarters is in the Masonic Building on Blythe Street in Paris.

Ever since its impoundment in 1944, Kentucky Lake has attracted impressive numbers of waterfowl, waterbirds, and shorebirds. A good place for observing a fine variety is Pace Point in the Big Sandy Unit. Here from October to April appear regularly transient or wintering Canada Geese, American Black Ducks, Gadwalls, Common Pintails, Green-winged and Blue-winged Teal, American Wigeons, Northern Shovelers, Ring-necked Ducks, Lesser Scaups, and Buffleheads. Mallards, Wood Ducks, and Hooded Mergansers are present all year long, although the Hooded Merganser is less common in summer.

During late summer and early fall there should be Great Blue Herons, Little Blue Herons, Great Egrets, and at least a few Forster's, Common, Little, and Black Terns. In winter there are large numbers of gulls and usually a good representation of Bonaparte's Gulls. Among shorebirds that appear commonly are Semipalmated Plovers, Common Snipes, Solitary Sandpipers, Greater and Lesser Yellowlegs, Pectoral Sandpipers, Least Sandpipers, and Semipalmated Sandpipers. Other shorebirds include Lesser Golden Plovers (in spring only), Black-bellied Plovers, Upland Sandpipers, White-rumped Sandpipers, Baird's Sandpipers, Short-billed and Long-billed Dowitchers, Stilt Sandpipers, and Sanderlings.

To reach Pace Point from Paris, drive south on State 69; 1.0 mile after it crosses the intersection with US 79, turn off left on an unnumbered road with directional signs to the town of Big Sandy, 14 miles distant. At the corner service station in Big Sandy, turn left (north) and, immediately after crossing the railroad, turn right; at the fork, bear right on a blacktop road that soon passes old brushy fields, with summer-resident Prairie Warblers and Field Sparrows, and then crosses a creek bordered by bottomland woods, worth exploring for summer-resident Northern Parula, Kentucky, and Hooded Warblers.

Backtrack to Lick Creek Road and continue north for 4 miles to a gravel road where there is a Refuge sign on the left. Pick up a self-guiding tour map here and continue on to Pace Point, where the flooded waters of the Tennessee and Big Sandy Rivers merge. Here and at Bennetts and Robbins Creeks, in view on the way to Pace Point, are the best places for waterbirds, waterfowl, and shorebirds. Just beyond an old church, near Tour Marker 2, Wild Turkeys browse in a clear area on the left. In the oak-hickory woods between Tour Markers 2 and 3, the Scarlet Tanager as well as the more common Summer Tanager are summer residents.

PIKEVILLE
Fall Creek Falls State Park

Fall Creek Falls State Park (15,777 acres) in central east Tennessee, a beautiful, rugged spot in the Cumberland Plateau, is splendid for birds of prey and smaller woodland birds. From Pikeville, drive west on State 30 for about 15 miles and turn south on a road marked with a directional sign to the Park.

The spectacular scenic features of the Park are the deep ravines, or gulfs, cut by the main waterway, Cane Creek, and its tributaries, Fall Creek and Pine Creek. Each has beautiful waterfalls and awesome escarpments. Fall Creek Falls, which drops 257 feet, is said to be the highest falls east of the Rocky Mountains. The gulf cut by Cane Creek is 2 miles long and 600 feet deep. Second-growth oak and shortleaf pine cover the bluffs; fine stands of hemlock, draped with *Usnea,* and cucumber tree grow in the gulfs along the Creeks. Scattered here and there are clumps of rhododendron and mountain laurel.

The inaccessible shelf on one prominent escarpment was once the site of a Peregrine Falcon aerie. Turkey Vultures, Sharp-shinned Hawks, Red-tailed Hawks, Broad-winged Hawks, Great Horned Owls, and Barred Owls breed regularly. Other breeding species are the Whip-poor-will, Pileated Woodpecker, Red-bellied Woodpecker, Great Crested Flycatcher, Carolina Chickadee, Tufted Titmouse, Wood Thrush, Black-and-white Warbler, Northern Parula Warbler, Black-throated Green Warbler, Yellow-throated Warbler, Pine Warbler, Prairie Warbler, Ovenbird, Louisiana Waterthrush,

Kentucky Warbler, Hooded Warbler, Scarlet Tanager, Summer Tanager, Indigo Bunting, and Field Sparrow.

May and the first two weeks of June constitute the best period for visits, since the birds are then in full song. Various foot trails lead along the tops of the cliffs.

ROAN MOUNTAIN

Roan Mountain | Roan Mountain State Park

For mountain bird finding in Tennessee beyond Great Smoky Mountains National Park, **Roan Mountain** in the Unaka Range of northeastern Tennessee is strongly recommended. Rather than one peak, Roan Mountain is actually a complex of lofty knobs, the highest reaching 6,385 feet, an altitude exceeded in eastern United States only by peaks in the Great Smoky and Black Mountains. Above 5,500 feet the Mountain is to a great extent densely covered with spruce and fir, interrupted on the slopes near the summits and in the saddles between by many acres of grassy areas, or 'balds.' Dotting and bordering them are natural 'gardens' of rhododendrons that produce showy flowers from purple to red. When the gardens are in full bloom, from about 20 June through 4 July, the nesting season for birds is at its height. Hence the time is ideal for the following trip.

From the intersection of State 143 and 19E in the village of Roan Mountain, take State 143 southward. At 1.8 miles, stop at the parking area on the left and explore the narrow meadow on the right. At the south end of the meadow is a hollow with deciduous shrubs and small trees. Listen here for the Golden-winged Warbler. Other summer-resident birds here, or in the woods across the highway, are the Least Flycatcher, Black-throated Blue Warbler, Black-throated Green Warbler, Ovenbird, Louisiana Waterthrush, Kentucky Warbler, Hooded Warbler, Scarlet Tanager, Rose-breasted Grosbeak, and Indigo Bunting.

Continue on State 143, at 4.4 miles entering **Roan Mountain State Park** (3,500 acres), with headquarters in the valley to the right. At 10.6 miles, stop near a roadside table on the left and listen for the songs of the Winter Wren, Solitary Vireo, and Canada Warbler.

Continue up to Carver's Gap (approximately 5,500 feet) at 12.6 miles on the Tennessee-North Carolina boundary. Around the picnic areas, explore the spruce-fir forest for Red-breasted Nuthatches, Golden-crowned Kinglets, and Slate-colored Juncos. The Gap is especially good for birds during their migrations. East of the Gap are the balds of Little Roan Mountain (5,800 feet) and Grassy Bald Mountain (6,000 feet), nesting habitat for both the (Prairie) Horned Lark and Vesper Sparrow. Water Pipits and Snow Buntings have appeared on the balds in some winters. The Appalachian Trail, easily reached from Carver's Gap, crosses the balds of both knobs.

At Carver's Gap, turn right (west) off State 143 onto a Park road, paved at its beginning. At 13.7 miles from the start of the trip, Balsam Road (for Forest Service use and closed to traffic) takes off to the left through dense spruce-fir woods and dead-ends after a couple of miles. During late fall and winter in years with a good cone crop, a walk along Balsam Road is likely to yield Evening Grosbeaks, Purple Finches, Pine Siskins, and sometimes Red Crossbills. In the next 1.6 miles after the Balsam Road turnoff, the Park road passes rhododendron gardens.

At 15 miles from the start of the trip, the Park road ends in a turn-around. Leave the car and walk west about a mile up to an observation platform on Roan High Bluff at an elevation of about 6,200 feet. While enjoying the magnificent view, watch for Northern Ravens in flight below. On the way up through the conifers one may hear or see the Veery, Magnolia Warbler, and most of the other high-country birds that may have been missed during earlier stops on the trip.

TIPTONVILLE

Reelfoot Lake | Reelfoot National Wildlife Refuge | Reelfoot Lake State Park

In 1811–1812 a series of violent earthquakes shook northwestern Tennessee, causing the terrain to crack open in some places, to rise up or subside in others. The forested land along Reelfoot Creek sank below the country around it and became covered with water, forming **Reelfoot Lake,** as it is known today. Stumps of the submerged forest may still be seen through the clear water.

Reelfoot comprises several basins of open water. Blue Basin, the largest, at the south end, has a depth of about 20 feet, but elsewhere on 'islands,' or towheads, between the basins and near the shore, tall cypresses emerge from much shallower water. Cow lilies, water chinquapins, pondweeds, and pickerel weeds form a luxuriant surface vegetation. From this area of emergent vegetation to the farming country around the Lake is a succession of plant communities: stretches of giant cutgrass, then a belt of buttonbush and black willow, and finally a great wooded swamp—except at the south and east ends—where there are stands of immense, thickly growing bald cypress, pecan, sweet gum, water locust, downy poplar, red maple, and white ash.

Birds nesting in the parts of the Reelfoot where there are aquatic grasses and other emergent plants include Pied-billed Grebes, Least Bitterns, King Rails, Purple Gallinules (a few), Common Gallinules, and American Coots. Great Blue Herons, Green Herons, Little Blue Herons, Cattle Egrets, and Great Egrets frequently visit the Lake from nesting colonies elsewhere, as do Little Terns.

From mid-October to April, Reelfoot is a haven for transient and wintering waterfowl. Canada Geese, Mallards, American Black Ducks, Gadwalls, American Wigeons, Ring-necked Ducks, Lesser Scaups, and Ruddy Ducks are usually the most abundant. Others appearing regularly are Common Pintails, Green-winged Teal, Northern Shovelers, Wood Ducks, Redheads, Canvasbacks, Common Goldeneyes, Buffleheads, Hooded Mergansers, and Common Mergansers. Although Wood Ducks and Hooded Mergansers breed in the Reelfoot area, they are more numerous in fall and winter.

Late in fall come hordes of Red-winged Blackbirds and European Starlings, sometimes a million or more of each species, and hundreds of thousands of Common Grackles and Brown-headed Cowbirds. Among them are Rusty Blackbirds, very few comparatively, and eclipsed by the multitude. By winter, when the entire aggregation has reached its maximum, the evening gathering of these birds to roost in the cutgrasses, willows, and buttonbushes is nothing short of spectacular.

Apart from the blackbird invasion, the wintertime ornithological

WOOD DUCK

feature is the assemblage of Bald Eagles. The number observed in four consecutive Christmas bird counts (1971–1974) ranged from 41 to 66 individuals.

Reelfoot Lake, including the islands, wooded swamps, and most of the shoreline, altogether 23,000 acres, is owned by the State of Tennessee. Under agreement with the United States Fish and Wildlife Service, 9,586 acres of the northern parts of Reelfoot Lake have been leased as the **Reelfoot National Wildlife Refuge,** with headquarters in Samburg on State 22, about 9 miles east of Tiptonville. **Reelfoot Lake State Park** (750 acres) lies along the western and southern parts of the Lake. Among its facilities are camping and picnic areas and a museum. Most of the remainder of the State property is the Reelfoot Wildlife Management Area, administered by the Tennessee Wildlife Resources Agency.

For bird finding, here are some suggestions. From Tiptonville, drive east on State 22 for 2.6 miles to the Park's museum. Stop here and take the walkway that leads back from the museum among the bald cypresses, where there are Prothonotary Warblers and other swamp-dwelling passerines. From the museum, except in winter, the Park operates a 3.5-hour boat cruise, with a stop at Caney Island for a short nature hike. The trip is an enjoyable way to become acquainted with Reelfoot's natural qualities and to see herons, egrets, and other birds.

Continue on State 22 around the south side of the Lake and northeastward. In winter, watch for Bald Eagles, often perched in some of the tallest trees. At Samburg, call in at Refuge headquarters, on State 22, one block north of the village's only stop sign; ask for permission to enter the Refuge's Grassy Island Unit (*see below*). Continue again on State 22 for 5 miles, turn left on State 157 for 2.3 miles, then left again on a paved backroad for 1.0 mile through the community of Walnut Log to a bridge over the Walnut Log Ditch. At Walnut Log one may hire a boat for a trip out of the Ditch into Reelfoot's Upper Blue Basin, which is excellent for grebes, bitterns, rails, gallinules, and other birds commonly nesting in the aquatic vegetation.

The bridge over the Ditch marks the entry into the Grassy Island Unit and is usually chained by the Refuge management. From this point a good gravel road passes 2.7 miles through wooded swampland to a turn-around. Just short of the turn-around an observation walkway juts out into the Basin. Bird finding along this road and walkway is rewarding in any season. Barred Owls are noisy during the day as well as at night. Red-shouldered Hawks, Pileated Woodpeckers, Red-bellied Woodpeckers, Acadian Flycatchers, Fish Crows, Carolina Wrens, Wood Thrushes, Prothonotary Warblers, Northern Parula Warblers, Cerulean Warblers, Yellow-throated Warblers, Kentucky Warblers, Hooded Warblers, and Summer Tanagers are common nesting birds. Less common, more often heard than seen, are Swainson's Warblers and Louisiana Waterthrushes. At two open places, Mississippi Kites are often in view overhead. During winter there is no better vantage point at Reelfoot than the walkway for viewing Bald Eagles, geese, ducks, and scores of American Coots.

Vermont

(BICKNELL'S) GRAY-CHEEKED THRUSH

This little state has among its 9,609 square miles probably more mountain land than the areas of the White Mountains, Adirondacks, and Catskills put together. As essentially a mountain state, its mountain ranges alternating with valleys, Vermont's birdlife is limited accordingly.

Although all the mountains of Vermont are frequently referred to as the Green Mountains, they actually comprise two different ranges, both running north and south. The Green Mountain Range, the more prominent, extends the length of the state's interior from Massachusetts to Quebec, gaining altitude northward. Its highest peak—and also the highest peak in the state—is Mt. Mansfield, with an elevation of 4,393 feet. The Taconic Range, extending north from Massachusetts, lies to the southwest. A number of isolated peaks are to the east.

Mt. Mansfield and a few other peaks rise above the timber line, presenting exposed ledges and boulder-cluttered summits. Hardy shrubs and herbaceous plants gain a foothold in every available niche and fissure; stunted firs, spruces, and birches form thick, matted tangles in the heads of ravines reaching up from the valleys. Of birds inhabiting the Green Mountains in late spring and summer, only the Slate-colored Junco seems able to endure these exposed conditions, often making itself at home on the very summits.

From the timber line down to 3,000 feet, most of the trees are fir and spruce, mixed occasionally with white and yellow birches. Dwarfed and dense at the timber line, they become taller at successively lower altitudes. Among the dwarfed conifers of the timber line the Bicknell's race of the Gray-cheeked Thrush breeds in abundance. Indeed, its center may be right here in the timber line of the Green Mountains. Sharing the timber line with the (Bicknell's) Gray-cheeked Thrush in equal if not greater abundance are the Blackpoll Warbler, Slate-colored Junco, and White-throated Sparrow. Here, too—though they are more numerous farther down, among the taller conifers—are the Yellow-bellied Flycatcher, Red-breasted Nuthatch, Brown Creeper, Winter Wren, Swainson's Thrush, Golden-crowned Kinglet, Solitary Vireo, several warblers including the Magnolia, Myrtle, and Blackburnian, Purple Finch, Pine Siskin, and, occasionally, the Red Crossbill and White-winged Crossbill. The Olive-sided Flycatcher, Boreal Chickadee, and Ruby-crowned Kinglet breed sparingly.

Between 2,000 and 3,000 feet the mountain forests are transitional in character, linking the coniferous above with the deciduous below. Here, spruces and firs meet and become intermixed with beeches and maples; and hemlocks are quite common. Birds from the coniferous and deciduous associations are also intermingled.

Most of the forests of the mountain valleys and lower slopes below 2,000 feet are deciduous, with beech, maple, birch, and basswood predominating; there are, however, a few forests where white pine prevails.

The bird finder who has enough time and energy to investigate the birdlife of all or parts of the Vermont mountains from valley to summit can do no better than to follow the famous Long Trail, a well-marked, 261-mile wilderness footpath beginning at Blackinton,

Massachusetts, and passing over mountains and through valleys the whole length of the Green Mountain Range to Quebec. Every few miles there are shelters and cabins, all of which have bunks for overnight rest. The bird finder need carry only blankets, food supplies, an axe, and cooking utensils. It is better to take the Trail from south to north, as a climax is achieved in northern Vermont, where the country is wildest and where there is an unexcelled view from Jay Peak.

The Long Trail is maintained for the public by the Green Mountain Club. Before hiking the Trail, or part of it, communicate with this organization's headquarters at Rutland, Vermont, and purchase the guidebook, which gives detailed descriptions of the route, valuable suggestions on equipment, and the rules of the Trail.

Between the forested mountains and rolling hills of Vermont are numerous valleys with deciduous woods and farmlands (fields, wet meadows, brushy lands, orchards, and dooryards) where the following are characteristic breeding birds:

DECIDUOUS WOODS

Broad-winged Hawk
Ruffed Grouse
Black-billed Cuckoo
Barred Owl
Whip-poor-will
Yellow-shafted Flicker
Pileated Woodpecker
Hairy Woodpecker
Downy Woodpecker
Great Crested Flycatcher
Least Flycatcher
Eastern Pewee
Blue Jay
Black-capped Chickadee
Tufted Titmouse (*southern Vermont*)
White-breasted Nuthatch
Wood Thrush
Yellow-throated Vireo
Red-eyed Vireo
Warbling Vireo
Ovenbird
American Redstart
Baltimore Oriole
Scarlet Tanager
Rose-breasted Grosbeak

FARMLANDS

Eastern Kingbird
Eastern Phoebe
Tree Swallow
Barn Swallow

FARMLANDS (*cont.*)

Gray Catbird
Brown Thrasher
Eastern Bluebird
Yellow Warbler
Chestnut-sided Warbler
Common Yellowthroat
Bobolink
Eastern Meadowlark
Common Grackle
Northern Cardinal (*southern Vermont*)
Indigo Bunting
American Goldfinch
Savannah Sparrow
Vesper Sparrow
Chipping Sparrow
Field Sparrow
Song Sparrow

Vermont's waterways do not usually attract large numbers of breeding waterfowl, waterbirds, or marsh birds. A few Herring Gulls nest on islands in Lake Champlain. Adjacent to Lake Champlain, which forms the state's western boundary for nearly a hundred miles, and adjacent to the southern end of Lake Memphramagog on the northern boundary, there are a few marshy areas of sufficient extent and composition for nesting ducks, Common Gallinules, and Black Terns. There are also a few small marshes, such as that north of West Rutland, where cattails and other aquatic vegetation provide conditions suitable for a typical bittern-rail association.

Except for American Woodcock and a few other species that begin nesting in April or earlier, the breeding season for most birds of the Vermont valleys begins in mid-May and is nearly ended by mid-July. In the mountains above 2,000 feet, however, the season begins and ends about two weeks later. Transient waterfowl and shorebirds stop regularly on Lakes Champlain and Memphremagog, but elsewhere in the state relatively few appear. Canada Geese are often seen in passage in late March, early April, late October, and November. Landbirds migrate in large numbers through the valleys between the big mountain ranges. Presumably larger numbers pass through the Connecticut River Valley along the eastern boundary than through the interior mountain valleys. The height of landbird migration comes in the first two weeks of May and the last two weeks of September.

During winter, Evening Grosbeaks, Common Redpolls, Pine Siskins, and American Tree Sparrows frequent towns and lowland countrysides. A few Northern Shrikes and occasionally Pine Grosbeaks and Red and White-winged Crossbills make an appearance, too. (Northern) Horned Larks and Snow Buntings and sometimes a Snowy Owl show up in windswept fields.

Authorities

George W. Davis, Thomas H. Davis, Jr., Elizabeth Downs, Richard D. Eldred, I. Scott Fillebrown, Kurt R. Groote, Richard M. Marble, Susan C. McClary, Ralph H. Minns, James C. Otis, Jr., L. Henry Potter, Barbara Rice, Lucretius H. Ross, Wendell P. Smith, Robert N. Spear, Jr., George J. Wallace.

References

Birds of Vermont. Compiled by Robert N. Spear, Jr. Green Mountain Audubon Society (P.O. Box 33, Burlington, Vt. 05401), 1976.

Birds of East-Central Vermont. By Richard B. Farrar, Jr. Vermont Institute of Natural Science, Woodstock, 1973.

ARLINGTON
Stratton Mountain

From Arlington in southern Vermont, an unnumbered road to Wardsboro follows the Roaring Branch east through wildly beautiful woods, and, after 15 miles, crosses the Long Trail. Leave the car and walk the Long Trail north from the crossing for a steady, easy 3-mile climb to the summit of **Stratton Mountain** (3,859 feet). From here the Long Trail drops rather steeply to Stratton Pond and returns to the Arlington-Wardsboro Road on a 3.5-mile branch trail marked with blue blazes. The woods is mixed deciduous-coniferous, but chiefly coniferous on the Mountain.

Breeding birds in woods and near streams include the (Bicknell's) Gray-cheeked Thrush and the following: Broad-winged Hawk, Yellow-bellied Sapsucker, Yellow-bellied Flycatcher, Traill's Flycatcher, Olive-sided Flycatcher, Boreal Chickadee, Red-breasted Nuthatch, Brown Creeper, Winter Wren, Hermit Thrush, Swainson's Thrush, Veery, Golden-crowned Kinglet, Solitary Vireo, Nashville Warbler, Magnolia Warbler, Blackburnian Warbler, Blackpoll Warbler, Northern Waterthrush, Canada Warbler, Slate-colored Junco, and White-throated Sparrow.

BRANDON
Mt. Horrid | Cape Lookoff Mountain | White Rock Mountain

For the bird finder eager to see or hear summer-resident (Bicknell's) Gray-cheeked Thrushes along with Yellow-bellied Flycatchers, Red-breasted Nuthatches, Winter Wrens, Swainson's Thrushes, Golden-crowned Kinglets, Magnolia Warblers, Blackpoll Warblers, Mourning Warblers, and Canada Warblers, and does not mind some rather steep climbing, **Mt. Horrid** (3,216 feet) is perfect. Maples and birches with scattered conifers cover most of its lower slopes; conifers predominate on its upper slopes.

Drive east from Brandon in west-central Vermont on State 73, through Brandon Gap (2,170 feet) for about 8 miles to the Long Trail crossing. Walk north on the Long Trail up Mt. Horrid to the Great Cliffs, where a Peregrine Falcon formerly nested regularly and, regrettably, is seen no more. If time permits, continue northward on the Long Trail to **Cape Lookoff Mountain** (3,298 feet) and then to **White Rock Mountain** (3,307 feet), about 3.5 miles from State 73.

BURLINGTON
Intervale | Coast Guard Marina Center | Sand Bar State Waterfowl Area | Sand Bar Bridge

Transient waterfowl and shorebirds provide the notable opportunities for bird finding in this northern city, by virtue of its situation on Lake Champlain, one of few bodies of water in or adjacent to Vermont where these birds appear in appreciable numbers. Two locations within the city offer good prospects. One is **Intervale,** a large marshy area. When it floods in early spring, Canada and Snow Geese stop here. In May, Common Snipes, Solitary Sandpipers, and both Greater and Lesser Yellowlegs appear in the wet fields and pastures. By June, some of the birds nesting are Pied-billed Grebes, American Bitterns, American Black Ducks, Blue-winged Teal, Wood Ducks, Soras, Virginia Rails, Common Gallinules, Marsh Wrens, and Swamp Sparrows.

To reach the area from I 89, take Exit 14W (US 2), go west on

Main Street almost to the Lake, turn right on Battery Street (which becomes North Street), then right on US 27 north, which crosses the marsh. Turn left on Manhattan Drive and continue a short distance to the entrance road to the city dump on the left. Park here and walk or, if the gate is open, drive down to a railroad spur line, and walk along the track beside the marsh. This track and an adjacent sand knoll are excellent points for observation, especially with a spotting scope.

The other location within the city is the Burlington waterfront at the **Coast Guard Marina Center.** To reach it, drive west on Main Street to the Lake, right on Lake Street, and left into the Marina. In winter, after the Lake freezes over, ducks gather in a small stretch of open water in front of the municipal power plant nearby. Some species to be expected are Mallards, Common Pintails, Ring-necked Ducks, Canvasbacks, Greater and Lesser Scaups, Common Goldeneyes, Buffleheads, and Hooded Mergansers.

About 10 miles north of the city the **Sand Bar State Waterfowl Area** encompasses a large marsh where American Black Ducks, Wood Ducks, and Common Goldeneyes nest in May and June. To reach the Area, drive north on I 89 to Exit 17, then west on US 2 for about 3 miles to headquarters on the left. Farther on, US 2 crosses a part of Lake Champlain to Grand Isle on **Sand Bar Bridge,** from which one may see many transient waterfowl, including Redheads, White-winged Scoters, and Common Mergansers.

GROTON
Groton State Forest

Northwest of Groton in northern Vermont lies **Groton State Forest** (15,607 acres) of mixed, primarily deciduous, growth. The area supports a larger than average population of Ruffed Grouse and at least one pair of Northern Ravens. There are many other species such as Broad-winged Hawks, Yellow-bellied Sapsuckers, Eastern Pewees, Wood and Hermit Thrushes, White-breasted and Red-breasted Nuthatches, Warbling and Red-eyed Vireos, Black-and-white Warblers, Nashville Warblers, Black-throated Green Warblers, Chestnut-sided Warblers, Ovenbirds, Common Yellowthroats, Canada Warblers, Slate-colored Juncos, and White-throated Sparrows. Scarlet Tanagers and Rose-breasted Grosbeaks are present but not common.

The Forest is reached from Groton by driving northwest on US 302 for 1.7 miles, then turning right on an unnumbered road that crosses the Forest from south to north and joins US 2 just east of Marshfield. Within the Forest are six lakes; the largest is Lake Groton in the southwest section. The highest point (1,958 feet), at Owls Head Mountain, is almost in the center. Along the road crossing the Forest are campgrounds, picnic sites, and access to some 30 miles of foot trails. The north-central quarter of the Forest is a game refuge.

NEWPORT

Sand Bank Bluff | South Bay Marsh | Morgan Area

Much of the country around this northeastern town at the southern end of Lake Memphramagog remains relatively unsettled and wild. Thus, for the bird finder, besides the Lake itself, there are vast forests and bogs and many small lakes, ponds, and marshes with connecting streams, all attracting a wide variety of species. Among areas yielding particularly wide varieties are the following three:

Sand Bank Bluff, north of Newport, the best observation point for birds on Lake Memphramagog. To reach it, drive east from Newport on State 105, turn north at the edge of town and continue north when State 105 goes east; turn left on the first road, then right to its end; left again, then the first right and the first left on a dirt road leading to the Bluff. Among waterbirds and waterfowl to be looked for on the Lake below, from October to freeze-up and in spring after the ice leaves, are the Common Loon, Horned Grebe, Canada Goose, Greater Scaup, Common Goldeneye, Bufflehead, and Common and Red-breasted Mergansers. During May, late August, and September, the narrow, sandy, and muddy shore directly below the Bluff attracts numerous shorebirds including the Semipalmated and Black-bellied Plovers, Ruddy Turnstones, Least and Semipalmated Sandpipers, and Sanderlings.

South Bay Marsh, south of Newport. Drive east on State 105 to the edge of town. When State 105 turns north, continue straight ahead, cross a bridge, and take the first right on a road that follows the east side of South Bay and soon crosses the Marsh. Among summer-resident birds to be looked for are the Pied-billed Grebe, Green Heron, Least and American Bitterns, American Black Duck, Blue-winged Teal, Ring-necked Duck, Common Gallinule, Common Snipe, Black Tern, and Marsh Wren.

Morgan Area, east of Newport. Drive east on State 105 to Derby, turn left on State 111 for about 15 miles, through spruce and fir woods with occasional spruce and tamarack bogs, to State 114; turn left on State 114 for about 1.25 miles to a pond on the right with a spruce-tamarack bog on its south side. This is worth exploring in nesting season for Yellow-bellied Sapsuckers, Ruby-crowned Kinglets, Tennessee Warblers, Northern Parula Warblers, Myrtle Warblers, Blackburnian Warblers, and Bay-breasted Warblers.

Continue north 0.25 mile to another pond on the right, then walk for about a mile along a logging road that begins on the left side of the road opposite the pond and crosses a spruce forest, passes a beaver dam and pond, traverses a bog, some second-growth deciduous woods, an alder swale, and a weedy field, and ends in a cutover spruce forest. Among breeding birds in the spruce forest are the Yellow-bellied Flycatcher, Boreal Chickadee, Winter Wren, Red-breasted Nuthatch, Swainson's Thrush, Magnolia Warbler, Pine Siskin; at the beaver pond, the Rusty Blackbird; in the second-growth woods, the Broad-winged Hawk, Black-billed Cuckoo, Solitary Vireo, Chestnut-sided Warbler; in the alder swale, the Traill's Flycatcher, Wilson's Warbler.

RUTLAND
West Rutland Marsh | Killington Peak | Pico Peak

Commencing at the northern edge of West Rutland, in central Vermont, and extending northward, is **West Rutland Marsh,** approximately 2 miles long and 0.5 mile wide. Cattails predominate near town, but farther north there are stretches of sedges with islands of willows and alders. A sluggish stream, the beginning of the Castleton River, meanders the entire length of the Marsh. On the eastern side stretch cultivated fields, meadows, pastures, and woodland; on the western side rise steep wooded hills.

In late May, June, and early July a trip along the roads around the Marsh, with side trips into the several different habitats, will yield a long list of birds. Some of the species are: *in the Marsh,* American Bittern, American Black Duck, occasionally Northern Harrier, Virginia Rail, Sora, Marsh Wren, Swamp Sparrow; *in the fields and wet meadows,* Sedge Wren, Bobolink; *in the small trees and shrubs,* Traill's Flycatcher, Yellow Warbler, Chestnut-sided Warbler, Common Yellowthroat; *in the woodlands,* Wood Thrush, Veery, Yellow-throated Vireo, Warbling Vireo. To reach the Marsh, drive west from Rutland on US 4 to West Rutland, then north on an unnumbered road to Florence. The Marsh begins at the north edge of West Rutland.

Most people acquainted with Vermont's birdlife agree that a trip to the top of **Killington Peak** (4,235 feet), the state's second highest mountain, will yield the greatest variety of montane birds. Proceed northeast from Rutland (500 feet) on US 4, which ascends the Green Mountain Range to Sherburne Pass (2,190 feet). On this 10-mile drive, first through farmland and then up through the deciduous forests of the lower slopes, one will see nearly all the species common to the Vermont countryside. Leave the car at Long Trail Lodge, on the Long Trail at Sherburne Pass, a popular stopping place not only for tourists but also for hikers. The Lodge is in a superb forest, amid fern-clad cliffs, where stands of white and yellow birches, maple, beech, and hemlock mix with scattered spruce and fir. Being transitional in character, the forest has an avifauna showing a mixture of deciduous- and coniferous-forest species. From the Lodge, take the Long Trail south over **Pico Peak** (3,957 feet) to Killington Peak, 5.5 miles distant.

Pico and Killington Peaks are both cloaked with a stunted growth of spruce and fir; the very summit of Killington is bare ledge. (Bicknell's) Gray-cheeked Thrushes occupy both Peaks, but on Killington the population descends as far down as 3,200 feet and is probably the larger and second in size only to the population on Mt. Mansfield (*see under* **Stowe**). Sharing the upper levels with the Gray-cheeked Thrushes are Winter Wrens, Myrtle Warblers, Blackpoll Warblers, Slate-colored Juncos, and White-throated Sparrows. At the lower levels, one may find Yellow-bellied Flycatchers, Brown Creepers, Swainson's Thrushes, and several warblers: Nashville, Magnolia, Black-throated Blue, Black-throated Green, Blackburnian, Mourning, and Canada.

For the bird finder not wishing to climb the Peaks, there are a number of short, pleasant trails from the Lodge through the largely hardwood forest.

ST. JOHNSBURY
Moose River Valley

Drive east from St. Johnsbury on US 2 for about 10 miles to North Concord and turn north on an unnumbered road that parallels the alder-lined Moose River in **Moose River Valley.** Between Victory and Gallup Mills the road passes through approximately 3,000 acres of spruce and fir, interrupted by marshes, meadows, and old fields. In early spring, American Woodcock perform their crepuscular flight displays above these old fields. A visit to almost any such field after sundown or before sunup between mid-April and mid-May is consequently rewarding.

STOWE
Mt. Mansfield

Vermont's highest peak, **Mt. Mansfield** (4,393 feet), is readily accessible and also affords the best sampling of birdlife in the higher altitudes of the Green Mountains.

From Stowe, drive northwest on State 108 and turn left on a toll road to the Mt. Mansfield Hotel, 544 feet from the summit. At about 3,000 feet the road passes from the deciduous forests at the base of the mountain, where there are Yellow-bellied Sapsuckers, Hermit Thrushes, Veeries, Solitary Vireos, Black-throated Blue Warblers, Black-throated Green Warblers, and Scarlet Tanagers, to stunted stands of spruce and fir where (Bicknell's) Gray-cheeked Thrushes reside during the summer and where Winter Wrens, Blackpoll Warblers, Slate-colored Juncos, and White-throated Spar-

BLACKPOLL WARBLER

rows abound. The hotel, close to the tree line, is literally surrounded by their habitat.

From the Hotel's porch in late afternoon of any June or early-July day the bird finder may hear the low-voiced, veery-type song of (Bicknell's) Gray-cheeked Thrushes in the nearby evergreens. The chorus usually begins by 3:00 or 4:00 P.M., increases gradually as evening approaches, and reaches a climax in a short period of from 15 to 30 minutes after sunset. A regular feature of the climax is the spectacular display of flight songs, each consisting of aerial spirals and tumbles accompanied by musical, reedy notes and lasting fully half a minute, sometimes longer. With the onset of total darkness the birds become silent.

At least six foot trails lead to the summit of Mt. Mansfield. Of the two most convenient, the one on the east constitutes a portion of the Long Trail from State 108 in Smugglers Notch to Taft Lodge; the other, on the west extends from Underhill Center to Butler Lodge. Both Lodges usually offer overnight shelter for hikers. Check the Green Mountain Club guidebook for current information. With sweeping views of northwestern New England, the Lodges are delightful headquarters for the bird finder who, not objecting to a leisurely climb, wishes to observe the transition of birds from mountain base to summit, to hear the sundown concert at tree line, and to search for the elusive songsters on the morrow.

SWANTON
Missisquoi National Wildlife Refuge | Mud Creek State Access Area

The bird finder looking for waterfowl and marsh birds in Vermont can do no better than go to the **Missisquoi National Wildlife Refuge** (4,680 acres) in the extreme northwestern part of the state, with headquarters on State 78, 2.5 miles northwest of Swanton.

The Refuge lies in the delta of the Missisquoi River, only a few feet above Lake Champlain, which bounds it on the north and west. Each spring the Refuge is almost completely inundated. A little more than half is swampy woodland, with red maple predominating, followed by silver maple, white ash, and oak. Except for a little grassy meadowland, the rest of the Refuge is marsh in which bulrushes, wild rice, sedges, spike rushes, bur reeds, and pickerel weeds are the predominant emergent aquatic plants, although many shrubs, such as willows, alders, dogwoods, and buttonwoods have gained a foothold in shallow sections.

Altogether, 186 bird species have been seen at the Refuge. American Black Ducks, Blue-winged Teal, Wood Ducks, and Common Goldeneyes nest regularly. Marsh birds breeding here include the American Bittern, Virginia Rail, Common Gallinule, Black Tern, and Red-winged Blackbird. The Killdeer, American Woodcock, Common Snipe, and Spotted Sandpiper are common in summer. Other nesting species include the Northern Harrier, Black Tern, Whip-poor-will, Belted Kingfisher, Yellow-shafted Flicker, Bank Swallow, Red-eyed Vireo, Common Yellowthroat, Bobolink, Rusty Blackbird, Purple Finch, and American Goldfinch.

Canada Geese, Snow Geese, Lesser Scaups, and Common Mergansers are regular transients in both spring and fall. Common Pintails, Green-winged Teal, and Ring-necked Ducks are more common in fall. Occasionally a Snow Goose, Northern Shoveler, Redhead, or Ruddy Duck shows up in fall. In fall also, American Black Ducks, Wood Ducks, Blue-winged Teal, and Common Goldeneyes increase, their numbers being swelled by transient individuals. Nearly all the geese and ducks appear in spring as early as the ice break-up—somewhere between mid-March and the third week of April, and stay in fall until freeze-up—anytime from early November to the middle of December.

About 7.5 miles farther west on State 78 from Refuge headquarters is the entrance road to the **Mud Creek State Access Area,** marked by a sign. Cross the dam and follow the access road to a railroad right-of-way, park, and walk out onto a wooded peninsula. The parking area, right-of-way, and peninsula are all good observation points for nesting waterbirds and waterfowl, such as the Great Blue Heron, Black-crowned Night Heron, Green Heron, American Bittern, American Black Duck, Mallard, and Blue-winged Teal. From the parking area, watch especially for Hooded Mergansers.

VERGENNES
Dead Creek Waterfowl Area

The **Dead Creek Waterfowl Area** (2,578 acres) in west-central Vermont consists mostly of wetlands, formed by a series of dams and dikes on Dead Creek under management of the Vermont Fish and Game Service. Marshes with cattails and other emergent plants provide excellent cover for waterfowl and marsh birds. The list of summer residents in the marsh area includes the Great Blue Heron, Wood Duck, and Red-winged Blackbird as common, with lesser numbers of the Black Tern, Mallard, American Black Duck, Common Pintail, Green-winged Teal, Blue-winged Teal, Northern Harrier, and Common Gallinule. In farming country surrounding the marsh should be Red-tailed Hawks, Red-shouldered Hawks, Killdeers, Eastern Kingbirds, Barn Swallows, Purple Martins, Eastern Bluebirds, Bobolinks, Eastern Meadowlarks, American Goldfinches, Vesper Sparrows, Field Sparrows, and Song Sparrows. Migrating Canada Geese, Snow Geese, and ducks stop at the marsh; northern shorebirds appear in August and September.

Drive south from Vergennes on State 22A for 6 miles to Addison, turn right on State 17, turn left on a narrow road to the Waterfowl Area marked by a sign. Just before crossing Dead Creek, stop at an 'observation area' where a controlled flock of Canada Geese invariably attracts a number of wild geese and ducks. Cross the bridge. From here there are two possibilities. (1) Turn right and take each successive right thereafter to the Farrel Access Area, the best spot for observing migrating geese, and, if one has a canoe, the best area for shorebirds in migration. (2) Turn left on a narrow road that leads into the main part of the marsh.

WATERBURY
Camel's Hump

The **Camel's Hump** (4,083 feet) is an impressive mountain, comparable to Mt. Mansfield in the vertical distribution of birdlife. Of the four trails leading to the summit, the one most convenient for bird finders traveling by car is from Crouching Lion Farm on the east side. To reach the Farm, cross the Winooski River in Waterbury and turn right on an unnumbered road to North Duxbury, then turn south to the Farm, the starting point of a 3-mile trail to the summit.

The coniferous forests adjacent to the summit contain a varied assortment of birds, the characteristic species being the (Bicknell's) Gray-cheeked Thrush. Both Myrtle and Blackpoll Warblers are abundant, and there is a better chance of finding Boreal Chickadees here than on any other Vermont peak.

In September the Bamforth Ridge Trail is very good for watching migrating hawks. Close by the Trail, where it tops out on the first spruce ridges about 2 miles from the starting point, an open ledge overlooks a broad valley—a fine outlook for hawks soaring over the valley. To reach Bamforth Trail, drive west from Waterbury on US 2 to Jonesville, cross the Winooski River, take the first left on a minor road for 2.5 miles, and look for the trail sign on the right.

WINDSOR
Mt. Ascutney

Ascutney State Park (1,530 acres), a public recreation area in southern Vermont, has several trails leading to the summit of **Mt. Ascutney** (3,144 feet), and a scenic 4-mile automobile road nearly to the summit. The woods on the slopes, though primarily mixed deciduous-coniferous, has good stands of conifers here and there. Despite crowded conditions in the Park during summer, the bird finder has a good chance of observing on the higher slopes such species as the Yellow-bellied Flycatcher, Boreal Chickadee, Red-breasted Nuthatch, Winter Wren, and Blackpoll Warbler. Birds on the lower slopes consist largely of those appearing commonly in the adjacent Connecticut River Valley. The Park is reached from Windsor by driving south on US 5 for about 4 miles, turning right on State 44 (Brownsville Road) and then left to the Park entrance.

WOODSTOCK
Mt. Tom

For an easy climb on a well-maintained trail as well as productive bird finding through a heavily deciduous-wooded area, **Mt. Tom** (1,300 feet) in southeast-central Vermont is ideal. Besides being excellent for spring and fall transients, the woods hold a variety of breeding species including the Ruffed Grouse, Great Horned Owl, Barred Owl, Pileated Woodpecker, Great Crested Flycatcher, Eastern Pewee, Brown Creeper, Wood Thrush, Hermit Thrush, Veery, Warbling Vireo, Black-throated Green Warbler, Scarlet Tanager, Rose-breasted Grosbeak, and Slate-colored Junco. Mt. Tom may be reached from Woodstock Inn in Woodstock by driving over the Ottauquechee River on the Green and the Middle Bridges to Mountain Avenue and the entrance to Falkner and French Memorial Parks, where the trail begins.

Virginia

BLACK SKIMMERS

There are sharply contrasting environments in Virginia: the shell-strewn beaches, grass-studded sand dunes, and lush salt marshes of Assateague and Cobb Islands on the Atlantic Ocean; the idyllic wilderness of the Great Dismal Swamp, with its mirrored waterways and magnificent gray-bearded cypresses; and, overlooking the rolling Piedmont Plateau and the placid Shenandoah Valley, the refreshing crests of the Blue Ridge Mountains, adorned in spring by a succession of blooms—lacy white shadblows in April, pink and flame azaleas, dogwoods, and redbuds in May, mountain laurels and pink and white rhododendrons in June. Birdlife is just as sharply contrasting: Black Skimmers nesting in the glaring sun on island beaches; Prothonotary and Hooded Warblers in the shaded recesses of the Great Dismal Swamp; and Northern Ravens patrolling their nesting cliffs high up in the Skyland area of the Blue Ridge. Bird finding in the Old Dominion is never monotonous.

The wide Coastal Plain, or 'Tidewater,' of Virginia, east of the Fall Line, is a low, undulating area cut by sluggish rivers into long peninsulas. In the north and central parts of the Plain, the eastern extremities of the peninsulas, together with the Eastern Shore or Delmarva Peninsula extending south from Maryland, form lower Chesapeake Bay. Where the Coastal Plain fronts directly on the ocean—on the eastern side of Delmarva Peninsula and in the extreme southeast—there are barrier islands with wide beaches and sand dunes; behind these lie salt and brackish marshes, shallow bays, and channels. Several of these islands and nearby marshes off the Eastern Shore Peninsula attract large numbers of nesting birds, including the following species:

Clapper Rail
American Oystercatcher
Wilson's Plover
Willet
Laughing Gull
Gull-billed Tern
Forster's Tern
Common Tern
Little Tern
Royal Tern
Black Skimmer
Marsh Wren
Seaside Sparrow

Nearly two-thirds of the Coastal Plain is forested. There are vast stretches of loblolly and Jersey pines, sometimes growing in pure stands, sometimes interspersed with silver maples, sweet gums, and various oaks. On the flood plains bordering the streams and rivers, sycamores and black birches predominate, except on drier soils, where there are also aspens, gums, and oaks. In the Great Dismal Swamp of the southeast and in similar wet areas, cypress becomes a prominent forest type. Almost all the forests are second growth.

The Piedmont Plateau, extending between the Coastal Plain and the Blue Ridge Mountains, is narrow in the north but widens gradually toward the North Carolina line, extending all the way from Emporia westward. Somewhat rolling in the east, with elevations of about 300 feet, it rises imperceptibly toward the Blue Ridge, the terrain becoming more rugged, with hills or ridges attaining heights of 1,000 to 1,200 feet, and occasionally higher. About two-thirds of the Piedmont is forested, almost entirely with second-growth timber. The prevalence of oaks, chiefly white oak, is characteristic of the forests, although there are scattered stands of shortleaf and Jersey pines and, less frequently, loblolly pine.

Among landbirds characteristic of both the Coastal Plain and the Piedmont are Fish Crows, in the forests near the coast, the bays, and tidewater rivers; Brown-headed Nuthatches, Pine Warblers, and a few Bachman's Sparrows in pinelands; Yellow-throated Warblers either in pinelands or in nearby swampy places; and Chuck-will's-widows in lowland woods, chiefly on the Coastal Plain.

The picturesque Blue Ridge Mountains extend southwest across the state from West Virginia and Maryland to North Carolina. In the north they comprise a distinct ridge about 10 miles wide and up to 4,000 feet in elevation, with spurs and subordinate ridges on each side. In the south the Mountains broaden and increase their elevations, producing near the North Carolina line a high, 70-mile-wide plateau of rolling uplands, deep valleys, and peaks. The highest point in the Blue Ridge, Mt. Rogers (5,719 feet), is also the highest point in the state.

Paralleling the Blue Ridge Mountains on the west is the Shenandoah Valley (Valley of Virginia), whose deciduous forests have for the most part been cleared for agriculture. Between the Valley and the West Virginia and Kentucky boundaries are the ridges of the Allegheny Mountains, whose maximum elevations approximate 4,000 feet.

The Blue Ridge and the Allegheny Mountains have similar second-growth forests, principally hardwoods—oak, maple, basswood, hickory, walnut, black locust, black gum—intermixed occasionally with white pine, red spruce, fir, and hemlock. Birdlife of these ranges is also similar, but is more readily accessible on the Blue Ridge because of a scenic parkway—Skyline Drive, in Shenandoah National Park, and Blue Ridge Parkway, its continuation south into North Carolina—which passes the highest crests from north to south. Characteristic breeding species are the Ruffed Grouse, Yellow-bellied Sapsucker, Least Flycatcher, Northern Raven, Veery, Solitary Vireo, Black-and-white Warbler, Golden-winged Warbler, Black-throated Blue Warbler, Black-throated Green Warbler, Blackburnian Warbler, Chestnut-sided Warbler, Canada Warbler, Scarlet Tanager, Rose-breasted Grosbeak, and Slate-colored Junco. Most of these birds may be observed along Skyline Drive in Shenandoah National Park, which embraces the loftiest peaks of the northern Blue Ridge. On the highest peaks of the southern Blue Ridge (e.g., Mt. Rogers), where spruce and fir predominate, there are, in addition, the Black-capped Chickadee, Red-breasted Nuthatch, Brown Creeper, Winter Wren, and Golden-crowned Kinglet.

In the deciduous woods of the Coastal Plain, the Piedmont, and the Shenandoah Valley, and in the adjoining farmlands (fields, wet meadows, brushy lands, orchards, and dooryards), the following birds breed regularly:

DECIDUOUS WOODS

Turkey Vulture
Black Vulture
Red-shouldered Hawk
Yellow-billed Cuckoo
Black-billed Cuckoo
Common Screech Owl
Barred Owl
Whip-poor-will
Yellow-shafted Flicker
Pileated Woodpecker
Red-bellied Woodpecker
Hairy Woodpecker
Downy Woodpecker
Great Crested Flycatcher
Acadian Flycatcher
Eastern Pewee
Blue Jay
Carolina Chickadee
Tufted Titmouse
White-breasted Nuthatch
Wood Thrush
Blue-gray Gnatcatcher
Yellow-throated Vireo
Red-eyed Vireo
Warbling Vireo
Prothonotary Warbler
Worm-eating Warbler
Northern Parula Warbler
Cerulean Warbler
Ovenbird
Louisiana Waterthrush
Kentucky Warbler
Hooded Warbler
American Redstart
Baltimore Oriole
Summer Tanager

FARMLANDS

Common Bobwhite
Mourning Dove
Eastern Kingbird
Eastern Phoebe
Tree Swallow (*local*)
Barn Swallow

House Wren
Bewick's Wren
Carolina Wren
Northern Mockingbird
Gray Catbird
Brown Thrasher
Eastern Bluebird
Loggerhead Shrike (*except near coast*)
White-eyed Vireo
Yellow Warbler
Prairie Warbler
Common Yellowthroat
Yellow-breasted Chat
Eastern Meadowlark
Orchard Oriole
Common Grackle
Northern Cardinal
Blue Grosbeak
Indigo Bunting
American Goldfinch
Rufous-sided Towhee
Grasshopper Sparrow
Henslow's Sparrow
Vesper Sparrow
Chipping Sparrow
Field Sparrow
Song Sparrow

Migration movements in Virginia are heavily coastwise. Such birds as Red-throated Loons, Northern Gannets, Double-crested Cormorants, Brant, Snow Geese, White-winged Scoters, and Red-breasted Mergansers follow the outer coast; Whistling Swans, Canvasbacks, and Ruddy Ducks pass through Chesapeake Bay; Common Loons, Canada Geese, and American Black Ducks tend to use both routes. Shorebirds frequent all beaches, although Ruddy Turnstones, Red Knots, and Sanderlings appear more often on ocean beaches, Whimbrels on mud flats and oyster bars of the outer bays. Tree Swallows, hawks, and other landbirds 'bunch up' in fall at Cape Charles on the southern tip of Delmarva Peninsula, much as they do on the Cape May Peninsula in New Jersey.

The following timetable indicates when to expect the heaviest migration movements:

Waterfowl: 20 February–1 April; 20 October–10 December
Shorebirds: 25 April–25 May; 5 August–10 October
Landbirds: 10 April–10 May; 5 September–1 November

The coastal bays and tidewater rivers attract large populations of waterfowl in winter. Such ducks as Common Goldeneyes, White-winged Scoters, and Red-breasted Mergansers, as well as many waterbirds, stay near the mouth of Chesapeake Bay and can be seen from the fishing pier on the South Island of the Chesapeake Bay Bridge–Tunnel. Two stellar places for wintering waterfowl are the Chincoteague National Wildlife Refuge on the northeast coast, reached from the little town of Chincoteague, and the Back Bay National Wildlife Refuge on the southeast coast, reached from Norfolk. Both have impressive concentrations of Canada and Snow Geese; in addition, Chincoteague is excellent for Brant, as is Back Bay for Whistling Swans.

Authorities

Curtis S. Adkisson, John W. Aldrich, Ray J. Beasley, Almond O. English, Paul G. Favour, Jr., John H. Grey, Jr., C. O. Handley, Sr., Norman D. Hellmers, D. Ralph Hostetter, Raymond K. Long, Jr., Brooke Meanley, James J. Murray, James J. Murray, Jr., Jack E. Perkins, Catherine W. Reed, Frederic R. Scott, Charles E. Stevens, Jr., Henry M. Stevenson, Robert J. Watson, Alexander Wetmore.

BLACKSBURG

Virginia Polytechnic Institute and State University | New River | Poverty Hollow | Mountain Lake

The campus of **Virginia Polytechnic Institute and State University,** covering nearly a square mile just west of Blacksburg in the Allegheny Mountain foothills in southwestern Virginia, has small ponds about 3 acres in area on its western periphery. The best way to reach the ponds is to turn west off Main Street onto Washington Street, proceed to its end, and go right until they come into view. Resident Canada Geese and Mallards serve as decoys to attract a winter population of waterfowl surprisingly large for such small bodies of water. The birds often become rather tame, thus affording an unusual opportunity to study them at close range. From October to April there are, off and on, such species as the American Black Duck, American Wigeon, Green-winged and Blue-winged Teal, Ring-necked Duck, and Lesser Scaup.

Other ducks, particularly diving species, gather on **New River** in winter and early spring. The best places for viewing them begin at the little town of McCoy, 12 miles west of Blacksburg. Turn west off Main Street near the center of town onto Prices Fork Road, which leads to State Secondary 685, and follow this to the first small hamlet, Prices Fork; turn right on State Secondary 652 and stay on it to the River at McCoy. From here one may follow the River northward along a road that parallels it closely for about 2 miles. Common Goldeneyes, Buffleheads, and Hooded Mergansers may be expected. Look for Ospreys in spring.

For a variety of birds in the Allegheny Mountains west of Blacksburg, drive north from the city on either Bypass or Business US 460; high on a 2,500-foot ridge, 3.9 miles from the point where the routes join outside town, turn left on a gravel road of the United States Forest Service which passes through a second-growth forest of oak, hickory, and pine, gradually descending 500 feet in 2 miles to **Poverty Hollow.** Here white pine and hemlock are the dominant trees, with thickets of rhododendron and mountain laurel flanking Poverty Creek.

Among summer-resident birds one may see or hear along the road and in Poverty Hollow are the Worm-eating Warbler, Pine Warbler, Ovenbird, Kentucky Warbler, Hooded Warbler, American Redstart, and Rose-breasted Grosbeak. Where the forest on the road's right has been recently cut are Golden-winged Warblers and, occasionally, Blue-winged Warblers—and the possibility of a hybrid of the two species. At night, Whip-poor-wills are vociferous and often may be in the road, their eyes reflecting the car's headlights. Ruffed Grouse, Wild Turkeys, and Pileated Woodpeckers are year-round residents. Red Crossbills are known to appear during nearly every month of the year.

Backtrack to US 460 and continue west; 2 miles beyond the village of Newport, turn north onto State Secondary 700, which ascends Salt Pond Mountain to **Mountain Lake** at an elevation of 3,873 feet. Here the terrain, drained by clear, cool streams, is largely forested with oak, hickory, and pine on the slopes, and beech and hemlock, with a dense undergrowth of mountain laurel and rhododendron, in the ravines.

In this setting bird finding is ideal. During late May, June, and early July, when the birds are in full song, one may list the following species, selected to indicate the opportunities: Yellow-bellied Sapsucker, Great Crested Flycatcher, Least Flycatcher, Northern Raven, White-breasted Nuthatch, Winter Wren, Wood Thrush, Veery, Cedar Waxwing, Solitary Vireo, Black-and-white Warbler,

Black-throated Blue Warbler, Black-throated Green Warbler, Blackburnian Warbler, Chestnut-sided Warbler, Louisiana Waterthrush, Canada Warbler, Scarlet Tanager, Indigo Bunting, and Slate-colored Junco.

CHARLOTTESVILLE

Observatory Hill | Monticello | Henley's Lake | Beaver Creek Reservoir | Whitehall Gravel Pits | Humpback Rocks

Lying among low mountains just east of the Blue Ridge, this lovely town, home of the University of Virginia, offers opportunities for bird finding in a pleasant setting.

Observatory Hill, on the University grounds, is excellent for following spring migration very close to town. From the intersection of Alderman and McCormick Roads, take McCormick Road west as it winds up the Hill through mature oak-hickory forest to the Observatory. In early May, transient warblers abound along the Road and the foot trails leading from it. Barred Owls reside in the forest the year round.

Two warblers—the Cerulean and Worm-eating—are summer residents in the woods near the entrance to **Monticello,** Thomas Jefferson's home, 2.5 miles south of Charlottesville just off State 20.

Outstanding for transient waterfowl and waterbirds is **Henley's Lake,** a 25-acre artificial pond at the eastern foot of the Blue Ridge. Take US 250 from Charlottesville to Mechum River, turn right on State 240 to Crozet, right on County 810 for 2 miles, and right again on County 811 at the 'Rose Valley Farm' entrance. A farm gate at 0.9 mile leads to the pond on the right. Surrounded by open fields, the Lake is attractive to waterfowl and waterbirds, with the result that many unusual records have been obtained through the years.

American Black Ducks, American Wigeons, Redheads, Ring-necked Ducks, Lesser Scaups, and Hooded Mergansers are common from the second week in February to the third week in April; Blue-winged Teal, Ruddy Ducks, and Red-breasted Mergansers become common by late March; Wood Ducks are frequent only in fall. After rainy days in March and April, look for Horned and Pied-billed Grebes, Great Blue Herons, Green Herons, and American

Bitterns. Such waterbirds as Herring Gulls, Ring-billed Gulls, Laughing Gulls, Bonaparte's Gulls, Forster's Terns, and Common Terns appear more or less regularly in April and early May.

For wintering waterbirds and waterfowl, including Common Loons and Canada Geese as well as some of the species in view during migration at Henley's Lake, **Beaver Creek Reservoir,** a flood-control lake, is productive. Take US 250 west from Charlottesville to Mechum River; just before the junction with State 240, turn right on County 682; at a Y junction, keep right on County 680. When at 0.9 mile from US 250, the road crosses the Reservoir dam, providing good views.

The **Whitehall Gravel Pits** offer the best chance for seeing shorebirds. Drive west from Charlottesville to Crozet (*as above*), then right on County 810 to a river crossing at 0.5 mile beyond Whitehall. The pits are on the right. Merely gravel diggings with small intermittent pools and one small cattail community, they attract Spotted Sandpipers, Solitary Sandpipers, Greater and Lesser Yellowlegs, Least Sandpipers, and Semipalmated Sandpipers in April and May and in late summer and early fall.

For watching the fall hawk migration along the Blue Ridge, drive west from Charlottesville on either I 64 or US 250 to the crest of the Blue Ridge, then south on the Blue Ridge Parkway for 6 miles to a parking area on the left below **Humpback Rocks.** From here walk up the steep trail for 0.75 mile to the summit overlooking the Valley of Virginia. In late September and early October, on any clear, windy day just after a cold front, chances are good for seeing southward flights of hawks—Sharp-shinned, Red-tailed, Red-shouldered, and Broad-winged—and a few Ospreys.

CHERITON
Cobb Island

In their large size and the great variety of their nesting sea birds, the colonies on **Cobb Island** have few rivals on the Atlantic coast between New England and Florida.

Like Assateague and other long narrow islands paralleling the ocean side of the Delmarva Peninsula, Cobb Island has a wide ocean beach, back of which the sand mounts into a ridge of low dunes, then subsides into a salt marsh interrupted by brackish ponds and tidal channels. Separating Cobb Island from the mainland, 8 miles distant, is the Broadwater, a shallow sound with marshy islands called Little Eastward, Big Eastward, Gull Marsh, and Cedar. Close to the southern tip of Cobb Island is tiny Cardwell Island—Cobb Island in miniature.

When Cobb Island was settled in about 1840 by Nathan Cobb, it was approximately 7 miles long and a half-mile wide. A fertile soil capped the dry surfaces behind the dunes. On the southern tip, Nathan Cobb and his sons ran a prosperous farm, planted trees, built a hotel, and sold a strip of land to the Federal Government for a Coast Guard Station.

Since the area teemed with transient and nesting birds, it became a paradise for pot-hunters, eggers, and sportsmen. But at the turn of the century and in the years that followed, merciless storms accompanied by high tides washed away the top soil and the trees, razed the Cobb buildings, and changed the Island's configuration. Finally, during a severe storm in 1933, the last of the buildings was destroyed, and George W. Cobb, one of Nathan's grandsons and sole occupant of the building at the time, was drowned. Today, Cobb Island is 2 miles shorter and half as wide, and only beach grasses and sand-loving dune plants comprise the vegetation above the level of the salt marsh. The Coast Guard Station, all but demolished in the 1933 storm, was later rebuilt.

Distributed at intervals along the 5-mile beach are colonies with varying numbers of Gull-billed Terns, Common Terns, Little Terns, Royal Terns, and Black Skimmers. All the nests are on the open, flat, shell-strewn sand above the normal high-tide mark. Also nesting in the same situation and not infrequently close to the colonies are a few pairs of American Oystercatchers, Piping Plovers, and Wilson's Plovers. Between the beach and the marsh, usually in grasses growing between the dunes or on the edge of the marsh, are the Willets' nesting sites.

The marsh colonies contain Laughing Gulls, and sometimes small numbers of Forster's Terns, if there is a sufficient accumulation of tidal debris on which they may deposit their eggs. Generally the colonies are in widely separated portions of the marsh. Many Clapper

Rails and Seaside Sparrows hide their nests in the grasses but not necessarily near the colonies. All the colonies usually have young by the last of June or first of July.

To reach Cobb Island, drive east 1.0 mile from US 13 in Cheriton to the little seaside town of Oyster, and negotiate with a fisherman for transportation.

CHINCOTEAGUE
Chincoteague National Wildlife Refuge

One of the best spots on the coast for transient and wintering waterfowl and transient shorebirds is the **Chincoteague National Wildlife Refuge** on Assateague Island, a long barrier island paralleling the ocean side of Delmarva Peninsula from Virginia north. The Refuge occupies all the Virginia part of Assateague, roughly one-third of the Island (9,439 acres), and several salt-marsh islands (417 acres) in Chincoteague Bay on the mainland side of Assateague in Maryland. The following account applies only to the Virginia portion of the Refuge.

On the ocean front is a wide beach rising gradually to a ridge of low dunes sparsely covered with grasses and various sand-loving plants. On the opposite side are extensive salt marshes. Here and there, water from rains has been impounded by dikes to form seven shallow fresh-water pond and marsh units. On high ground are stands of pine and oak, with underlying thickets of myrtle, bayberry, and other shrubs.

About November, Canada Geese, Brant, and Snow Geese appear on the Refuge. While a good many continue south to areas on the coast of North Carolina, large flocks of Brant and Snow Geese frequently pass the winter, sometimes remaining until late March. A few Whistling Swans occasionally stop by in migration but are unusual in winter. The Snow Geese tend to stay in the salt marshes at Ragged Point near the north end of the Refuge, the Brant in Assateague Cove, a nearly enclosed harbor near the southern end. At times the birds fly over the Refuge or outlying waters as they move from one feeding ground to another.

From October through December and from March through April

the duck migrations are at their height. American Black Ducks are the most abundant; the Common Pintails and American Wigeons rank next. Other ducks on the Refuge marshes and ponds are Mallards, Gadwalls, Green-winged Teal, Blue-winged Teal, Northern Shovelers, Wood Ducks, Redheads, Ring-necked Ducks, Canvasbacks, Greater Scaups, Lesser Scaups, Ruddy Ducks, and Hooded Mergansers. Offshore there are great numbers of White-winged Scoters, Surf Scoters, and Black Scoters, together with fewer Common Goldeneyes, Buffleheads, Oldsquaws, and Red-breasted Mergansers. Although the duck populations decline during winter, thousands nonetheless winter on the Refuge itself and offshore.

In August and September and again in April the shorebird migrations are under way. In suitable places—the shores of Tom's Cove (the inner part of Assateague Cove), the ocean beach, the mud flats, and the banks of marsh creeks—the bird finder can compile a list that will include the Semipalmated Plover, Black-bellied Plover, Ruddy Turnstone, Common Snipe, Whimbrel, Spotted Sandpiper, Greater and Lesser Yellowlegs, Red Knot, Pectoral Sandpiper, Least Sandpiper, Dunlin, Semipalmated Sandpiper, Western Sandpiper, and Sanderling.

Among birds nesting on the Refuge in June and early July are Mallards, American Black Ducks, Clapper Rails, American Oystercatchers, Piping Plovers, Wilson's Plovers, Killdeers, Willets, Laughing Gulls, Gull-billed Terns, Forster's Terns, Common Terns, occasionally Little and Royal Terns, Forster's Terns, and Black Skimmers. Gray Catbirds, both Pine and Prairie Warblers, Common Yellowthroats, Yellow-breasted Chats, and Rufous-sided Towhees are common in forested areas and shrubby thickets.

When coming to the Refuge from the north, leave US 13 about 4 miles south of the Maryland-Virginia line on State 175 eastward. After about 6 miles the highway begins crossing the causeway and bridges to Chincoteague Island over tidal marshes, mud flats, oyster bars, and channels, excellent for waterbirds, including herons, egrets, and Glossy Ibises, and shorebirds. From time to time, stop the car off the pavement to scan this vast area. If there are shrubs and high grasses near the causeway, be alert for Seaside Sparrows. Boat-tailed Grackles will be conspicuous.

After crossing Chincoteague Channel, the highway enters the

town of Chincoteague on Chincoteague Island. In summer months here, take a side trip by turning immediately right on Main Street and proceeding out of town to the end at a turn-around at South Point. The road follows the marshy edges of Chincoteague Channel, where Clapper Rails and Willets abound; and there should be a few Black-crowned and Yellow-crowned Night Herons among a variety of wading birds. Inland from the turn-around is a sandy area where Wilson's Plovers begin nesting in June.

Backtrack on Main Street past the bridge for 7 blocks and turn right on Maddocks Boulevard, which leads to a bridge across Assateague Channel into the Refuge on Assateague Island. From the bridge the Refuge road goes through pine woods to headquarters and beyond toward the beach, passing Ponds A and F on the left and Tom's Cove on the right. On reaching the traffic circle back from the beach, take the road south between the beach and Tom's Cove. During early summer, look for Piping Plovers that nest in adjacent sandy areas well above high-tide line.

FRONT ROYAL
Shenandoah National Park

Seventy-five miles west of Washington, D.C., in the heart of the Blue Ridge Mountains, is **Shenandoah National Park** (331 square miles). An elongated strip running in a northeast-southwest direction for 75 miles, it embraces the highest and most scenic section of the northern Blue Ridge.

The Mountains form a distinct ridge, with the higher peaks reaching altitudes exceeding 4,000 feet. On the east side the incline is gradual for the most part, and there are many spur ridges with deep coves; on the west the rise from the floor of the Shenandoah Valley is sharp in numerous places, producing escarpments. Hawksbill Mountain, highest peak in the Park, has an elevation of 4,049 feet.

About 85 per cent of the Park, including the summits, is forested; the remainder is chiefly meadowland or old fields. Hardwoods—largely oaks, with scattered stands of maples, hickories, walnuts, black locusts, black gums, and basswoods—comprise the principal tree growth. There are, however, certain places where conifers such as pines, hemlocks, firs, and spruces prevail. Junipers are scattered at lower elevations. Azaleas, mountain laurels, rhododendrons, and other shrubs form a luxuriant undergrowth in many of the forests. When these are blooming in May and June, the mountainsides are a riot of color.

Spectacular Skyline Drive, a 105-mile parkway with long, sweeping curves, follows the ridge from north to south near the crest and commands majestic vistas here and there: on the east the tree-covered foothills of the Piedmont; on the west the Shenandoah Valley, with its patchwork of farms cut by winding streams. Along this highway many bird-finding areas are readily accessible. Three, especially recommended for birds in June and July, are the Elkwallow picnic area, Skyland area, and Big Meadows area.

Skyline Drive is accessible at four points: at Front Royal, on the north, where US 522, US 340, and State 55 converge; at Thornton Gap on US 211; at Swift Run Gap on US 33; and at Rockfish Gap, on the south, where I 64 and US 340 intersect southeast of Waynesboro. Both Skyland and Big Meadows are between Thornton Gap and Swift Run Gap.

The Elkwallow picnic area (2,500 feet), 24 miles south from Front Royal, supports mature white pines surrounded by second-growth forest—a year-round habitat for Wild Turkeys.

The Skyland area (3,680 feet), 41 miles south of Front Royal, is largely wooded and outstanding for birds characteristic of the Virginia high country. The Stony Man Nature Trail of easy grade leads for 1.0 mile to the summit of Stony Man (4,010 feet), from whose rocky brow there is an excellent opportunity to observe Northern Ravens. One pair usually nests on the perpendicular cliff just below the observation point. Along the Trail one should find Ruffed Grouse, Wood Thrushes, Veeries, Solitary Vireos, Black-and-white Warblers, Chestnut-sided Warblers, Canada Warblers, Rose-breasted Grosbeaks, and Slate-colored Juncos.

Ten miles farther south, in the central section of the Park, is the Big Meadows area (elevation 3,500 feet), where there are 200 acres of open fields with brushy spots; adjoining woods, mostly deciduous, containing trees in various stages of growth; and the Big Meadows Swamp, supporting shrubs and partly surrounded by an extensive stand of gray birches.

This area, accessible by the Swamp Nature Trail, is notable for a variety of birds not necessarily characteristic of the high country. In addition to several species of warblers, the common nesting species in the area include: Common Bobwhite, House Wren, Gray Catbird, Brown Thrasher, Eastern Meadowlark, Scarlet Tanager, American Goldfinch, Rufous-sided Towhee, Vesper Sparrow, Field Sparrow, Song Sparrow. The American Woodcock, also nesting here, performs its courtship flights in early evening from mid-March to mid-April.

Within the Park are lodges, restaurants, five campgrounds, and seven picnic grounds, plus 200 miles of foot trails, including a 95-mile section of the Appalachian Trail that parallels closely the Skyline Drive. The Park is open to visitors all year, but most of the lodges and restaurants are closed during winter and early spring months. From May to October, Park naturalists conduct nature walks and guided trips and give illustrated evening talks. Obtain a map and information concerning these activities at any of the entrances. Park headquarters (address: Luray, Virginia 22835) is midway between Luray and Thornton Gap on US 211.

LEXINGTON
Big Spring | Lime Kiln Bridge | Goshen Pass

The Lexington area, in the Valley of Virginia in the western part of the state, features rolling hills, steep ridges, isolated peaks, narrow valleys, and elevations varying between 800 and 4,000 feet. There are no large bodies of water; rivers are few, but there are many streams that are usually swift and rocky. Over half the country is forested, principally with second-growth hardwoods. Patches of pine stand on some of the lower slopes, and hemlock along the banks of shaded ravines; the remainder of the country is open farmland. Landbirds are abundant and diversified.

Big Spring, a pond of 3 acres bordered by sycamores, boxelders, and willows, has a marshy shore with cattail beds. Bewick's Wrens, Blue-gray Gnatcatchers, Warbling Vireos, Orchard Orioles, and Baltimore Orioles are summer residents in the immediate area. Barn Swallows nest in a barn at the south end; Cliff Swallows plaster their nests on a barn a quarter-mile north. To reach Big Spring, drive west from Lexington on US 60 for 7 miles, then right on State Secondary 631 for half a mile. The pond is on the left.

An area where one may find breeding birds characteristic of open, mixed hardwoods with thick undergrowth and grassy clearings is in the vicinity of the **Lime Kiln Bridge** over the Maury River. To reach the Bridge, drive west from US 11 in Lexington on State Secondary 631—at the north end of the Maury River Bridge—for 2 miles. Among the birds here are Wood Ducks, Pileated Woodpeckers, Acadian Flycatchers, Wood Thrushes, Black-and-white Warblers, Worm-eating Warblers, Northern Parula Warblers, Cerulean Warblers, Prairie Warblers, Louisiana Waterthrushes, Summer Tanagers, Blue Grosbeaks, and Indigo Buntings.

Goshen Pass, northwest of Lexington, is a winding gap 4 miles long, cut through the mountains by the Maury River and flanked on both sides by a heavy growth of hemlock and an undergrowth of rhododendron. State 39 west from Lexington passes through the gap. Near the entrance, about 15 miles from Lexington, Summer Tanagers, Blue Grosbeaks, and other low-country birds are summer residents; high up in the Pass are Black-throated Blue Warblers, Black-throated Green Warblers, Blackburnian Warblers, Scarlet Tanagers, and other mountain birds. The month of June, when the rhododendrons are blooming and the birds are in full song, is the ideal time to make the trip.

LYNCHBURG
Apple Orchard Mountain

Excellent for mountain birds is an area near **Apple Orchard Mountain** (4,275 feet) on the Blue Ridge Parkway. From Lynchburg, drive northwest on US 501 about 15 miles and turn left onto the Parkway. Follow it through Petit Gap and over Thunder Ridge (3,485 feet). Shortly after the observation tower on Apple Orchard Mountain comes into view, a woods road—a 1.0-mile trail to the summit—turns right from the Parkway. On the opposite side of the road and about 100 yards into the forest is a spring near which Veeries, Solitary Vireos, Black-throated Blue Warblers, Chestnut-sided Warblers, Canada Warblers, Rose-breasted Grosbeaks, and Slate-colored Juncos reside in late May, June, and early July.

MARION
White Top Mountain | Mt. Rogers

In southwestern Virginia, 5 miles north of the North Carolina border, rise **White Top Mountain** (5,520 feet) and **Mt. Rogers** (5,719 feet), the state's two highest peaks. The deciduous woods clothing their lower slopes are gradually succeeded by the dense growth of conifers that caps their summits. A paved road leading to the summit of White Top makes it possible to drive through this change of vegetation and see, at the same time, the corresponding change in birdlife. The best time to make the trip is between late May and early July, when the birds are in full song.

To reach White Top from I 81 at Marion, take State 16 east and south to Troutdale; turn right on State Secondary 603 and drive almost to Konnarock, 12 miles distant; turn left on State Secondary 600, which climbs to a high saddle, with a parking area, between the two mountains. The forests passed through during the ascent are of mixed hardwoods, with shrubby thickets along streams and near the roadsides. Among breeding birds are Ruffed Grouse, Whip-poor-wills, Acadian Flycatchers, Wood Thrushes, Black-and-white Warblers, Swainson's Warblers, Yellow-throated Warblers, Kentucky Warblers, Yellow-breasted Chats, Hooded Warblers, Scarlet Tanagers, and Rose-breasted Grosbeaks.

On the saddle is an extensive 'bald' with a thick turf of grass and dwarf heather, where birds such as (Prairie) Horned Larks and Vesper Sparrows are summer residents.

From the saddle the road bears right and proceeds to the summit of White Top. In the red spruces at the summit, expect to find the following breeding species of northern affinities: Saw-whet Owl, Red-breasted Nuthatch, Brown Creeper, Winter Wren, Swainson's Thrush, Golden-crowned Kinglet, Magnolia Warbler, Blackburnian Warbler, Canada Warbler, Purple Finch, Red Crossbill (irregular). Other species of northern affinities, although appearing on the summit, are apt to be more common in the mixed coniferous and deci-

NORTHERN RAVEN

duous forest directly below. Some of these species are the Yellow-bellied Sapsucker, Black-capped Chickadee, Hermit Thrush, Veery, Solitary Vireo, Black-throated Blue Warbler, Black-throated Green Warbler, and Slate-colored Junco.

A road-trail, closed to cars, bears left from the saddle to the summit of Mt. Rogers, 3 miles distant. The birds on the summit are identical with those on White Top, but the denser and more extensive coniferous growth, consisting of both red spruce and Fraser's fir, attracts a larger nesting population. The Northern Raven may be seen regularly from both mountains.

NORFOLK

Chesapeake Bay Bridge–Tunnel | Seashore State Park | Lynnhaven Bridge | Back Bay National Wildlife Refuge

During spring migration the bird finder using the **Chesapeake Bay Bridge–Tunnel** (US 13), which extends 17.5 miles over and under lower Chesapeake Bay from Norfolk to Cape Charles at the southern tip of Virginia's part of Delmarva Peninsula, should look for small north-bound landbirds on South Island. Of huge loose rocks, and barren of vegetation except for seaweed at the water's edge, this totally man-made bit of land, 6 miles out in the Bay from Norfolk, is at the terminus of the first bridge where the highway disappears into the tunnel under Thimble Shoal Channel. Landbirds moving north, confronted with a long flight over water, often hold up and wait here, feeding and resting for a day or more.

In April or early May, South Island sometimes abounds in a variety of passerines—flycatchers, wrens, thrashers, thrushes, kinglets, vireos, warblers, and many kinds of fringillids. Seaside Sparrows are often very numerous. These birds, many of which normally inhabit dense woods or thickets, perch in full view on guard rails around the parking area and dart in and out among the rock slabs below or hop about on the seaweed, foraging presumably for insects, spiders, and small crustaceans.

Seashore State Park (2,770 acres) lies between the Bridge-Tunnel and Cape Henry and stretches from Chesapeake Bay inland for 3 miles. In addition to a long sand beach and low sand dunes on the Bay, the area includes large plots of pines and hardwoods, many lakes and creeks, and several small cypress pools.

To reach the Park, take US 60 east from the Bridge-Tunnel terminus. Just before entering the Park, US 60 crosses **Lynnhaven Bridge.** A stop here may be worthwhile, especially at low tide, when shorebirds gather on the mud flats. Marsh Wrens and Seaside Sparrows nest in the marshes at the west end of the Bridge. The marshes on the island at the east end of the Bridge may have Clapper Rails.

Upon entering the Park (through which US 60 passes), the bird finder may see, in summer, Common Terns and Little Terns from the beach on the left of the road. During migration he should look for other terns—Forster's, Royal, Caspian, and Black—and, in winter and early spring, Northern Gannets offshore.

The wooded section, on the right of the highway, is best approached by the eastern entrance (nearest Cape Henry). Bear left past the cabins to the beginning of the trails. Although it is possible to drive down the main trail, it is best to walk so as to explore the surroundings of two cypress pools, only 100 yards from the entrance, for breeding Prothonotary and Northern Parula Warblers and broods of Wood Ducks in summer. A hundred yards beyond these pools, the first trail to the left leads to another cypress pool, where Black-crowned Night Herons nest and where there may be broods of Blue-winged Teal and Wood Ducks. When walking along the trails, watch for permanent-resident Pileated Woodpeckers in the surrounding hardwoods.

The **Back Bay National Wildlife Refuge** on the southeastern Virginia coast, southeast of Norfolk, comprises 4,589 acres of sand beach, dunes, and brackish marshes, together with about 4,500 acres of bay waters. Having relatively few nesting birds, the Refuge is of interest chiefly as a place for transient and wintering waterfowl and transient shorebirds.

The eastern half of the Refuge consists of a 4.5-mile barrier beach backed by a low ridge of dunes, which falls away westward to brackish marshes. Some of the dunes are partly grass-covered, and between them, in sheltered places, are patches of bayberry, myrtle, and other shrubs. The western half of the Refuge has insular brackish marshes separated by channels, coves, and bays. In the marshes are numerous ponds with fresh to brackish water.

Waterfowl begin arriving in numbers during October, and by December they have passed the peak of abundance. In spring the peak is reached in March, and by late April all have disappeared, save a few American Black Ducks that stay to nest. The winter population, though smaller than that in spring or fall, is nonetheless impressive in both numbers and species. A trip over the Refuge any day in December, January, or February will almost certainly yield the following: Whistling Swan, Canada Goose, Snow Goose, American Black Duck, American Wigeon, Common Pintail, Green-winged Teal, Redhead, Ring-necked Duck, Canvasback, Greater Scaup, Lesser Scaup, Common Goldeneye, Bufflehead, White-winged Scoter, Surf Scoter, Black Scoter, Ruddy Duck, Hooded Merganser, Red-breasted Merganser.

In winter the bird finder should be further rewarded by seeing Common Loons, Northern Gannets, Double-crested Cormorants, American Coots (usually abundant), Greater Black-backed Gulls, Herring Gulls, Ring-billed Gulls, Laughing Gulls, and (occasionally) Bonaparte's Gulls. On land he will probably find (Northern) Horned

Larks, Fish Crows, Water Pipits, Myrtle Warblers, Eastern Meadowlarks, Red-winged Blackbirds, American Goldfinches, Savannah Sparrows, American Tree Sparrows, White-throated Sparrows, Song Sparrows, and (occasionally) Snow Buntings.

To reach the Refuge, drive southeast on State 165 to Princess Ann; then east on State 149, State 615, and State 629 to the settlement of Sandbridge Beach; and south at a cafe for 1.25 miles to the end of the hard-surfaced road. The Refuge and subheadquarters are 4 miles farther south on the beach and cannot be approached with confidence except in a vehicle equipped with a 4-wheel drive. The bird finder planning to visit the Refuge is advised to communicate with the manager (P.O. Box 6128, Virginia Beach, Virginia 23456) well in advance, and arrange to be taken to the Refuge from Sandbridge Beach by Refuge personnel on one of their routine trips.

RICHMOND

Byrd Park | Curles Neck

On the west side of this capital city, Belt Boulevard (State 161) crosses 450-acre **Byrd Park** with three fairly good-sized lakes attractive to many transient and wintering waterfowl as well as to waterbirds. The middle one, Swan Lake, is the largest and always the most productive. Surrounding it are roads and paths, permitting easy access. In winter one may expect many American Wigeons, Ring-necked Ducks, American Coots, and Ring-billed Gulls. Feeding them on Sunday afternoons is a popular Richmond diversion; hence they have become quite tame.

Other birds of regular appearance in winter are Pied-billed Grebes, Mallards, American Black Ducks, Redheads, Lesser Scaups, Ruddy Ducks, Common Mergansers, and Herring Gulls. Waterfowl are usually present from the last week of October to the first week in April, with greater numbers after the beginning and just before the end of this period, when migrations are under way. In summer one or two pairs of American Coots remain to nest.

Curles Neck, a dairy farm of about 4,000 acres, lies in a wide meander of the James River 10 miles southeast of the city limits, on State 5. When one is coming from Richmond, the entrance, marked by two stone pillars, is on the right. Drive through the entrance and proceed through pine woods and farmland until the road makes a sharp turn left, 2.5 miles beyond State 5. A short distance to the right is the office, where one may obtain permission to look for birds. Here the road makes a sharp right turn and continues for another mile through farmland before it ends. Ahead, at the tip of the meander, lies a large wooded swamp. A dirt road to the left goes to the entrance of a marsh; a road to the right leads to the River.

Curles Neck is best known for its large wintering population of Canada Geese, often feeding in the fields along the roads. Look carefully for Snow Geese. In the marsh itself large numbers of Mallards, American Black Ducks, Common Pintails, and other ducks winter; fewer appear on the River. In summer the pine woods throng with Pine and Yellow-throated Warblers, and the wooded swamp echoes to the lively song of the Prothonotary Warbler. In late summer, Great Egrets line the River banks and are especially common in the marsh. Gulls and terns of at least six species also appear in late summer and fall. Large groups of Black Vultures reside in the area the year round.

ROANOKE
Murray's Pond

Spring-fed **Murray's Pond,** sometimes called Hollins Pond, consisting of about 9 acres of open water and marsh, is bordered by a hard-surfaced road. To reach it, drive north from Roanoke on US 11 for 7 miles, making a right turn at a sign marked 'Hollins Station.' The Pond is 0.2 mile from the highway on a secondary road. Transient waterfowl are present almost daily in spring and fall, and they sometimes linger into December. A few transient shorebirds appear in late April and May. The Least Bittern has been known to nest in the marsh, and a few pairs of Rough-winged Swallows nest regularly in a nearby abandoned rock quarry.

SALTVILLE
Saltville Ponds

An outstanding place for waterfowl and shorebirds in southwestern Virginia is the group of small, shallow **Saltville Ponds,** at the southwestern edge of Saltville, off State 91. Among many kinds of ducks that appear here in spring and fall are Gadwalls, American Wigeons, Green-winged Teal, Northern Shovelers, Lesser Scaups, Buffleheads, Hooded Mergansers, and Red-breasted Mergansers. The transient shorebirds that feed on the extensive mud flats around some of the Ponds are probably the chief attraction. Among species regularly recorded are the Semipalmated Plover, Common Snipe, Solitary Sandpiper, Greater Yellowlegs, Lesser Yellowlegs, Pectoral Sandpiper, Least Sandpiper, and Semipalmated Sandpiper.

SUFFOLK
Great Dismal Swamp | Dismal Swamp National Wildlife Refuge

Lying across the Virginia–North Carolina line in southeastern Virginia, only 25 miles from the sea, is the **Great Dismal Swamp,** an area 35 miles long from north to south and 13 miles wide. The Swamp begins about 5 miles south of Norfolk; more than half extends into North Carolina.

Contrary to notions suggested by its name, the Dismal is no grim morass, no stronghold of harmful creatures, but rather a romantic wilderness of tall trees and reflecting waters, a trackless haunt of fascinating wildlife. In the 1760s George Washington made at least six visits to the Dismal, which in correspondence he called 'a glorious paradise.' No visiting bird finder can fail to share his enthusiasm.

Much of the Dismal is true swamp, for the water is shallow, and a tree growth—principally bald cypresses, gums, tupelos, maples, and oaks—is dominant; but there are areas where the tree growth stands on fairly dry land, and still others where the water is present and the tree growth is replaced by 'lights'—strangely open, reedy, vine-matted jungles. White cedars, formerly an abundant forest type, now exist sparingly. As in so many true swamps, the water is color-stained by vegetation: deep reds and purples by cypresses and tupelos, orange-yellows by maples and cedars.

In the very heart of the Dismal rests oval-shaped Lake Drummond, the famous 'Lake of the Dismal Swamp.' Roughly 3.75 by 3 miles in diameter, it is surrounded by giant cypresses supporting beards of Spanish moss. Old stumps and even live cypresses stand in the water far out from shore. Although the Lake is only 6 feet or so deep, the sandy bottom is barely discernible through the wine-colored water. Somewhat sterile as it is, the Lake is mainly a resting place rather than a feeding area for waterbirds and waterfowl.

Long ago, to facilitate drainage and to provide a means of carrying out lumber, canals or 'ditches' were cut from Lake Drummond to the Dismal's edge, and locks were constructed to control the water level. The Washington Ditch, dug under the direction of the great surveyor himself, passes 5.5 miles northwest from the Lake and was designed to get timber from the Dismal to the Nansemond River near Suffolk. The main waterway is the Dismal Swamp Canal on the east side. This passes across the Swamp east of Lake Drummond from the village of Deep Creek on the northeast side to the village of South Mills, North Carolina, on the southeast side, a distance of 22 miles. As the Canal proceeds south, it is joined by the Feeder Ditch from Lake Drummond. US 17, south from Portsmouth, parallels the Dismal Swamp Canal all the way from Deep Creek to South Mills.

Although the Dismal has been altered through the years by the digging of canals, removal of high-grade timber, and disastrous fires, it still remains a wilderness abounding with wildlife. Much of the Virginia portion—altogether 49,097 acres—now comprises the **Dismal Swamp National Wildlife Refuge.**

A trip today along the canals—straight, dark waterways beneath cypress-tupelo arches—into beautiful Lake Drummond is a rewarding experience. The best time to make the trip into the Dismal is late April and in May before the mosquitoes and ticks become obnoxious. A good entry where boats may be rented is Arbuckle's Landing at the junction of the Dismal Swamp and the Feeder Ditch along US 17, 10 miles south of Deep Creek. The Feeder Ditch, the Waste Weir—a clearing on high ground about 0.25 mile east of Lake Drummond—and Lake Drummond will produce characteristic birds. After crossing the Canal by boat at Arbuckle's Landing, it is possible to walk along the rim of the Feeder Ditch to the locks of the Waste Weir and beyond into the Swamp.

Another good entry that will produce characteristic birds is along Jericho Ditch, which starts from near Suffolk and leads into the northeastern part of the Swamp to Lake Drummond. This is paralleled by a road called Jericho Ditch Lane, which is accessible to cars, though permission to drive it must first be obtained at Refuge headquarters (200 North Main Street in Suffolk). To reach the Lane from Suffolk, take Washington Street east to White Marsh Road, then south for 0.75 mile; after passing under the power line, turn east into the Swamp.

Among breeding birds in the Dismal, Red-eyed Vireos and Prothonotary Warblers are extraordinarily common, Ovenbirds and Hooded Warblers nearly so. (Wayne's) Black-throated Green Warblers nest high up in stands of tupelos, and Swainson's Warblers low in deciduous thickets. Barred Owls make themselves evident in daytime by their cackling and laughing sounds. Other birds breeding regularly include Great Blue Herons, Green Herons, Little Blue Herons, Black-crowned Night Herons, Turkey Vultures, Black Vultures, Red-shouldered Hawks, Chimney Swifts (apparently nesting in hollow tree stubs), Pileated Woodpeckers, Acadian Flycatchers, both American and Fish Crows, Northern Parula Warblers (nesting in Spanish moss), and Prairie Warblers (mainly in the lights). Wood Ducks, often nesting in old Pileated Woodpecker holes, appear with their broods along the ditches or near the shore of Lake Drummond.

Large numbers of Eastern Phoebes, Gray Catbirds, Brown Thrashers, American Robins, Rufous-sided Towhees, Fox Sparrows, and other passerines pass the winter in the Dismal, taking advantage of the mild climate, protection afforded by the dense forest cover, and abundance of food. A truly spectacular sight in winter is the roosting of blackbirds near the east side of the Dismal traversed by the Carapeake Ditch on the Virginia-North Carolina line. As evening approaches, if one parks his car on US 17 where it crosses the state line, he will see clouds of blackbirds sweeping in from surrounding country. Except for a few Brown-headed Cowbirds, Rusty Blackbirds, and European Starlings, Common Grackles and Red-

HOODED WARBLER

winged Blackbirds make up the gathering aggregation. The roost has been used since at least the turn of this century. In December 1961 the number of birds was estimated to be 31 million!

Reference

The Great Dismal Swamp. By Brooke Meanley. Audubon Naturalist Society of the Central Atlantic States, 1973. Available from the Society's Bookshop, 1621 Wisconsin Avenue N.W., Washington, D.C. 20007.

WACHAPREAGUE
Parramore Island | Cedar Island

Some Norfolk bird watchers claim that the islands off this village on Delmarva Peninsula, particularly the two named **Parramore** and **Cedar,** rival Cobb Island (*see under* **Cheriton**) for sea birds. And the comfortable inn at Wachapreague is often a gathering place for bird enthusiasts. One can easily charter a boat here for the outer islands with their beaches and grass-studded dunes backed by salt marshes. Between Parramore Island and the mainland are smaller islands with nesting Laughing Gulls and Forster's Terns. These usually have young by the last of June or the first of July.

WILLIAMSBURG
Battlefield Area of the Colonial National Historical Park

The **Battlefield Area of the Colonial National Historical Park** (7,233 acres) extends for approximately 2.75 miles along the south side of the York River, and almost the same distance inland. About three-quarters of the Area embraces a mixed deciduous and coniferous forest; about 700 acres take in the village of Yorktown and open farmlands, and about 400 acres comprise marsh and wooded swamp.

Bird finding is good at all seasons, although at no time are there great concentrations of any particular species. Among landbirds presumably nesting in the area are Red-bellied Woodpeckers, Rough-winged Swallows, Carolina Chickadees, Tufted Titmice, Brown-headed Nuthatches, Carolina Wrens, Northern Mockingbirds, Blue-gray Gnatcatchers, White-eyed Vireos, Yellow-throated Warblers, Pine Warblers, Prairie Warblers, Hooded Warblers, Orchard Orioles, Summer Tanagers, and Rufous-sided Towhees. Common winter visitants include Golden-crowned Kinglets, Myrtle Warblers, Slate-colored Juncos, and White-throated Sparrows. From October to April the following usually appear on the adjacent York River: Horned Grebes, Pied-billed Grebes, Gadwalls, American Wigeons, Common Pintails, Lesser Scaups, Common Goldeneyes, Buffleheads, and Ruddy Ducks.

Colonial Parkway, which goes southeast from Williamsburg to Yorktown, 13 miles distant, passes through part of the Area between the York River on one side and a fine woods on the other. Bird finding is invariably rewarding near the highway.

West Virginia

MAURICE BROOKS

CANADA WARBLER

George Washington, in his journal for 1770, recorded a great flight of waterfowl along the Ohio and Great Kanawha Rivers in what is now West Virginia. Systematic ornithological observation in the state began, however, with John James Audubon and Alexander Wilson. Audubon writes of collecting a number of Carolina Parakeets in 1831 near the mouth of the Great Kanawha River, which is now Point Pleasant, West Virginia. An earlier record of this species, but published the same year, was made by Wilson, in Wood County.

Despite the fact that West Virginia ranks forty-first among the forty-eight contiguous states in point of area, it occupies a remarkable geographic position. The tip of its northern panhandle, largely within the Shenandoah Valley, extends to within 50 miles of Washington, D.C. Its southern extension is farther south than Richmond, and at Kenova, where the Ohio River leaves the state, the longitude is almost exactly that of Port Huron, Michigan. This geographical diversity, coupled with the range in altitude, contributes a fascinating mixture of eastern/western and northern/southern plants and animals. There is a popular toast that says it all:

> Here's to West Virginia: the most northern of the southern states, the most southern of the northern states, the most western of the eastern states, and the most eastern of the western states; but here's to West Virginia, a mighty good state for the shape she's in!

Thus, in West Virginia we sometimes find Sedge Wrens and Bewick's Wrens nesting at the edges of the spruce belt; Yellow-breasted Chats and Carolina Wrens appearing above 4,000 feet elevation; Swainson's Warblers breeding in dry, open, upland woods; and Northern Parula Warblers building nests in hardwood trees, without benefit of Spanish moss or *Usnea.* West Virginia is also at the extremity of a number of eastern North American bird ranges. The state includes the southernmost breeding areas for Hermit and Swainson's Thrushes, Nashville and Mourning Warblers, Northern Waterthrush, Purple Finch, and Savannah Sparrow; it may also include the southern breeding limits of the Northern Goshawk and Common Snipe, both of which have been observed in summer. Fewer species are at the northern limits of their breeding range in the state, but the list does include Swainson's Warbler.

A number of species and well-marked races reach or approach in West Virginia their northern breeding limits in the Appalachian region, although they may extend farther northward at lower elevations to east and west. Hence they may exhibit locally the aberrations in behavior that characterize so many species at their breeding limits. The state probably does not include the eastern or western breeding limits of any species, but such birds as the Dickcissel and Lark Sparrow are near their eastern limits here. Bird finders from other regions will find many species nesting in unusual situations or behaving according to unfamiliar patterns, and in this is to be found much of the state's ornithological fascination.

The state lies wholly within the unglaciated Appalachian Mountain region, and most of its boundaries are determined by mountain ridges or winding rivers. Elevations range from 242 feet above sea level at Harpers Ferry to 4,860 feet at the summit of Spruce Knob, highest point in the state, with a mean of 1,500 feet—higher than that of any other state east of the Mississippi.

The Allegheny Backbone, in places a secondary continental divide separating waters that drain to the Atlantic seaboard from those that reach the Gulf of Mexico, bisects West Virginia unequally from northeast to southwest. East of the divide are the Potomac and the James River systems, and to the west the Ohio and its tributaries carry away an abundant rainfall and snowfall. The latter waterways drain roughly four-fifths of the state's area—approximately 25,000 square miles—and the flora and fauna of the state partake more of the Mississippi Valley than of the Atlantic seaboard. The area has no natural lakes, but a series of large artificial impoundments and the intricate patterns of the rivers do something to offset this lack.

Originally the whole of West Virginia was forested. As might be expected in a region of such broken terrain, forest types tended to become badly mixed, but there was, and still is, a fairly definite pattern. In the broader river valleys, forests were of the central-hardwoods type: largely oak, hickory, and chestnut, with elm, sycamore, and black walnut as abundant secondary species. In rich coves lying back from the streams there were areas of tulip tree, black walnut, white oak, white ash, and hemlock. Breeding birds of the central-hardwoods forest are usually those whose centers of abundance are southward. Species typical of this forest type and of the farmlands (fields, wet meadows, brushy lands, orchards, and dooryards), which are interspersed, are the following:

DECIDUOUS FOREST

Turkey Vulture
Black Vulture
Sharp-shinned Hawk
Red-shouldered Hawk
Broad-winged Hawk
Ruffed Grouse
Yellow-billed Cuckoo
Common Screech Owl
Barred Owl
Whip-poor-will
Yellow-shafted Flicker
Pileated Woodpecker
Red-bellied Woodpecker
Hairy Woodpecker
Downy Woodpecker
Great Crested Flycatcher

Acadian Flycatcher
Eastern Pewee
Blue Jay
Carolina Chickadee
Tufted Titmouse
White-breasted Nuthatch
Wood Thrush
Blue-gray Gnatcatcher
Red-eyed Vireo
Prothonotary Warbler
Swainson's Warbler
Worm-eating Warbler
Cerulean Warbler
Yellow-throated Warbler
Ovenbird
Louisiana Waterthrush
Kentucky Warbler
Hooded Warbler
American Redstart
Baltimore Oriole
Scarlet Tanager
Summer Tanager

FARMLANDS

Common Bobwhite
Mourning Dove
Red-headed Woodpecker
Eastern Kingbird
Eastern Phoebe
(Prairie) Horned Lark
Barn Swallow
House Wren
Bewick's Wren
Carolina Wren
Northern Mockingbird
Gray Catbird
Brown Thrasher
Eastern Bluebird
Loggerhead Shrike
White-eyed Vireo
Yellow Warbler
Prairie Warbler
Common Yellowthroat
Yellow-breasted Chat
Eastern Meadowlark
Orchard Oriole
Common Grackle
Northern Cardinal
Blue Grosbeak
Indigo Bunting
American Goldfinch
Rufous-sided Towhee
Grasshopper Sparrow
Henslow's Sparrow
Vesper Sparrow
Bachman's Sparrow
Chipping Sparrow
Field Sparrow
Song Sparrow

Perhaps the most extensive forest of the state was of a type known as 'Appalachian hardwoods.' In this there are no clear dominants, but rather a mixture of twenty or twenty-five common species, among them oaks, hickories, maples, birches, chestnut, ash, basswood, cherry, black locust, buckeye, magnolias, tulip tree, walnut, beech, and others. Higher up the mountain slopes there were almost pure stands of birch-beech-maple forest. Just west of the main Allegheny crest was an extensive belt of white pine. The Appalachian hardwoods region is essentially an area of transition. Plants and animals characteristic of the lower river valleys meet and mingle with species of more northern associations.

Most of the birds listed above for the central hardwoods appear at least locally, and they are joined by such species as the Yellow-throated Vireo, Black-and-white Warbler, Golden-winged Warbler, Black-throated Green Warbler, Least Flycatcher, and Rose-breasted Grosbeak. There is a nice gradient in bird populations. At the lower elevations the breeding species are predominantly of southern association, the more northern birds being restricted to favorable local niches. At higher elevations, southern species drop out one by one, and the population becomes more boreal, until, at the upper reaches of the forest type, species from the valleys are restricted to local areas.

Near the higher mountain summits, hardwoods and conifers, principally yellow birch, red maple, hemlock, and red spruce, became mixed. Atop the highest ridges were pure stands of red spruce, and in a few mountain bogs this species became mixed with balsam-fir. On the drier eastern slopes, oak-pine, with some red cedar was the dominant forest type. Subject to changes that inevitably follow lumbering and agriculture, coupled with the almost total disappearance of the American chestnut, these same forest types persist. The spruce-balsam forest of West Virginia is now an enclave, an island of boreal forest separated from its kind both to the north and to the south. Originally the state's spruce belt included some 700,000 acres, and there remain some spruce-clad summits that are unbroken for 30 to 40 miles. Characteristic species are listed below. Species recorded throughout the year, though without actual nesting data, are marked with an asterisk.

* Golden Eagle
Saw-whet Owl
Yellow-bellied Sapsucker
Olive-sided Flycatcher
Northern Raven
Black-capped Chickadee
Red-breasted Nuthatch
Brown Creeper
Winter Wren
Hermit Thrush

Swainson's Thrush
Golden-crowned Kinglet
Solitary Vireo
Nashville Warbler
Magnolia Warbler
Black-throated Blue Warbler
Blackburnian Warbler
Northern Waterthrush
Mourning Warbler
Canada Warbler
Purple Finch
* Pine Siskin
* Red Crossbill
Slate-colored Junco

West Virginia lies between two of the major North American flyways—the Atlantic and the Mississippi—and thus lacks many of the spectacular migration phenomena that characterize territories to both east and west—great flights of waterfowl and waves of warblers, for example. Nonetheless there is much of interest. The northern portion of the state is in the path of a heavy autumnal waterfowl flight that crosses from Lake Erie to Chesapeake Bay and has traditionally passed over this lakeless area. Since the construction of artificial impoundments such as Cheat Lake (*see under* **Morgantown**) and Grafton Reservoir, however, waterfowl sometimes alight in astonishing numbers and variety, as, for example, when a flight meets an incoming easterly storm at the crest of the Alleghenies. At such times there are flocks of several hundred Common Loons and Pied-billed Grebes, even on small ponds, and flocks of ducks containing as many as twenty species: all three eastern species of scoters, the Redhead, Canvasback, Oldsquaw, and others. Meteorological conditions also produce occasional concentrations of shorebirds, which alight in flooded meadows during September storms.

During both spring and fall migrations there are heavy crossover movements of certain species through the West Virginia Alleghenies. Bonaparte's Gulls are regularly seen, sometimes in large numbers, and often many miles from any large body of water. Palm Warblers appear in very large numbers in October. There are heavy flights of White-crowned Sparrows in both spring and fall, at even the highest mountain crests. Autumn brings large flights of hawks, which may be observed advantageously from many of the higher Allegheny crests. Favorite spots for such observations are at the Allegheny Backbone, along US 219 and at Roaring Plains, reached by Forest Service road from Petersburg on US 220. During the last two weeks of September, Broad-winged Hawks ordinarily make up the bulk of such flights, but later in the season there are often heavy movements of Sharp-shinned, Red-tailed, Red-shouldered, and other hawks.

In winter the study of birds in West Virginia is not ordinarily highly productive. Most of the species leave the higher crests, moving southward or to lower altitudes. Save in the lower river valleys, the bird finder will have to work very hard to make a list of twenty-five to thirty species on any winter day. Many persons imagine that high mountain areas in the state should be excellent places to look for northern finches and other unusual winter visitants. Actually, most of the irregular winter visitants, casual or strays, that have been recorded have been found in the valleys. It is pleasant to visit the ice- and snow-clad spruces in the winter, but the ornithological rewards are likely to be few.

Reference

The List of West Virginia Birds. By George A. Hall. Reprinted from *The Redstart,* vol. 38, no. 2 (1971). Available from the Brooks Bird Club, 707 Warwood Avenue, Wheeling, W. Va. 26003.

CHARLESTON
Kanawha River Valley

Charleston, capital city of the state, lies in the **Kanawha River Valley,** a convenient place for observing birdlife characteristic of the warmer river valleys, notably the large numbers of Swainson's Warblers that nest in the vicinity.

To find the breeding grounds of the Swainson's Warbler, follow US 60 and US 119 eastward from the city to the Kanawha City Bridge. Cross the Bridge and continue south on US 119. A half-mile beyond the Bridge a series of cross streets leads to the right toward a group of low, wooded hills. Dirt roads and trails cross a railroad 0.25 mile off US 119 and lead into ravines between the hills. Follow one of these cross streets, leave the car at the end of the pavement, and explore these ravines and hills around them on foot.

Swainson's Warblers nest in most of the ravines, and, during breeding season, will be found most easily by listening for their loud ringing songs. In this locality they build their rather bulky nests in tangles of grapevines, spicebush, and other low vegetation. One observer found eleven such nests in a single season; many species listed above as common in river valleys will be found in Swainson's Warbler territory: Acadian Flycatchers, Carolina Wrens, Blue-gray Gnatcatchers, White-eyed Vireos, Worm-eating, Kentucky, and Hooded Warblers, Summer Tanagers, and others. In pine stands near the ridge tops there is a chance of finding Yellow-throated Warblers.

DAVIS
Canaan Valley

Davis, on the Blackwater River just above spectacular Blackwater Falls, is a central point from which to visit the extensive mountain sphagnum and cranberry bogs in **Canaan Valley** (pronounced *ka-náne* to rhyme with *inane*).

To reach Canaan Valley, follow State 32 southeastward from Davis for 4 miles to the top of Canaan Mountain, at 3,700 feet elevation. Spread below is the 25,000-acre Valley, almost completely surrounded by mountains. Destructive lumbering and fire have removed the magnificent stands of spruce and balsam-fir that originally covered the Valley, but it is now regenerating in patches. Much of the south end of the Valley has been drained and is now crop and grazing land, but the north end is overgrown with shrubs, bracken, and scattered patches of fir, spruce, and quaking aspen. There are still extensive areas of sphagnum and cranberry.

Brown Creepers, Winter Wrens, Hermit and Swainson's Thrushes, Veeries, Golden-crowned Kinglets, Magnolia, Black-throated Blue, Blackburnian, Mourning, and Canada Warblers, Purple Finches, and Slate-colored Juncos are locally abundant. Along the streams Traill's Flycatchers and Swamp Sparrows breed. Northern Ravens are numerous in the Valley and on surrounding mountains. On timbered slopes above the Valley, Wild Turkeys and Pileated Woodpeckers range. Although no nest has been found, juvenile Common Snipes have been sighted in the Valley.

State 32 crosses the Valley, and a network of dirt roads and trails permits almost unlimited exploration. The bird finder should follow these roads and head for the nearest timber. One place is about as likely to yield results as another, although the largest populations are often at the Valley margins, where the mountain slopes begin.

In the development stage are plans for an impoundment that will create a lake of about 7,800 acres in the northern, lower end of the Canaan Valley. From this lake, water will be pumped to a smaller and higher impoundment on Cabin Mountain overlooking Canaan. Gravity flow will turn turbines generating electric power. This construction, flooding all the lower valley, will radically alter the biota of Canaan Valley, inundating most of the breeding grounds for American Woodcock and Common Snipe. Plans include the management of the impoundment areas by the Division of Wildlife Resources, West Virginia Department of Natural Resources. There is hope that the project will provide a breeding ground for waterfowl and waterbirds.

ELKINS
Cheat Mountains

Take US 250 south from Elkins. Five miles beyond Huttonsville, where US 250 nears the Riffle School, the road begins to climb over the **Cheat Mountains.** The 15-mile drive between the school and the village of Durbin will give a good idea of the Cheat ranges. On the Greenbrier River, just east of the main Cheat Mountain ranges and west of the Allegheny Backbone, at the village of Bartow, is a modern motel and restaurant, a convenient headquarters for visits to the Cheat Mountain area.

The Cheat ranges include West Virginia's finest mountains. There are twin ranges, with an elevated plateau between them, through which Shavers Fork of the Cheat River flows. For 30 miles the ranges are almost entirely above 4,000 feet, with points above 4,800 feet. Most summits are clothed with a dense second-growth forest, with a few stands of virgin spruce. For 20 miles south of US 250, there is not a road or a trail that crosses or follows the ridges. This is a great wilderness area, of infinite interest to the naturalist. With the possible exception of the Nashville Warbler, every bird species that breeds in the West Virginia spruce belt is present. One can see or hear most of them from the highway, but there are rewarding side roads and trails to be followed.

From the foot of the mountain on the west, at the Riffle School, to its summit, twenty-two species of warblers nest:

Black-and-white Warbler	Prairie Warbler
Worm-eating Warbler	Ovenbird
Golden-winged Warbler	Northern Waterthrush
Northern Parula Warbler	Louisiana Waterthrush
Yellow Warbler	Kentucky Warbler
Magnolia Warbler	Mourning Warbler
Black-throated Blue Warbler	Common Yellowthroat
Black-throated Green Warbler	Yellow-breasted Chat
Cerulean Warbler	Hooded Warbler
Blackburnian Warbler	Canada Warbler
Chestnut-sided Warbler	American Redstart

The bird finder will begin his climb by car in forest that is predominantly oak-hickory, with birds of the lowland species; he will ascend through rich Appalachian hardwood forest where Veeries, Chestnut-sided Warblers, Rose-breasted Grosbeaks, and other more northern birds reside; then he will pass a zone of almost pure birch-beech-maple forest, with Slate-colored Juncos and many northern warblers. Where the highway reaches the mountain summit, the bird finder will be in a mixed forest of yellow birch, red maple, hemlock, and spruce, with rhododendron, mountain laurel, and other heaths, as well as dense growths of viburnums, deciduous holly, mountain ash, and other fruiting shrubs with typical spruce-belt birds.

Two miles eastward from the first summit is a side trail to the right. At a bridge across Red Run, 200 yards up this road, are dense stands of mountain long-stemmed holly, found on only a few Appalachian peaks. In October these plants may be loaded with showy cherry-red fruit, much loved by birds. Above the bridge is a beaver dam, and just beyond is a fringe of tall spruces, a favorite area of Winter Wrens and Swainson's Thrushes. US 250 continues eastward from this side road, through a spruce forest and along Shavers Fork of the Cheat, which it crosses just before the village of Cheat Bridge. Some of the best bird finding on the mountain is along the lumber railroad grade to the right, upstream, from the highway. Here Magnolia, Black-throated Blue, Mourning, and Canada Warblers are abundant; Traill's Flycatchers and Swamp Sparrows nest along the River; and Purple Finches sing in the spruces.

Eastward beyond Cheat Bridge, US 250 passes, on the left, a dense tangle of balsam and spruce known locally as the 'Blister-pine Swamp,' where Yellow-bellied Sapsuckers, Winter Wrens, Hermit and Swainson's Thrushes, and Northern Waterthrushes, as well as many other warblers, nest. One mile from Cheat Bridge is a forest opening in which American Woodcock nest, and from which they begin their 'sky-dancing' on April and May evenings just as the thrushes are closing their chorus.

Two miles east of Cheat Bridge, near the Randolph-Pocahontas county line, is a forest road to the left marked 'Gaudineer Recreation Area, 2 miles.' This is the most rewarding side trip in the Cheat Mountains. One can easily drive to the summit, but he will see more birds if he walks. The road to Gaudineer skirts the head of Blister-pine Swamp through a stand of second-growth beech. Presently it enters a forest composed for the most part of yellow birch, under which alder-leaved viburnum, mountain holly, and red-fruited elder grow. The ground is carpeted by club mosses and ferns, and there is a fine show of spring wildflowers, among which painted trillium and northern Clinton's lily are prominent.

At 1.5 miles from US 250 the forest road passes through a stand of giant spruces, where Red-breasted Nuthatches, Brown Creepers, and Golden-crowned Kinglets nest. Just beyond this spruce stand is a forest-access road to the right, which leads, after 1.25 miles, to one of the few stands of virgin spruce on the Mountain. A recent census here revealed 482 pairs of breeding birds per 100 acres—highest ever counted in a virgin-spruce forest.

The road to Gaudineer summit inclines to the left, through a wonderfully dense stand of young spruces. At the top is a parking area at 4,445 feet elevation. Along the road and the trails, mountain laurel blooms in dense masses in late June.

Pine Siskins and Red Crossbills have been observed more

frequently at Gaudineer Knob than at any other point in the Cheat ranges. Chances of finding them are best in mid-June, when the spruces are in blossom. Crossbills feast on the waxy carpels of these blossoms. During one June there were hundreds of White-winged Crossbills in this area. Winter Wrens and Purple Finches sing from the spruces beneath the tower. The mountain's outstanding bird feature, however, is its evening chorus of Hermit and Swainson's Thrushes.

The Gaudineer road is as good a place as any in the mountains for Wild Turkeys. It is the type locality for the Appalachian Ruffed Grouse, and nearby are the type localities for the big, gray West Virginia flying squirrel and the little gold-flecked Netting's salamander. Snowshoe hares feed along the trails, and black bears destroy the Forest Service signs.

Gaudineer Knob has long been topped by a United States Forest Service lookout tower, a famous one for bird students. Along with most other such towers in eastern National Forests, Gaudineer Lookout is being phased out. It is cheaper and faster to use planes and helicopters to scout for fires, and the few towers left are maintained for scenic purposes.

In earlier years a major attraction of the Cheat Mountains area was breeding Olive-sided Flycatchers, which also resided regularly in Canaan Valley (*see under* **Davis**), Dolly Sods Area (*see under* **Petersburg**), and Cranberry Glades (*see under* **Marlinton**). Apparently these flycatchers have completely disappeared in summer in the West Virginia mountains; at least there have been no reports in recent years. So far as is known, this is the only bird of boreal association that has been lost as a nesting species.

MARLINTON
Cranberry Glades

Marlinton, on the Greenbrier River in a transition area where Bewick's Wrens, Blue-gray Gnatcatchers, and White-eyed Vireos meet Least Flycatchers, Solitary Vireos, and Rose-breasted Grosbeaks, is a fine point of departure for a visit to **Cranberry Glades.** The route to Cranberry Glades leads southward along US 219 to the village of

Mill Point, where State 39 goes right up Stamping Creek to the summit of Cranberry Mountain. There, on the left side for travelers going west, the United States Forest Service maintains an information center from May to October, with a theater for film presentations, exhibits, and, on its staff, knowledgeable naturalists who conduct scheduled tours of the Cranberry Glades.

Directly opposite the information center is the southern terminus of the West Virginia Scenic Highway (State 150), about 15 miles of which are completed. This highway leads through scenic wild country, with fine blooming of rhododendron, mountain laurel, and other heaths, and with autumn displays of native mountain ash, and passes through the designated Cranberry Back Country, a sanctuary for West Virginia's state mammal, the black bear.

A half-mile west of the information center, on State 39, a road bears right to the entrance of Cranberry Glades, where there is a parking area and the beginning of a boardwalk through two of the Glades. To give some protection to the flora and fauna of the Glades, the Forest Service has constructed the half-mile circular boardwalk leading through Round and Flag Glades, past masses of cranberries growing over the peat mosses, through borders of swampy woods, and through orchid areas—at their best in late June and early July. Trailside markers call attention to features of particular interest. The other major glades—Big and Long Glades—are closed to the public, although serious investigators may secure permission to enter from the District Ranger, Gauley Ranger District, United States Forest Service, Richwood (West Virginia 26261), a town on State 39 about 36 miles west of Mill Point.

Cranberry Glades is a high mountain bog (3,300 feet) of 600 acres, surrounded by much higher mountains. The Glades proper contain four major openings, true northern muskegs. Over a thick carpet of mosses grow masses of cranberries, with many northern herbs and shrubs. In early July there are orchids of a half-dozen species. The Glade openings are surrounded by dense thickets of alder, spruce, aspen, and rhododendron. Cranberry Glades is the known southern breeding limit in eastern United States for the Hermit and Swainson's Thrushes, Northern Waterthrush, Mourning Warbler, and Purple Finch. It has most of the nesting birds of the spruce belt and, in addition, such species as the Blue-gray Gnatcatcher, White-eyed

YELLOW-BREASTED CHAT

Vireo, Worm-eating Warbler, Louisiana Waterthrush, and Yellow-breasted Chat.

MARTINSBURG
Federal Fish Hatchery at Leetown | Altona Marshes

Just as Charleston is a representative river-valley area west of the Allegheny crest, so Martinsburg, in the heart of the Shenandoah Valley, serves eastern West Virginia. To the north is the Potomac River and, to the east, the Shenandoah. The town is surrounded by orchards, but there are wooded ridges close by.

The region is ornithologically famous as the collecting locality of the only known specimens of 'Sutton's Warbler.' It is virtually impossible to give any directions for searching for this bird—whether a valid species or race or merely a chance hybrid between Yellow-throated and Northern Parula Warblers. The best hint is to investigate thoroughly all situations in which Northern Parula Warblers breed. Attention was called to the collected male because it was singing a *double* Parula song, as were other birds reported twice from the general region of its discovery and from a few other scattered localities from Virginia to Florida. The two known specimens, a male and a female, both in breeding condition, were collected on different days and 16 miles apart. They were in wholly different habitats, the male on a dry oak-pine ridge, and the female in dense river-bottom forest.

The theory that Sutton's Warbler is a hybrid between the Northern Parula and Yellow-throated Warblers has been strengthened in recent years by the discovery of sparse and local Yellow-throated Warblers in the region. At the time of the discovery of Sutton's, Yellow-throated Warblers had not been recorded from this section of West Virginia.

Even if no other Sutton's Warblers are found, there is much of ornithological interest in the region. Blue Grosbeaks occasionally nest near the brushy margins of apple orchards. Prothonotary Warblers have been found in the river-bottom forest across the Shenandoah River from Millville, near Harpers Ferry. Bachman's Sparrows sing in the cedar and hawthorn-dotted fields. To reach Millville, drive southeast from Martinsburg on State 9, then east on US 340 for about 5 miles, and right (south) for about a mile.

The **Federal Fish Hatchery at Leetown** has extensive ponds that afford one of the few good breeding places in the state for Pied-billed Grebes, ducks, rails, Common Gallinules, and other marsh-dwelling species. In late summer this is one of the state's best areas for shorebirds. From Martinsburg, drive southeast on State 9 for about 7 miles to Kearneysville, then right on State 48 for 4 miles.

Near Charles Town (not to be confused with Charleston), 15 miles southeast of Martinsburg on State 9, are the **Altona Marshes,** an extensive wet area quite unusual in the Shenandoah Valley. Here King and Virginia Rails nest, as do occasional Common Gallinules. Traill's Flycatchers are common, and in surrounding fields Upland Sandpipers may breed. During some years at least, this area also has surprising winter populations of the Common Snipe and American Woodcock.

To reach the Marshes from Charles Town, go south on State 51 for about 1.5 miles and turn right on a road marked 'Piedmont.' At the railroad crossing the Marshes are along the right-of-way.

MORGANTOWN
Cheat Lake | Coopers Rock State Forest

Seven miles east of Morgantown on State 73 is **Cheat Lake,** an artificial impoundment that often becomes a resting place for large flocks of waterfowl in spring and fall. Lying at the foot of the westernmost Allegheny ridge, it is a meeting place for northern and southern plant and animal species. Near the Lake are stands of redbud, pawpaw, persimmon, yellow oak, and buckeye; and common birds include Blue-gray Gnatcatcher, White-eyed Vireo, Worm-eating Warbler, Prairie Warbler, Kentucky Warbler, and Hooded Warbler. Higher up the slopes are the Least Flycatcher, Solitary Vireo, Black-throated Blue Warbler, Black-throated Green Warbler, Chestnut-sided Warbler, and Slate-colored Junco.

Three miles beyond Cheat Lake on State 73 is **Coopers Rock State Forest,** more than 12,000 acres in extent, where Wild Turkeys reside, and there is a fine population of Ruffed Grouse. Pileated Woodpeckers are fairly common, and Northern Ravens nest sparingly on some of the bold cliffs.

PETERSBURG
Dolly Sods Area | Bear Rocks | Seneca Rocks-Spruce Knob National Recreation Area

Petersburg, in the scenic and historic South Branch of the Potomac Valley, is a convenient stopping place for visitors to the **Dolly Sods Area,** a center of much ornithological activity.

Dolly Sods, on a plateau atop the Alleghenies with elevations above 4,000 feet, constitutes a secondary continental divide, with waters to the west reaching the Mississippi, and to the east joining the Potomac. Once clothed with a dense red spruce forest, destructive lumbering, followed by fire, has greatly reduced the forested area. In place of the forest there are now very large areas of heath 'balds' and several thousand acres of blueberries, huckleberries, and other shrubs.

In poorly drained places there are peat bogs that have saved important elements of original biota, including plants and animals of boreal association. Around these openings were areas where grasses and other forage plants flourished; these by local terminology were the 'sods,' and since the early owners were named 'Dolly,' they became Dolly Sods.

Dolly Sods is now within the United States Forest Service's Monongahela National Forest. With the help and encouragement of the West Virginia Chapter of Nature Conservancy, the mineral rights under 15,800 acres of this area have recently been purchased, and the Forest Service, with local cooperation, has placed the whole Area under long-range recreation and conservation management. A new spruce forest is establishing itself, but there are still large bald openings.

Dolly Sods holds as breeding birds all species listed as characteristic of the red spruce forest, although, as with other mountain areas, there are no recent summer records of Olive-sided Flycatchers. Along the well-marked Northlands Trail are especially good spots for enjoying the evening thrush chorus in June, with Veeries and Hermit and Swainson's Thrushes much in evidence. Nashville Warblers also nest here, and Winter Wrens and Purple Finches are common. During some seasons Red Crossbills are abundant and, more rarely, there are summering White-winged Crossbills.

During autumn, Dolly Sods is a center of the Operation Recovery bird-banding program by enthusiasts who often spend weeks netting and banding thousands of migrants. In late September these workers are joined by many persons attracted by the hawk flights which, under favorable conditions, sometimes pass the Bear Rocks overlook at the north end of Dolly Sods. On days of good flights, counts of several hundred hawks are frequent, and occasional days yield 3,000 to 5,000 birds. At other times, with different winds, there may be large flights of Blue Jays, Red-breasted Nuthatches, and diurnal wood warblers.

Dolly Sods is reached from Petersburg by following State 28 westward for about 5 miles, going right on Jordan Run Road for 1.0 mile, and left on the well-marked road to the summit for 9 miles. There is always a chance of seeing Wild Turkeys along this road as well as other game such as grouse and deer.

At the top of the mountain the road joins Forest Service Road 75, a remarkable drive of several miles along the crest. Its northern terminus is at **Bear Rocks,** the favored spot for hawk flights. Along this road are two campgrounds, marked nature trails, several cranberry bogs, a fire-lookout tower, and extensive beaver dams. From the beginning of blossoming, the road presents a succession of showy plants in flower and fruit.

The **Seneca Rocks–Spruce Knob National Recreation Area,** under joint management of the National Park Service and the Forest Service, includes as special features spectacular Seneca Rocks and West Virginia's highest point, Spruce Knob, at 4,860 feet. In its 1,000 square miles are widely variant plants and animals. Drive west and south from Petersburg on State 28 for about 18 miles to the intersection with US 33 at Mouth of Seneca. Seneca Rocks are east of here, across the river; the Spruce Knob Unit is west along US 33.

POINT PLEASANT
McClintic Wildlife Station

Ten miles north of Point Pleasant on State 62 is the **McClintic Wildlife Station,** an experimental area maintained by the West Virginia Department of Natural Resources. With a number of shallow ponds, the facility is designed primarily for waterfowl study, but wildlife plantings and good cover make it an excellent place for observing a variety of birds. Being near the Ohio River, it is in the path of sizable waterfowl migrations. In spring and fall the Station is one of the better places in the upper Ohio Valley for a great variety of waterfowl as well as waterbirds. A resident flock of Canada Geese is successfully raising young, and these free-flying individuals are returning in some numbers as breeding birds. The Station also supports a large number of breeding Wood Ducks and a lesser number of American Black Ducks.

WHEELING
Oglebay Park

City-owned **Oglebay Park,** long a center of nature activities, is 5 miles north of Wheeling on State 88. With 700 acres of woodland and open fields, the Park has some miles of nature trails, a nature center, a small museum, and resident naturalists who gladly assist visitors. Local birdlife includes species typical of the upper Ohio Valley, with a great variety of wood warblers—Blue-winged, Kentucky, Hooded, Yellow-breasted Chat, Worm-eating, and others—among them.

Wisconsin

WHISTLING SWANS

In Wisconsin, waterfowl make the headlines. When Whistling Swans gather each spring in vast numbers on Lake Winnebago, when thousands of Canada Geese congregate each fall on Horicon Marsh, local newspapers, from Kenosha to Superior and from Prairie du Chien to Green Bay, give the events a front-page spread, with pictures. But waterfowl receive more than publicity in Wisconsin: much is being undertaken for their welfare. Excellent refuges have been created for their protection, management, and study. In Milwaukee the Juneau Park Lagoon is aerated to prevent freezing, thus encouraging ducks to remain all winter. For the bird finder there is more to Wisconsin than waterfowl—much more—but he must plan to be a waterfowl finder part of the time.

Because of its relation to various water areas, Wisconsin has many attractions for waterbirds and shorebirds as well as waterfowl; in the east and north, the 500-mile shoreline of Lake Michigan and Lake Superior; in the west, the St. Croix and Mississippi Rivers, with the hundreds of floodplain lakes on their bottomlands; in the interior, about eight thousand lakes and numberless marshes, rivers, and streams.

In general, Wisconsin is a plain, with an average elevation of only 1,050 feet. In the extreme northwestern part, which adjoins Lake Superior, is a stretch of lowland only 150 to 350 feet above lake level. South of this, however—in north-central Wisconsin—are the Northern Highlands, of which Rib Mountain near Wausau reaches an elevation of 1,940 feet—highest point in the state. Other high areas worthy of mention are the Eastern Highlands along Lake Michigan (the most densely populated section of the state) and the Western Upland near the St. Croix and Mississippi Rivers. The shores of Lake Superior and Lake Michigan, with their gradual slope and beaches of loose rock and gravel, are relatively unspectacular, but the high bluffs along the St. Croix and Mississippi Rivers present some of the finest scenery to be found in the state, if not in the entire Midwest.

Wisconsin was originally covered with forest except in the southern part, where there were prairie openings. The coniferous forest was characteristic of the northern third of the state: white and red pines, white spruce, and fir on the uplands, black spruce, tamarack, and white cedar in the bogs and wet lowlands. Fires and lumbering operations destroyed nearly all these forests, and a secondary growth of such trees as aspen and birch replaced them. In a few areas, particularly in lowlands, the original forest conditions still prevail and maintain a distinctive association of birds. Although an intensive study of the coniferous-forest birds of Wisconsin has yet to be made, the following are probably typical:

Spruce Grouse
Yellow-bellied Sapsucker
Black-backed Three-toed Woodpecker
Olive-sided Flycatcher
Gray Jay
Boreal Chickadee
Red-breasted Nuthatch
Brown Creeper
Winter Wren
Hermit Thrush
Swainson's Thrush

Golden-crowned Kinglet
Ruby-crowned Kinglet
Solitary Vireo
Nashville Warbler
Northern Parula Warbler
Magnolia Warbler
Cape May Warbler
Black-throated Blue Warbler
(*northeastern Wisconsin*)
Myrtle Warbler
Black-throated Green Warbler
Blackburnian Warbler
Northern Waterthrush
Mourning Warbler
Canada Warbler
Purple Finch
Pine Siskin
Red Crossbill
White-winged Crossbill
Slate-colored Junco
White-throated Sparrow

Through the middle third of the state, the original forests were of white pine, red pine, jack pine, and oak, but these have almost disappeared. In their place are farmlands or cutover lands now grown to aspen and new oak and jack pine. Deciduous hardwood forests predominated in the southern third of the state and along the western border. Here were sugar maple, white, red, and black oaks, elm, basswood, shellbark hickory, honey locust, black walnut, and butternut. On the bottomlands of the Mississippi and lower Wisconsin Rivers were river birch, swamp white oak, and Kentucky coffee tree—all more characteristic of localities in southern states. Today, a few of these forests still stand, more or less modified in form, on the bluffs and bottomlands of the St. Croix and Mississippi Rivers, the bottomlands of the lower Wisconsin River, and a few other areas; but most of them have been entirely removed for farming space, or cut over so extensively that they have lost their original character. Very few prairie openings are recognizable now, grazing and cultivation having made them indistinguishable from farmlands. Characteristic breeding birds of the present-day deciduous hardwoods and farmlands (fields, wet meadows, brushy lands, orchards, and dooryards) in the southern third of the state are listed below:

DECIDUOUS HARDWOODS

Red-shouldered Hawk
Broad-winged Hawk
Ruffed Grouse
Yellow-billed Cuckoo
Black-billed Cuckoo
Common Screech Owl

DECIDUOUS HARDWOODS (*cont.*)

Barred Owl
Whip-poor-will
Yellow-shafted Flicker
Red-bellied Woodpecker
Hairy Woodpecker
Downy Woodpecker
Great Crested Flycatcher
Least Flycatcher
Eastern Pewee
Blue Jay
Black-capped Chickadee
White-breasted Nuthatch
Wood Thrush
Blue-gray Gnatcatcher
Yellow-throated Vireo
Red-eyed Vireo
Warbling Vireo
Cerulean Warbler
Ovenbird
American Redstart
Baltimore Oriole
Scarlet Tanager
Rose-breasted Grosbeak

FARMLANDS

Common Bobwhite
Mourning Dove
Red-headed Woodpecker
Eastern Kingbird
Eastern Phoebe
(Prairie) Horned Lark
Tree Swallow
Barn Swallow
House Wren
Gray Catbird
Brown Thrasher
Eastern Bluebird
Yellow Warbler
Common Yellowthroat
Bobolink
Eastern Meadowlark
Western Meadowlark
Brewer's Blackbird
Common Grackle
Northern Cardinal
Indigo Bunting
Dickcissel
American Goldfinch
Rufous-sided Towhee
Savannah Sparrow
Grasshopper Sparrow
Henslow's Sparrow
Vesper Sparrow
Chipping Sparrow
Clay-colored Sparrow
Field Sparrow
Song Sparrow

Marshes containing cattails, bulrushes, and sedges, mingled with many less conspicuous aquatic plants, are scattered throughout the state, but nowhere so abundantly as in the central and southeastern sectors. Some of these marshes contain Yellow-headed Blackbird colonies; a few have a pair or two of Wilson's Phalaropes nesting in the grasses along the damp outskirts. The larger marshes have King Rails and Common Gallinules nesting commonly. In marshes throughout the state the following breed more or less regularly:

Pied-billed Grebe
American Bittern
Least Bittern
Virginia Rail
Sora
American Coot
Black Tern
Marsh Wren
Red-winged Blackbird
Swamp Sparrow

Wisconsin has three prominent migration routes: first, the Lake Michigan shore, used by waterbirds, waterfowl, and shorebirds; second, the Mississippi River, used by a similar variety of waterbirds and waterfowl and, in addition, by small landbirds such as kinglets, vireos, warblers, and fringillids; third, the Wisconsin River, which flows south through the middle of the state and then swings west to the Mississippi, and is used by waterfowl, a few shorebirds, and many landbirds. Outstanding features of the Wisconsin migration are the spring swan concentrations in the Fox River Valley from Green Bay to Lake Winnebago, the great gatherings of geese in both spring and fall on certain marshes and lakes in the state's interior, and the fall flights of hawks on the Lake Michigan shore. The following timetable applies to the southern half of the state:

Waterfowl: 25 March–20 April; 5 October–15 November
Shorebirds: 1 May–1 June; 1 August–25 September
Landbirds: 20 April–25 May; 20 August–10 October

The Wisconsin winter is definitely not mild, and only in a limited degree does the nearness of Lake Superior and Lake Michigan modify the temperature extremes. In the Madison and Milwaukee regions it is not unusual to see Evening Grosbeaks, Common Redpolls, Pine Siskins, Red Crossbills, White-winged Crossbills, Lapland Longspurs, and Snow Buntings; Horned Larks (northern races), Slate colored Juncos, and American Tree Sparrows are common winter residents. Waterfowl remain through the winter only in a very few places where the water fails to freeze over.

Authorities

N. R. Barger, Wallace B. Grange, Donald V. Gray, Owen J. Gromme, Joseph J. Hickey, Clarence S. Jung, Harold A. Mathiak, Robert A. McCabe, Gordon Orians, Howard L. Orians, Carl H. Richter, Samuel D. Robbins, Jr., Mrs. Walter E. Rogers, Vern E. Rudolph, A. W. Schorger, Walter E. Scott, William E. Southern, Harold C. Wilson.

References

Wisconsin Birds. A Checklist with Migration Graphs. By N. R. Barger, Roy H. Lound, and Samuel D. Robbins, Jr. Wisconsin Society for Ornithology, 1975.

Wisconsin's Favorite Bird Haunts [A guide to locations for birds, with individual maps]. Compiled and edited by Darryl D. Tessen. Rev. ed. Wisconsin Society for Ornithology, 1976.

These two publications may be obtained from W.S.O. Supply Department, Hickory Hill Farm, Loganville, Wis. 53943.

GRANTSBURG
Crex Meadows Wildlife Area

The **Crex Meadows Wildlife Area** in northwestern Wisconsin, about 75 miles from Minneapolis–St. Paul, Minnesota, is well worth a visit from early April to late October. The following directions apply to a visit in June or early July. From the intersection of State 48 and State 70 in Grantsburg, drive north on County F for about a mile to the southwestern corner of the Area, where directional signs point the way into it.

Owned by the State Department of Natural Resources, the Crex Meadows Area embraces some 20,000 acres, principally lakes and marshes. These have been restored after being drained by white settlers for farmlands in about 1890. Dikes and water-control structures impound the flowage into the St. Croix River, with the result that the Area is a major stopover for waterfowl in spring and fall and the chief nesting site for waterfowl and marsh birds in northern Wisconsin.

On entering the Area from the southwestern corner, proceed to the Phantom Lake Flowage and drive along Dike 7, which overlooks it from the west side. Just offshore, notice an island with a Great Blue Heron colony as well as a few nesting Black-crowned Night Herons and a small colony of Double-crested Cormorants. On the water, look for summer-resident Common Loons and ducks, commonest being Mallards, Blue-winged Teal, and Ring-necks. Other waterfowl that may be expected are American Black Ducks, Wood Ducks, Green-winged Teal, and Hooded Mergansers.

Continue on Dike 7, passing the intersection with Dike 9 on the left, and then bear right on the Main Dike Road, which leads to Dike 4 and eventually Dike 3. Watch for Brewer's Blackbirds along the way.

The marsh on the right of Dike 3 is especially productive. Listen and look for such birds as Pied-billed Grebes, both American and Least Bitterns, Virginia Rails, Soras, American Coots, Black Terns (abundant), Marsh Wrens, and Yellow-headed Blackbirds. Carefully inspect the sedges for Le Conte's Sparrows. Difficult to see and flush, they are more easily discovered by listening for their peculiarly weak, buzzy, two-noted songs, which have an insect-like quality.

Continue on Dike 3 to its intersection with Dike 2 and bear left, taking the West Refuge Road, which, as its name indicates, passes along the west side of the refuge (where no hunting is allowed) to an observation area on the north side of the refuge. Sandhill Cranes are often in view from here, as well as a resident flock of Canada Geese.

MADISON
University of Wisconsin Arboretum | Vilas Park | University Bay

The **University of Wisconsin Arboretum** at Madison, in south-central Wisconsin, is remarkably fine for the warbler migration and for many nesting birds. Its 1,240 acres along the south shore of Lake Wingra, on the southwest outskirts of this state capital, embrace many diverse habitats for birds: open water, marshes, grassy meadows, open prairie-like fields, shrubby thickets, and woods of oak, hickory, and black cherry. From 5 to 20 May the warbler migration is at its best. A hundred or more species of birds nest in the Arboretum. In the open parts, American Woodcock flight-sing overhead during April evenings soon after sundown.

Elsewhere in habitats suited to them are the Green Heron, Least

Bittern, Wood Duck, Red-shouldered Hawk, Common Gallinule, Black-billed Cuckoo, Great Horned Owl, Red-headed Woodpecker, Traill's Flycatcher, Bewick's Wren (a few), Marsh Wren, Wood Thrush, Blue-gray Gnatcatcher, Yellow-throated Vireo, Hooded Warbler (a few), and Rose-breasted Grosbeak.

The Arboretum may be reached by driving southwest from the Capitol on West Washington Street for 11 blocks, then right on Regent Street for 11 blocks, and left on Monroe Street for 12 blocks. Noting the Arboretum on the left, continue on Monroe Street (which becomes Nakoma Road), then go left on Manitou Way for 2 blocks to the entrance, Arboretum Drive, which crosses the area from southwest to northeast.

During spring and fall migrations, at least two areas for water-loving birds should be visited. One is **Vilas Park** on the north shore of Lake Wingra, reached by turning west from the northeast end of Arboretum Drive onto Wingra Drive. The lagoons here attract gulls and terns, as well as surface-feeding ducks, their presence encouraged partly by exotic waterfowl belonging to the neighboring zoo.

The other is **University Bay** on Lake Mendota, reached from the Capitol by driving west on State Street to the University of Wisconsin campus, turning right to the Lake and left on Willow Drive (this becomes Mendota Drive), which follows the shore. Besides gulls and terns, look for other waterbirds, wading birds, and a variety of ducks, both surface-feeding and diving species. Late in summer a few Great Egrets occasionally appear and, later in fall, Common Loons, Horned Grebes, Common Mergansers, Red-breasted Mergansers, and sometimes White-winged Scoters and Surf Scoters.

MAZOMANIE
Wisconsin River Bottomlands

In south-central Wisconsin northwest of Madison, an exceptionally rewarding area for the bird finder borders the south side of the Wisconsin River between Mazomanie and Spring Green. Generally known as the **Wisconsin River Bottomlands,** sometimes locally as

Mazomanie Bottoms, here and there are floodplains extending nearly 0.5 mile inland from the River. Dense woods of green ash, red oak, river birch, silver maple, and basswood cover most of the plains, making the area attractive to a great number of forest-dwelling birds.

From the first pronounced migration in early April to the last big exodus in early October, this is invariably an exciting area for the bird finder. Red-shouldered Hawks and Pileated Woodpeckers nest where the timber is heavy; Lark Sparrows nest sparingly on sandy wastes where most trees fail to grow; Upland Sandpipers and Henslow's Sparrows nest in old pastures and grassy meadows. Both the migrating and breeding populations are large in number of species, but what makes the area exceptional is its proximity to the northernmost limits of the ranges of several forest and forest-edge species: Red-bellied Woodpecker, Tufted Titmouse, Blue-gray Gnatcatcher, Prothonotary Warbler, Blue-winged Warbler, Cerulean Warbler, Louisiana Waterthrush, Kentucky Warbler, and Yellow-breasted Chat.

For places where most, if not all, of the above-mentioned birds may be found, here are four sets of directions.

1. *For woodland birds,* drive west from the post office in Mazomanie for 1.5 miles and, after crossing Black Earth Creek, take the next right north. In the next mile the road eventually meanders down a hill and crosses Marsh Creek. Continue 0.3 mile to the point where the road turns sharply right. Park here and walk on a lane that leads into the woods and soon peters out. To go into deeper woods ahead, the bird finder is on his own. By bearing left, he will reach the Wisconsin River bank close to the confluence of Marsh and Dunlap Creeks.

2. *For the Lark Sparrow,* from the intersection of US 14 and County Y in Mazomanie drive west on US 14 for 3 blocks, then turn off south on a road that reaches a suitable sandy habitat after 0.6 mile.

3. *For Upland Sandpipers and Henslow's Sparrows,* drive north from Mazomanie on County Y, which, after 1.5 miles, goes down a hill to a broad meadow.

4. *For more woodland birds,* continue north on County Y. Bear left at each of the next three forks. After going left at the third fork, continue for 1.5 miles on an unimproved road (sometimes impassable in wet weather) that passes through a wooded bottomland where bird finding is good anywhere along the way.

MILWAUKEE

Juneau Park Lagoon | Bradford Beach | Schlitz Audubon Center

Wisconsin's largest city, occupying some 49 square miles along a bay on Lake Michigan's shore, offers ample opportunity for the observation of waterbirds and waterfowl. In the very heart of Milwaukee, on the bay just north of the Lincoln Memorial Bridge and east of Lincoln Memorial Drive, is **Juneau Park Lagoon,** where many surface-feeding ducks gather, beginning in October. Mallards, American Black Ducks, American Wigeons, Common Pintails, Green-winged Teal, and Northern Shovelers come in great numbers and stay for several weeks. At times they are joined by a few diving ducks: Redheads, Ring-necked Ducks, Canvasbacks, and Lesser Scaups. Many of the ducks remain in the Lagoon all winter, since the water, agitated from below by an air pump, is kept free of ice.

The bay waters of Lake Michigan immediately offshore from the Park are a stopping point for many waterbirds and waterfowl in their migratory journeys. This is particularly true in late October and November, after a cold spell with north winds, or in late March and April, after the ice in the bay has broken up. Hundreds, and sometimes thousands, of diving ducks can be seen, among them Redheads, Ring-necked Ducks, Canvasbacks, Lesser Scaups, Common Goldeneyes, Buffleheads, Oldsquaws, White-winged Scoters, Ruddy Ducks, and all three species of mergansers. A great many stay through winter. The waters offshore are also on the migratory flyway of thousands of Bonaparte's Gulls, some of which linger in fall until the bay begins to freeze over in late December. Other migrating birds to look for are Common Loons, Red-throated Loons, Horned Grebes, Pied-billed Grebes, Double-crested Cormorants, Gadwalls, American Coots, Forster's Terns, Common Terns, a few Caspian Terns, and Black Terns.

In late summer and fall until mid-October, the shorebird migration is worth observing on **Bradford Beach,** north of Juneau Park and paralleled by Lincoln Memorial Drive. The most common are Semipalmated Plovers, Ruddy Turnstones, Least Sandpipers, Semipalmated Sandpipers, and Sanderlings.

Nine miles north of downtown Milwaukee on a bluff overlooking Lake Michigan is the youth-oriented **Schlitz Audubon Center,** owned and operated by the National Audubon Society for environmental education. On the 185-acre property a deep woods slopes down to old grassy fields and an unspoiled half-mile of Lake Michigan beach. Along foot trails, in summer, one may observe many of the smaller landbirds common to southeastern Wisconsin. In September and early October, shorebirds stop on the beach during their southward passage, and from late August to December southbound migrating hawks pass overhead, much as they do farther north on Lake Michigan (*see under* **Port Washington**). The Center may be reached from Milwaukee by proceeding north on US 141, exiting east on East Brown Deer Road (State 32), then going south on North Lake Shore Drive (still State 32) to the entrance on the right at 8566 North Lake Shore Drive.

NECEDAH
Necedah National Wildlife Refuge

The **Necedah National Wildlife Refuge** in west-central Wisconsin is notable for the large numbers of waterfowl that congregate on its marshes, sloughs, and artificial pools, especially in fall. At the peak of migration, 14–31 October, many thousands of Canada Geese plus small aggregations of Snow Geese and sometimes a few Whistling Swans are present. Besides a resident flock of Canada Geese, there are Mallards, Blue-winged Teal, Ruddy Ducks, and Hooded Mergansers nesting on the Refuge. During fall migration many more of these same species arrive, along with many Gadwalls, Green-winged Teal, American Wigeons, Ring-necked Ducks, Lesser Scaups, Common Goldeneyes, Buffleheads, and Common Mergansers. A small number of Sandhill Cranes usually shows up. In late August and September a few transient shorebirds—commonly Common Snipes, Solitary Sandpipers, Greater and Lesser Yellowlegs, and Pectoral Sandpipers—appear along the edges of the sloughs and pools. Where there are marshes, Virginia Rails, Soras, Marsh Wrens, Sedge Wrens, and Red-winged Blackbirds reside in summer, as do Common Yellowthroats and Swamp Sparrows on the brushy borders.

The 39,600 acres of the Refuge comprise, in addition to wet areas, an extensive upland terrain where jack pine and red oak grow abundantly, providing habitats for Ruffed Grouse, introduced Wild Turkeys, and a variety of passerines such as the Yellow-throated Vireo, Red-eyed Vireo, Pine Warbler, Ovenbird, Scarlet Tanager, and Rose-breasted Grosbeak. There are also grassy areas where American Woodcock establish their singing grounds in early spring, and Upland Sandpipers nest later.

The Refuge may be reached from Necedah by driving west on State 21 for 4.5 miles, then north at a directional sign for 2 miles to headquarters. The Rynearson Pools, immediately north and east of headquarters, offer good views of waterfowl, Sandhill Cranes, and shorebirds. At headquarters, inquire about foot trails that lead through a typical pine-oak woods in the southeastern part of the Refuge, and also obtain permission to walk on the dikes around the Sprague-Mather Pool.

To reach the Pool, return to Necedah and drive north on State 80 toward the town of Sprague, 8 miles distant. Shortly before reaching the town, turn west on the Sprague-Mather Road, which soon runs along the south side of the dike that impounds the Pool. In the ditch between the Road and dike, where there are cattails and other marsh vegetation, watch for rails and wrens. At a point about 3.5 miles from State 80, turn north on a road along the dike for excellent views of the Pool. Stay in the car so as not to disturb the birds close by.

OSHKOSH
Lake Winnebago

One of the most spectacular sights in Wisconsin is the annual spring concentration of Whistling Swans on **Lake Winnebago,** from about

25 March—usually as soon as the ice breaks up—to 10 April. Several hundred of these handsome great white birds tarry here briefly during their northward migration. They can best be seen from the west shore near the highways paralleling it: US 45 south from Oshkosh to Fond du Lac, or County A north from Oshkosh to Neenah.

PARK FALLS
Clam Lake–Teal Lake Area

Parts of the Chequamegon National Forest in northwestern Wisconsin have a rich coniferous-forest avifauna. Obtain a map of the Forest from the supervisor's office in Park Falls, then drive northward for 21 miles on State 13 and turn west into the Forest on State 77 for 14 miles to **Clam Lake.** From Clam Lake, continue southwestward about 15 miles to **Teal Lake** or turn south on County GG. Between the Lakes or along the first 10 miles of County GG are upland woods consisting of sugar maple, aspen, and white birch, together with some stands of white spruce, white pine, and red pine; and also bogs with thick stands of black spruce, white cedar, and tamarack.

The best time to look for birds along the road between the Lakes is June and early July at the height of the nesting season, when most of the species are in full song. Stop to listen and search at places along the way where the forest cover, or lack of it, seems particularly suited to the following species: *edges of bogs with mixed stands of hardwoods and conifers,* Sharp-shinned Hawk, Yellow-bellied Sapsucker, Yellow-bellied Flycatcher, Gray Jay, Boreal Chickadee, Red-breasted Nuthatch, Winter Wren, Swainson's Thrush, Veery, Solitary Vireo, Black-and-white Warbler, Northern Parula Warbler (spruces with *Usnea* lichen), Magnolia Warbler, Cape May Warbler, Myrtle Warbler, Black-throated Green Warbler, Blackburnian Warbler, and Northern Waterthrush; *edges of woods and clearings with shrubby growth,* Chestnut-sided Warbler, Mourning Warbler, and Canada Warbler; *upland pines,* Pine Warbler, Hermit Thrush, and Slate-colored Junco; *blow-downs with still-standing stubs,* Black-backed Three-toed Woodpecker; *open bogs with scattered trees of spruce and tamarack,* Olive-sided Flycatcher, Purple Finch, and

MOURNING WARBLER

Nashville Warbler. Anywhere along the way, look for Northern Ravens, sometimes flying across the road or scavenging around the Lakes.

PORT WASHINGTON
Harrington Beach State Park

From the last of August to the first of December, hundreds of diurnal birds of prey migrate from northern regions southward over the west shore of Lake Michigan. Northern Harriers and Broad-winged Hawks are usually in the vanguard, the Broad-wings moving in spectacular numbers by mid-September. Meanwhile come the

Sharp-shinned and Red-tailed Hawks, but not in great numbers until October. In passage almost any time from September on, although never in such great numbers, are Cooper's Hawks, Ospreys, Merlins, American Kestrels, and an occasional Red-shouldered Hawk, Bald Eagle, or Peregrine Falcon. Tending to draw up the rear in November are Rough-legged Hawks and Northern Goshawks.

While a few predators are likely to be moving southward any day, the big flights depend on weather conditions: clearing sky and strong westerly winds just after a cold front, and preceded one to three days earlier by storms and lowering temperatures in regions to the north. To determine upon a good day for flights, begin studying weather maps several days in advance.

To see the hawk flights, select any vantage point along Lake Michigan offering a wide view of shore, water, and sky. One good position is at the northern end of **Harrington Beach State Park,** which is reached from Port Washington (30 miles north of Milwaukee) by driving north on US 141 and State 32 for 5.5 miles and, at the village of Lake Church, turning off east on County D to the parking area, then walking to the shore. Another good spot is on the shore in the Schlitz Audubon Center (*see under* **Milwaukee**).

SARONA
Hunt Hill Sanctuary

The **Hunt Hill Sanctuary,** maintained by the National Audubon Society on the south shore of Devil's Lake in northwestern Wisconsin, has 330 acres of varied habitat: two small, deep lakes, a spruce-tamarack bog, mature and secondary deciduous woods, meadowlands, and fields. The two little lakes, marsh-fringed and undisturbed, attract Black Terns for nesting, as well as an occasional pair of American Bitterns and Soras; Ospreys nest in the immediate environs. Devil's Lake usually hosts a pair of Common Loons. Bobolinks and Clay-colored Sparrows are common summer residents of the open uplands.

The buildings of the Audubon Workshop, where in July and August the Society centers its teaching of natural history and ecology to adults, are surrounded by a magnificent forest of sugar maple, basswood, and hemlock in which Cerulean Warblers are among regular breeding birds. Cliff Swallows plaster their nests under the eaves of one of the Workshop buildings. A feeding station nearby is patronized all summer by Purple Finches and five kinds of woodpeckers: Hairy, Downy, Red-bellied, Red-headed, and Yellow-bellied Sapsucker.

Sarona is on US 53 between Spooner and Rice Lake. To reach the Sanctuary, drive east from Sarona on County D for 4 miles.

SOLON SPRINGS

Solon Springs–Drummond Pine Barrens | Brule River State Forest

Solon Springs, in northwestern Wisconsin, is on State 53, 37 miles south of Superior. Eastward for 30 miles from Solon Springs to Drummond are the **Solon Springs–Drummond Pine Barrens,** in which commercially planted jack pines are the predominant tree growth. Explore this area in June and early July for breeding Connecticut Warblers. From Solon Springs, drive northeast on County A. After its junction with County P, 3 miles from Solon Springs at the north end of Lake St. Croix, County A turns abruptly south for 1.7 miles and then takes an easterly direction. For the next 6 miles, County A traverses good Connecticut Warbler habitat.

County S, which intersects County A at the end of the fifth mile and goes north, also passes through good habitat for the first 2 miles. Stop and listen for the singing males wherever there are stands of jack pine 15 to 30 feet tall, either with their lower branches still present or with an understory of bracken fern and scrub oak.

Northeast of Solon Springs is the southwestern part of the **Brule River State Forest,** which borders the Brule River to its mouth on Lake Superior. After exploring the Pine Barrens for Connecticut Warblers, backtrack on County A and go north on County P. For 1.5 miles, until the road crosses the Brule River, a fine stand of black spruce and white cedar attracts a nesting 'colony' of Cape May Warblers and a host of other northern passerines including the Yellow-bellied Flycatcher, Olive-sided Flycatcher, possibly the

Boreal Chickadee, both Golden-crowned and Ruby-crowned Kinglets, and the Northern Parula, Magnolia, Myrtle, and Blackburnian Warblers.

WAUPUN

Horicon National Wildlife Refuge | Horicon Marsh Wildlife Area

The stellar place in Wisconsin for seeing spring goose concentrations and a rich variety of nesting marshbirds is Horicon Marsh, an area of well over 36,000 acres in the southeastern part of the state, surrounded by gently rolling farmland and cut through by the Rock River. The northernmost 20,900 acres are included in the **Horicon National Wildlife Refuge**; the 15,000 acres adjoining them to the south comprise the **Horicon Marsh Wildlife Area,** belonging to the State Department of Natural Resources.

To reach National Refuge headquarters, drive east from Waupun on State 49, across the northern end of the Refuge, for 6.5 miles, then south on County Z for 3.5 miles to a sign indicating the direction of headquarters on the east side of the Marsh. To reach State Area headquarters, just north of Horicon, drive east from Waupun on State 49 for 0.1 mile, south on State 26 for 14 miles, and east on State 33 for 11 miles to Horicon. From here follow directional signs north to headquarters on the southern end of the Marsh.

About 8,000 acres of the Marsh are flooded by a dike across the Rock River, thus providing suitable resting and feeding areas for waterfowl, waterbirds, and shorebirds. Flock after flock of geese stop here during March and April, the greatest number appearing from 5 to 15 April. Well over 50,000 individuals may be present at one time—and a spectacular sight it is. Early each morning the birds rise in flocks from the Marsh, go into flight formation, and disappear outside the Marsh to forage on the farmlands; later in the day they return, again in flight formation, and break ranks to settle on the Marsh for the night. The spring population consists largely of Canada Geese, with lesser numbers of Snow Geese.

The fall goose migration, beginning in late September and continuing through November, is even more impressive. The greatest concentrations occur from mid-October to mid-November, when peak numbers may reach 200,000, largely Canada Geese and usually no more than a few thousand Snow Geese.

The best vantage points for viewing the spectacle: (1) from State 49 as it crosses the northern part of the National Refuge east of Waupun; (2) near both the National and State headquarters; (3) from the east side of the Main Dike between the National and State parts of the Marsh, reached by continuing south from National headquarters on County Z and following directional signs.

The goose concentrations tend to eclipse the presence of many other transient waterfowl, such as Mallards, American Black Ducks, Blue-winged Teal, American Wigeons, and Redheads, whose numbers are impressive.

Many shallower parts of Horicon Marsh have an abundance of cattails, bulrushes, and sedges that consequently provide nesting habitat for Pied-billed Grebes, American Bitterns, perhaps as many as a dozen species of ducks, King Rails, Virginia Rails, Soras, Common Gallinules, American Coots, Black Terns, Marsh Wrens, Yellow-headed Blackbirds, and Red-winged Blackbirds. A few Forster's Terns nest on muskrat houses. At the time of year when these birds are nesting (June and early July), high water may make access roads impassable. Inquire at either headquarters about reaching the best habitats.

On Four Mile Island in the State Area is a colony of Great Blue Herons and Black-crowned Night Herons, together with a few Great Egrets. From a distance, after mid-June, one can see young in the nests. The Island is usually accessible only by boat.

In late August and September, when mud flats are exposed by the low water level, shorebirds are well represented by Solitary Sandpipers, Greater and Lesser Yellowlegs, Pectoral Sandpipers, Long-billed Dowitchers, and others.

Index

Abbreviations used in the Index

AL	Alabama	MA	Massachusetts	OH	Ohio
CT	Connecticut	MD	Maryland	PA	Pennsylvania
DC	District of Columbia	ME	Maine	RI	Rhode Island
DE	Delaware	MI	Michigan	SC	South Carolina
FL	Florida	MS	Mississippi	TN	Tennessee
GA	Georgia	NC	North Carolina	VA	Virginia
IL	Illinois	NH	New Hampshire	VT	Vermont
IN	Indiana	NJ	New Jersey	WI	Wisconsin
KY	Kentucky	NY	New York	WV	West Virginia
		NYC	New York City		

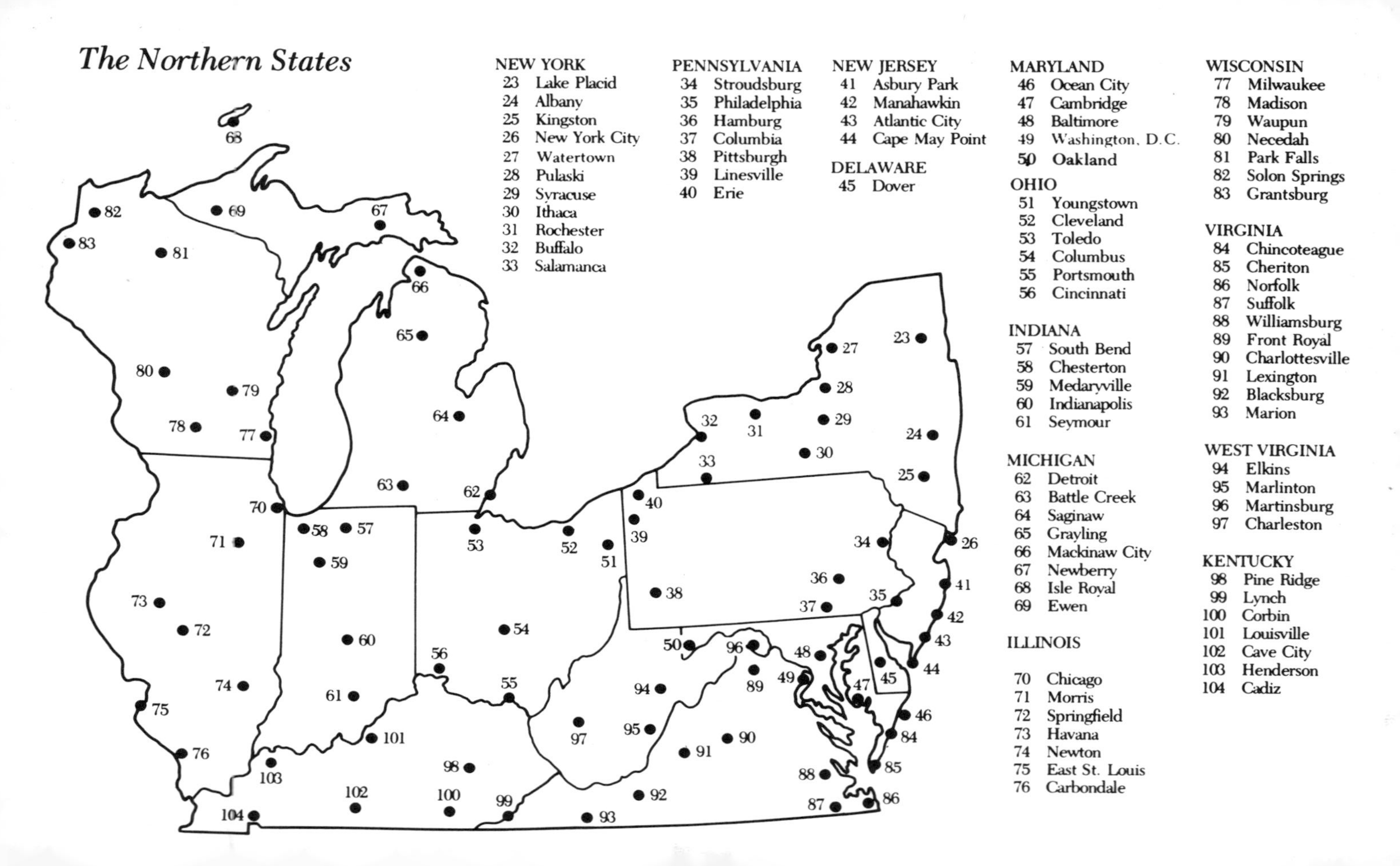
The Northern States
NEW YORK
23 Lake Placid
24 Albany
25 Kingston
26 New York City
27 Watertown
28 Pulaski
29 Syracuse
30 Ithaca
31 Rochester
32 Buffalo
33 Salamanca
PENNSYLVANIA
34 Stroudsburg
35 Philadelphia
36 Hamburg
37 Columbia
38 Pittsburgh
39 Linesville
40 Erie
NEW JERSEY
41 Asbury Park
42 Manahawkin
43 Atlantic City
44 Cape May Point
DELAWARE
45 Dover
MARYLAND
46 Ocean City
47 Cambridge
48 Baltimore
49 Washington, D.C.
50 Oakland
OHIO
51 Youngstown
52 Cleveland
53 Toledo
54 Columbus
55 Portsmouth
56 Cincinnati
INDIANA
57 South Bend
58 Chesterton
59 Medaryville
60 Indianapolis
61 Seymour
MICHIGAN
62 Detroit
63 Battle Creek
64 Saginaw
65 Grayling
66 Mackinaw City
67 Newberry
68 Isle Royal
69 Ewen
ILLINOIS
70 Chicago
71 Morris
72 Springfield
73 Havana
74 Newton
75 East St. Louis
76 Carbondale
WISCONSIN
77 Milwaukee
78 Madison
79 Waupun
80 Necedah
81 Park Falls
82 Solon Springs
83 Grantsburg
VIRGINIA
84 Chincoteague
85 Cheriton
86 Norfolk
87 Suffolk
88 Williamsburg
89 Front Royal
90 Charlottesville
91 Lexington
92 Blacksburg
93 Marion
WEST VIRGINIA
94 Elkins
95 Marlinton
96 Martinsburg
97 Charleston
KENTUCKY
98 Pine Ridge
99 Lynch
100 Corbin
101 Louisville
102 Cave City
103 Henderson
104 Cadiz